Ergebnisse der Mathematik   Volume 36
und ihrer Grenzgebiete

3. Folge

A Series of Modern Surveys
in Mathematics

Editorial Board

E. Bombieri, Princeton   S. Feferman, Stanford
M. Gromov, Bures-sur-Yvette   J. Jost, Leipzig
J. Kollár, Salt Lake City, Utah   H.W. Lenstra, Jr., Berkeley
P.-L. Lions, Paris   R. Remmert (Managing Editor), Münster
W. Schmid, Cambridge, Mass.   J.Tits, Paris

Springer
*Berlin
Heidelberg
New York
Barcelona
Budapest
Hong Kong
London
Milan
Paris
Singapore
Tokyo*

Jacek Bochnak
Michel Coste
Marie-Françoise Roy

Real Algebraic Geometry

 Springer

Jacek Bochnak
Mathematisch Instituut
Vrije Universiteit
De Boelelaan 1081
NL-1081 HV Amsterdam
The Netherlands
e-mail: bochnak@cs.vu.nl

Michel Coste
Institut Mathématique de Rennes
Université de Rennes 1
Campus de Beaulieu
F-35042 Rennes cedex
France
e-mail: michel.coste@univ-rennes1.fr

Marie-Françoise Roy
Institut Mathématique de Rennes
Université de Rennes 1
Campus de Beaulieu
F-35042 Rennes cedex
France
e-mail: marie-francoise.coste-roy@univ-rennes1.fr

The present edition is a substantially revised and expanded English version of the book "Géometrie algébrique réelle", originally published in French as *Ergebnisse der Mathematik und ihrer Grenzgebiete, 3. Folge,* Vol. 12 (ISBN 3-540-16951-2)

Library of Congress Cataloging-in-Publication Data applied for
Die Deutsche Bibliothek - CIP-Einheitsaufnahme
Bochnak, Jacek:
Real algebraic geometry/Jacek Bochnak; Michel Coste; Marie-Françoise Roy. - Berlin; Heidelberg; New York; Barcelona; Budapest; Hong Kong; London; Milan; Paris; Singapore; Tokyo: Springer, 1998
(Ergebnisse der Mathematik und ihrer Grenzgebiete; Folge 3, Vol. 36)
ISBN 3-540-64663-9

Mathematics Subject Classification (1991):
Primary: 14Pxx. Secondary: 03C10, 11Exx, 12J15, 14F05, 19E99, 19G12, 32C05, 55R50, 57N80, 58A07, 58A35

ISSN 0071-1136
ISBN 3-540-64663-9 Springer-Verlag Berlin Heidelberg New York

This work is subject to copyright. All rights are reserved, whether the whole or part of the material is concerned, specifically the rights of translation, reprinting, reuse of illustrations, recitation, broadcasting, reproduction on microfilms or in any other ways, and storage in data banks. Duplication of this publication or parts thereof is permitted only under the provisions of the German Copyright Law of September 9, 1965, in its current version, and permission for use must always be obtained from Springer-Verlag. Violations are liable for prosecution under the German Copyright Law.

© Springer-Verlag Berlin Heidelberg 1998
Printed in Germany

Typesetting: Camera-ready copy produced by the authors' output file using a Springer T$_E$X macro package
SPIN 10497071   44/3143 - 5 4 3 2 1 0 - Printed on acid-free paper

Preface

The present volume is a translation, revision and updating of our book (published in French) with the title "Géométrie Algébrique Réelle". Since its publication in 1987 the theory has made advances in several directions. There have also been new insights into material already in the French edition. Many of these advances and insights have been incorporated in this English version of the book, so that it may be viewed as being substantially different from the original.

We wish to thank Michael Buchner for his careful reading of the text and for his linguistic corrections and stylistic improvements. The initial LaTeX file was prepared by Thierry van Effelterre.

The three authors participate in the European research network "Real Algebraic and Analytic Geometry". The first author was partially supported by NATO Collaborative Research Grant 960011.

April 1998
Jacek Bochnak
Michel Coste
Marie-Françoise Roy

Table of Contents

Preface V

Introduction 1

1. **Ordered Fields, Real Closed Fields** 7
   1.1 Ordered Fields, Real Fields 7
   1.2 Real Closed Fields 9
   1.3 Real Closure of an Ordered Field 14
   1.4 The Tarski-Seidenberg Principle 17

2. **Semi-algebraic Sets** 23
   2.1 Algebraic and Semi-algebraic Sets 23
   2.2 Projection of Semi-algebraic Sets. Semi-algebraic Mappings .. 26
   2.3 Decomposition of Semi-algebraic Sets 30
   2.4 Connectedness 34
   2.5 Closed and Bounded Semi-algebraic Sets. Curve-selection Lemma 35
   2.6 Continuous Semi-algebraic Functions. Lojasiewicz's Inequality 42
   2.7 Separation of Closed Semi-algebraic Sets 46
   2.8 Dimension of Semi-algebraic Sets 50
   2.9 Some Analysis over a Real Closed Field 54

3. **Real Algebraic Varieties** 59
   3.1 Real and Complex Algebraic Sets 59
   3.2 Real Algebraic Varieties 62
   3.3 Nonsingular Points 65
   3.4 Projective Spaces and Grassmannians 70
   3.5 Some Useful Constructions 76

4. **Real Algebra** 83
   4.1 The Artin-Lang Homomorphism Theorem and the Real Nullstellensatz 83
   4.2 Cones, Convex Ideals 86
   4.3 Prime Cones 88
   4.4 The Positivstellensatz 90
   4.5 Real Principal Ideals 94

5. The Tarski-Seidenberg Principle as a Transfer Tool 97
  5.1 Extension of Semi-algebraic Sets 97
  5.2 The Full Strength of the Tarski-Seidenberg Principle 98
  5.3 Further Results on Extension of Semi-algebraic Sets and Mappings 100

6. Hilbert's 17th Problem. Quadratic Forms 103
  6.1 Solution of Hilbert's 17th Problem 103
  6.2 The Equivariant Version of Hilbert's 17th Problem 106
  6.3 Hilbert's Theorem about Positive Forms 111
  6.4 Quantitative Aspects of Hilbert's 17th Problem 114
  6.5 A Bound on the Number of Inequalities 122
  6.6 Bibliographic and Historical Notes 128

7. Real Spectrum 133
  7.1 Definition and General Properties of the Real Spectrum 133
  7.2 Real Spectrum of a Ring of Polynomial Functions 142
  7.3 Semi-algebraic Functions on the Real Spectrum 146
  7.4 Semi-algebraic Families of Sets and Mappings 149
  7.5 Semi-algebraically Connected Components. Dimension 154
  7.6 Orderings and Central Points 157

8. Nash Functions 161
  8.1 Germs of Nash Functions and Algebraic Power Series 161
  8.2 Local Properties of Nash Functions 167
  8.3 Approximation of Formal Solutions of a System of Nash Equations 171
  8.4 The Artin-Mazur Description of Nash Functions 172
  8.5 The Substitution Theorem. The Positivstellensatz for Nash Functions 175
  8.6 Nash Sets, Germs of Nash Sets 178
  8.7 Henselian Properties. Noetherian Property 184
  8.8 Efroymson's Approximation Theorem 192
  8.9 Tubular Neighbourhood. Extension Theorem 197
  8.10 Families of Nash Functions 202

9. Stratifications 207
  9.1 Stratifying Families of Polynomials 207
  9.2 Triangulation of Semi-algebraic Sets 216
  9.3 Semi-algebraic Triviality of Semi-algebraic Mappings 221
  9.4 Triangulation of Semi-algebraic Functions 227
  9.5 Half-branches of Algebraic Curves 232
  9.6 The Theorems of Sard and Bertini 235
  9.7 Whitney's Conditions a and b 236

10. Real Places 245
10.1 Real Places and Orderings 245
10.2 Real Places and Specialization in the Real Spectrum 249
10.3 Half-branches of Algebraic Curves Again 254
10.4 Fans and Basic Semi-algebraic Sets 256

11. Topology of Real Algebraic Varieties 263
11.1 Combinatorial Properties of Algebraic Sets 264
11.2 Local Euler-Poincaré Characteristic of Algebraic Sets 266
11.3 Fundamental Class of a Real Algebraic Variety. Algebraic Homology 271
11.4 Injective Regular Self-Mappings of an Algebraic Set......... 278
11.5 Upper Bound for the Sum of the Betti Numbers of an Algebraic Set 281
11.6 Nonsingular Algebraic Curves in the Real Projective Plane .. 285
11.7 Appendix: Homology of Semi-algebraic Sets over a Real Closed Field 290

12. Algebraic Vector Bundles 297
12.1 Algebraic Vector Bundles 297
12.2 Algebraic Line Bundles and the Divisor Class Group 306
12.3 Approximation of Continuous Sections by Algebraic Sections. 308
12.4 Algebraic Approximation of $\mathcal{C}^\infty$ Hypersurfaces 312
12.5 Vector Bundles over Algebraic Curves and Surfaces 320
12.6 Algebraic $\mathbb{C}$-vector Bundles 325
12.7 Nash Vector Bundles and Semi-algebraic Vector Bundles 331

13. Polynomial or Regular Mappings with Values in Spheres . 339
13.1 Polynomial Mappings from S^n into S^k 339
13.2 Hopf Forms and Nonsingular Bilinear Forms 346
13.3 Approximation of Mappings with Values in S^1, S^2 or S^4 352
13.4 Homotopy Classes of Mappings into S^n 361
13.5 Mappings from a Product of Spheres into a Sphere 368

14. Algebraic Models of $\mathcal{C}^\infty$ Manifolds 373
14.1 Algebraic Models of $\mathcal{C}^\infty$ Manifolds 373
14.2 More about the Topology of Real Algebraic Sets............ 380

15. Witt Rings in Real Algebraic Geometry 383
15.1 K_0 and the Witt Ring 383
15.2 Separation of Connected Components by Signatures 392
15.3 Comparison between $W(\mathcal{P}(V))$ and $W(\mathcal{S}^0(V))$ 399

Bibliography 407

Index of Notation 421

Index 427

Introduction

In simplest terms, algebraic geometry is the study of the set of solutions of a system of polynomial equations. The main goal of real algebraic geometry is the study of real algebraic sets i.e. subsets of $\mathbb{R}^n$ defined by polynomial equations. By means of a simple example one can see some features which point up the difference between real and complex algebraic geometry. Let us consider the intersection of the straight line $x = t$, depending on the parameter t, with the cubic $y^2 = x^3 - x$. For $t = -1, 0, 1$ the straight line is tangent to the cubic. In the complex plane, when t is different from $-1, 0, 1$, the straight line always intersects the cubic in two points. In the real plane the situation is more intricate.

a) The intersection may be empty because the field of real numbers is not algebraically closed. At first glance this appears to be a glaring defect. Just the same, this defect has some positive aspects, as we shall see.

b) While the intersection may be empty there is, nevertheless, an invariant, namely, the parity of the number of intersections (when the straight line is not tangent). This result can be traced back to complex conjugation, and for this reason one frequently encounters in real algebraic geometry invariants modulo 2.

c) The set of parameters t for which there is nonempty intersection is the union of two intervals. This set cannot be described by means of only polynomial equations and their negations ($\neq$). One has to make use of inequalities (viz. $x^3 - x \geq 0$). One is thus led to consider semi-algebraic sets, which are the subsets of $\mathbb{R}^n$ defined by a finite number of polynomial equations and inequalities. The ordering of the reals, which plays an important role in this example, is also closely related to the euclidean topology of $\mathbb{R}^n$, and is much more important for some real phenomena than the Zariski topology.

As the above example shows, real algebraic geometry is concerned not only with the zeros of polynomials but also with domains where the polynomials have constant sign. A famous example of this type is the 17th problem of Hilbert, which asks whether a polynomial which is nonnegative on $\mathbb{R}^n$ is a sum of squares of rational functions. The solution of this problem given by Artin has a remarkable feature: one is not able to resolve this question by considering only points of $\mathbb{R}^n$. One has to consider points belonging to other fields, which contain the field of rational functions $\mathbb{R}(X_1, ..., X_n)$ and have the algebraic properties of $\mathbb{R}$. More precisely, the fields in question are real closed fields i.e. fields which admit a unique ordering, such that every positive element has a square root and every polynomial of odd degree has a root. When these fields strictly contain $\mathbb{R}$ they necessarily have elements which are positive and less than all positive real numbers. Other real closed fields are strictly contained in $\mathbb{R}$, as, for instance, the field of real algebraic numbers, which is not complete (Cauchy sequences do not always converge).

The introduction of the theory of real closed fields is therefore seen to be indispensable for resolving algebraic problems, even if they are initially posed only over $\mathbb{R}$. The theory of real closed fields is also relevant to geometry: most of the geometric results and their proofs concerning the study of algebraic and semi-algebraic subsets of $\mathbb{R}^n$ carry over to any real closed field. The fact that these proofs require only the axioms of a real closed field sheds a different light on certain results over the reals (for instance, the theorem of Sard or the combinatorial study of the algebraic topology of real algebraic sets). It is nevertheless necessary to emphasize that certain questions require the use of transcendental methods which are only valid for $\mathbb{R}$. For example, the comparison between the algebraic and differentiable categories is based for the most part on the Stone-Weierstrass theorem, which is a special feature of the real numbers.

Special features and tools of real algebraic geometry

Algebraic sets and polynomial and regular functions are studied in algebraic geometry over any field, but in the case of the real field they have rather special behaviour. For example, an irreducible algebraic set can have several connected components with respect to the euclidean topology, the set of non-singular points is not necessarily dense, the dimension may decrease locally, etc. The fact that there are nonconstant polynomials without zeros gives rise to rational functions which are everywhere defined and are not polynomials. One can also replace a finite number of polynomials defining a real algebraic set by a single polynomial (the sum of squares of the original polynomials). This imparts to real algebraic sets a flexibility which is sometimes astonishing. One can show that a number of the constructions of classical geometry (projective space, grassmannians, blowing-up) can be given without leaving the affine setting. The Alexandrov compactification of an algebraic set can be made algebraic. Regular mappings between real algebraic sets are less rigid than polynomial mappings and this leads to new ways of approximating topological situations algebraically.

Semi-algebraic sets and semi-algebraic functions (i.e. functions having semi-algebraic graph) are objects which are truly a special feature of real algebraic geometry. This class of sets has remarkable stability properties, of which the most important is stability under projection. Practically all the useful constructions with semi-algebraic sets do not leave the realm of semi-algebraic sets. Semi-algebraic sets also have a very pleasant topological structure: they have good stratifications. Furthermore, semi-algebraic functions grow in a very well controlled way.

Another family of functions which is of interest in real algebraic geometry is the family of Nash functions, which are real algebraic analytic fuctions (for example the function $\sqrt{1+x^2}$). They can be defined and studied over any real closed field, by nontranscendental methods. They retain the good algebraic properties of polynomials and the flexibility of analytic functions (implicit function theorem, preparation theorem and division theorem), which allows them to be used in the study of smooth semi-algebraic phenomena (tubular neighbourhoods, approximation of differentiable functions).

The development of real algebraic geometry was accompanied by that of real algebra. The basic notion is that of a prime cone of a ring, which simultaneously generalizes the notion of a point of a real algebraic set and that of a total ordering of a field. The set of prime cones of a ring, equipped with a suitable topology, constitutes the real spectrum, which has general properties analogous to those of the spectrum of a ring (the prime ideals of the ring). An example of this is the compactness of the real spectrum. The real spectrum of the ring of polynomial functions on a real algebraic set is sufficiently close to the set itself, with its euclidean topology, that it provides significant geometric information. The real spectrum thus serves as a kind of dictionary between algebraic properties and topological or geometric properties.

The algebraic theory of quadratic forms and real algebraic geometry have deep connections with each other. We have already been given some indication of the importance of sums of squares. The study of quadratic forms over the field of rational functions gives quantitative information concerning Hilbert's 17th problem and allows one to bound the number of inequalities needed to define a semi-algebraic set. Incidentally, the results of semi-algebraic geometry can be used to give information about quadratic forms over the ring of polynomial functions.

Algebraic vector bundles constitute a very important tool in the study of real algebraic sets. They provide, for example, information about $\mathbb{Z}/2$-homology classes represented by real algebraic varieties, factoriality of the ring of regular functions, approximation of smooth mappings by regular ones etc. On the other hand, one does not have a good cohomological theory of coherent algebraic sheaves in real algebraic geometry.

This book does not deal with a certain number of topics which the reader would perhaps have wished to find. We confine ourselves strictly to the algebraic setting and we do not deal with real analytic geometry, which is a very active field. We have also omitted the recent results concerning real algebraic surfaces (in particular, their classification), real abelian varieties, enumerative geometry over $\mathbb{R}$ and the moduli spaces of real algebraic curves. Neither do we give an account of a number of works concerning the 16th problem of Hilbert. Of course this does not mean that we consider these subjects to be unimportant or not interesting; we have simply had to limit ourselves in order to remain within the confines of one volume and to finish the work in a reasonable amount of time.

Some remarks concerning the development of real algebraic geometry

Real algebraic geometry shares many of the basic concepts with algebraic geometry over an algebraically closed field. Nevertheless, it has its own problems and methods. The results that have been obtained have a measure of coherence and aesthetic of their own; they are not necessarily analogous to results of complex algebraic geometry. (A typical example of such a result is the fact that real projective varieties are affine). The development of real algebraic geometry is connected with a number of other areas of mathematics: differential topology, including the theory of singularities, commutative algebra, analytic geometry and also algebraic topology, quadratic forms, model theory and analysis. A start has been made in the algorithmic applications of real algebraic geometry, and they are potentially very important, especially in connection with robotics, computer vision, and computer aided design. Real algebraic geometry has only recently become an independent discipline. Mathematicians working in this field have a wide variety of backgrounds, such as differential topology, complex algebraic geometry, quadratic forms, model theory.

It is still somewhat mysterious why a systematic study of real algebraic geometry was not undertaken earlier. Nevertheless, results which are important for real algebraic geometry punctuated the development of mathematics in the 19th and 20th centuries: Sturm's theorem which gives a procedure for computing the number of real roots of a polynomial in an interval (1835), Harnack's theorem on the maximum number of connected components of a real curve (1876), many works, since 1891, on Hilbert's 16th problem concerning the position of ovals of a nonsingular curve in the real projective plane, the solution of Hilbert's 17th problem by Artin (1927), the Tarski-Seidenberg principle (formulated in the early 1930's but published later) or Comessatti's investigation of real algebraic surfaces and real abelian varieties in the period 1912-1928. But the notion of a semi-algebraic set, which seems so natural to us now, came into existence much later and rather as a byproduct of the study of semi-analytic sets towards the end of the 1950's. The Real Nullstellensatz, whose proof presents hardly more technical difficulties than Hilbert's Nullstellensatz, was proved only in the 1970's. A systematic study of real algebraic varieties started seriously only in 1973 after Tognoli's surprising discovery (based on earlier work of John Nash) that every compact smooth manifold is diffeomorphic to a nonsingular real algebraic set. The extraordinary success of complex analysis and complex geometry is certainly one of the causes of this situation. If one looks at the first chapters of a text on algebraic geometry, the assumption that one is going to work over an algebraically closed field is considered to be so obvious and so inevitable that usually no argument is given to justify this assumption. Occasionally justifications are given: for example, the existence of curves with real equations and without real points.

The real realm is thus labeled as one which lacks the sort of regular phenomena which are necessary for the satisfactory development of a theory. A different and interesting point of view was expressed by R. Thom in his book [324] : "It might be argued that the importance accorded by analysis to the complex field and the theory of analytic functions during the last century has had an unfortunate effect on the orientation of mathematics. By allowing the construction of a beautiful (even too beautiful) theory which was in perfect harmony with the equally successful quantification of physical theories, it has led to a neglect of the real and qualitative nature of things. Now, well past the middle of the twentieth century, it has taken the blossoming of topology to return mathematics to the direct study of geometrical objects, a study which, however, has barely begun; compare the present neglected state of real algebraic geometry with the degree of sophistication and formal perfection of complex algebraic geometry. In the case of any phenomenon governed by an algebraic equation it is of paramount importance to know whether this equations has solutions, *real solutions*, and precisely this question is suppressed when complex scalars are used. As examples of situations in which the reality plays an essential role, we have the following: the characteristic values of a linear differential system, the index of critical points of a function, and the elliptic or hyperbolic character of a differential operator."

Real algebraic geometry is still in its infancy, but it is no longer in the state of neglect deplored by R. Thom. We hope that this book will reflect some of the recent advances in the subject.

1. Ordered Fields, Real Closed Fields

Abstract. The first three sections of this chapter briefly review Artin-Schreier theory: ordered fields, real fields, real closed fields and the real closure of an ordered field. The fourth section is devoted to the Tarski-Seidenberg principle, which is an essential tool for real algebraic geometry.

1.1 Ordered Fields, Real Fields

Definition 1.1.1. *An* ordering *of a field F is a total order relation $\leq$ satisfying:*
  (i) $x \leq y \Rightarrow x + z \leq y + z$,
  (ii) $0 \leq x$, $0 \leq y \Rightarrow 0 \leq xy$.
An ordered field $(F, \leq)$ *is a field F, equipped with an ordering $\leq$.*

We are familiar with $\mathbb{Q}$ and $\mathbb{R}$ with their natural orderings. We now describe the orderings of the field of rational functions $\mathbb{R}(X)$.

Example 1.1.2. There is exactly one ordering of $\mathbb{R}(X)$ such that X is positive and smaller than any positive real number. If

$$P(X) = a_n X^n + a_{n-1} X^{n-1} + \cdots + a_k X^k \quad \text{with } a_k \neq 0,$$

then $P(X) > 0$ for this ordering if and only if $a_k > 0$, and $P(X)/Q(X) > 0$ if and only if $P(X)Q(X) > 0$. It is easy to verify that this indeed defines an ordering of $\mathbb{R}(X)$. Note that, with this ordering, the field $\mathbb{R}(X)$ is not archimedean. It contains infinitely small elements (i.e. positive and smaller than $1/n$, for every $n \in \mathbb{N}$ with $n \neq 0$), such as X, and also infinitely large elements (i.e. bigger than n, for every $n \in \mathbb{N}$) such as $1/X$.

Given any ordering of $\mathbb{R}(X)$, X determines a *cut* (I, J) in $\mathbb{R}$ where $I = \{x \in \mathbb{R} \mid x < X\}$ and $J = \{x \in \mathbb{R} \mid X < x\}$. The cuts $(\emptyset, \mathbb{R})$, $(]-\infty, a[, [a, \infty[)$, $(]-\infty, a],]a, +\infty[)$ and $(\mathbb{R}, \emptyset)$ are respectively denoted by $-\infty$, a_-, a_+ and $+\infty$. Performing, respectively, the change of variables $Y = -1/X$, $Y = a - X$, $Y = X - a$ and $Y = 1/X$, we get an ordering of $\mathbb{R}(Y)$ such that Y is positive and smaller than any positive real number. We have just seen that there is exactly one such ordering. We conclude that there is a bijection between the set of orderings of $\mathbb{R}(X)$ and the set of cuts of $\mathbb{R}$, which is $\{a_+ \mid a \in \mathbb{R}\} \cup \{a_- \mid a \in \mathbb{R}\} \cup \{-\infty, +\infty\}$.

By abuse of notation, we also denote by $a_+, a_-, -\infty, +\infty$ the orderings determined by these cuts. Note that the sign of $f \in \mathbb{R}(X)$ for the ordering a_- is the sign of f on some small open interval $]a - \varepsilon, a[$.

Definition 1.1.3. *A* cone *of a field F is a subset P of F such that:*
  (i) $x \in P$, $y \in P \Rightarrow x + y \in P$,
  (ii) $x \in P$, $y \in P \Rightarrow xy \in P$,
  (iii) $x \in F \Rightarrow x^2 \in P$.
  The cone P is said to be proper *if in addition:*
  (iv) $-1 \notin P$.

Definition 1.1.4. *Let $(F, \leq)$ be an ordered field. The subset*
$$P = \{x \in F \mid x \geq 0\}$$
is called the positive cone *of $(F, \leq)$.*

Proposition 1.1.5. *Let $(F, \leq)$ be an ordered field. The positive cone P of $(F, \leq)$ is a proper cone satisfying:*
  (v) $P \cup -P = F$ *(where $-P = \{x \in F \mid -x \in P\}$).*
  Conversely, if P is a proper cone of a field F satisfying (v), then F is ordered by
$$x \leq y \Leftrightarrow y - x \in P.$$

Notation 1.1.6. *The set of sums of squares of elements of F is denoted by $\sum F^2$.*

The set $\sum F^2$ is a cone, and is contained in every cone of F.

Lemma 1.1.7. *Let P be a proper cone of F.*
  (i) *If $-a \notin P$ then $P[a] = \{x + ay \mid x, y \in P\}$ is a proper cone of F.*
  (ii) *The cone P is contained in the positive cone of an ordering of F.*

Proof. (i) Let us show that $-1 \notin P[a]$: if $-1 = x + ay$, with $x, y \in P$, then either $y = 0$ and $-1 \in P$, or $-a = (1/y)^2 y(1+x) \in P$. Both cases lead to a contradiction.

(ii) Using Zorn's lemma, there exists a maximal proper cone Q containing P. It is enough to show that $Q \cup -Q = F$. Let $a \notin Q$. By (i), $Q[-a]$ is a proper cone and, hence, $Q = Q[-a]$, since Q is maximal. This implies that $-a \in Q$. □

Theorem 1.1.8. *Let F be a field. Then the following properties are equivalent:*
  (i) *F can be ordered.*
  (ii) *The field F has a proper cone.*

(iii) $-1 \notin \sum F^2$.
(iv) For every $x_1, \ldots, x_n$ in F
$$\sum_{i=1}^{n} x_i^2 = 0 \Rightarrow x_1 = \ldots = x_n = 0 \ .$$

Proof. (i) $\Rightarrow$ (ii) $\Rightarrow$ (iii) $\Leftrightarrow$ (iv) are easy. We show that (iii) $\Rightarrow$ (i) : if $-1 \notin \sum F^2$, it follows that $\sum F^2$ is a proper cone. Then use condition (ii) of Lemma 1.1.7. □

Definition 1.1.9. *A field satisfying the properties of the preceding theorem is called a* real field.

It is worth noting that a real field always has characteristic 0.

Proposition 1.1.10. *Let F be a field containing $\mathbb{Q}$ and P a cone of F. Then P is the intersection of the positive cones of orderings of F containing P (the intersection is F if the family of these positive cones is empty).*

Proof. The cone P is certainly contained in this intersection. If $a \notin P$, then P is proper since, if $-1 \in P$, then $a = \frac{1}{4}((1+a)^2 - (1-a)^2) \in P$. Hence, $P[-a]$ is proper by Lemma 1.1.7 (i). Lemma 1.1.7 (ii) gives a positive cone of an ordering containing $P[-a]$ and, hence, P but not a. □

Corollary 1.1.11. *Let F be a field containing $\mathbb{Q}$. Then $\sum F^2$ is the intersection of the positive cones of all orderings of F.*

1.2 Real Closed Fields

Definition 1.2.1. *A* real closed field *F is a real field that has no nontrivial real algebraic extension $F_1 \supset F$, $F_1 \neq F$.*

Theorem 1.2.2. *Let F be a field. Then the following properties are equivalent:*

(i) *The field F is real closed.*
(ii) *There is a unique ordering of F whose positive cone is the set of squares of F and every polynomial of $F[X]$, of odd degree, has a root in F.*
(iii) *The ring $F[i] = F[X]/(X^2+1)$ is an algebraically closed field.*

Proof. (i) $\Rightarrow$ (ii) Let $a \in F$. If a is not a square in F, then $F[\sqrt{a}] = F[X]/(X^2-a)$ is a nontrivial algebraic extension of F and, hence, $F[\sqrt{a}]$ is not real. So

$$-1 = \sum_{i=1}^{n}(x_i + \sqrt{a}\,y_i)^2, \quad \text{hence} \quad -1 = \sum_{i=1}^{n} x_i^2 + a\left(\sum_{i=1}^{n} y_i^2\right) \text{ in } F \ .$$

Since F is real, $-1 \neq \sum_{i=1}^{n} x_i^2$ and, hence, $\sum_{i=1}^{n} y_i^2 \neq 0$. So,

$$-a = \left(\sum_{i=1}^n y_i^2\right)^{-1}\left(1 + \sum_{i=1}^n x_i^2\right) \in \sum F^2 \ .$$

This shows, on the one hand, that $\sum F^2 \cup -\sum F^2 = F$, hence that there is only one possible ordering of F, with $\sum F^2$ as positive cone. On the other hand, if a is not a square, it is negative for this order. Hence, every positive element is a square.

It remains to show that, if $f \in F[X]$ has odd degree, then f has a root in F. If this is not the case, let f be a polynomial of odd degree $d > 1$ such that every polynomial of odd degree $< d$ has a root in F. Since a polynomial of odd degree has at least one odd irreducible factor, f is irreducible. The quotient $F[X]/(f)$ is a nontrivial algebraic extension of F and, hence,

$$-1 = \sum_{i=1}^n h_i^2 + fg \quad \text{with} \quad \deg(h_i) < d \ .$$

Since the term of highest degree in the expansion of $\sum_{i=1}^n h_i^2$ has a coefficient which is a sum of squares and F is real, $\sum_{i=1}^n h_i^2$ is a polynomial of even degree $\leq 2d - 2$. The polynomial g is, hence, of odd degree $\leq d - 2$ and has a root x in F. But then $-1 = \sum_{i=1}^n h_i(x)^2$, which contradicts the fact that F is real.

(ii) $\Rightarrow$ (iii) Let $f \in F[X]$ of degree $d = 2^m n$ with n odd. Let us show by induction on m that f has a root in $F[i]$. For $m = 0$, we know that f has a root in F. Let us suppose, now, that the result is true for $m - 1$. Take $y_1, \ldots, y_d$ to be the roots of f in an algebraic closure of F and define

$$g_h = \prod_{\lambda < \mu}(X - y_\lambda - y_\mu - h y_\lambda y_\mu), \quad \text{for } h \in \mathbb{Z} \ .$$

The polynomial g_h is symmetric in $y_1, \ldots, y_d$ and, hence, $g_h \in F[X]$. The degree of g_h is $d(d-1)/2 = 2^{m-1} n'$ with n' odd. By induction, g_h has a root in $F[i]$ and, hence, there exist λ and μ with $y_\lambda + y_\mu + h y_\lambda y_\mu \in F[i]$. Letting h range over $\mathbb{Z}$, we see that there exist λ and μ with $y_\lambda + y_\mu \in F[i]$ and $y_\lambda y_\mu \in F[i]$. These elements y_λ and y_μ are the solutions of a quadratic equation with coefficients in $F[i]$, which has its two solutions in $F[i]$ (proceed as for $\mathbb{C}$). The polynomial f has, thus, a root in $F[i]$.

Suppose now that $f \in F[i][X]$. Let $\overline{f}$ be the polynomial obtained by replacing the coefficients of f with their conjugates. Since $f\overline{f} \in F[X]$, $f\overline{f}$ has a root x in $F[i]$. Then either x is a root of f, or it is a root of $\overline{f}$, and in this case, its conjugate $\overline{x}$ is a root of f.

(iii) $\Rightarrow$ (i) The field F is real. We already know that -1 is not a square in F, since $F[i]$ is a field. It is enough to show that in F, a sum of squares is still a square: let $a, b \in F$ and $c, d \in F$ such that $a + ib = (c + id)^2$; then $a^2 + b^2 = (c^2 + d^2)^2$.

To conclude, note that $F[i]$ is the only nontrivial algebraic extension of F. $\square$

Example 1.2.3. The field $\mathbb{R}$ is, of course, real closed. The real algebraic numbers (the real numbers algebraic over $\mathbb{Q}$) form a real closed field denoted by $\mathbb{R}_{\text{alg}}$. Let $\mathbb{R}(X)^\wedge$ (resp. $\mathbb{C}(X)^\wedge$) be *the field of Puiseux series* with real (resp. complex) coefficients, i.e. the set of expressions

$$\sum_{i=k}^{+\infty} a_i X^{i/q} \text{ with } k \in \mathbb{Z},\ q \in \mathbb{N} \setminus \{0\},\ a_i \in \mathbb{R} \text{ (resp. } \mathbb{C}).$$

It is known ([334], p.98) that $\mathbb{C}(X)^\wedge$ is algebraically closed. Since $\mathbb{C}(X)^\wedge = \mathbb{R}(X)^\wedge[i]$, $\mathbb{R}(X)^\wedge$ is real closed. A positive element of $\mathbb{R}(X)^\wedge$ is a Puiseux series of the form $\sum_{i=k}^{+\infty} a_i x^{i/q}$ with $a_k > 0$.

We now extend to polynomials in one variable with coefficients in a real closed field R some properties that are well known for differentiable functions over $\mathbb{R}$. Intervals in R will be denoted in the following way:

$$[a,b] = \{x \in R \mid a \le x \le b\},\quad]a,b[= \{x \in R \mid a < x < b\},\ \text{etc...}$$

Proposition 1.2.4. *Let R be a real closed field, $f \in R[X]$, $a, b \in R$ with $a < b$. If $f(a)f(b) < 0$, then there exists x in $]a,b[$ such that $f(x) = 0$.*

Proof. By property (iii) of Theorem 1.2.2, the irreducible factors of f are linear, or have the form $(X-c)^2 + d^2 = (X - c - id)(X - c + id)$. If $f(a)$ and $f(b)$ have opposite signs, then $g(a)$ and $g(b)$ have opposite signs for some linear factor g of f. Hence the root of g is in $]a,b[$. $\square$

Proposition 1.2.5. *Let R be a real closed field, $f \in R[X]$, $a, b \in R$ with $a < b$ and $f(a) = f(b) = 0$. Then the derivative polynomial f' has a root in $]a,b[$.*

Proof. We can suppose that a and b are two consecutive roots of f, i.e. that f never vanishes in $]a,b[$. Then

$$f = (X-a)^m (X-b)^n g,$$

where g never vanishes in $[a,b]$. Hence, by Proposition 1.2.4, g has constant sign on $[a,b]$. Then

$$f' = (X-a)^{m-1}(X-b)^{n-1} g_1$$

where

$$g_1 = m(X-b)g + n(X-a)g + (X-a)(X-b)g'.$$

Thus, $g_1(a) = m(a-b)g(a)$ and $g_1(b) = n(b-a)g(b)$, hence $g_1(a)$ and $g_1(b)$ have opposite signs. By Proposition 1.2.4, g_1 has a root in $]a,b[$ and so does f'. $\square$

Corollary 1.2.6. *Let R be a real closed field, $f \in R[X]$, $a, b \in R$ with $a < b$. There exists $c \in]a,b[$ such that $f(b) - f(a) = (b-a)f'(c)$.*

Corollary 1.2.7. *Let R be a real closed field, $f \in R[X]$, $a, b \in R$ with $a < b$. If the derivative f' is positive (resp. negative) on $]a, b[$, then f is strictly increasing (resp. strictly decreasing) on $[a, b]$.*

The following results concern root counting.

Definition 1.2.8. *Let R be a real closed field, and let f and g be in $R[X]$. The Sturm sequence of f and g is the sequence of polynomials $(f_0, \ldots, f_k)$ defined as follows:*

$f_0 = f$, $\quad f_1 = f'g$,

$f_i = f_{i-1}q_i - f_{i-2}$ with $q_i \in R[X]$ and $\deg(f_i) < \deg(f_{i-1})$ for $i = 2, \ldots, k$,

f_k *is a greatest common divisor of f and $f'g$.*

Given a sequence $(a_0, \ldots, a_k)$ of elements of R with $a_0 \neq 0$, we define the number of sign changes in the sequence $(a_0, \ldots, a_k)$ as follows: count one sign change if $a_i a_\ell < 0$ with $\ell = i + 1$ or $\ell > i + 1$ and $a_j = 0$ for every j, $i < j < \ell$.

If $a \in R$ is not a root of f and $(f_0, \ldots, f_k)$ is the Sturm sequence of f and g, we define $v(f, g; a)$ to be the number of sign changes in $(f_0(a), \ldots, f_k(a))$.

Theorem 1.2.9 (Sylvester's Theorem). *Let R be a real closed field and let f and g be two polynomials in $R[X]$. Let $a, b \in R$ be such that $a < b$ and neither a nor b are roots of f. Then the difference between the number of roots of f in the interval $]a, b[$ for which g is positive and the number of roots of f in the interval $]a, b[$ for which g is negative, is equal to $v(f, g; a) - v(f, g; b)$.*

Proof. First note that the Sturm sequence $(f_0, \ldots f_k)$ is (up to signs) equal to the sequence obtained from the euclidean algorithm. Define a new sequence $(g_0, \ldots, g_k)$ by $g_i = f_i/f_k$ for $i \in \{0, \ldots, k\}$. Note that the number of sign changes in $(f_0(x), f_1(x))$ (resp. $(f_{i-1}(x), f_i(x), f_{i+1}(x))$) and the number of sign changes in $(g_0(x), g_1(x))$ (resp. $(g_{i-1}(x), g_i(x), g_{i+1}(x))$) coincide for any x which is not a root of f. Note also that the roots of g_0 are exactly the roots of f which are not roots of g. Observe that for $i \in \{0, \ldots, k\}$, g_{i-1} and g_i are relatively prime.

We consider, now, how $v(f, g; x)$ behaves when x passes through a root c of a polynomial g_i.

If c is a root of g_0, then it is not a root of g_1. We write $f'(c_-) > 0$ (resp. < 0) if f' is positive (resp. negative) immediately to the left of c. The sign of $f'(c_+)$ is defined similarly. According to the signs of $g(c)$, $f'(c_-)$ and $f'(c_+)$ we have, by Corollary 1.2.7, the following eight cases:

$g(c) > 0$, $f'(c_-) > 0$, $f'(c_+) > 0$ $\qquad$ $g(c) < 0$, $f'(c_-) > 0$, $f'(c_+) > 0$

	c	
f	$-\ \ 0$	$+$
$f'g$	$+$	$+$

	c	
f	$-\ \ 0$	$+$
$f'g$	$-$	$-$

$g(c) > 0,\ f'(c_-) < 0,\ f'(c_+) > 0$ $\qquad$ $g(c) < 0,\ f'(c_-) < 0,\ f'(c_+) > 0$

$$\begin{array}{c|ccc} & & c & \\ \hline f & + & 0 & + \\ f'g & - & & + \end{array} \qquad \begin{array}{c|ccc} & & c & \\ \hline f & + & 0 & + \\ f'g & + & & - \end{array}$$

$g(c) > 0,\ f'(c_-) > 0,\ f'(c_+) < 0$ $\qquad$ $g(c) < 0,\ f'(c_-) > 0,\ f'(c_+) < 0$

$$\begin{array}{c|ccc} & & c & \\ \hline f & - & 0 & - \\ f'g & + & & - \end{array} \qquad \begin{array}{c|ccc} & & c & \\ \hline f & - & 0 & - \\ f'g & - & & + \end{array}$$

$g(c) > 0,\ f'(c_-) < 0,\ f'(c_+) < 0$ $\qquad$ $g(c) < 0,\ f'(c_-) < 0,\ f'(c_+) < 0$

$$\begin{array}{c|ccc} & & c & \\ \hline f & + & 0 & - \\ f'g & - & & - \end{array} \qquad \begin{array}{c|ccc} & & c & \\ \hline f & + & 0 & - \\ f'g & + & & + \end{array}$$

In every case as x passes through c, the number of sign changes in $(f_0(x), f_1(x))$ decreases by 1 if $g(c) > 0$, and increases by 1 if $g(c) < 0$.

If c is a root of g_i with $i = 1, \ldots, k$, then it is neither a root of g_{i-1} nor a root of g_{i+1}, and $g_{i-1}(c)g_{i+1}(c) < 0$, by the definition of the sequence. Passing through c does not lead to any modification of the number of sign changes in $(f_{i-1}(x), f_i(x), f_{i+1}(x))$ in this case.

The conclusion of the theorem is now obvious. $\square$

Corollary 1.2.10 (Sturm's Theorem). *Let R be a real closed field and $f \in R[X]$. Let $a, b \in R$ be such that $a < b$ and neither a nor b are roots of f. Then the number of roots of f in the interval $]a, b[$ is equal to $v(f, 1; a) - v(f, 1; b)$.*

Proof. Apply Theorem 1.2.9 with $g = 1$. $\square$

Lemma 1.2.11. *Let R be a real closed field, $f = a_n X^n + \cdots + a_0 \in R[X]$ with $a_n \neq 0$. Let $M = 1 + |a_{n-1}/a_n| + \cdots + |a_0/a_n|$ (where $|c|$ denotes c if $c \geq 0$ and $-c$ if $c < 0$). Then f never vanishes on $[M, +\infty[$ (resp. $]-\infty, -M]$), and its sign is the sign of a_n (resp. $(-1)^n a_n$).*

Proof. Let $x \in R$ with $|x| \geq M$. Then, setting $b_i = a_i/a_n$,

$$f(x) = a_n x^n (1 + b_{n-1} x^{-1} + \cdots + b_0 x^{-n})$$

and, since $|b_{n-1} x^{-1} + \cdots + b_0 x^{-n}| \leq (|b_{n-1}| + \cdots + |b_0|) M^{-1} < 1$, $f(x)$ has the same sign as $a_n x^n$. $\square$

Corollary 1.2.12. *Let R be a real closed field, f and g in $R[X]$, $(f_0,\ldots,f_k)$ the Sturm sequence of f and g. Let $v(f,g;+\infty)$ (resp. $v(f,g;-\infty)$) denote the number of sign changes in the sequence of coefficients of the highest degree terms of $(f_0,\ldots,f_k)$ (resp. $(f_0(-X),\ldots,f_k(-X))$). Then the difference between the total number of roots of f in R for which g is positive and the total number of roots of f in R for which g is negative, is equal to $v(f,g;-\infty) - v(f,g;+\infty)$.*

Proof. Lemma 1.2.11 shows that we can choose $M \in R$ large enough so that all the roots of f in R are in the interval $]-M,+M[$ and that $v(f,g;M)$ (resp. $v(f,g;-M)$) is equal to $v(f,g;+\infty)$ (resp. $v(f,g;-\infty)$). The corollary now follows from Sylvester's Theorem 1.2.9. □

Remark 1.2.13. It is easy to see, using the preceding corollary, that a monic squarefree polynomial f (i.e. such that f and its derivative f' are relatively prime) of degree n with coefficients in R, has its n roots in R if and only if the Sturm sequence of f and 1 consists of $n+1$ polynomials $f_0 = f$, $f_1 = f'$, $f_2,\ldots,f_n$ and all the leading coefficients of these polynomials are positive.

The next result is of a different nature: it gives a simple upper bound on the number of roots.

Proposition 1.2.14 (Descartes's Lemma). *Let R be a real closed field, $f = a_n X^n + a_{n-1} X^{n-1} + \cdots + a_k X^k \in R[X]$ with $a_n a_k \neq 0$. Then the number of positive roots of f is less than or equal to the number of sign changes in the sequence of coefficients $a_n,\ldots,a_k$.*

Proof. Proceed by induction on n. The result is clear for $n = 1$. So assume it to be true for $n - 1$, with $n > 1$. We may also assume that X does not divide f.

We have $f = a_n X^n + \cdots + a_q X^q + a_0$ and $f' = na_n X^{n-1} + \cdots + qa_q X^{q-1}$ with a_n, a_q, a_0 different from 0. By induction, the number of sign changes in the sequence $a_n,\ldots,a_q$ bounds the number of positive roots of f'. Take for c the smallest positive root of f' (take $c = +\infty$ if there is none). Then f' has the same sign on $]0,c[$ as a_q. Since $f(0) = a_0$, we see from the variation of f that it may have roots in $]0,c[$ only if $a_q a_0 < 0$, which is the case if the number of sign changes in $a_n,\ldots,a_0$ exceeds by 1 the number of sign changes in $a_n,\ldots,a_q$. Application of Proposition 1.2.5 completes the proof. □

1.3 Real Closure of an Ordered Field

Definition 1.3.1. *An algebraic extension R of an ordered field $(F, \leq)$ is called a* real closure *of F if R is real closed and its unique ordering extends the ordering of F (i.e. the inclusion $F \hookrightarrow R$ preserves the ordering).*

1.3 Real Closure of an Ordered Field

Theorem 1.3.2. *Every ordered field $(F, \leq)$ has a real closure. If R and R' are two real closures of $(F, \leq)$, then there exists a unique F-isomorphism $\Phi : R \to R'$.*

Proof. Choose an algebraic closure $\overline{F} \supset F$ of F, and let $\mathcal{E}$ be the family of ordered sub-extensions $(K, \leq)$ with $F \subset K \subset \overline{F}$, the inclusion $F \subset K$ being order-preserving. The family $\mathcal{E}$ is ordered by the relation $(K, \leq) \prec (K', \leq)$, defined by requiring that $K \subset K'$ and the inclusion is order-preserving. By Zorn's lemma, $\mathcal{E}$ has a maximal element $(R, \leq)$. We show, now, that R is real closed. First we show that every positive element of $(R, \leq)$ is a square: if $a \in R$ is positive and is not a square in R, let P be the subset of $R(\sqrt{a}) \subset \overline{F}$ of elements of the form
$$\sum_{i=1}^{n} b_i(c_i + d_i\sqrt{a})^2$$
with $c_i, d_i \in R$ and b_i in the positive cone of $(R, \leq)$. Then P is a cone and is proper because, if
$$-1 = \sum_{i=1}^{n} b_i(c_i + d_i\sqrt{a})^2 ,$$
then $-1 = \sum_{i=1}^{n} b_i(c_i^2 + ad_i^2)$ would be in the positive cone of $(R, \leq)$. By Lemma 1.1.7, a maximal proper cone containing P is the positive cone of an ordering of $R(\sqrt{a})$, extending the ordering of R. This contradicts the maximality of $(R, \leq)$. Hence the field R has a unique ordering, whose positive elements are squares of R. This implies that if K is a real field with $R \subset K \subset \overline{F}$, then any ordering of K extends the ordering of R, whence $R = K$ by maximality. The field R is, thus, real closed.

We now show the second statement of the theorem. Let R be a real closure of $(F, \leq)$ and R' a real closed field containing F, such that the ordering of R' extends the ordering of F. Let $\mathcal{F}$ be the family of homomorphisms $\varphi : K \to R'$, with $F \subset K \subset R$, such that φ is order-preserving (K is ordered by restricting the ordering of R). We define a partial order relation on $\mathcal{F}$: $\varphi_1 \prec \varphi_2$ if there is a commutative diagram

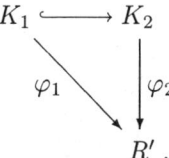

By Zorn's lemma, we can choose $\Phi : L \to R'$ to be maximal in $\mathcal{F}$. We show that $L = R$. If there exists $a \in R \setminus L$, let $f = \sum_{i=0}^{q} c_i X^i \in L[X]$ be its minimal polynomial over L. The polynomial f has no multiple roots since L has characteristic zero. Let $a_1 < \ldots < a_n$ be the roots of f in R and say $a = a_j$. The Sturm sequence of f and 1 is a sequence of polynomials in $L[X]$ and, since Φ is order-preserving, we see, by using Corollary 1.2.12, that

$f_\Phi = \sum_{i=0}^{q} \Phi(c_i)X^i$ has n roots $b_1 < \ldots < b_n$ in R' as well. We thus have a homomorphism $\Psi : L(a) \to R'$ extending Φ, such that $\Psi(a) = b_j$.

Our previous proof and the primitive element theorem give the following result that will be useful later:

Lemma 1.3.3. *Let L_1 be an extension of finite degree of L contained in R. Then there exists a homomorphism $\Phi_1 : L_1 \to R'$ extending Φ.*

We now show that $\Psi : L(a) \to R'$ defined above is order-preserving. Let $y \in L(a)$, $y \geq 0$, and choose $x_1, \ldots, x_{n-1}, z$ in R with $x_i^2 = a_{i+1} - a_i$ and $z^2 = y$. Using Lemma 1.3.3 for $L_1 = L(a_1, \ldots, a_n, y, x_1, \ldots, x_{n-1}, z)$, we have $\Phi_1 : L_1 \to R'$. The $\Phi_1(a_i)$ are roots of f_Φ in R' and, since $\Phi_1(a_{i+1}) - \Phi_1(a_i) = (\Phi_1(x_i))^2 \geq 0$, we have $\Phi_1(a_i) = b_i$ and, in particular, $\Phi_1(a) = b_j$. Hence $\Phi_1|_{L(a)} = \Psi$. Then $\Psi(y) = \Phi_1(y) = (\Phi_1(z))^2$, which shows that Ψ is order-preserving. This contradicts the maximality of Φ.

Conclusion: we have $L = R$ and Φ is an F-homomorphism from R into R'. Such a Φ is unique, for if $a \in R$ is the jth root in R of its minimal polynomial f over F, the proof above shows that $\Phi(a)$ is the jth root of f_Φ in R'. Let us state the result that we have obtained:

Proposition 1.3.4. *Let $(F, \leq)$ be an ordered field, R a real closure of $(F, \leq)$, and R', a real closed extension of F whose ordering extends that of F. Then there exists a unique F-homomorphism $\Phi : R \to R'$.*

Now we can complete easily the proof of Theorem 1.3.2. If R and R' are two real closures of $(F, \leq)$, then, by Proposition 1.3.4, there are two unique F-homomorphisms $\Phi : R \to R'$ and $\Phi' : R' \to R$. By uniqueness, we have $\Phi' \circ \Phi = \mathrm{Id}_R$ and $\Phi \circ \Phi' = \mathrm{Id}_{R'}$. $\square$

Remark 1.3.5. The uniqueness of the real closure of an ordered field is stronger than the uniqueness of the algebraic closure of a field in the following sense: a real closure of $(F, \leq)$ has no other F-automorphism than the identity.

By abuse of language we shall speak, from now on, of *the* real closure of an ordered field.

Example 1.3.6.
   a) The real closure of $\mathbb{Q}$ is $\mathbb{R}_{\mathrm{alg}}$.
   b) We describe, now, the real closure of $\mathbb{R}(X)$ with the ordering 0_+ (see Example 1.1.2). The field $\mathbb{R}(X)$ is canonically a subfield of the field of Puiseux series $\mathbb{R}(X)^\wedge$, which is real closed (Example 1.2.3). The ordering of $\mathbb{R}(X)^\wedge$ makes X positive (since $X = (X^{1/2})^2$) and smaller than every positive real number (since, if $a \in \mathbb{R}$, $a > 0$, $a - X$ is a square in $\mathbb{R}(X)^\wedge$). Hence it extends the ordering considered on $\mathbb{R}(X)$. Thus, the real closure of $\mathbb{R}(X)$ for the ordering 0_+ is the field $\mathbb{R}(X)^\wedge_{\mathrm{alg}}$ of Puiseux series algebraic over $\mathbb{R}(X)$.

Proposition 1.3.7. *Let $(F, \leq)$ be an ordered field, $F(a)$ a finite algebraic extension of F generated by a and let f be the minimal polynomial of a over F. The number of orderings of $F(a)$, extending the ordering of F, is equal to the number of roots of f in the real closure of $(F, \leq)$. This number is congruent to the degree of the extension $[F(a) : F]$ modulo 2.*

Proof. Choose, if there exists one, an ordering $\leq$ of $F(a)$ extending the ordering of F. The real closure of $(F(a), \leq)$ is isomorphic (up to a unique F-isomorphism) to the real closure of $(F, \leq)$, and the image of a in this real closure is a root of f. Conversely, if b is a root of f in the real closure R of $(F, \leq)$, then the F-homomorphism $\Phi : F(a) \to R$ defined by $\Phi(a) = b$ induces an ordering of $F(a)$ extending the ordering of F. The last statement is a consequence of the facts that f has $[F(a) : F]$ roots in $R[i]$ and the roots of f in $R[i] \setminus R$ are pairwise conjugate. □

1.4 The Tarski-Seidenberg Principle

Notation 1.4.1. *Let R be a real closed field, $a \in R$. Define*

$$\begin{aligned} \operatorname{sign}(a) &= 0 & \text{if} & \quad a = 0, \\ \operatorname{sign}(a) &= 1 & \text{if} & \quad a > 0, \\ \operatorname{sign}(a) &= -1 & \text{if} & \quad a < 0. \end{aligned}$$

The Tarski-Seidenberg principle is the following result.

Theorem 1.4.2. *Let $f_i(X, Y) = h_{i,m_i}(Y) X^{m_i} + \cdots + h_{i,0}(Y)$ for $i = 1, \ldots, s$ be a sequence of polynomials in $n + 1$ variables with coefficients in $\mathbb{Z}$, where $Y = (Y_1, \ldots, Y_n)$. Let ϵ be a function from $\{1, \ldots, s\}$ to $\{-1, 0, 1\}$. Then there exists a boolean combination $\mathcal{B}(Y)$ (i.e. a finite composition of disjunctions, conjunctions and negations) of polynomial equations and inequalities in the variables Y with coefficients in $\mathbb{Z}$ such that for every real closed field R and for every $y \in R^n$, the system*

$$\begin{cases} \operatorname{sign}(f_1(X, y)) = \epsilon(1) \\ \quad \vdots \\ \operatorname{sign}(f_s(X, y)) = \epsilon(s) \end{cases}$$

has a solution x in R if and only if $\mathcal{B}(y)$ holds true in R.

The proof of the theorem will use the following notation.

Notation 1.4.3. *Let $f_1, \ldots, f_s$ be a sequence of polynomials in $R[X]$ and let $x_1 < \ldots < x_N$ be the roots in R of all f_i that are not identically zero. By convention we define $x_0 = -\infty$, $x_{N+1} = +\infty$. If $I_k =]x_k, x_{k+1}[$, $\operatorname{sign}(f_i(x))$ is constant for $x \in I_k$, and is denoted $\operatorname{sign}(f_i(I_k))$.*
The matrix with s rows and $2N + 1$ columns whose i^{th} row is

$$\text{sign}(f_i(I_0)), \text{sign}(f_i(x_1)), \text{sign}(f_i(I_1)), \ldots, \text{sign}(f_i(x_N)), \text{sign}(f_i(I_N))$$

is denoted $\text{SIGN}_R(f_1, \ldots, f_s)$. Note that $\text{SIGN}_R(f_1, \ldots, f_s)$ is a matrix with entries in $\{-1, 0, 1\}$.

If $m = \max(\{\deg(f_i) \mid i = 1, \ldots, s\})$ then $N \leq sm$. The disjoint union of the sets of matrices with entries in $\{-1, 0, 1\}$ having s rows and $2\ell + 1$ columns, for $\ell = 0, \ldots, sm$, is denoted $W_{s,m}$.

Lemma 1.4.4. *Let ϵ be a function from $\{1, \ldots, s\}$ to $\{-1, 0, +1\}$. Then there exists a subset $W(\epsilon)$ of $W_{s,m}$ such that for every real closed field R and every sequence $f_1, \ldots, f_s$ of polynomials in $R[X]$ of degrees $\leq m$, the system*

$$\begin{cases} \text{sign}(f_1(X)) = \epsilon(1) \\ \quad \vdots \\ \text{sign}(f_s(X)) = \epsilon(s) \end{cases}$$

has a solution x in R if and only if $\text{SIGN}_R(f_1, \ldots, f_s) \in W(\epsilon)$.

Proof. $W(\epsilon)$ is the subset of $W_{s,m}$ whose elements are matrices having one of their columns coinciding with the sequence $\epsilon(1), \ldots, \epsilon(s)$. □

The importance of the concept of "SIGN_R" is that the "SIGN_R" of a sequence of polynomials $f_1, \ldots, f_s$ is completely determined by the "SIGN_R" of a new and simpler sequence.

Lemma 1.4.5. *There exists a mapping φ from $W_{2s,m}$ to $W_{s,m}$ such that for every real closed field R and every sequence $f_1, \ldots, f_s$ of polynomials in $R[X]$ of degrees $\leq m$, with f_s nonconstant and none of the $f_1, \ldots, f_{s-1}$ identically zero, we have:*

$$\text{SIGN}_R(f_1, \ldots, f_s) = \varphi(\text{SIGN}_R(f_1, \ldots, f_{s-1}, f'_s, g_1, \ldots, g_s)),$$

where f'_s is the derivative of f_s, and $g_1, \ldots, g_s$ are the remainders of the euclidean division of f_s by $f_1, \ldots, f_{s-1}, f'_s$, respectively.

Proof. Let $x_1 < \ldots < x_N$, with $N \leq 2sm$, be the roots in R of those polynomials among $f_1, \ldots, f_{s-1}, f'_s, g_1, \ldots, g_s$ that are not identically zero. Extract from these roots the subsequence $x_{i_1} < \ldots < x_{i_M}$ of the roots of the polynomials $f_1, \ldots, f_{s-1}, f'_s$. The sequence $i_1, \ldots, i_M$ depends only on $w = \text{SIGN}_R(f_1, \ldots, f_{s-1}, f'_s, g_1, \ldots, g_s)$. By convention, let $i_0 = 0$ with $x_0 = -\infty$ and let $i_{M+1} = N + 1$ with $x_{N+1} = +\infty$. For $k = 1, \ldots, M$ one of the polynomials $f_1, \ldots, f_{s-1}, f'_s$ vanishes at x_{i_k}. It is enough to know w in order to choose a function $\theta : \{1, \ldots, M\} \to \{1, \ldots, s\}$ such that $f_s(x_{i_k}) = g_{\theta(k)}(x_{i_k})$. We show that the existence of a root of f_s in an interval $]x_{i_k}, x_{i_{k+1}}[$, for $k = 0, \ldots, M$, depends only on w. The polynomial f_s has a root

– in $]x_{i_k}, x_{i_{k+1}}[$, for $k = 1, \ldots, M - 1$, if and only if

$$\text{sign}(g_{\theta(k)}(x_{i_k})) \, \text{sign}(g_{\theta(k+1)}(x_{i_{k+1}})) = -1,$$

- in $]-\infty, x_{i_1}[$ (if $M \neq 0$) if and only if
$$\text{sign}(f'_s(]-\infty, x_1[))\,\text{sign}(g_{\theta(1)}(x_{i_1})) = 1\,,$$
- in $]x_{i_M}, +\infty[$ (if $M \neq 0$) if and only if
$$\text{sign}(f'_s(]x_N, +\infty[))\,\text{sign}(g_{\theta(M)}(x_{i_M})) = -1\,,$$
- in $]-\infty, +\infty[$ always if $M = 0$.

Now let $y_1 < \ldots < y_L$ (with $L \leq sm$) be the roots in R of the polynomials $f_1, \ldots, f_s$. As before, let $y_0 = -\infty$, $y_{L+1} = +\infty$. Define the function
$$\rho : \{0, \ldots, L+1\} \longrightarrow \{0, \ldots, M+1\} \cup \{(k, k+1) \mid k = 0, \ldots, M\}$$
$$\ell \longmapsto \begin{cases} k & \text{if } y_\ell = x_{i_k}\,, \\ (k, k+1) & \text{if } y_\ell \in]x_{i_k}, x_{i_{k+1}}[\,. \end{cases}$$

From what we have seen before, the number L and the function ρ depend only on w. We are now ready to verify that $\text{SIGN}_R(f_1, \ldots, f_s)$ depends only on w.

For $j = 1, \ldots, s-1$, we have
- if $\rho(\ell) = k$ $\quad\quad\quad\quad \text{sign}(f_j(y_\ell)) = \text{sign}(f_j(x_{i_k}))\,,$
- if $\rho(\ell) = (k, k+1)$ $\quad \text{sign}(f_j(y_\ell)) = \text{sign}(f_j(]x_{i_k}, x_{i_{k+1}}[))\,.$

We also have
- if $\rho(\ell) = k$ or $(k, k+1)$ $\quad \text{sign}(f_j(]y_\ell, y_{\ell+1}[)) = \text{sign}(f_j(]x_{i_k}, x_{i_{k+1}}[))\,.$

We now deal with the case $j = s$. We have
- if $\rho(\ell) = k$ $\quad\quad\quad\quad \text{sign}(f_s(y_\ell)) = \text{sign}(g_{\theta(k)}(x_{i_k}))\,,$
- if $\rho(\ell) = (k, k+1)$ $\quad \text{sign}(f_s(y_\ell)) = 0\,.$

The most delicate case concerns $\text{sign}(f_s(]y_\ell, y_{\ell+1}[))$:
- if $\ell \neq 0$, $\rho(\ell) = k$ $\quad\quad \text{sign}(f_s(]y_\ell, y_{\ell+1}[)) = \text{sign}(g_{\theta(k)}(x_{i_k}))$
  if this is nonzero,
  $\quad\quad\quad\quad\quad\quad\quad\quad\quad \text{sign}(f_s(]y_\ell, y_{\ell+1}[)) = \text{sign}(f'_s(]x_{i_k}, x_{i_{k+1}}[))$
  otherwise,
- if $\ell \neq 0$, $\rho(\ell) = (k, k+1)$
  $\quad\quad\quad\quad\quad\quad\quad\quad\quad \text{sign}(f_s(]y_\ell, y_{\ell+1}[)) = \text{sign}(f'_s(]x_{i_k}, x_{i_{k+1}}[))\,,$
- if $\ell = 0$ $\quad\quad\quad\quad\quad\quad\quad \text{sign}(f_s(]-\infty, y_1[)) = -\text{sign}(f'_s(]-\infty, x_1[))\,.$

$\square$

By Lemma 1.4.4, the proof of the Tarski-Seidenberg principle can be reduced to the following result.

Proposition 1.4.6. *Let $f_i(X, Y) = h_{i,m_i}(Y)X^{m_i} + \cdots + h_{i,0}(Y)$, for $i = 1, \ldots, s$, be a sequence of polynomials in $n+1$ variables with coefficients in $\mathbb{Z}$, where $Y = (Y_1, \ldots, Y_n)$, and let $m = \max(\{m_i \mid i = 1, \ldots, s\})$. Let W' be a subset of $W_{s,m}$. Then there exists a boolean combination $\mathcal{B}(Y)$ of polynomial equations and inequalities in the variables Y with coefficients in $\mathbb{Z}$, such that, for every real closed field R and every $y \in R^n$, one has*
$$\text{SIGN}_R(f_1(X, y), \ldots, f_s(X, y)) \in W' \Leftrightarrow \mathcal{B}(y) \text{ is satisfied in } R\,.$$

Proof. Without loss of generality, we may assume that none of the polynomials $f_1, \ldots, f_s$ is identically zero and that $h_{i,m_i}(Y)$ is not identically zero for $i = 1, \ldots, s$. We associate to the sequence of polynomials $(f_1, \ldots, f_s)$ the sequence $(m_1, \ldots, m_s)$ of their degrees in X. To compare these finite sequences of integers, define a strict order as follows:

$$\sigma = (m'_1, \ldots, m'_t) \prec \tau = (m_1, \ldots, m_s)$$

if there exists $p \in \mathbb{N}$ such that, for every $q > p$, the number of times q appears in σ is equal to the number of times q appears in τ, and the number of times p appears in σ is smaller than the number of times p appears in τ. This gives a well-ordering of the set of sequences of integers: there is no infinite chain $\sigma_1 \succ \sigma_2 \succ \sigma_3 \succ \ldots$. We proceed now by induction with respect to the order $\prec$.

Let $m = \max(\{m_1, \ldots, m_s\})$. If $m = 0$, then the result is straightforward, since $\text{SIGN}_R(f_1(X,y), \ldots, f_s(X,y))$ is the list of signs of "constant terms" $h_{1,0}(y), \ldots, h_{s,0}(y)$.

Suppose that $m \geq 1$ and $m_s = m$. Let $W'' \subset W_{2s,m}$ be the inverse image of $W' \subset W_{s,m}$ under the mapping φ defined in Lemma 1.4.5. By this lemma, for every real closed field R and for every $y \in R^n$ such that $h_{i,m_i}(y) \neq 0$ for $i = 1, \ldots, s$, the property

$$\text{SIGN}_R(f_1(X,y), \ldots, f_s(X,y)) \in W'$$

is equivalent to the property

$$\text{SIGN}_R(f_1(X,y), \ldots, f_{s-1}(X,y), f'_s(X,y), g_1(X,y), \ldots, g_s(X,y)) \in W'',$$

where f'_s is the derivative of f_s with respect to X and $g_1, \ldots, g_s$ are the remainders in the euclidean division (with respect to X) of f_s by $f_1, \ldots, f_{s-1}, f'_s$, respectively, multiplied by appropriate even powers of $h_{1,m_1}, \ldots, h_{s,m_s}$, respectively, in order to clear the denominators. Now, the sequence of degrees in X of $f_1, \ldots, f_{s-1}, f'_s, g_1, \ldots, g_s$ is smaller than $(m_1, \ldots, m_s)$ with respect to the order $\prec$. On the other hand, if at least one among the $h_{i,m_i}(y)$ is zero, we can truncate the corresponding polynomial f_i and obtain a sequence of polynomials, whose sequence of degrees in X is smaller than $(m_1, \ldots, m_s)$ with respect to the order $\prec$. This completes the proof of Proposition 1.4.6 and proves the Tarski-Seidenberg principle as well.
□ □

The Tarski-Seidenberg principle will also be useful in the following form:

Corollary 1.4.7. *Let F be a real field, $f_1(X, Y), \ldots, f_s(X, Y)$ a sequence of polynomials in $n+1$ variables with coefficients in F, where $Y = (Y_1, \ldots, Y_n)$. Let ϵ be a function from $\{1, \ldots, s\}$ to $\{-1, 0, +1\}$. Then there exists a boolean combination $\mathcal{B}(Y)$ of polynomial equations and inequalities in the variables Y with coefficients in F, such that, for every real closed field R containing F and every $y \in R^n$, the system*

$$\begin{cases} \text{sign}(f_1(X,y)) = \epsilon(1) \\ \quad\vdots \\ \text{sign}(f_s(X,y)) = \epsilon(s) \end{cases}$$

has a solution x in R if and only if $\mathcal{B}(y)$ holds true in R.

Proof. Note that we can write $f_i(X,Y) = G_i(X,Y,a)$, where a is the sequence of coefficients of the f_i (so $a \in F^m$ for some appropriate m) and $G_i \in \mathbb{Z}[X,Y,T]$. Then we can apply apply Theorem 1.4.2 to the polynomials G_i, with the y in 1.4.2 replaced by $(y,a) \in R^{n+m}$. □

Bibliographic Notes. The content of this chapter is almost completely classical. The first three sections present the theory developed by Artin and Schreier [16] for the solution of Hilbert's 17th problem (cf. Chap. 6). This can be found in many textbooks of algebra, for example [210]. Sturm's theorem [311] is well known. Sylvester's extension of Sturm's theorem [313], presented here, is less well known. It plays an important role in recent algorithms for sign determination [281]. The classical approach of the Artin-Schreier theory, presented here, is highly nonconstructive. For a constructive approach, see [164, 218, 285]. Descartes's Lemma (Proposition 1.2.14, [113]) is somewhat different; it may be viewed as the precursor of Khovansky's result [184, 185, 272] saying that the number of nondegenerate solutions of a system of polynomial equations $f_1 = \ldots = f_n = 0$ in $\{(x_1,\ldots,x_n) \in R^n \mid x_1 > 0, \ldots, x_n > 0\}$ is bounded by a function of n and the number of different monomials occuring in $f_1,\ldots,f_n$.

The Tarski-Seidenberg principle [318, 290] is a more recent result. It was first announced without proof in [317]. The idea of our proof is taken from Hörmander [167].

2. Semi-algebraic Sets

Abstract. This chapter deals with semi-algebraic sets over a real closed field R. These are the sets defined by a boolean combination of polynomial equations and inequalities. This class of sets has a remarkable property: stability under projection. Several applications of this basic property are investigated. The study of semi-algebraic sets is based mainly on the "slicing" technique, which makes it possible to decompose them into a finite number of subsets semi-algebraically homeomorphic to open hypercubes. Using this decomposition, we show that a semi-algebraic set has a finite number of semi-algebraically connected components. The notions of connectedness and compactness over a real closed field, other than $\mathbb{R}$, require some care. Nevertheless, closed and bounded semi-algebraic subsets of R^n preserve several of the properties known in the case $R = \mathbb{R}$. They are proved using the curve-selection lemma. All this is the subject of the first five sections of this chapter. In Section 6, we study continuous semi-algebraic functions and we show Łojasiewicz's inequality. Section 7 deals with the separation of disjoint closed semi-algebraic sets. Section 8 introduces the notion of dimension for a semi-algebraic set and establishes its expected properties. Finally, the last section contains essentially an implicit function theorem in the semi-algebraic framework (this result is well known over $\mathbb{R}$ but it is also useful over real closed fields other than $\mathbb{R}$).

Throughout this chapter R is a fixed real closed field.

2.1 Algebraic and Semi-algebraic Sets

Algebraic sets in R^n are defined as for any field:

Definition 2.1.1. *Let B be a subset of $R[X_1, \ldots, X_n]$. Denote*

$$\mathcal{Z}(B) = \{x \in R^n \mid \forall f \in B \;\; f(x) = 0\} \,.$$

The elements of $\mathcal{Z}(B)$ are the zeros of B. An algebraic subset of R^n is the set of zeros of some $B \subset R[X_1, \ldots, X_n]$.

Notation 2.1.2. *Given a subset S of R^n, denote by*

$$\mathcal{I}(S) = \{f \in R[X_1,\ldots,X_n] \mid \forall x \in S \; f(x) = 0\}$$

the ideal of $R[X_1,\ldots,X_n]$ of polynomials vanishing on S.

Note that each algebraic subset of R^n can be given by a single equation.

Proposition 2.1.3. *Given an algebraic subset V of R^n, there exists f in $R[X_1,\ldots,X_n]$ such that $V = \mathcal{Z}(f)$.*

Proof. Take $f = f_1^2 + \cdots + f_m^2$, where $f_1,\ldots,f_m$ generate $\mathcal{I}(V)$. □

We shall study the special properties of algebraic sets in other chapters. We are interested, here, in the properties they share with a larger class of subsets of R^n, the semi-algebraic sets.

Definition 2.1.4. *A semi-algebraic subset of R^n is a subset of the form*

$$\bigcup_{i=1}^{s} \bigcap_{j=1}^{r_i} \{x \in R^n \mid f_{i,j} *_{i,j} 0\},$$

where $f_{i,j} \in R[X_1,\ldots,X_n]$ and $*_{i,j}$ is either < or =, for $i = 1,,\ldots,s$ and $j = 1,\ldots,r_i$.

Note that the semi-algebraic subsets of R^n form the smallest family of subsets containing all sets of the form

$$\{x \in R^n \mid f(x) > 0\}, \quad \text{where } f \in R[X_1,\ldots,X_n],$$

and closed under taking finite intersections, finite unions and complements. Alternatively, if we call the conditions $f(x) > 0$, $f(x) < 0$ or $f(x) = 0$, *sign conditions on the polynomial f*, then a semi-algebraic subset of R^n is defined by a boolean combination (obtained by disjunction, conjunction and negation) of sign conditions involving a finite number of polynomials.

Example 2.1.5.

(a) An algebraic set is, of course, semi-algebraic.

Figure 2.1.

2.1 Algebraic and Semi-algebraic Sets   25

(b) Semi-algebraic sets can take various and pleasant shapes, like the one of Figure 2.1 which is defined by

$$\{(x,y) \in \mathbb{R}^2 \mid x^2/25 + y^2/16 < 1 \text{ and } x^2 + 4x + y^2 - 2y > -4$$
$$\text{and } x^2 - 4x + y^2 - 2y > -4 \text{ and } (x^2 + y^2 - 2y \neq 8 \text{ or } y > -1)\}.$$

c) We shall see later in this chapter that many sets turn out to be semi-algebraic, even if they do not appear so at first. For example, the closure of a semi-algebraic set, the set of points equidistant from two given semi-algebraic sets, etc., are semi-algebraic.

It is, of course, not the case that every set is semi-algebraic.

d) The set $\{(x,y) \in \mathbb{R}^2 \mid y = e^x\}$ is not semi-algebraic.

e) The set $\{(x,y) \in \mathbb{R}^2 \mid \exists n \in \mathbb{N} \ \ y = nx\}$ is not semi-algebraic.

f) The set

$$\{(x,y) \in \mathbb{R}^2 \mid y = \lfloor x \rfloor \text{ or } (x \in \mathbb{Z} \text{ and } x \leq y \leq x + 1)\}$$

is not semi-algebraic (where $\lfloor x \rfloor$ denotes the greatest integer function).

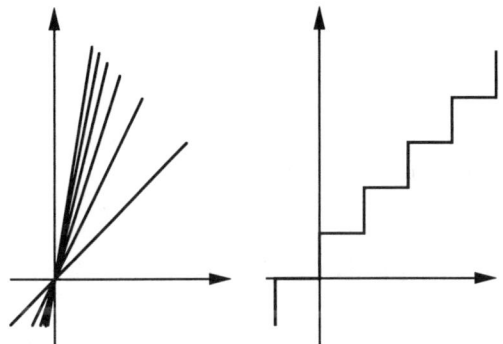

Figure 2.2. Example e (left) and Example f (right)

Remark 2.1.6. The infinite fan (Example e) and the infinite staircase (Example f) are different in the following sense: every point of $\mathbb{R}^2$ has a neighbourhood whose intersection with the infinite staircase is semi-algebraic, while this is not true for the infinite fan (look at the origin). The infinite staircase is said to be locally semi-algebraic.

One sees immediately that

Proposition 2.1.7. *Semi-algebraic subsets of R are exactly the finite unions of points and open intervals (bounded or unbounded).*

Proposition 2.1.8. *Every semi-algebraic subset of R^n can be written as a finite union of semi-algebraic sets of the form:*

$$\{x \in R^n \mid f_1(x) = \cdots = f_\ell(x) = 0, \ g_1(x) > 0, \ldots, \ g_m(x) > 0\},$$

where $f_1, ,\ldots, f_\ell, g_1, \ldots, g_m$ are in $R[X_1, \ldots, X_n]$.

Proof. The family of these finite unions is clearly closed under taking finite intersections and finite unions. We only have to verify that the complement of a semi-algebraic set of the form described in the proposition, can be written as the union of semi-algebraic sets of the same form. We leave this easy task to the reader. □

We shall consider two topologies on R^n: the *Zariski topology*, whose closed subsets are the algebraic sets and the finer *euclidean topology* coming from the ordering structure on R.

Definition 2.1.9. *Let* $x = (x_1, \ldots, x_n) \in R^n$, $r \in R$, $r > 0$. *Denote*

$$\begin{aligned} \|x\| &= \sqrt{x_1^2 + \cdots + x_n^2}, \\ B_n(x,r) &= \{y \in R^n \mid \|y - x\| < r\} \quad \text{(open ball)}, \\ \overline{B}_n(x,r) &= \{y \in R^n \mid \|y - x\| \leq r\} \quad \text{(closed ball)}, \\ S^{n-1}(x,r) &= \{y \in R^n \mid \|y - x\| = r\} \quad \text{($(n-1)$-sphere)}. \end{aligned}$$

By convention, S^{n-1} will denote $S^{n-1}(0,1)$, where 0 is the origin of R^n.

The euclidean topology *on R^n is the topology for which open balls form a basis of open subsets.*

In what follows, R^n will always be considered with its euclidean topology, unless stated otherwise.

The sets $B_n(x,r)$, $\overline{B}_n(x,r)$ and $S^{n-1}(x,r)$ are semi-algebraic. Polynomials are continuous with respect to the euclidean topology.

2.2 Projection of Semi-algebraic Sets. Semi-algebraic Mappings

By their very definition, semi-algebraic sets are stable under finite union, finite intersection and taking complements. They are also stable under projection.

Theorem 2.2.1. *Let S be a semi-algebraic subset of R^{n+1}, $\Pi : R^{n+1} \to R^n$ the projection on the space of the first n coordinates. Then $\Pi(S)$ is a semi-algebraic subset of R^n.*

Proof. By Proposition 2.1.8, it is enough to prove the theorem for a semi-algebraic set of the form:

$$\{(y,x) \in R^n \times R \mid f_i(y,x) = 0,\ i = 1, \ldots, \ell,\quad g_j(y,x) > 0,\ j = 1, \ldots, m\}.$$

By the Tarski-Seidenberg principle (more specifically, its Corollary 1.4.7), there exists a boolean combination of polynomial equations and inequalities $\mathcal{B}(Y)$ in the variables Y with coefficients in R such that, for every y in R^n, the system

2.2 Projection of Semi-algebraic Sets. Semi-algebraic Mappings 27

$$\begin{cases} f_1(y, X) = \cdots = f_\ell(y, X) = 0 \\ g_1(y, X) > 0 \\ \quad \vdots \\ g_m(y, X) > 0 \end{cases}$$

has a solution x in R if and only if $\mathcal{B}(y)$ is satisfied. Since the set of y in R^n, satisfying $\mathcal{B}(y)$, is semi-algebraic, the theorem is proved. □

Their stability under projection makes semi-algebraic sets particularly interesting. For example, the projection of an algebraic set is a semi-algebraic set (it is not, in general, an algebraic set). Conversely, a semi-algebraic subset of R^n can be easily written as the projection of an algebraic subset of R^{n+m} for some m:

$$\{x \in R^n \mid f_1(x) = \cdots = f_\ell(x) = 0, \ g_1(x) > 0, \ldots, \ g_m(x) > 0\}$$

is the projection of

$$\{(x, y) \in R^{n+m} \mid f_1(x) = \cdots = f_\ell(x) = 0, \ y_1^2 g_1(x) = 1, \ldots, \ y_m^2 g_m(x) = 1\}.$$

Motzkin [241] has shown that every semi-algebraic subset of R^n is, in fact, the projection of an algebraic subset of R^{n+1}.

The following is a pleasant consequence of stability under projection.

Proposition 2.2.2. *The closure and the interior of a semi-algebraic set are semi-algebraic.*

Proof. Let S be a semi-algebraic subset of R^n. The closure of S

$$\text{clos}(S) = \{x \in R^n \mid \forall t \in R \ \exists y \in S \ (\|y - x\|^2 < t^2 \text{ or } t = 0)\}$$

can also be written as

$$R^n \setminus \Pi_2 \left[R^{n+1} \setminus \Pi_1(\{(x, y, t) \in R^{2n+1} \mid y \in S \text{ and } (\|y - x\|^2 < t^2 \text{ or } t = 0)\}) \right]$$

where $\Pi_1 : R^{2n+1} \to R^{n+1}$ is the projection given by $\Pi_1(x, y, t) = (x, t)$ and $\Pi_2 : R^{n+1} \to R^n$ the one given by $\Pi_2(x, t) = x$. By Theorem 2.2.1, the closure of S is semi-algebraic. By taking complements, one sees that the interior is also semi-algebraic. □

It would be wrong to believe that the closure of a semi-algebraic set is always obtained just by relaxing the strict inequalities describing it. For example, the closure of the set

$$A = \{(x, y) \in R^2 \mid x^3 - x^2 - y^2 > 0\}$$

is not the set

$$B = \{(x, y) \in R^2 \mid x^3 - x^2 - y^2 \geq 0\}.$$

The closure of A is obtained by removing the point $(0, 0)$ from B, and can be described as

$$\text{clos}(A) = \{(x,y) \in R^2 \mid x^3 - x^2 - y^2 \geq 0,\ x \geq 1\}.$$

It can be seen in the proof of Proposition 2.2.2 that the description of a semi-algebraic set is easier to read with quantifiers than with projections. Thus it will be useful to become familiar with a few concepts from logic.

Definition 2.2.3. *A first-order formula of the language of ordered fields with parameters in R is a formula written with a finite number of conjunctions, disjunctions, negations, and universal or existential quantifiers on variables, starting from atomic formulas which are formulas of the kind $f(x_1, \ldots x_n) = 0$ or $g(x_1, \ldots, x_n) > 0$, where f and g are polynomials with coefficients in R. The free variables of a formula are those variables of the polynomials appearing in the formula, which are not quantified.*

By definition, semi-algebraic sets are described by quantifier free first-order formulas of the language of ordered fields with parameters in R. The properties of stability of semi-algebraic sets under finite intersections and unions, taking complements and projections can be expressed in the following way.

Proposition 2.2.4. *Let $\Phi(x_1, \ldots, x_n)$ be a first-order formula of the language of ordered fields, with parameters in R, with free variables $x_1, \ldots, x_n$. Then $\{x \in R^n \mid \Phi(x)\}$ is a semi-algebraic set.*

Proof. By induction on the construction of the formula, starting from atomic formulas. The cases of conjunction, disjunction and negation are obvious. If $\Phi(x)$ has the form $\exists y\ \Psi(x,y)$, where $S = \{(x,y) \in R^{n+1} \mid \Psi(x,y)\}$ is semi-algebraic, then $\{x \in R^n \mid \Phi(x)\}$ is the projection of S and thus is semi-algebraic by Theorem 2.2.1. Universal quantifiers are reduced to existential quantifiers since "$\forall y \ldots$" is equivalent to "not $\exists y$ not $\ldots$". □

We emphasize the fact that *quantifiers are only allowed on variables that range over R*. We have already seen (Example 2.1.5 e) that

$$\{(x,y) \in \mathbb{R}^2 \mid \exists n \in \mathbb{N}\ \ y = nx\}$$

is not semi-algebraic.

We now define semi-algebraic mappings.

Definition 2.2.5. *Let $A \subset R^m$ and $B \subset R^n$ be two semi-algebraic sets. A mapping $f : A \to B$ is semi-algebraic if its graph is semi-algebraic in R^{m+n}.*

Proposition 2.2.6.
  (i) *The composition $g \circ f$ of semi-algebraic mappings $f : A \to B$ and $g : B \to C$ is semi-algebraic.*
  (ii) *The R-valued semi-algebraic functions on a semi-algebraic set A form a ring.*

2.2 Projection of Semi-algebraic Sets. Semi-algebraic Mappings

Proof. (i) Let $F \subset R^{m+n}$ be the graph of f and $G \subset R^{n+p}$ the graph of g. The graph of $g \circ f$ is the projection of $(F \times R^p) \cap (R^m \times G)$ onto R^{m+p} and, hence, is semi-algebraic by Theorem 2.2.1.

(ii) follows from (i), by noting for instance that $f + g$ is the composition of $(f, g) : A \to R^2$ with $+ : R^2 \to R$. □

Proposition 2.2.7. *Let $f : A \to B$ be a semi-algebraic mapping. If $S \subset A$ is semi-algebraic, then its image $f(S)$ is semi-algebraic. If $T \subset B$ is semi-algebraic, then its inverse image $f^{-1}(T)$ is semi-algebraic.*

Proof. The set $f(S)$ (resp. $f^{-1}(T)$) is the image of $(S \times B) \cap \mathrm{Graph}(f)$ (resp. $(A \times T) \cap \mathrm{Graph}(f)$) by the projection $A \times B \to B$ (resp. $A \times B \to A$). □

General semi-algebraic functions are not very interesting. Most of the time, we shall work with some subclasses, such as continuous semi-algebraic functions. Here is an interesting example of a continuous semi-algebraic function:

Proposition 2.2.8. *Let $A \subset R^n$ be a nonempty semi-algebraic set.*
(i) *For every x in R^n, the distance between x and A*

$$\mathrm{dist}(x, A) = \inf(\{\|x - y\| \mid y \in A\})$$

is well-defined.
(ii) *The function $x \mapsto \mathrm{dist}(x, A)$ from R^n to R is continuous semi-algebraic, vanishing on $\mathrm{clos}(A)$ and positive elsewhere.*

Proof. (i) The set $\{\|x - y\| \mid y \in A\}$ is the image of A by the semi-algebraic function $y \mapsto \|x - y\|$. Hence, it is semi-algebraic. Moreover, being bounded from below, it has a lower bound (by Proposition 2.1.7).

(ii) The graph of this function is

$$\{(x, t) \in R^{n+1} \mid t \geq 0 \text{ and } \forall y \in A \quad t^2 \leq \|x - y\|^2$$
$$\text{and } \forall \epsilon \in R \quad \epsilon > 0 \Rightarrow \exists y \in A \quad t^2 + \epsilon > \|x - y\|^2\}$$

and hence is semi-algebraic by Proposition 2.2.4. The other properties are straightforward. □

The next proposition will often be useful.

Proposition 2.2.9. *Let A be a semi-algebraic subset of R^n, which is locally closed (i.e. which is the intersection of an open set with a closed set). Then there exists a semi-algebraic homeomorphism from A onto a closed subset of R^{n+1}.*

Proof. The set A is the intersection of $\mathrm{clos}(A)$ with $U = (R^n \setminus \mathrm{clos}(A)) \cup A$ which is an open semi-algebraic set. Unless A is already closed, $R^n \setminus U$ is nonempty and the function $x \mapsto \mathrm{dist}(x, R^n \setminus U)$ is semi-algebraic and continuous (by Proposition 2.2.8). The semi-algebraic homeomorphism

$$x \mapsto (x, (\mathrm{dist}(x, R^n \setminus U))^{-1})$$

maps A onto the closed semi-algebraic set

$$\{(x,y) \in R^{n+1} \mid x \in \mathrm{clos}(A) \text{ and } y\,\mathrm{dist}(x, R^n \setminus U) = 1\}.$$

□

2.3 Decomposition of Semi-algebraic Sets

The word "decomposition" means that we shall decompose semi-algebraic sets into simpler subsets that are semi-algebraically homeomorphic to open hypercubes $]0,1[^d$. The decomposition of a semi-algebraic set $A \subset R^n$ is obtained by induction on n. The tool allowing us to pass from n to $n+1$ is the following theorem.

Let X denote $(X_1, \ldots, X_n)$.

Theorem 2.3.1. *Let $f_1(X,Y), \ldots, f_s(X,Y)$ be polynomials in $n+1$ variables with coefficients in R. There exist a partition of R^n into a finite number of semi-algebraic sets $A_1, \ldots, A_m$ and, for $i = 1, \ldots, m$, a finite number (possibly zero) of continuous semi-algebraic functions $\xi_{i,1} < \ldots < \xi_{i,\ell_i}$, $\xi_{i,j} : A_i \to R$, such that:*

(i) For every x in A_i, $\{\xi_{i,1}(x), \ldots, \xi_{i,\ell_i}(x)\}$ is the set of roots of those polynomials among $f_1(x,Y), \ldots, f_s(x,Y)$ which are not identically zero.

(ii) For every x in A_i, the signs of $f_k(x,y)$, $k = 1, \ldots, s$, depend only on the signs of $y - \xi_{i,j}(x)$, $j = 1, \ldots, \ell_i$.

In particular, the graph of each $\xi_{i,j}$ is contained in the set of zeros of some f_k, with k depending on i and j.

Proof. We may assume that the family $f_1, \ldots, f_s$ is stable under derivation with respect to the variable Y, by adding the missing derivatives, since we can remove the functions $\xi_{i,j}$ that do not give the roots of polynomials belonging to the initial family. We now apply the following corollary of Proposition 1.4.6.

Lemma 2.3.2. *Let $f_1(X,Y), \ldots, f_s(X,Y)$ be polynomials in $n+1$ variables (X,Y) with coefficients in R and let q be the maximum degree in Y of the f_k's. Let $w \in W_{s,q}$ (with the same notations as in 1.4.3). Then there exists a boolean combination $\mathcal{B}_w(X)$ of polynomial equations and inequalities in the variables X with coefficients in R such that, for every x in R^n, we have $\mathrm{SIGN}_R(f_1(x,Y), \ldots, f_s(x,Y)) = w$ if and only if $\mathcal{B}_w(x)$ is satisfied.*

2.3 Decomposition of Semi-algebraic Sets   31

Proof. Let $a \in R^p$ be the list of coefficients of the polynomials f_k, $k = 1,\ldots,s$. Then $f_k(X,Y) = G_k(a,X,Y)$, where $G_k(T,X,Y)$ is a polynomial in $p+n+1$ variables with coefficients in $\mathbb{Z}$. By Proposition 1.4.6, there exists a boolean combination $\mathcal{B}'_w(T,X)$ of polynomial equations and inequalities in the variables (T,X) with coefficients in $\mathbb{Z}$ such that, for every $(t,x) \in R^{p+n}$, one has $\mathrm{SIGN}_R(G_1(t,x,Y),\ldots,G_s(t,x,Y)) = w$ if and only if $\mathcal{B}'_w(t,x)$ is verified. Finally, take $\mathcal{B}_w(X) = \mathcal{B}'_w(a,X)$. $\square$

We now return to the proof of Theorem 2.3.1. To each $w \in W_{s,q}$ corresponds a semi-algebraic set $A_w = \{x \in R^n \mid \mathcal{B}_w(x)\}$. Let $A_1,\ldots,A_m$ be those sets among the A_w which are nonempty. Of course, they form a partition of R^n, and $\mathrm{SIGN}_R(f_1(x,Y),\ldots,f_s(x,Y))$ is constant on each A_i: there is a number $\ell_i \leq sq$ such that, for each $x \in A_i$, the polynomials among $f_1(x,Y),\ldots,f_s(x,Y)$ which are not identically zero have altogether ℓ_i roots $\xi_{i,1}(x) < \ldots < \xi_{i,\ell_i}(x)$, and, for every $k = 1,\ldots,s$, the signs $\mathrm{sign}(f_k(x,\xi_{i,j}(x)))$, $j = 1,\ldots,\ell_i$, and $\mathrm{sign}(f_k(x,]\xi_{i,j}(x),\xi_{i,j+1}(x)[))$, $j = 0,\ldots,\ell_i$, do not depend on $x \in A_i$ (with the convention $\xi_{i,0}(x) = -\infty$, $\xi_{i,\ell_i+1}(x) = +\infty$). It only remains to show that the $\xi_{i,j}$ are semi-algebraic and continuous. The graph of $\xi_{i,j}$ is

$$\{(x,y) \in A_i \times R \mid \exists (y_1,\ldots,y_{\ell_i}) \in R^{\ell_i} \, (\prod_k f_k(x,y_1) = \cdots = \prod_k f_k(x,y_{\ell_i}) = 0$$
$$\text{and } y_1 < \ldots < y_{\ell_i} \text{ and } y = y_j)\}$$

(where k ranges over the set of subscripts of the polynomials that are not identically zero on A_i), which shows, by Proposition 2.2.4, that the function $\xi_{i,j}$ is semi-algebraic. Let us fix $x' \in A_i$. Then $y_j = \xi_{i,j}(x')$ is a simple root of at least one among the $f_k(x',Y)$, say of $f_1(x',Y)$. For $\epsilon \in R$ small enough, we have $f_1(x',y_j - \epsilon)f_1(x',y_j + \epsilon) < 0$. Hence, in a neighbourhood U of x' in R^n, we have

$$\forall x \in U \quad f_1(x,y_j - \epsilon)f_1(x,y_j + \epsilon) < 0$$

and $f_1(x,Y)$ has a root between $y_j - \epsilon$ and $y_j + \epsilon$. As this can be done for all the j simultaneously, the root of $f_1(x,Y)$ between $y_j - \epsilon$ and $y_j + \epsilon$ is $\xi_{i,j}(x)$. This proves that $\xi_{i,j}$ is continuous. $\square$

Remark 2.3.3. It was indeed necessary to add the derivatives with respect to Y. Consider

$$f(X,Y) = (X - (Y-1)^2)^2(X + (Y+1)^2)^2 \ .$$

Then $\mathrm{SIGN}_R f(x,Y)$ is constant for $x \in R$, but it is impossible to find two continuous semi-algebraic functions $\xi_1 < \xi_2 : R \to R$, giving the roots of $f(x,Y)$. In this case, it is necessary to partition R into three subsets: $]-\infty,0[$, $\{0\}$ and $]0,+\infty[$.

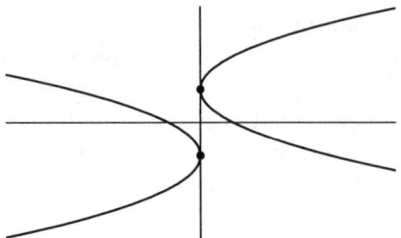

Figure 2.3.

We return to the statement of Theorem 2.3.1. We can see the cylinders $A_i \times R$ as "salamis" cut into "slices" by the graphs of the functions $\xi_{i,j}$, in such a way that the signs of the polynomials $f_1, \ldots, f_s$ are constant on the slices. For instance, the Figure 2.4 shows what happens for the polynomial $(Y - 7)^3 - (Y - 7)X_1 - X_2$, which is the equation of a Whitney cusp.

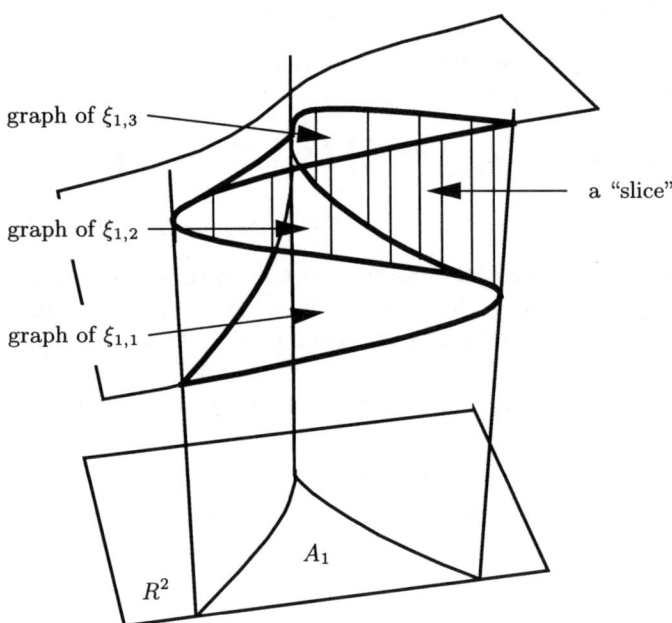

Figure 2.4.

We turn this rather unsophisticated, but illuminating, image into a more formal description:

Definition 2.3.4. *Let $f_1(X, Y), \ldots, f_s(X, Y)$ be polynomials in $n + 1$ variables with coefficients in R. A partition of R^n into semi-algebraic sets $A_1, \ldots, A_m$, together with continuous semi-algebraic functions*

$$\xi_{i,1} < \ldots < \xi_{i,\ell_i} : A_i \to R$$

satisfying properties (i) and (ii) of Theorem 2.3.1 is called a slicing of $f_1, \ldots, f_s$ and is denoted by $(A_i, (\xi_{i,j})_{j=1,\ldots,\ell_i})_{i=1,\ldots,m}$.

If the $A_1, \ldots, A_m$ are given by boolean combinations of sign conditions on the polynomials $g_1, \ldots, g_t \in R[X_1, \ldots, X_n]$, we say that the $g_1, \ldots, g_t$ slice the $f_1, \ldots, f_s$.

Lemma 2.3.5. *Let $f_1(X, Y), \ldots, f_s(X, Y)$ be polynomials in $R[X, Y]$ and $(A_i, (\xi_{i,j})_{j=1,\ldots,\ell_i})_{i=1,\ldots,m}$ a slicing of $f_1, \ldots, f_s$. Then for every i, $1 \leq i \leq m$, and every j, $0 \leq j \leq \ell_i$, the slice*

$$]\xi_{i,j}, \xi_{i,j+1}[= \{(x, y) \in R^{n+1} \mid x \in A_i \text{ and } \xi_{i,j}(x) < y < \xi_{i,j+1}(x)\}$$

is semi-algebraic and semi-algebraically homeomorphic to $A_i \times]0, 1[$ (with the convention $\xi_{i,0} = -\infty$ and $\xi_{i,\ell_i+1} = +\infty$).

Proof. Each slice is semi-algebraic, since A_i and the functions $\xi_{i,j}$, $j = 1, \ldots, \ell_i$, are semi-algebraic. We now give, explicitly, the semi-algebraic homeomorphism

$$h :]\xi_{i,j}, \xi_{i,j+1}[\to A_i \times]0, 1[\, .$$

For $j = 1, \ldots, \ell_i - 1$ define:

$$h(x, y) = (x, (y - \xi_{i,j}(x))/(\xi_{i,j+1}(x) - \xi_{i,j}(x))) \, .$$

For $j = 0$, $\xi_{i,0} = -\infty$, define (if $\ell_i \neq 0$):

$$h(x, y) = (x, (1 + \xi_{i,1}(x) - y)^{-1}) \, .$$

For $j = \ell_i \neq 0$, $\xi_{i,\ell_i+1} = +\infty$, define:

$$h(x, y) = (x, (y - \xi_{i,\ell_i}(x) + 1)^{-1}) \, .$$

Finally, if $\ell_i = 0$, $\xi_0 = -\infty$ and $\xi_1 = +\infty$, define:

$$h(x, y) = (x, (y + \sqrt{1 + y^2})/2\sqrt{1 + y^2}) \, .$$

$\square$

Theorem 2.3.6. *Every semi-algebraic subset of R^n is the disjoint union of a finite number of semi-algebraic sets, each of them semi-algebraically homeomorphic to an open hypercube $]0, 1[^d \subset R^d$, for some $d \in \mathbb{N}$ (with $]0, 1[^0$ being a point).*

Proof. By induction on n. For $n = 1$, we already know that every semi-algebraic subset of R is the union of a finite number of points and open intervals (Proposition 2.1.7). Open intervals are clearly semi-algebraically homeomorphic to $]0, 1[$. We now assume the result proved for n. Let S be a semi-algebraic subset of R^{n+1}, given by a boolean combination of sign conditions on the polynomials $f_1, \ldots, f_s$, and let $(A_i, (\xi_{i,j})_{j=1,\ldots,\ell_i})_{i=1,\ldots,m}$

be a slicing of $f_1, \ldots, f_s$. By induction, we may assume that all A_i are semi-algebraically homeomorphic to open hypercubes. Moreover, it is clear that S is the union of a finite number of semi-algebraic sets that are either the graph of a function $\xi_{i,j}$, or a slice $]\xi_{i,j}, \xi_{i,j+1}[$, as in Lemma 2.3.5. The graph of $\xi_{i,j}$ is semi-algebraically homeomorphic to A_i, while, by Lemma 2.3.5, the slice $]\xi_{i,j}, \xi_{i,j+1}[$ is semi-algebraically homeomorphic to $A_i \times]0,1[$. □

2.4 Connectedness

Some price has to be paid for full generality: an arbitrary real closed field is not connected (in the euclidean topology). It is even possible to prove that $\mathbb{R}$ is the only real closed field that is connected. We shall give only two counter-examples.

Example 2.4.1.
  a) The intersection of $]-\infty, \pi[$, ($\pi = 3.14\ldots$) with $\mathbb{R}_{\text{alg}}$ is a closed and open subset of the field of real algebraic numbers $\mathbb{R}_{\text{alg}}$.
  b) In $\mathbb{R}(X)^\wedge$ (the field of Puiseux series, as in Example 1.2.3) the set $\{f \in \mathbb{R}(X)^\wedge \mid \exists r \in \mathbb{R} \quad r > 0 \text{ and } f > r\}$ is a closed and open set.

The two closed and open sets given in the example are not semi-algebraic (in (b), the variable r ranges over $\mathbb{R}$ and not over $\mathbb{R}(X)^\wedge$!). According to the description of semi-algebraic subsets of R (Proposition 2.1.7), it is not possible to find a closed and open semi-algebraic subset of R different from both R and $\emptyset$.

Definition 2.4.2. *A semi-algebraic subset A of R^n is* semi-algebraically connected *if for every pair of semi-algebraic sets F_1 and F_2 closed in A, disjoint and satisfying $F_1 \cup F_2 = A$, one has $F_1 = A$ or $F_2 = A$.*

Proposition 2.4.3. *An open hypercube $]0,1[^d \subset R^d$ is semi-algebraically connected.*

Proof. Suppose that $]0,1[^d$ is not semi-algebraically connected. Then we can find two nonempty closed semi-algebraic sets F_1 and F_2 that are the complement of each other in $]0,1[^d$. Let x_i be a point of F_i, $1 = 1, 2$, and S the segment joining x_1 and x_2. Considering the intersections of the F_i with S, it would follow that S is not semi-algebraically connected. But it is clear that a segment is semi-algebraically connected (by Proposition 2.1.7). □

Theorem 2.4.4. *Every semi-algebraic subset A of R^n is the disjoint union of a finite number of semi-algebraically connected semi-algebraic sets $C_1, \ldots, C_s$ which are both closed and open in A. The $C_1, \ldots, C_s$ are called the* semi-algebraically connected components *of A.*

Proof. By Theorem 2.3.6, we know that A is the disjoint union of a finite number of semi-algebraic sets A_i, semi-algebraically homeomorphic to open hypercubes $]0,1[^d$ and, hence, semi-algebraically connected by Proposition 2.4.3. Consider the equivalence relation $\mathcal{R}$ on the set of the A_i generated by the relation " $A_i \cap \mathrm{clos}(A_j) \neq \emptyset$ ". Let $C_1, \ldots, C_s$ be the unions of the sets in the same equivalence class for $\mathcal{R}$. The C_k are semi-algebraic, disjoint, closed in A and their union is A. Suppose that $C_k = F_1 \cup F_2$, with F_1 and F_2 disjoint, semi-algebraic and closed in C_k. Since each A_i is semi-algebraically connected, $A_i \subset C_k$ implies that $A_i \subset F_1$ or $A_i \subset F_2$. Since F_1 (resp F_2) is closed in C_k, if $A_j \subset F_1$ (resp. F_2) and $A_i \cap \mathrm{clos}(A_j) \neq \emptyset$ then $A_i \subset F_1$ (resp. F_2). According to the definition of the C_k, $C_k = F_1$ or $C_k = F_2$. □

All we have used above is the decomposition of semi-algebraic sets (Theorem 2.3.6) and the fact that an open hypercube is semi-algebraically connected.

Since for $R = \mathbb{R}$ an open hypercube is connected, we have:

Theorem 2.4.5. *A semi-algebraic subset A of $\mathbb{R}^n$ is semi-algebraically connected if and only if it is connected. Every semi-algebraic set (and, in particular, every algebraic subset of $\mathbb{R}^n$) has a finite number of connected components, which are semi-algebraic.*

2.5 Closed and Bounded Semi-algebraic Sets. Curve-selection Lemma

The notion of compactness also causes difficulties in the case of a real closed field other than $\mathbb{R}$.

Example 2.5.1.
 a) The interval $[0,1]$ is not compact in $\mathbb{R}_{\mathrm{alg}}$. The family $([0, r[\cup]s, 1])$ for $0 < r < \pi/4 < s < 1$, $r, s \in \mathbb{R}_{\mathrm{alg}}$ ($\pi = 3.14\ldots$), is an open cover of $[0,1]$ by semi-algebraic subsets of $\mathbb{R}_{\mathrm{alg}}$ and it is impossible to extract a finite subcover from it.
 b) The interval $[0,1]$ is not compact in $\mathbb{R}(X)^\wedge$ either. The family consisting of all $[0, f[$ for f positive and smaller than every positive real number and of all $]r, 1]$ for $r \in \mathbb{R}$, $0 < r < 1$, is an open cover of $[0,1]$ by semi-algebraic subsets of $\mathbb{R}(X)^\wedge$ and it is impossible to extract a finite subcover from it.

So we have to be very careful. Closed and bounded semi-algebraic sets are not necessarily compact. In what follows, we shall study the properties of *closed and bounded* semi-algebraic sets and forget about compactness. In the course of this study, we shall establish interesting results also when $R = \mathbb{R}$, so that this section will be useful even for a person interested only in this special case.

Lemma 2.5.2. *Let $A \subset R$ be a semi-algebraic set and $\varphi : A \to R$ a semi-algebraic function. There exists a nonzero polynomial $f \in R[X,Y]$ such that for every x in A, $f(x, \varphi(x)) = 0$.*

Proof. By Proposition 2.1.8, the graph of φ is the finite union of semi-algebraic sets of the form

$$\{(x,y) \in R \times R \mid f_i(x,y) = 0, \ i = 1, \ldots, \ell, \ g_j(x,y) > 0, \ j = 1, \ldots, m\}$$

with at least one among the f_i nonzero, since, otherwise, the graph of φ would contain a nonempty open subset of R^2. We can then take f to be the product of these nonzero polynomials. $\square$

Proposition 2.5.3. *Let $\varphi :]0,r] \to R$ be a continuous bounded semi-algebraic function defined on an interval $]0,r] \subset R$. Then φ can be continuously extended to 0.*

Proof. By Lemma 2.5.2, there is a nonzero polynomial $f \in R[X,Y]$ such that $f(x, \varphi(x)) = 0$ for every $x \in]0,r]$. We shall proceed by induction on the degree d of f with respect to Y.

If $d = 1$, we have $\varphi(x) = N(x)/D(x)$ where N and D are relatively prime polynomials and X does not divide D, since the absolute value of φ is bounded. We can take $\varphi(0) = N(0)/D(0)$.

We assume now the result proved for polynomials of degree $< d$ with respect to Y. We can always suppose that f is not divisible by X and, also, (choosing r small enough) that there is a slicing $(A_i, (\xi_{i,j})_{j=1,\ldots,\ell_i})_{i \in I}$ of $(f, \frac{\partial f}{\partial Y})$ with $A_1 =]0,r]$ and $\varphi = \xi_{1,j_0}$ for some j_0 (here we use the fact that $]0,r]$ is semi-algebraically connected in order to see that φ coincides with one of the $\xi_{1,j}$). If φ is also a root of $\frac{\partial f}{\partial Y}$, the desired result follows, because $\frac{\partial f}{\partial Y}$ is of degree $d-1$. Otherwise, we have, for instance, $\frac{\partial f}{\partial Y}(x, \varphi(x)) > 0$ for every x in $]0,r]$. We now choose two continuous semi-algebraic functions ρ and θ from $[0,r]$ to R such that, for every x in $]0,r]$, $\rho(x) < \varphi(x) < \theta(x)$ and $\frac{\partial f}{\partial Y}(x,y) > 0$ for every y in $]\rho(x), \theta(x)[$. For ρ take the function ξ_{1,j_0-1} if $j_0 > 1$ (it is necessarily a root of $\frac{\partial f}{\partial Y}$, and, therefore, it can be continuously extended to 0), or, if $j_0 = 1$, the constant function $-M-1$ (where M bounds $|\varphi|$ from above). Proceed for θ in a similar way.

If $\rho(0) = \theta(0)$, defining $\varphi(0) = \rho(0)$ gives a continuous extension of φ to 0.

Otherwise, $\rho(0) < \theta(0)$ and $\frac{\partial f}{\partial Y}(0,y)$ is never < 0 on the interval $[\rho(0), \theta(0)]$, by continuity. If $f(0,Y)$ were constant, it would be identically zero since $f(0, \rho(0)) \leq 0 \leq f(0, \theta(0))$, and this is impossible, since X does not divide f. Hence, the function $f(0,Y)$ is strictly increasing and has one and only one root y_0 in $[\rho(0), \theta(0)]$. Define $\varphi(0) = y_0$. It remains to show that φ is continuous at 0.

If $\rho(0) < y_0 < \theta(0)$, then

$$f(0, y_0 - \epsilon) < 0 \, , \ f(0, y_0 + \epsilon) > 0 \, , \ \rho(0) < y_0 - \epsilon < y_0 < y_0 + \epsilon < \theta(0)$$

2.5 Closed and Bounded Semi-algebraic Sets. Curve-selection Lemma

for every $\epsilon \in R$ with $\epsilon > 0$ small enough. Hence, there exists $\eta \in R$, $\eta > 0$, such that, for every $x \in \,]0, \eta[$,
$$f(x, y_0 - \epsilon) < 0 \,, \; f(x, y_0 + \epsilon) > 0 \,, \; \rho(x) < y_0 - \epsilon \,, \; y_0 + \epsilon < \theta(x) \,.$$
This implies that, for every $x \in \,]0, \eta[$, $\varphi(x) \in \,]y_0 - \epsilon, y_0 + \epsilon[$.

If $\rho(0) = y_0$, we have $f(0, y_0 + \epsilon) > 0$ for every $\epsilon \in R$ with $\epsilon > 0$ small enough. But then there exists $\eta \in R$, $\eta > 0$, such that, for every $x \in \,]0, \eta[$, $f(x, y_0 + \epsilon) > 0$, $y_0 - \epsilon < \rho(x) < y_0 + \epsilon$. This implies that for every $x \in \,]0, \eta[$, $\varphi(x) \in \,]y_0 - \epsilon, y_0 + \epsilon[$.

We proceed in a similar way if $\theta(0) = y_0$, and this completes the proof. □

In what follows, a *family of polynomials in R[X] stable under derivation* is a family $\mathcal{F}$ that does not contain the zero polynomial and such that, if $f \in \mathcal{F}$, then $f' \in \mathcal{F}$ or $f' = 0$.

Proposition 2.5.4 (Thom's Lemma). *Let $f_1, \ldots, f_s$ be a family of polynomials in $R[X]$ stable under derivation. Let ϵ be a function from $\{1, \ldots, s\}$ to $\{-1, 0, 1\}$. Let $A_\epsilon \subset R$ be the semi-algebraic set*
$$A_\epsilon = \bigcap_{k=1}^{s} \{x \in R \mid \mathrm{sign}(f_k(x)) = \epsilon(k)\} \,.$$
Denote by $A_{\overline{\epsilon}} \subset R$ the semi-algebraic set obtained by relaxing the strict inequalities in A_ϵ:
$$A_{\overline{\epsilon}} = \bigcap_{k=1}^{s} \{x \in R \mid \mathrm{sign}(f_k(x)) \in \overline{\epsilon(k)}\} \,,$$
where $\overline{0} = \{0\}$, $\overline{-1} = \{-1, 0\}$, $\overline{1} = \{0, 1\}$. Then
  (i) *either A_ϵ is empty, or A_ϵ is a point, or A_ϵ is an open interval,*
  (ii) *if A_ϵ is nonempty, then its closure is $A_{\overline{\epsilon}}$,*
  (iii) *if A_ϵ is empty, then $A_{\overline{\epsilon}}$ either is empty or reduces to a point.*

Proof. By induction on s. There is nothing to prove if $s = 0$. We assume now the result proved for s and let f_{s+1} be of maximal degree in the family $f_1, \ldots, f_{s+1}$ which is stable under derivation. The family $f_1, \ldots, f_s$ is still stable under derivation. Let ϵ' be a function from $\{1, \ldots, s+1\}$ to $\{-1, 0, 1\}$ and ϵ its restriction to $\{1, \ldots, s\}$. If A_ϵ is either a point, or empty, then $A_{\epsilon'} = A_\epsilon \cap \{x \in R \mid \mathrm{sign}(f_{s+1}(x)) = \epsilon'(s+1)\}$ satisfies the properties (i), (ii) and (iii). If A_ϵ is an open interval, the derivative of f_{s+1} (which is one of the f_k, with $1 \le k \le s$) has a constant nonzero sign on A_ϵ, (except if f_{s+1} is a constant, which is a trivial case). Then f_{s+1} is strictly monotone on $A_{\overline{\epsilon}}$ and hence the properties (i), (ii) and (iii) are satisfied for $A_{\epsilon'}$. □

Theorem 2.5.5 (Curve-selection Lemma). *Let A be a semi-algebraic subset of R^n and $x \in R^n$ a point belonging to the closure of A. Then there exists a continuous semi-algebraic mapping $f : [0,1] \to R^n$ such that $f(0) = x$ and $f(\,]0,1]) \subset A$.*

We shall see in Chap. 8 that we can actually choose f to be a Nash mapping (see Definition 2.9.3 and Proposition 8.1.13).

We shall say that a polynomial $f \in R[X_1,\ldots,X_n,Y]$ is *quasi-monic* with respect to Y if
$$f = a_d Y^d + g_{d-1}(X_1,\ldots,X_n)Y^{d-1} + \cdots + g_0(X_1,\ldots,X_n),$$
where a_d is a nonzero element of R.

Lemma 2.5.6. *Let $f_1,\ldots,f_s$ be a family of polynomials in $n+1$ variables (X,Y), $X = (X_1,\ldots,X_n)$, with coefficients in R. Suppose that the family is stable under derivation with respect to Y and that all f_k are quasi-monic with respect to Y. Let $(A_i,(\xi_{i,j})_{j=1,\ldots,\ell_i})_{i=1,\ldots,m}$ be a slicing of $f_1,\ldots,f_s$. Then every function $\xi_{i,j}$ can be continuously extended to $\mathrm{clos}(A_i)$.*

Proof. We prove Theorem 2.5.5 and Lemma 2.5.6 simultaneously, by proceeding in the following way:

(i) Theorem 2.5.5 is obviously true for $n = 1$,

(ii) Theorem 2.5.5 for n implies Lemma 2.5.6 for n,

(iii) Theorem 2.5.5 for n and Lemma 2.5.6 for n imply Theorem 2.5.5 for $n+1$.

We prove (ii). We fix i,j and define, for $k = 1,\ldots,s$,
$$\epsilon(k) = \mathrm{sign}(f_k(x,\xi_{i,j}(x))) \quad \text{for } x \in A_i.$$

Let x' belong to the closure of A_i. It is enough to show that $\xi_{i,j}$ can be continuously extended to $A_i \cup \{x'\}$. By Theorem 2.5.5 for n, we can find $f : [0,1] \to R^n$ continuous and semi-algebraic such that $f(0) = x'$, $f(\,]0,1]) \subset A_i \cap \overline{B}_n(x',1)$. Define $\varphi = \xi_{i,j} \circ (f|_{]0,1]})$. If $\xi_{i,j}(x)$ is a root of
$$f_k(x,Y) = Y^d + g_{d-1}(x)Y^{d-1} + \cdots + g_0(x),$$
then by Lemma 1.2.11
$$|\xi_{i,j}(x)| \leq 1 + |g_{d-1}(x)| + \cdots + |g_0(x)|.$$

We choose $a \in R$ such that, for every x in $A_i \cap \overline{B}_n(x',1)$, $|g_\lambda(x)| \leq a$ for $\lambda = 0,\ldots,d-1$ (it is clear that a polynomial is bounded on a bounded set). The function $|\varphi|$ is then bounded by $1 + da$. By Proposition 2.5.3, φ can be continuously extended to 0. We define $\xi_{i,j}(x') = \varphi(0)$ and show that $\xi_{i,j}$ is continuous at x'. If it is not continuous, then
$$\exists \mu \in R \ \ \mu > 0 \ \ \forall \eta \in R \ \ \eta > 0 \ \ \exists x \in A_i \ \ (\|x - x'\| < \eta$$
$$\text{and } |\xi_{i,j}(x) - \varphi(0)| \geq \mu).$$

2.5 Closed and Bounded Semi-algebraic Sets. Curve-selection Lemma

We define
$$C_\mu = \{x \in A_i \mid |\xi_{i,j}(x) - \varphi(0)| \geq \mu\} \cap \overline{B}_n(x', 1).$$
Since x' belongs to the closure of C_μ, another application of Theorem 2.5.5 for n gives $g : [0,1] \to R^n$, continuous and semi-algebraic, with $g(0) = x'$ and $g(\,]0,1]) \subset C_\mu$. The function $\psi = \xi_{i,j} \circ (g|_{]0,1]})$ can be continuously extended to 0 (same proof as above). By continuity, $|\varphi(0) - \psi(0)| \geq \mu$ and also $\operatorname{sign}(f_k(x', \varphi(0))) \in \overline{\epsilon(k)}$ and $\operatorname{sign}(f_k(x', \psi(0))) \in \overline{\epsilon(k)}$. By Thom's lemma (Proposition 2.5.4), since at least one of the $\epsilon(k)$ is 0, the set
$$\{y \in R \mid \operatorname{sign}(f_k(x', y)) \in \overline{\epsilon(k)}, \text{ for } k = 1, \ldots, s\}$$
either is empty, or reduces to a point. This gives a contradiction and completes the proof of (ii).

Let us now prove (iii). Let A be a semi-algebraic subset of R^{n+1} given by a boolean combination of sign conditions on $f_1, \ldots, f_s \in R[X, Y]$. We may assume that $f_1, \ldots, f_s$ satisfy the assumption of Lemma 2.5.6. First, a finite number of polynomials in $R[X, Y]$ can be made simultaneously quasi-monic with respect to Y by a linear substitution of variables $X_1 = X_1' + a_1 Y, \ldots, X_n = X_n' + a_n Y$ with suitably chosen $a_1, \ldots, a_n \in R$. Next, we can add the derivatives, since the derivative with respect to Y of a polynomial which is quasi-monic with respect to Y is itself quasi-monic.

Let $(A_i, (\xi_{i,j})_{j=1,\ldots,\ell_i})_{i=1,\ldots,m}$ be a slicing of $f_1, \ldots, f_s$. The set A is the union of the graphs of some functions $\xi_{i,j}$ and some slices $]\xi_{i,j}, \xi_{i,j+1}[$. Let (x, y) be a point in the closure of A.

– If (x, y) belongs to the closure of a graph of $\xi_{i,j}$ contained in A, choose $\varphi : [0, 1] \to R^n$ continuous and semi-algebraic such that $\varphi(0) = x$ and $\varphi(\,]0, 1]) \subset A_i$ (Theorem 2.5.5 for n). The function $\xi_{i,j}$ can be continuously extended at x (Lemma 2.5.6 for n) and, of course, $y = \xi_{i,j}(x)$. Define $\psi = \xi_{i,j} \circ \varphi$. Then $f = (\varphi, \psi)$ is continuous, semi-algebraic, $f(0) = (x, y)$ and $f(\,]0, 1]) \subset A$.

– If (x, y) belongs to the closure of a slice $]\xi_{i,j}, \xi_{i,j+1}[$, $1 \leq j < \ell_i$, contained in A, take, as above, $\varphi : [0, 1] \to R^n$ continuous and semi-algebraic, such that $\varphi(0) = x$ and $\varphi(\,]0, 1]) \subset A_i$. The functions $\xi_{i,j}$ and $\xi_{i,j+1}$ can be continuously extended to x. Define
$$t = \begin{cases} \frac{1}{2} & \text{if } \xi_{i,j}(x) = \xi_{i,j+1}(x) \\ (y - \xi_{i,j}(x))/(\xi_{i,j+1}(x) - \xi_{i,j}(x)) & \text{otherwise.} \end{cases}$$
Let $\psi = [(1-t)\xi_{i,j} + t\xi_{i,j+1}] \circ \varphi$. Then $f = (\varphi, \psi)$ is again continuous, semi-algebraic, $f(0) = (x, y)$ and $f(\,]0, 1]) \subset A$.

– The case where one of the bounds of the slice is infinite is left to the reader.

This completes the proof. □

Lemma 2.5.6 is false without the assumption that the family $f_1, \ldots, f_s$ is stable under derivation with respect to the variable Y. Consider for example the family consisting of the single polynomial

$$f(X_1, X_2, Y) = 36 \left(X_1^2 + \left(\frac{12X_2 + 5Y}{13} \right)^2 \right) - (X_1^2 + X_2^2 + Y^2 + 8)^2 .$$

The polynomial f is the equation of a "tilted torus" and we get a slicing of f above the partition of R^2 depicted in Figure 2.5.

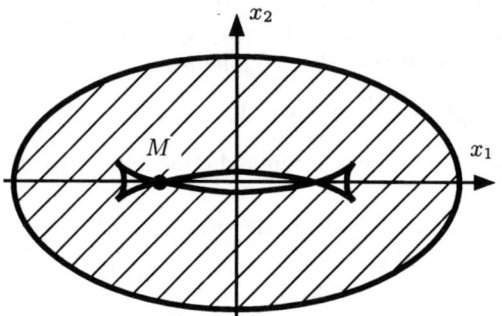

Figure 2.5.

Above the hatched subset, f has two roots that are given by two continuous functions ξ_1 and ξ_2. But neither ξ_1 nor ξ_2 can be continuously extended to the point M since for $i = 1, 2$ the limit of ξ_i approaching M "from above" (with $x_2 > 0$) is different from the limit of ξ_i approaching M "from below" (with $x_2 < 0$).

Proposition 2.5.7. *Let A be a closed and bounded semi-algebraic subset of R^{n+1} and $\Pi : R^{n+1} \to R^n$ the projection on the space of the first n coordinates. Then $\Pi(A)$ is a closed and bounded semi-algebraic set.*

Proof. It is obvious that $\Pi(A)$ is bounded. Let $x \in R^n$ be a point in the closure of $\Pi(A)$. The set A is given by a boolean combination of sign conditions on $f_1, \ldots, f_s$. Let $(A_i, (\xi_{i,j})_{j=1,\ldots,\ell_i})_{i=1,\ldots,m}$ be a slicing of $f_1, \ldots, f_s$. The point x belongs to the closure of one of the A_i contained in $\Pi(A)$, say A_1. Since A is closed and bounded, A contains the graph of at least one function $\xi_{1,j}$.

Take $\varphi : [0,1] \to R^n$ continuous and semi-algebraic with $\varphi(0) = x$ and $\varphi(]0,1]) \subset A_1$ (Theorem 2.5.5). Choose j such that the graph of $\xi_{1,j}$ is contained in A, and take $\psi = \xi_{1,j} \circ (\varphi|_{]0,1]})$. The function ψ is bounded, since A is bounded, and hence ψ can be continuously extended to 0. But then $(x, \psi(0)) \in A$, since A is closed, and thus $x \in \Pi(A)$. □

Theorem 2.5.8. *Let $A \subset R^n$ be a closed and bounded semi-algebraic set, $f : A \to R^p$ a continuous semi-algebraic mapping. Then $f(A)$ is a closed and bounded semi-algebraic set.*

Proof. The set $\mathrm{Graph}(f) \subset R^n \times R^p$ is closed since A is closed. We now show that $\mathrm{Graph}(f)$ is bounded. For this, it is necessary and sufficient to show that

2.5 Closed and Bounded Semi-algebraic Sets. Curve-selection Lemma 41

$\|f\|$ is bounded on A. Let

$$A' = \{((1 + \|f(x)\|)^{-1}x, (1 + \|f(x)\|)^{-1}) \in R^{n+1} \mid x \in A\} .$$

Take the semi-algebraic homeomorphism $h : A' \to A$ given by $h(x,t) = (x_1/t, \ldots, x_n/t)$. If $\|f\|$ is not bounded on A, the origin $(0,0)$ is a point in the closure of A' and, by Theorem 2.5.5, there exists $\varphi : [0,1] \to R^{n+1}$ continuous and semi-algebraic, such that $\varphi(0) = (0,0)$ and $\varphi(]0,1]) \subset A'$. Since $\psi = h \circ (\varphi|_{]0,1]})$ is bounded, by Proposition 2.5.3 it can be continuously extended to 0. Then $\psi(0) \in A$ and $h^{-1}(\psi(0)) = \varphi(0) = (0,0) \notin A'$, which is impossible. Graph(f) is thus closed and bounded and, hence, by using Proposition 2.5.7 several times, $f(A)$ is closed and bounded. □

We now define the Alexandrov compactification of a locally closed semi-algebraic set S and show that it is unique up to a semi-algebraic homeomorphism.

Proposition 2.5.9. *Let S be a locally closed semi-algebraic set that is not closed and bounded. Then there exists a pair $(\dot{S}, \eta)$ such that:*
  (i) *$\dot{S}$ is a closed and bounded semi-algebraic set,*
  (ii) *$\eta : S \to \dot{S}$ is a continuous semi-algebraic mapping, which is a homeomorphism onto its image,*
  (iii) *$\dot{S} \setminus \eta(S)$ consists of a single point.*
    If $(\dot{S}', \eta')$ is another pair with the same properties, then there exists a unique semi-algebraic homeomorphism $h : \dot{S} \to \dot{S}'$ such that $\eta' = h \circ \eta$.

Proof. By Proposition 2.2.9, we may assume S to be closed in R^n, unbounded and not containing the origin 0 of R^n. We define η by $\eta(x) = x/\|x\|^2$ and $\dot{S} = \eta(S) \cup \{0\}$. The uniqueness up to a semi-algebraic homeomorphism is a consequence of the following lemma.

Lemma 2.5.10. *Let $B \subset A$ be two closed and bounded semi-algebraic sets, such that $S = A \setminus B$ is not closed. Let $(\dot{S}, \eta)$ be a pair satisfying properties* (i), (ii) *and* (iii) *of Proposition 2.5.9. Define $\dot{S} \setminus \eta(S) = \{y\}$. Then the semi-algebraic mapping $\varphi : A \to \dot{S}$, defined by $\varphi|_S = \eta$ and $\varphi(B) = \{y\}$, is continuous.*

Proof. Suppose that φ is not continuous. Then, by Theorem 2.5.5, there exist $\epsilon > 0$ and a continuous semi-algebraic mapping $\gamma : [0,1] \to A$ with

$$\gamma(0) \in B \quad \text{and} \quad \gamma(]0,1]) \subset \{x \in S \mid \|\varphi(x) - y\| \geq \epsilon\} .$$

Since $E_\epsilon = \{z \in \dot{S} \mid \|z - y\| \geq \epsilon\}$ is closed and bounded, the mapping $\varphi \circ \gamma|_{]0,1]} :]0,1] \to E_\epsilon$ can be continuously extended to 0. Let t be the value of this extension at 0. Then $t \in \eta(S)$ and, by continuity, $\eta^{-1}(t) = \gamma(0)$. This is a contradiction, since $\gamma(0) \in B$. Hence φ is continuous. □ □

Definition 2.5.11. *The pair* $(\dot{S}, \eta)$, *defined up to a semi-algebraic homeomorphism, as in Proposition 2.5.9, is called the* semi-algebraic Alexandrov compactification *of S.*

The curve-selection lemma (Theorem 2.5.5) allows us to prove another result concerning connectedness.

Definition 2.5.12. *A semi-algebraic subset A of R^n is said to be semi-algebraically path connected if, for every x, y in A, there exists a continuous semi-algebraic mapping $\varphi : [0, 1] \to A$ such that $\varphi(0) = x$ and $\varphi(1) = y$.*

Proposition 2.5.13. *A semi-algebraic set is semi-algebraically connected if and only if it is semi-algebraically path connected.*

Proof. Since $[0, 1]$ is semi-algebraically connected, it is clear that semi-algebraic path connectedness implies semi-algebraic connectedness. We prove the converse by using Theorem 2.3.6 (decomposition of semi-algebraic sets) and the proof of Theorem 2.4.4. It is obvious that an open hypercube is semi-algebraically path connected. It is then enough to show that if A_i and A_j are semi-algebraically homeomorphic to open hypercubes, with $A_i \cap \mathrm{clos}(A_j) \neq \emptyset$, then $A_i \cup A_j$ is semi-algebraically path connected. But this is an immediate consequence of Theorem 2.5.5. $\square$

2.6 Continuous Semi-algebraic Functions. Łojasiewicz's Inequality

The rate of growth of a continuous semi-algebraic function with values in R is bounded by a polynomial.

Proposition 2.6.1. *Let f be a semi-algebraic function (not necessarily continuous) from $]a, +\infty[\subset R$ to R. There exist $r, c \in R$, $r > a$, and $p \in \mathbb{N}$, such that, for every $x \geq r$, we have $|f(x)| \leq cx^p$. Moreover, if $h \in R[X, Y]$ is a nonzero polynomial, such that $h(x, f(x)) = 0$ on $]a, +\infty[$, we can take p to be the degree of h with respect to X.*

Proof. Lemma 2.5.2 says that there exists a nonzero polynomial $h(X, Y) = q_m(X)Y^m + \cdots + q_0(X)$ such that $h(x, f(x)) = 0$ on $]a, +\infty[$. The polynomial q_m is supposed to be not identically zero, and thus $q_m(x)$ does not vanish for x large enough. Then, for such an x,

$$|f(x)| \leq 1 + |q_m(x)|^{-1}(|q_{m-1}(x)| + \cdots + |q_0(x)|) .$$

If p is the maximum of the degrees of $q_{m-1}, \ldots, q_0$, we can find a constant $c \in R$ such that $|f(x)| \leq cx^p$ for all x large enough. $\square$

2.6 Continuous Semi-algebraic Functions. Łojasiewicz's Inequality

Proposition 2.6.2. *Let $f : F \to R$ be a continuous semi-algebraic function, with F a closed semi-algebraic subset of R^n. Then $|f|$ is bounded by a polynomial: there exist $c \in R$ and $p \in \mathbb{N}$ such that for every $x \in F$,*

$$|f(x)| \leq c(1 + \|x\|^2)^p.$$

Proof. Let F_t be the semi-algebraic set $\{x \in F \mid \|x\| = t\}$. It is closed and bounded. Hence, by Theorem 2.5.8, we can define a function $v : R \to R$ by taking $v(t) = \sup(|f|(F_t))$ if $F_t \neq \emptyset$ and $v(t) = 0$ otherwise. The function v is semi-algebraic, since, by Proposition 2.2.4, its graph

$$\{(t,u) \in R \times R \mid (\exists x \in F_t \ \ u = |f(x)| \ \text{ and } \ \forall y \in F_t \ \ |f(y)| \leq u)$$
$$\text{or } (F_t = \emptyset \text{ and } u = 0)\}$$

is semi-algebraic. By Proposition 2.6.1, there exist $r \in R$, $p \in \mathbb{N}$ and $c_1 \in R$ such that $v(t) \leq c_1 t^p$ for $t \geq r$. Let c_2 be the maximum of $|f(x)|$ for $x \in F$ and $\|x\| \leq r$ (Theorem 2.5.8), and $c = \max(c_1, c_2)$. We have $|f(x)| \leq c(1 + \|x\|^2)^p$ on F. $\square$

We shall now prove Łojasiewicz's inequality, after a few technical preliminaries.

Lemma 2.6.3. *Let $f : A \to R$ be a semi-algebraic function on a semi-algebraic subset A of R^n. There exist a partition of A into a finite number of semi-algebraic sets A_i, $i = 1, \ldots, m$, and for every i a polynomial $g_i(X,Y)$ in $n+1$ variables, such that, for every x in A_i, $g_i(x,Y)$ is not identically zero and $g_i(x, f(x)) = 0$.*

Proof. The graph of f is given by a boolean combination of sign conditions on polynomials $g_1(X,Y), \ldots, g_s(X,Y)$. Let $(A_i, (\xi_{i,j})_{j=1,\ldots,\ell_i})_{i=1,\ldots,k}$ be a slicing of $g_1, \ldots, g_s$ such that A is the union of the A_i, $i = 1, \ldots, m$. Then the graph of $f|_{A_i}$ necessarily coincides with the graph of one of the $\xi_{i,j}$, which proves the result. $\square$

Proposition 2.6.4. *Let $f : A \to R$ be a semi-algebraic function on a locally closed semi-algebraic set. Let $g : \{x \in A \mid f(x) \neq 0\} \to R$ be a continuous semi-algebraic function. Then there exists an integer $N > 0$, such that the function $f^N g$, extended by 0 when $f(x) = 0$, is continuous on A.*

Proof. By Proposition 2.2.9, we may assume that A is a closed semi-algebraic subset of R^n. For $x \in A$ and $u \in R$, define

$$A_{x,u} = \{y \in A \mid \|y - x\| \leq 1 \text{ and } u|f(y)| = 1\}.$$

The semi-algebraic set $A_{x,u}$ is closed and bounded. Define

$$v(x,u) = \begin{cases} 0 & \text{if } A_{x,u} = \emptyset, \\ \sup\{|g(y)| \mid y \in A_{x,u}\} & \text{otherwise.} \end{cases}$$

The function $v : A \times R \to R$ is well-defined and semi-algebraic. By Lemma 2.6.3, $A \times R$ is the union of a finite number of semi-algebraic sets B_i, which have the following property: for each i, there is a polynomial $h_i(X, U, V)$, such that, for every (x, u) in B_i, $h_i(x, u, V)$ is not identically zero and $h_i(x, u, v(x, u)) = 0$.

Fix $x' \in A$ with $f(x') = 0$. Let $h_{x'}$ be the product of those h_i such that $B_i \cap (\{x'\} \times R)$ is nonempty. The polynomial $h_{x'}(x', U, V)$ is not identically zero and $h_{x'}(x', u, v(x', u)) = 0$. By Proposition 2.6.1, there is an integer p (which we can take independent of x', for instance p equal to the sum of the degrees in U of all the h_i) and numbers $r(x'), c(x') \in R$, such that

$$|v(x', u)| \leq c(x') u^p \quad \text{for every } u \geq r(x') \;.$$

This means that

$$|f(y)|^p |g(y)| \leq c(x') \quad \text{on} \quad \{y \in A \mid f(y) \neq 0 \text{ and } \|y - x'\| \leq 1\}$$

for $|f(y)|$ sufficiently small. The function $f^{p+1} g$, extended by 0, is thus continuous at x'. Since p is independent of x', this completes the proof. □

Remark 2.6.5. The fact that A is locally closed is essential. For instance, for

$$A = \{(x, y) \in R^2 \mid y > 0\} \cup \{(0, 0)\} \;, \; f(x, y) = x^2 + y^2 \;, \; g(x, y) = 1/y \;,$$

the conclusion of Proposition 2.6.4 is no longer true.

Theorem 2.6.6. *Let A be a locally closed semi-algebraic set and f and g two continuous semi-algebraic functions from A to R, such that $f^{-1}(0) \subset g^{-1}(0)$. Then there exist an integer $N > 0$ and a continuous semi-algebraic function $h : A \to R$, such that $g^N = hf$ on A.*

Proof. The function $1/f$ is continuous semi-algebraic on $\{x \in A \mid g(x) \neq 0\}$. By Proposition 2.6.4, there exists an integer N, such that the function $h : A \to R$ defined by

$$h(x) = \begin{cases} 0 & \text{if } f(x) = 0, \\ g^N(x)/f(x) & \text{if } f(x) \neq 0, \end{cases}$$

is continuous, semi-algebraic, and $g^N = hf$. □

Corollary 2.6.7 (Łojasiewicz's inequality). *Let A be a closed and bounded semi-algebraic set and f and g two continuous semi-algebraic functions from A to R such that $f^{-1}(0) \subset g^{-1}(0)$. Then there exist an integer $N > 0$ and a constant $c \in R$, such that $|g|^N \leq c|f|$ on A.*

Proof. Use Theorem 2.6.6, with $c = \sup\{|h(x)| \mid x \in A\}$. □

Given a semi-algebraic set A, we denote by $\mathcal{S}^0(A)$ the ring of continuous semi-algebraic functions from A into R.

2.6 Continuous Semi-algebraic Functions. Lojasiewicz's Inequality

Proposition 2.6.8. *Let $U \subset A \subset R^n$ be semi-algebraic sets, A locally closed, U open in A. Let f be a function in $\mathcal{S}^0(A)$, such that*

$$U = \{x \in A \mid f(x) \neq 0\}.$$

Then $\mathcal{S}^0(U)$ is isomorphic to the ring of fractions $\mathcal{S}^0(A)_f$.

Proof. The restriction mapping $\mathcal{S}^0(A) \to \mathcal{S}^0(U)$ induces a canonical homomorphism $\mathcal{S}^0(A)_f \to \mathcal{S}^0(U)$. This homomorphism is surjective, by Proposition 2.6.4. It is also injective, since the kernel of the restriction mapping $\mathcal{S}^0(A) \to \mathcal{S}^0(U)$ is the ideal consisting of those g in $\mathcal{S}^0(A)$ satisfying $fg = 0$. □

The Tietze-Urysohn theorem has a semi-algebraic version.

Proposition 2.6.9. *Let A be a locally closed semi-algebraic set, F a semi-algebraic set, closed in A, and $f : F \to R$ a continuous semi-algebraic function. Then there exists a continuous semi-algebraic function $\overline{f} : A \to R$, such that $\overline{f}|_F = f$.*

Proof. By decomposing f into $f = f^+ - f^-$ with $f^+ = \frac{1}{2}(f + |f|)$ and $f^- = \frac{1}{2}(|f| - f)$, we may assume that f is ≥ 0. Moreover, by Proposition 2.2.9, we may assume that A is closed in R^n, and hence we can take $A = R^n$. Applying Theorem 2.6.6 to the functions $|f(x) - f(y)|$ and $\|x - y\|$ on $F \times F$, we get

$$|f(x) - f(y)|^N = h(x, y) \|x - y\|$$

for some integer N and some nonnegative continuous semi-algebraic function $h : F \times F \to R$. By Proposition 2.6.2, there exists a polynomial $g(X, Y)$, such that $h(x, y) \leq g(x, y)$ on $F \times F$. Define

$$\Delta(x, y) = (g(x, y) \|x - y\|)^{1/N} + \|x - y\|$$

and

$$\overline{f}(x) = \inf(\{\Delta(x, y) + f(y) \mid y \in F\}).$$

If $(x, y) \in F \times F$, then $\Delta(x, y) \geq |f(x) - f(y)|$ and, hence, $\Delta(x, y) + f(y) \geq f(x)$. Therefore $\overline{f}|_F = f$. The function $\overline{f}$ is semi-algebraic. It remains to show that it is continuous. Let $x' \in R^n$ and $a = \overline{f}(x')$. Then

$$a = \inf(\{\Delta(x', y) + f(y) \mid y \in F \cap \overline{B}_n(x', a)\})$$

and, since $F \cap \overline{B}_n(x', a)$ is closed and bounded, there exists $y' \in F \cap \overline{B}_n(x', a)$, such that $a = \Delta(x', y') + f(y')$. Fix $\epsilon \in R$, $\epsilon > 0$. There exists $\eta \in R$, $\eta > 0$ such that $\|x - x'\| < \eta \Rightarrow |\Delta(x, y') - \Delta(x', y')| < \epsilon$. Thus,

$$\|x - x'\| < \eta \Rightarrow \overline{f}(x) < a + \epsilon.$$

Suppose that, for every μ, $0 < \mu < 1$, there is (x, y), such that $\|x - x'\| < \mu$, $y \in F$ and $\Delta(x, y) + f(y) \leq a - \epsilon$. We then have $y \in \overline{B}_n(x', 1 + a)$. Define

$$K = \{x \in R^n \mid \exists y \in F \cap \overline{B}_n(x', 1+a) \quad \Delta(x,y) + f(y) \leq a - \epsilon\}.$$

Then x' belongs to the closure of K, and K is closed and bounded, since it is the projection of

$$\{(x,y) \in R^n \times R^n \mid y \in F \cap \overline{B}_n(x', 1+a) \text{ and } \Delta(x,y) + f(y) \leq a - \epsilon\},$$

which is a closed and bounded semi-algebraic set. Thus $x' \in K$, and this contradicts the fact that $\overline{f}(x') = a$. Hence, there exists μ such that

$$\|x - x'\| < \mu \Rightarrow \overline{f}(x) > a - \epsilon,$$

which completes the proof. □

2.7 Separation of Closed Semi-algebraic Sets

Let F and G be two disjoint closed semi-algebraic subsets of R^n. Is it possible to find a function which is positive on F and negative on G? The answer, of course, depends on the class of functions we consider. The answer is easy for continuous semi-algebraic functions: we can take the function with value 1 on F and value -1 on G, and extend it to R^n using Proposition 2.6.9. However, this is not always possible for polynomials. Let

$$\begin{aligned} F &= \{(x,y) \in R^2 \mid x \leq 0 \text{ or } y \leq 0\} \\ G &= \{(x,y) \in R^2 \mid x \geq 1 \text{ and } y \geq 1\}. \end{aligned}$$

Suppose that a polynomial $f \in R[X,Y]$ is positive on F and negative on G. On account of its signs, the polynomial $f(X, tX) \in R[X]$ should be of odd degree for $t > 0$ and of even degree for $t < 0$, which is impossible.

Therefore polynomials do not suffice. But, if we take square roots of polynomials positive on R^n, then we can separate disjoint closed semi-algebraic sets. This result will be used in Chap. 8 and 15. Before showing this, let us establish a result which is important in its own right.

Definition 2.7.1. *A basic open semi-algebraic subset of R^n is a set of the form*

$$\{x \in R^n \mid f_1(x) > 0, \ldots, f_s(x) > 0\},$$

where $f_1, \ldots, f_s \in R[X_1, \ldots, X_n]$.

A basic closed semi-algebraic subset of R^n is a set of the form

$$\{x \in R^n \mid f_1(x) \geq 0, \ldots, f_s(x) \geq 0\},$$

where $f_1, \ldots, f_s \in R[X_1, \ldots, X_n]$.

Theorem 2.7.2 (Finiteness Theorem). *Let $A \subset R^n$ be an open (resp. closed) semi-algebraic set. Then A is a finite union of basic open (resp. basic closed) semi-algebraic sets.*

2.7 Separation of Closed Semi-algebraic Sets

Proof. It is enough to prove the result if A is an open set, since the case where A is closed then follows by taking the complement. The set A is a finite union of semi-algebraic sets of the form

$$B = \{x \in R^n \mid f_1(x) = \ldots = f_\ell(x) = 0, \; g_1(x) > 0, \ldots, g_m(x) > 0\},$$

where the f_i and the g_i are polynomials.

Let $f = f_1^2 + \cdots + f_\ell^2$ and $g(x) = \prod_{i=1}^m (|g_i(x)| + g_i(x))$. On the complement of A, $g(x) = 0$ if $f(x) = 0$ and, hence, by Theorem 2.6.6, there exist an integer N and a continuous semi-algebraic function h on $R^n \setminus A$, such that $g^N = hf$ on $R^n \setminus A$. By Proposition 2.6.2, there is $c \in R$ and $p \in \mathbb{N}$, such that $|h(x)| \leq c(1 + \|x\|^2)^p$ on $R^n \setminus A$. Define

$$B_1 = \left\{ x \in R^n \;\middle|\; f(x)c(1 + \|x\|^2)^p < \left(2^m \prod_{i=1}^m g_i(x)\right)^N, \right.$$

$$\left. g_1(x) > 0, \ldots, g_m(x) > 0 \right\}.$$

We have $B \subset B_1 \subset A$, and we see that A can be written as a finite union of basic open semi-algebraic sets by replacing B with B_1. □

Theorem 2.7.3. *Let F and G be two disjoint closed semi-algebraic subsets of R^n. Then there exists a function $f : R^n \to R$ of the form*

$$f = \sum_{i=1}^m P_i \sqrt{1 + \sum_{j=1}^{\ell_i} Q_{i,j}^2},$$

where the P_i and the $Q_{i,j}$ are polynomials, such that f is positive on F and negative on G.

Proof. By the finiteness theorem 2.7.2, G has the form

$$G = \bigcup_{\lambda=1}^p \left(\bigcap_{\mu=1}^{q_\lambda} \{x \in R^n \mid S_{\lambda,\mu}(x) \geq 0\} \right),$$

where the $S_{\lambda,\mu}$ are polynomials. Define $h_\lambda = \sum_{\mu=1}^{q_\lambda}(|S_{\lambda,\mu}| - S_{\lambda,\mu})$. The function $h = \prod_{\lambda=1}^p h_\lambda$ is continuous semi-algebraic, vanishes on G and is positive elsewhere. By Proposition 2.6.2, the function $1/h$ can, thus, be bounded from above on F by a polynomial $c(1 + \|x\|^2)^r$. Let $\epsilon(x) = c^{-1}(1 + \|x\|^2)^{-r}$. Now, let $\delta : R^n \to]0, 1]$ be a continuous semi-algebraic function. Define

$$f_1 = \prod_{\lambda=1}^p \left(\sum_{\mu=1}^{q_\lambda} (\sqrt{S_{\lambda,\mu}^2 + \delta^2} - S_{\lambda,\mu}) \right).$$

On F, $f_1 > h$ and, hence, $f_1 > \epsilon$. Let x be a point of G. We know that one of the h_λ vanishes at x, and, hence, we have on G

$$f_1 \leq \prod_{\lambda=1}^{p}(h_\lambda + q\delta) \leq q\delta \prod_{\lambda=1}^{p}(h_\lambda + q) \text{ , where } q = \max(q_1, \ldots, q_p) \text{ .}$$

Choose for δ the function $d^{-1}(1+\|x\|^2)^{-s}$, where $d(1+\|x\|^2)^s$ bounds from above the function $\max\left(2\epsilon^{-1}q\prod_{\lambda=1}^{p}(h_\lambda+q), 1\right)$. Then $f_1 \leq \epsilon/2$ on G. Define $f = \delta^{-p}(f_1\epsilon^{-1} - 1)$. The function f is positive on F, negative on G, and has the desired form. □

We shall need a relative separation theorem, that is, a version of 2.7.3 valid for disjoint closed semi-algebraic subsets of an open semi-algebraic set U. First we introduce the functions needed for this separation.

Definition 2.7.4. *Let U be an open semi-algebraic subset of R^n. Denote by $\mathcal{A}(R^n; U)$ the smallest subring of the ring of continuous semi-algebraic functions from R^n to R containing the polynomials and such that, if f is a sum of squares of functions of this subring, positive on U, then $\sqrt{f}$ belongs to the subring.*

After defining Nash functions (Definition 2.9.3), it will be clear that, for $f \in \mathcal{A}(R^n; U)$, the restriction $f|_U$ is Nash on U.

Proposition 2.7.5. *Let U be an open semi-algebraic subset of R^n. Then there exists a function in $\mathcal{A}(R^n; U)$ which is positive on U and vanishes on $R^n \setminus U$.*

Proof. Let $F = R^n \setminus U$. By the finiteness theorem 2.7.2, F is a finite union of closed semi-algebraic sets of the form

$$\{x \in R^n \mid P_0(x) = 0, \ P_1(x) \geq 0, \ldots, \ P_k(x) \geq 0\}$$

(with possibly P_0 identically zero). It is enough to prove the proposition for an F of the form above, since, for a finite union, we can take the product of the corresponding functions. We then proceed by induction on k. For $k = 0$, we take $f = P_0^2$. Given $k > 0$, assume that the proposition holds true for $k - 1$. Define

$$F' = \bigcup_{i=1}^{k}\{x \in R^n \mid P_0^2(x) + P_i^2(x) = 0,$$

$$P_1(x) \geq 0, \ldots, \ \widehat{P_i(x) \geq 0}, \ldots, \ P_k(x) \geq 0\} \text{ ,}$$

where the hat denotes the omission of the corresponding inequality. By induction, there exists a function $h \in \mathcal{A}(R^n; R^n \setminus F')$ vanishing on F' and positive elsewhere. Let

$$G_1 = \{(x,y) \in R^{n+1} \mid x \in F \text{ and } yh(x) = 1\}$$
$$G_2 = \{(x,y) \in R^{n+1} \mid P_0(x) = 0 \text{ and } (P_1(x) \leq 0 \text{ or } \ldots \text{ or } P_k(x) \leq 0)$$
$$\text{and } yh(x) = 1\} \text{ .}$$

2.7 Separation of Closed Semi-algebraic Sets

G_1 and G_2 are closed semi-algebraic subsets of R^{n+1}, and they are disjoint since

$$F' = F \cap \{x \in R^n \mid P_0(x) = 0 \text{ and } (P_1(x) \le 0 \text{ or } \ldots \text{ or } P_k(x) \le 0)\}.$$

Lemma 2.7.6. *Let F' be a closed semi-algebraic subset of R^n, h a function in $\mathcal{A}(R^n; R^n \setminus F')$, $h = 0$ on F', $h > 0$ on $R^n \setminus F'$, G_1 and G_2 two disjoint closed semi-algebraic subsets of $\{(x,y) \in R^{n+1} \mid yh(x) = 1\}$. Then there exists a function $g_1 \in \mathcal{A}(R^n; R^n \setminus F')$ positive on the projection of G_1 onto R^n, negative on the projection of G_2, and vanishing on F'.*

Proof. By the separation theorem 2.7.3, we can find a function $g : R^{n+1} \to R$ of the form

$$g = \sum_{i=1}^{m} Q_i \sqrt{1 + \sum_{j=1}^{\ell_i} S_{i,j}^2},$$

where the Q_i and the $S_{i,j}$ are polynomials in $n+1$ variables (X, Y), which is positive on G_1 and negative on G_2. Let d be the maximum of the degrees of the Q_i and the $S_{i,j}$ with respect to Y. Then

$$g_1(x) = h^{2d+1}(x)\, g\left(x, \frac{1}{h(x)}\right)$$

$$= h(x) \left[\sum_{i=1}^{m} h^d(x) Q_i\left(x, \frac{1}{h(x)}\right) \sqrt{h^{2d}(x) + \sum_{j=1}^{\ell_i} h^{2d}(x) S_{i,j}^2\left(x, \frac{1}{h(x)}\right)} \right]$$

is a function in $\mathcal{A}(R^n; R^n \setminus F')$ with the expected properties. $\square$

We return to the proof of Proposition 2.7.5. The function $f = \sqrt{P_0^2 + g_1^2} - g_1$ belongs to $\mathcal{A}(R^n; R^n \setminus F')$ and, hence, a fortiori, to $\mathcal{A}(R^n; U)$. It vanishes if and only if $P_0 = 0$ and $g_1(x) \ge 0$, i.e. exactly on F. Elsewhere, it is positive.
$\square$

Theorem 2.7.7. *Let U be an open semi-algebraic subset of R^n and F and G two disjoint closed semi-algebraic subsets of U. Then there exists a function in $\mathcal{A}(R^n; U)$ which is positive on F and negative on G.*

Proof. By Proposition 2.7.5, there exists a function $h \in \mathcal{A}(R^n; U)$ which is positive on U and vanishes on $R^n \setminus U$. Define

$$\begin{aligned} F_1 &= \{(x,y) \in R^{n+1} \mid x \in F \text{ and } yh(x) = 1\} \\ G_1 &= \{(x,y) \in R^{n+1} \mid x \in G \text{ and } yh(x) = 1\}. \end{aligned}$$

Lemma 2.7.6 gives a function $f \in \mathcal{A}(R^n; U)$ separating F and G, as stated in the theorem. $\square$

2.8 Dimension of Semi-algebraic Sets

The decomposition of semi-algebraic sets (Theorem 2.3.6) clearly shows what the dimension of a semi-algebraic set should be. If A is a semi-algebraic set which is the union of semi-algebraic sets A_i, each homeomorphic to an open hypercube $]0,1[^{d_i}$, then the dimension of A should be the maximum of the d_i. We shall give an algebraic definition of the dimension and show that we recover this property. The advantage of the algebraic definition is that it is intrinsic and coincides with the classical definition in the case of an algebraic set.

Definition 2.8.1. *Let $A \subset R^n$ be a semi-algebraic set. Denote by $\mathcal{P}(A) = R[X_1, \ldots, X_n]/\mathcal{I}(A)$ the ring of polynomial functions on A. The dimension of A, denoted $\dim(A)$, is the dimension of the ring $\mathcal{P}(A)$, i.e. the maximal length of chains of prime ideals of $\mathcal{P}(A)$.*

Proposition 2.8.2. *Let $A \subset R^n$ be a semi-algebraic set. Then*

$$\dim(A) = \dim(\operatorname{clos}(A)) = \dim(\operatorname{clos}_{\operatorname{Zar}}(A)) ,$$

where $\operatorname{clos}_{\operatorname{Zar}}(A) = \mathcal{Z}(\mathcal{I}(A))$ is the Zariski closure of A.

Proof. We have $\mathcal{I}(A) = \mathcal{I}(\operatorname{clos}(A)) = \mathcal{I}(\operatorname{clos}_{\operatorname{Zar}}(A))$. □

In order to develop the theory of dimension of semi-algebraic sets, we shall use results about algebraic sets. These results are classical, but are often stated only for algebraic sets over an algebraically closed field; so we prefer to recall them. The fact that R is real closed plays no role for the results of the next theorem. They are true for any base field.

Theorem 2.8.3.

(i) *An algebraic set $V \subset R^n$ is said to be* irreducible, *if, whenever $V = F_1 \cup F_2$, where F_1 and F_2 are algebraic sets, then $V = F_1$ or $V = F_2$. Every algebraic set V is the union – in a unique way – of a finite number of irreducible algebraic sets $V_1, \ldots, V_p$, such that $V_i \not\subset \bigcup_{j \neq i} V_j$ for $i = 1, \ldots, p$. The V_i, $i = 1, \ldots, p$, are* the irreducible components of V. *We have $\dim(V) = \max(\dim(V_1), \ldots, \dim(V_p))$.*

(ii) *An algebraic set $V \subset R^n$ is irreducible if and only if the ideal $\mathcal{I}(V)$ is prime. Denote by $\mathcal{K}(V)$ the field of fractions of $\mathcal{P}(V)$. The dimension of V is equal to the transcendence degree of $\mathcal{K}(V)$ over R.*

(iii) *If $V \subset R^m$ and $W \subset R^n$ are two algebraic sets, then the product $V \times W \subset R^{m+n}$ is an algebraic set and $\mathcal{P}(V \times W) = \mathcal{P}(V) \otimes_R \mathcal{P}(W)$. If V and W are irreducible, then $V \times W$ is irreducible. We have $\dim(V \times W) = \dim(V) + \dim(W)$.*

Hints for the proof:

(i) The proofs in [150], Proposition 1.5 and Corollary 1.6, or in [295], Chap. 1, § 3, Theorems 1 and 2 can be used for any base field. The last statement follows from the equality $\mathcal{I}(V) = \bigcap_{i=1}^{p} \mathcal{I}(V_i)$.

(ii) Use the proof in [150], Corollary 1.4. For the statement about the dimension, we refer the reader to [228], Chap. 5, § 14.

(iii) There is a canonical homomorphism $\mathcal{P}(V) \otimes_R \mathcal{P}(W) \to \mathcal{P}(V \times W)$, and we can show that it is an isomorphism, as in [295], Chap. 1, § 2,2, Example 4. The fact that the product of irreducible algebraic sets is irreducible is proved in [295], Chap. 1, § 3, Theorem 3. The statement about the dimension of the product can be proved by working with the irreducible components and by using the normalization lemma of E. Noether, as in [295], Chap. 1, § 6, 1, Example 4. □

We now return to semi-algebraic sets.

Proposition 2.8.4. *Let U be a nonempty open semi-algebraic subset of R^n. Then $\dim(U) = n$.*

Proof. We prove the result by induction on n. If $n = 1$, then U is infinite (it contains an open interval) and, thus, $\mathcal{I}(U) = \{0\}$. Now we assume $n > 1$ and the result to hold true for $n-1$. Then U contains an open set $U' \times]a, b[$, with U' an open semi-algebraic subset of R^{n-1}. Let $P(X', X_n) = Q_d(X') X_n^d + \cdots + Q_0(X')$ (where $X' = (X_1, \ldots, X_{n-1})$) be a polynomial in $\mathcal{I}(U)$. For every $x' \in U'$ we have $P(x', X_n) \in \mathcal{I}(]a, b[)$ and, thus, $P(x', X_n)$ is identically zero. Hence, by induction, $Q_d = \cdots = Q_0 = 0$, which shows $\mathcal{I}(U) = \{0\}$. Hence, $\mathcal{P}(U) = R[X_1, \ldots, X_n]$, and the dimension of $R[X_1, \ldots, X_n]$ is n (cf. Theorem 2.8.3 (ii)). □

Proposition 2.8.5.
(i) Let $A = \bigcup_{i=1}^{p} A_i$ be a finite union of semi-algebraic sets. Then
$$\dim(A) = \max(\dim(A_1), \ldots, \dim(A_p)) .$$

(ii) Let A and B be two semi-algebraic sets. Then
$$\dim(A \times B) = \dim(A) + \dim(B) .$$

Proof. (i) We have $\mathcal{I}(A) = \bigcap_{i=1}^{p} \mathcal{I}(A_i)$.

(ii) This follows from Theorem 2.8.3 (iii) and from the fact that
$$\mathrm{clos}_{\mathrm{Zar}}(A) \times \mathrm{clos}_{\mathrm{Zar}}(B) = \mathrm{clos}_{\mathrm{Zar}}(A \times B) .$$

The inclusion $\supset$ is clear. To show the other inclusion, let $P(X, Y) \in \mathcal{I}(A \times B)$. Then $P(x, Y) \in \mathcal{I}(B)$ for every $x \in A$, and, hence, for every $y \in \mathrm{clos}_{\mathrm{Zar}}(B)$, $P(x, y) = 0$. This shows that $A \times \mathrm{clos}_{\mathrm{Zar}}(B) \subset \mathrm{clos}_{\mathrm{Zar}}(A \times B)$. By interchanging A and B, we conclude that $\mathrm{clos}_{\mathrm{Zar}}(A) \times \mathrm{clos}_{\mathrm{Zar}}(B) \subset \mathrm{clos}_{\mathrm{Zar}}(A \times B)$. □

Proposition 2.8.6. *Let A be a semi-algebraic subset of R^{n+1} and $\Pi : R^{n+1} \to R^n$ the projection onto the space of the first n coordinates. Then $\dim(\Pi(A)) \leq \dim(A)$.*

Proof. By Proposition 2.8.2 and Theorem 2.8.3 (i), we may assume that A is an irreducible algebraic set. Then $B = \text{clos}_{\text{Zar}}(\Pi(A))$ is also irreducible: if $B = F_1 \cup F_2$, where F_1 and F_2 are algebraic, then

$$A = (A \cap \Pi^{-1}(F_1)) \cup (A \cap \Pi^{-1}(F_2)),$$

and thus $A \subset \Pi^{-1}(F_1)$ or $A \subset \Pi^{-1}(F_2)$. Hence $B \subset F_1$ or $B \subset F_2$. The projection Π induces an injective homomorphism $\Pi^* : \mathcal{P}(B) \to \mathcal{P}(A)$ and, hence, an injective homomorphism $\mathcal{K}(B) \to \mathcal{K}(A)$. The proposition now follows by using Theorem 2.8.3. □

Proposition 2.8.7. *Let $A \subset R^n$ be a semi-algebraic set and $f : A \to R^p$ a semi-algebraic mapping whose graph is $G(f) \subset R^{n+p}$. Then $\dim(A) = \dim(G(f))$.*

Proof. First we prove the result for $p = 1$. By Lemma 2.6.3, there exist a finite number of semi-algebraic sets A_i whose union is A, and polynomials $P_i(X, Y)$, such that, for every x in A_i, $P_i(x, Y)$ is not identically zero and $P_i(x, f(x)) = 0$. We fix an i and choose an irreducible component V of $\text{clos}_{\text{Zar}}(A_i)$. Certainly, $V \cap A_i \neq \emptyset$, and, thus, there exists $x \in V$ such that $P_i(x, Y)$ is different from zero. Hence, $\mathcal{Z}(P_i) \cap (V \times R)$ is strictly contained in $V \times R$. Since, by Theorem 2.8.3 (iii), the set $V \times R$ is irreducible of dimension $\dim(V) + 1$, we have

$$\begin{aligned} \dim(\text{Graph}(f|_{A_i \cap V})) &\leq \dim(\mathcal{Z}(P_i) \cap (V \times R)) \\ &< \dim(V) + 1 \leq \dim(A_i) + 1 \,. \end{aligned}$$

Collecting together these inequalities for all irreducible components of all $\text{clos}_{\text{Zar}}(A_i)$ and using Proposition 2.8.5 (i), we get:

$$\dim(G(f)) \leq \dim(A) \,.$$

The inequality in the opposite direction follows from Proposition 2.8.6.

The result for an arbitrary p is proved by induction. For $p > 1$, we assume the result to hold true for $p - 1$. We have $f = (f', f_p)$ with $f' : A \to R^{p-1}$ and $f_p : A \to R$. Define $B = \text{Graph}(f') \subset R^{n+p-1}$ and $g : B \to R$ by $g(x, y') = f_p(x)$. Then $G(f) = \text{Graph}(g)$ and $\dim(\text{Graph}(g)) = \dim(B)$ by the result for $p = 1$. By induction, we obtain $\dim(B) = \dim(A)$. □

Theorem 2.8.8. *Let A be a semi-algebraic set and $f : A \to R^p$ a semi-algebraic mapping. Then $\dim(A) \geq \dim(f(A))$. If f is a bijection from A onto $f(A)$, then $\dim(A) = \dim(f(A))$.*

2.8 Dimension of Semi-algebraic Sets 53

Proof. The projection of the graph of f onto R^p is $f(A)$; hence, $\dim(A) \geq \dim(f(A))$, by Propositions 2.8.7 and 2.8.6. The second statement is obvious. $\square$

Corollary 2.8.9. *If $A = \bigcup_{i=1}^{p} A_i$ is a finite union of semi-algebraic sets, where each A_i is semi-algebraically homeomorphic to an open hypercube $]0,1[^{d_i}$, then $\dim(A) = \max(d_1, \ldots, d_p)$.*

Proof. Use Propositions 2.8.4, 2.8.5 (i) and Theorem 2.8.8. $\square$

We now introduce a notion of local dimension.

Proposition 2.8.10. *Let $A \subset R^n$ be a semi-algebraic set, and let x be a point of A. There exists a semi-algebraic neighbourhood U of x in A, such that, for any other semi-algebraic neighbourhood U' of x in A contained in U, one has $\dim(U) = \dim(U')$.*

Proof. The family of ideals $\mathcal{I}(U)$ of $R[X]$, for U a semi-algebraic neighbourhood of x in A, form a directed system with respect to inclusion. This family is stationary since $R[X]$ is noetherian. $\square$

Definition 2.8.11. *Let $A \subset R^n$ be a semi-algebraic set, and let x be a point of A. The* local dimension *of A at x, denoted $\dim(A_x)$, is $\dim(U)$, where U is as in Proposition 2.8.10.*

Proposition 2.8.12. *Let A be a semi-algebraic set of dimension d. Then $A^{(d)} = \{x \in A \mid \dim(A_x) = d\}$ is a nonempty closed semi-algebraic subset of A.*

Proof. We use the decomposition again (Theorem 2.3.6). The set A is a finite union of semi-algebraic sets A_i, each semi-algebraically homeomorphic to $]0,1[^{d_i}$. Let A' be the closure in A of the union of the A_i such that $d_i = d$ (there are such A_i, since $d = \max(d_i)$, by Corollary 2.8.9). Of course, $A' \subset A^{(d)}$. If $x \notin A'$, there is an open neighbourhood U of x, such that $A_i \cap U \neq \emptyset$ implies $d_i < d$, hence $x \notin A^{(d)}$. Therefore $A^{(d)} = A'$. $\square$

Proposition 2.8.13. *Let $A \subset R^n$ be a semi-algebraic set. Then*

$$\dim(\mathrm{clos}(A) \setminus A) < \dim(A) .$$

Proof. Let V be an irreducible component of $\mathrm{clos}_{\mathrm{Zar}}(A)$. Then V is the Zariski closure of $V \cap A$. Thus we may assume that $\mathrm{clos}_{\mathrm{Zar}}(A)$ is irreducible. We may also assume that

$$A = \{x \in R^n \mid P(x) = 0, \ Q_1(x) > 0, \ldots Q_m(x) > 0\} .$$

Then $\mathrm{clos}(A) \setminus A$ is contained in $\mathrm{clos}_{\mathrm{Zar}}(A) \cap \bigcup_{i=1}^{m} \{x \in R^n \mid Q_i(x) = 0\}$, which is an algebraic set strictly contained in $\mathrm{clos}_{\mathrm{Zar}}(A)$ and thus of dimension lower than $\dim(A)$, since $\mathrm{clos}_{\mathrm{Zar}}(A)$ is irreducible. $\square$

We conclude this section with the following result valid over $\mathbb{R}$.

Proposition 2.8.14. *Let $A \subset \mathbb{R}^n$ be a semi-algebraic set which is a C^∞ submanifold of $\mathbb{R}^n$ of dimension d. Then $\dim(A) = d$.*

Proof. Let $x \in A$ and $T_x(A)$ be the tangent space of A at x. The orthogonal projection $A \to T_x(A)$ is a semi-algebraic mapping and maps bijectively an open semi-algebraic neighbourhood of x in A onto an open semi-algebraic subset of $T_x(A)$. Since $T_x(A)$ is a vector space of dimension d, $\dim(A_x) = \dim(T_x(A)) = d$, by Theorem 2.8.8 and Proposition 2.8.4. By Proposition 2.8.12, $\dim(A) = \max(\{\dim(A_x) \mid x \in A\}) = d$. $\square$

2.9 Some Analysis over a Real Closed Field

It is possible to copy for an arbitrary real closed field R the usual notions of differentiation over $\mathbb{R}$. We shall do this only for semi-algebraic functions. For the functions of one variable, Theorem 2.5.8 implies that a semi-algebraic function, which is continuous on a closed and bounded interval, is bounded and reaches its bounds. Hence Rolle's theorem and the mean value theorem are valid for semi-algebraic continuous functions. We can proceed to higher order derivatives in view of:

Proposition 2.9.1. *Let $f :]a, b[\to R$ be a semi-algebraic function differentiable on the interval $]a, b[$. Then its derivative f' is a semi-algebraic function.*

Proof. Describe the graph of f' by a first-order formula of the language of ordered fields with parameters in R and use Proposition 2.2.4. Alternatively, if f is continuously differentiable, we see that $f'(x) = \overline{h}(x, x)$, where $\overline{h}$ is the continuous extension of $h : (x, y) \mapsto (f(x) - f(y))/(x - y)$ to the diagonal, and that the graph of $\overline{h}$, which is the closure of the graph of h in $]a, b[^2 \times R$, is semi-algebraic by Proposition 2.2.2. $\square$

The differentiability of semi-algebraic functions of several variables behaves in the usual way.

Notation 2.9.2. *Let U be an open semi-algebraic subset of R^n and B a semi-algebraic subset of R^p. Denote by $\mathcal{S}^k(U, B)$, for $k = 0, \ldots, \infty$, the set of semi-algebraic mappings from U to B of class C^k (i.e. all partial derivatives up to order k exist and are continuous). Denote by $\mathcal{S}^k(U)$ the ring of semi-algebraic functions from U to R of class C^k.*

We have Taylor's theorem in the following form: let $U \subset R^n$ be an open semi-algebraic neighbourhood of the origin 0, $f \in \mathcal{S}^k(U)$, for $k \in \mathbb{N}$. Let $\Phi(x)$ be the Taylor expansion of f at 0 up to order k. Then for every $\epsilon \in R$, $\epsilon > 0$, there exists $\delta \in R$, $\delta > 0$, such that

$$\|x\| < \delta \Rightarrow |f(x) - \Phi(x)| < \epsilon \|x\|^k .$$

The case of $k = \infty$ is of particular importance.

2.9 Some Analysis over a Real Closed Field 55

Definition 2.9.3. *A Nash function from an open semi-algebraic subset U of R^n to R is a semi-algebraic function of class C^∞. The ring of Nash functions on U is denoted by $\mathcal{N}(U)$.*

Thus the two notations $\mathcal{N}(U)$ and $\mathcal{S}^\infty(U)$ denote the same object.

Example 2.9.4. We have already encountered Nash functions in the preceding section. A function in $\mathcal{A}(R^n; U)$ (Definition 2.7.4) is Nash on U. A typical example of a Nash function on R is $\sqrt{1+x^2}$. Of course, Nash functions cannot, in general, be expressed as compositions of operations $+, -, \times, /, \sqrt[n]{\ }$.

Proposition 2.9.5. *Let $\mathcal{N}_{R^n,0} = \varinjlim_r \mathcal{N}(B_n(0,r))$ be the ring of germs of Nash functions at the origin of R^n. The homomorphism $\mathcal{N}_{R^n,0} \to R[[X_1, \ldots, X_n]]$ sending a germ to its Taylor series is injective.*

Proof. Let $f \in \mathcal{N}_{R^n,0}$, and assume its Taylor series is zero. Then, for every integer p, $\lim_{x \to 0} f(x)/\|x\|^p = 0$. Let $g(r) = \sup(\{|f(x)| \mid \|x\| \le r\})$. The function g is semi-algebraic and does not vanish on an interval $]0, \epsilon[$, if f is not zero. By Corollary 2.6.7, unless f is zero, there is an integer p and a constant $c \in R$ such that $r^p \le cg(r)$ for r small enough. The kernel of the homomorphism thus reduces to the zero germ. □

We shall consider an implicit function theorem and explain the proof in some detail, even though it follows closely the classical proof.

Given a linear mapping $F: R^n \to R^p$, denote by

$$\|F\| = \sup(\{\|F(x)\| \mid x \in S^{n-1}\}),$$

which exists since $x \mapsto \|F(x)\|$ is continuous semi-algebraic and S^{n-1} is a closed and bounded semi-algebraic set.

Proposition 2.9.6. *Let x and y be two points of R^n, U an open semi-algebraic subset of R^n containing the segment $[x,y]$, and let $f \in \mathcal{S}^1(U, R^p)$. Then*

$$\|f(x) - f(y)\| \le M\|x - y\|,$$

where $M = \sup(\{\|df_z\| \mid z \in [x,y]\})$.

Proof. Define $g(t) = f((1-t)x + ty)$ for $t \in [0,1]$. Then $\|g'(t)\| \le M\|x-y\|$ for $t \in [0,1]$. Let $\epsilon \in R$, $\epsilon > 0$, and define

$$A_\epsilon = \{t \in [0,1] \mid \|g(t) - g(0)\| \le M\|x-y\|t + \epsilon t\}.$$

This is a closed semi-algebraic subset of $[0,1]$ containing 0. It contains a maximal interval $[0, t_0]$. Suppose $t_0 \ne 1$. We have

$$\|g(t_0) - g(0)\| \le M\|x-y\|t_0 + \epsilon t_0.$$

Since $\|g'(t_0)\| \le M\|x-y\|$, we can find $\mu > 0$ in R such that, if $t_0 < t < t_0 + \mu$,

$$\|g(t) - g(t_0)\| \leq M\|x - y\|(t - t_0) + \epsilon(t - t_0) .$$

So, for $t_0 < t < t_0 + \mu$, we have

$$\|g(t) - g(0)\| \leq M\|x - y\|t + \epsilon t ,$$

which contradicts the maximality of t_0. Thus, $1 \in A_\epsilon$ for every ϵ, which implies the conclusion of the proposition. □

Proposition 2.9.7. *Let U' be an open semi-algebraic neighbourhood of the origin 0 of R^n, $f \in \mathcal{S}^k(U', R^n)$, $k \geq 1$. Assume $f(0) = 0$ and $df_0 : R^n \to R^n$ invertible. Then there exist open semi-algebraic neighbourhoods U and V of 0 in R^n, $U \subset U'$, such that $f|_U$ is a homeomorphism onto V and $(f|_U)^{-1} \in \mathcal{S}^k(V, U)$.*

Proof. We may assume that df_0 is the identity Id of R^n (by composing f with $(df_0)^{-1}$). Take $g = f - \text{Id}$. Then $dg_0 = 0$, and there is $\epsilon_1 > 0$ such that $\|dg_x\| \leq \frac{1}{2}$ if $x \in B_n(0, \epsilon_1)$. By Proposition 2.9.6, if $x, y \in B_n(0, \epsilon_1)$, then

$$\|f(x) - f(y) - (x - y)\| \leq \frac{1}{2}\|x - y\| ,$$

and thus

$$\frac{1}{2}\|x - y\| \leq \|f(x) - f(y)\| \leq \frac{3}{2}\|x - y\| .$$

This implies that f is injective on $B_n(0, \epsilon_1)$. We can find $\epsilon_2 < \epsilon_1$ with df_x invertible for $x \in B_n(0, \epsilon_2)$.

Let us prove $f(B_n(0, \epsilon_2)) \supset B_n(0, \epsilon_2/4)$. Choose y^0 with $\|y^0\| < \epsilon_2/4$ and define $h(x) = \|f(x) - y^0\|^2$. Then h reaches its minimum on $\overline{B}_n(0, \epsilon_2)$ and does not reach it on the boundary $S^{n-1}(0, \epsilon_2)$. Indeed, if $\|x\| = \epsilon_2$, then $\|f(x)\| \geq \epsilon_2/2$ and $h(x) > (\epsilon_2/4)^2 > h(0)$. This minimum is reached at a point $x^0 \in B_n(0, \epsilon_2)$. Then we have, for $i = 1, \ldots, n$,

$$\frac{\partial h}{\partial x_i}(x^0) = 0 , \quad \text{i.e.} \quad \sum_{j=1}^n (f_j(x^0) - y_j^0) \frac{\partial f_j}{\partial x_i}(x^0) = 0 .$$

Since df_{x^0} is invertible, we get $f(x^0) = y^0$.

We define $V = B_n(0, \epsilon_2/4)$, $U = f^{-1}(V) \cap B_n(0, \epsilon_2)$. The mapping f^{-1} is continuous, because $\|f^{-1}(x) - f^{-1}(y)\| \leq 2\|x - y\|$ for $x, y \in V$, and we easily get $d(f^{-1})_x = (df_{f^{-1}(x)})^{-1}$. □

Corollary 2.9.8 (Implicit Function Theorem). *Let $(x^0, y^0) \in R^{n+p}$, and let $f_1, \ldots, f_p$ be semi-algebraic functions of class C^k on an open neighbourhood of (x^0, y^0), such that $f_j(x^0, y^0) = 0$ for $j = 1, \ldots, p$ and the matrix*

$$\left[\frac{\partial f_j}{\partial y_i}(x^0, y^0)\right]$$

is invertible. Then there exist an open semi-algebraic neighbourhood U (resp. V) of x^0 (resp. y^0) in R^n (resp. R^p) and a mapping $\varphi \in \mathcal{S}^k(U,V)$, such that $\varphi(x^0) = y^0$ and

$$f_1(x,y) = \cdots = f_p(x,y) = 0 \Leftrightarrow y = \varphi(x)$$

for every $(x,y) \in U \times V$.

Proof. Apply the inverse function theorem 2.9.7 to the mapping $(x,y) \mapsto (x, f(x,y))$. □

Thus we have all the tools needed to develop "semi-algebraic differential geometry". The notion of Nash diffeomorphism between open semi-algebraic subsets of R^n is defined in the obvious way.

Definition 2.9.9. *A semi-algebraic subset M of R^n is said to be a Nash submanifold of R^n of dimension d if, for every point x of M, there exists a Nash diffeomorphism φ from an open semi-algebraic neighbourhood Ω of the origin in R^n onto an open semi-algebraic neighbourhood Ω' of x in R^n such that $\varphi(0) = x$ and $\varphi((R^d \times \{0\}) \cap \Omega) = M \cap \Omega'$.*

Let M be a Nash submanifold of R^n. A mapping $f : M \to R^p$ is a Nash mapping if it is semi-algebraic and, for every φ as above, $f \circ \varphi|_{R^d \cap \Omega}$ is a Nash mapping. We denote by $\mathcal{N}(M)$ the ring of Nash functions from M to R. Two Nash submanifolds M and M' are said to be Nash diffeomorphic if there exists a bijection $f : M \to M'$, such that f and f^{-1} are Nash.

The study of these objects will be developed in Chap. 8. We shall then show that in the case $R = \mathbb{R}$, Nash functions coincide with the analytic functions satisfying an algebraic equation.

Proposition 2.9.10. *Let $S \subset R^n$ be a semi-algebraic set. Then S is the disjoint union of a finite number of Nash submanifolds M_i, each Nash diffeomorphic to an open hypercube $]0,1[^{\dim(M_i)}$.*

Proof. We proceed by induction on n, referring the reader to the proof of Theorem 2.3.6. The result is obvious for $n = 1$. Assume that it holds for n. Then, slicing polynomials $P_1(X,Y), \ldots, P_s(X,Y)$ (where $X = (X_1, \ldots, X_n)$), we may assume, by induction, that the A_i of the partition of R^n are Nash submanifolds, Nash diffeomorphic to $]0,1[^{d_i}$. We may also assume that the family $P_1, \ldots, P_s$ is stable under derivation with respect to Y, so that each zero $\xi_{i,j}$ on A_i is a simple root of one of the $P_1, \ldots, P_s$. Hence, by the implicit function theorem 2.9.8, every $\xi_{i,j}$ is a Nash function on A_i. From this it follows that the graphs of $\xi_{i,j}$ and the slices $]\xi_{i,j}, \xi_{i,j+1}[$ are Nash submanifolds, Nash diffeomorphic to A_i and $A_i \times]0,1[$, respectively. □

Bibliographic Notes. To the best of our knowledge, the first systematic study of semi-algebraic sets and mappings is due to H. Brakhage [67]. He called them "Elementarmengen" and "Elementarabbildungen" and he developed the theory over an arbitrary real closed field. Unfortunately Brakhage's doctoral dissertation was never published and remained unknown to the mathematical community. We are indebted to G.-M. Greuel for providing us a copy of this dissertation after the publication of the French edition of this book.

The actual development of the theory of semi-algebraic sets (as well as the theory of semi-analytic sets) began with S. Lojasiewicz [215]. The extension of some results that were known over $\mathbb{R}$ to the case of an arbitrary real closed field can be found in [76, 108]. The smiling face of Example 2.1.5 b) was taken from [76]. In [215], Lojasiewicz also deals with locally semi-algebraic sets. The logical aspect of the Tarski-Seidenberg principle is only touched upon here (Definition 2.2.3 and Proposition 2.2.4); it plays an important role in the development of model theory, as an example of quantifier elimination [275]. The "slicing" technique (Theorem 2.3.1) is used in one form or another in any study of semi-algebraic sets, for instance in [215, 87]. It is also present in the cylindrical algebraic decomposition algorithm of Collins [88]. Several papers on this algorithmic aspect and a bibliography can be found in [11]. Before Lojasiewicz, Whitney [338] had proved that the difference of two real algebraic sets has a finite number of connected components. Proposition 2.5.4 is referred as Thom's lemma in [215]. The curve-selection lemma (Theorem 2.5.5) is a semi-algebraic version of the statements formulated in [74] and [233]. Lojasiewicz's inequality (Corollary 2.6.7) was first used in problems of the division of a distribution by a function [214, 166]. The proof of the semi-algebraic Tietze-Urysohn theorem is taken from [251]. The theorems of separation of closed semi-algebraic sets are given by Mostowski [238] (his proof of the relative version was completed in [43]). For recent advances concerning the problem of separation of semi-algebraic sets by signs of polynomials, see [3]. The finiteness theorem 2.7.2, whose local version can be found in [215], has been given several proofs [43, 99, 111, 265], including one of a logical nature [118]. The treatment of the dimension in Section 8 is close to that in [108]. The proof of the implicit function theorem for an arbitrary real closed field can be found in [76].

Many results presented in this chapter hold more generally for definable sets and functions in o-minimal structures [119]. This includes the sets and functions definable with the exponential function [341].

3. Real Algebraic Varieties

Abstract. In the first section, we show how real and complex algebraic sets exhibit strikingly different behavior. In the second section, we define real algebraic varieties. In fact, we shall be concerned almost exclusively with affine real algebraic varieties, i.e. real algebraic sets "up to a biregular isomorphism". The third section concerns the notion of nonsingularity. In addition to recalling some properties of varieties valid over an arbitrary field of characteristic zero, we stress a few special properties of the real case. The fourth section describes important examples of real algebraic varieties: projective spaces and grassmannians. These are *affine* real algebraic varieties, which explains why, in the real case, there is much less need to leave the affine framework (as compared to the complex case). In the fifth section, we conclude by giving a few useful constructions, such as blowing up and some constructions specific to the real case: the algebraic "Alexandrov compactification" and blowing down of a subvariety to a point.

3.1 Real and Complex Algebraic Sets

We shall illustrate the differences of behavior between real and complex algebraic sets. Let $V \subset \mathbb{C}^n$ be an algebraic set. Then V can be considered as an algebraic subset of $\mathbb{R}^{2n}$, by separating the real and imaginary parts in the equations of V.

Proposition 3.1.1. *Let $V \subset \mathbb{C}^n$ be an irreducible algebraic set, of complex dimension d, considered as an algebraic subset of $\mathbb{R}^{2n}$. Then*
  (i) *V is connected,*
  (ii) *V is not bounded (except if V is a point),*
  (iii) *$\dim(V_x) = 2d$ at every point x of V.*

Hints for the proof:
  (i) [295], Chap. 7, § 2.
  (ii) If $\overline{V} \subset \mathbb{P}_n(\mathbb{C})$ is the projectivization of V and if V is not a point, then $\overline{V} \setminus V \neq \emptyset$ and V is dense in $\overline{V}$ ([295], Chap. 7, § 2.1, Lemma 1). Therefore V cannot be compact.
  (iii) If x is a nonsingular point of V, then V is a $\mathcal{C}^\infty$ manifold of (real) dimension $2d$ in a neighbourhood of x and $\dim(V_x) = 2d$ (Proposition 2.8.14).

Since the set of nonsingular points is dense in V ([295], Chap. 7, § 2.1, Lemma 1), one has $\dim(V_x) = 2d$ at every point x of V. □

The following examples show that Proposition 3.1.1 is no longer true for real algebraic sets.

Example 3.1.2. a) The circle $\{(x,y) \in \mathbb{R}^2 \mid x^2 + y^2 = 1\}$ is a bounded algebraic set.

b) The cubic curve in $\mathbb{R}^2$ given by the equation $x^2 + y^2 - x^3 = 0$ has an isolated point at the origin.

c) The cubic curve in $\mathbb{R}^2$ given by the equation $x + y^2 - x^3 = 0$ has two connected components of dimension 1 (its projectivization also has two connected components). It is nonsingular (cf. Section3). The two cubic curves b) and c) are, of course, irreducible.

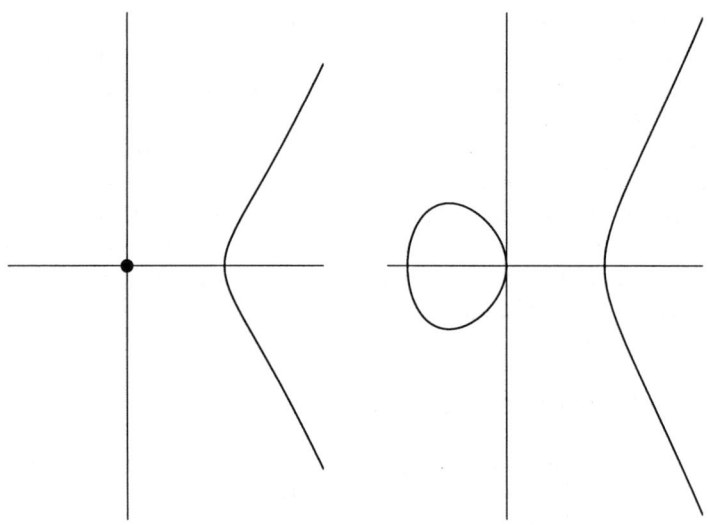

Figure 3.1. Examples b (left) and c (right)

d) The "Cartan umbrella" given by the equation $z(x^2 + y^2) - x^3 = 0$ is a connected irreducible surface in $\mathbb{R}^3$, with a "stick" (the z-axis) of dimension 1.

e) Another umbrella is the surface in $\mathbb{R}^3$ given by the equation $x^3 + zx^2 - y^2 = 0$. In this case, the "stick" (the z-axis) intersects the "cloth" (the set of points where the local dimension is 2) along a whole half-line.

f) The surface V in $\mathbb{R}^3$ given by the equation $x^2(1 - z^2) = x^4 + y^2$ is not bounded, but its subset $V^{(2)}$ of points where the local dimension is 2 is bounded.

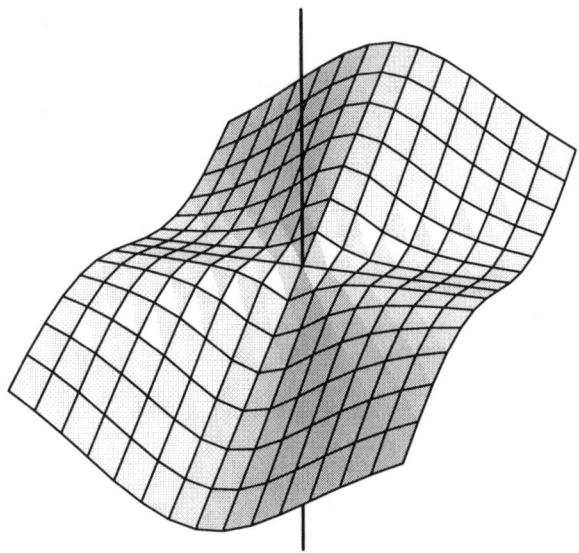

Figure 3.2. Example d

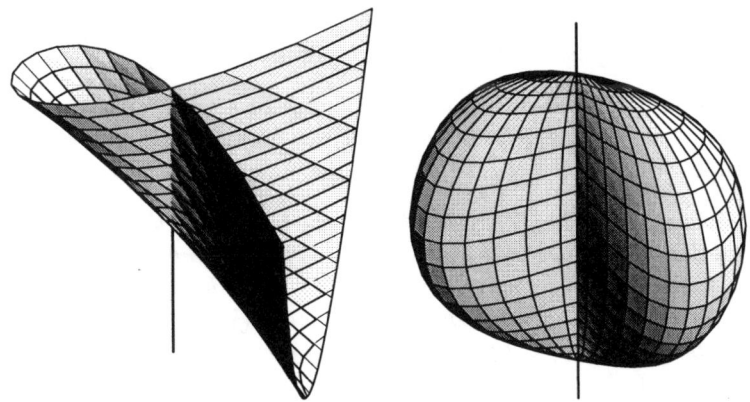

Figure 3.3. Examples e (left) and f (right)

3.2 Real Algebraic Varieties

In what follows R denotes a real closed field.

Let V be an algebraic subset of R^n. Recall that $\mathcal{P}(V) = R[X_1, \ldots, X_n]/\mathcal{I}(V)$ denotes the ring of polynomial functions from V to R. A *polynomial mapping* from V to a subset W of R^p is a mapping whose coordinate functions are polynomial. We denote by $\mathcal{P}(V,W)$ the set of polynomial mappings from V to W.

Definition 3.2.1. *Let U be a Zariski open subset of V. A regular function on U is the quotient $f = g/h$, where g, h are in $\mathcal{P}(V)$ and $h^{-1}(0) \cap U = \emptyset$. The regular functions on U form a ring denoted by $\mathcal{R}(U)$. A regular mapping from U into a subset W of R^p is a mapping whose coordinate functions are regular. We denote by $\mathcal{R}(U,W)$ the set of regular mappings from U to W.*

The ring $\mathcal{R}(U)$ is therefore the ring of fractions of $\mathcal{P}(V)$ for the multiplicative set
$$\{h \in \mathcal{P}(V) \mid h^{-1}(0) \cap U = \emptyset\}.$$

Notation 3.2.2. *Given a subset A of $\mathcal{P}(V)$ (resp. $\mathcal{R}(U)$), we denote by*
$$\mathcal{Z}_V(A) = \{x \in V \mid \forall P \in A \ P(x) = 0\}$$
$$(\text{resp.} \ \mathcal{Z}_U(A) = \{x \in U \mid \forall f \in A \ f(x) = 0\})$$
the set of zeros of A.

Given a subset X of V (resp. U), we denote by
$$\mathcal{I}_{\mathcal{P}(V)}(X) = \{P \in \mathcal{P}(V) \mid \forall x \in X \ P(x) = 0\}$$
$$(\text{resp.} \ \mathcal{I}_{\mathcal{R}(U)}(X) = \{f \in \mathcal{R}(U) \mid \forall x \in X \ f(x) = 0\})$$
the ideal of $\mathcal{P}(V)$ (resp. $\mathcal{R}(U)$) of functions vanishing on X.

The usual definition of regular functions is of a local nature. Here, the local nature (in the Zariski topology) of the notion of regular function is compatible with the existence of a global denominator.

Proposition 3.2.3. *Let $V \subset R^n$ be an algebraic set, $(U_i)_{i=1,\ldots,p}$ a finite family of Zariski open subsets of V, $U = \bigcup_{i=1}^p U_i$, f a function from U to R. If, for $i = 1, \ldots, p$, there exist $P_i, Q_i \in \mathcal{P}(V)$, Q_i vanishing nowhere on U_i, such that $f|_{U_i} = P_i/Q_i$, then there exist $P, Q \in \mathcal{P}(V)$, Q vanishing nowhere on U, such that $f = P/Q$ (in other words, if for $i = 1, \ldots, p$, $f|_{U_i} \in \mathcal{R}(U_i)$, then $f \in \mathcal{R}(U)$).*

Proof. Choose $S_i \in \mathcal{P}(V)$ such that $S_i^{-1}(0) = V \setminus U_i$. Then $Q = \sum_{i=1}^p S_i^2 Q_i^2$ vanishes nowhere on U, and $f = \left(\sum_{i=1}^p S_i^2 Q_i P_i\right)/Q$. □

Corollary 3.2.4. *Let $V \subset R^n$ be an algebraic set. Then $\mathcal{R} : U \mapsto \mathcal{R}(U)$ is a sheaf of rings on V for the Zariski topology.*

3.2 Real Algebraic Varieties   63

Proposition 3.2.5. *Let $V \subset R^n$, $V' \subset R^p$ be two algebraic sets, $U \subset V$ and $U' \subset V'$ Zariski open sets. A regular mapping $\varphi : U \to U'$ induces an R-algebra homomorphism $\varphi^* : \mathcal{R}(U') \to \mathcal{R}(U)$, $\varphi^*(f) = f \circ \varphi$. The mapping $\varphi \mapsto \varphi^*$ is a bijection from $\mathcal{R}(U, U')$ onto the set of R-algebra homomorphisms from $\mathcal{R}(U')$ to $\mathcal{R}(U)$.*

Proof. Let $\theta : \mathcal{R}(U') \to \mathcal{R}(U)$ be an R-algebra homomorphism. The mapping $\varphi = (\theta(X_1), \ldots, \theta(X_p))$ is a regular mapping from U to U', and $\varphi^* = \theta$. This proves that $\varphi \mapsto \varphi^*$ is a bijection. □

Definition 3.2.6. *Let $V \subset R^n$, $V' \subset R^p$ be two algebraic sets, and let $U \subset V$ and $U' \subset V'$ be Zariski open sets. A biregular isomorphism from U onto U' is a bijective regular mapping whose inverse is also regular.*

Remark 3.2.7. By Proposition 3.2.5, U and U' are biregularly isomorphic if and only if $\mathcal{R}(U)$ is isomorphic to $\mathcal{R}(U')$ as an R-algebra.

Example 3.2.8.

a) Let $S^1 \subset R^2$ be the circle given by the equation $x^2 + y^2 = 1$ and $T \subset R^3$ the torus given by the equation $16(x^2 + y^2) = (x^2 + y^2 + z^2 + 3)^2$, obtained by revolving the circle $S^1((2,0),1)$ of the (x,z)-plane around the z-axis. The polynomial mapping $\varphi : S^1 \times S^1 \to T$ defined by $\varphi(t, u, v, w) = (t(2+v), u(2+v), w)$ is bijective and φ^{-1} is regular:

$$\varphi^{-1}(x, y, z) = (x/\rho, y/\rho, \rho - 2, z) \text{ where } \rho = (x^2 + y^2 + z^2 + 3)/4 \, .$$

Hence, φ is a biregular isomorphism. Note that $\mathcal{P}(S^1 \times S^1)$ and $\mathcal{P}(T)$ are not isomorphic (the first ring is regular, while the second is not).

b) A bijective regular mapping ψ between real algebraic varieties is not necessarily biregular, even if the inverse is real analytic. Take for instance $V = \{(x, y) \in R^2 \mid y^3 + y - x = 0\}$, W the x-axis, and $\psi : V \to W$ the projection onto the x-axis.

We shall see in Chap. 13 that the set of regular mappings from a compact connected nonsingular real algebraic curve V into the circle S^1 is dense in the space of smooth mappings. Hence, for any such curve V, there exists a bijective regular mapping $\psi : V \to S^1$ with ψ^{-1} analytic, but not regular.

We now define real algebraic varieties.

Definition 3.2.9. *An affine real algebraic variety (over R) is a topological space X equipped with a sheaf $\mathcal{R}_X$ of functions with values in R, isomorphic to an algebraic set $V \subset R^n$ with its Zariski topology, equipped with its sheaf of regular functions $\mathcal{R}_V$. The sheaf $\mathcal{R}_X$ is called the sheaf of regular functions on X.*

Proposition 3.2.10. *Let $V \subset R^n$ be a real algebraic set and U a Zariski open subset of V. Then $(U, \mathcal{R}_V|_U)$ is an affine real algebraic variety.*

Proof. Take $P \in \mathcal{P}(V)$, such that $P^{-1}(0) = V \setminus U$. Then U is biregularly isomorphic to the algebraic set $W = \{(x, y) \in R^{n+1} \mid x \in V \text{ and } yP(x) = 1\}$, and thus $(U, \mathcal{R}_V|_U)$ is isomorphic to $(W, \mathcal{R}_W)$. □

Definition 3.2.11. *A real algebraic variety (over R) is a topological space X equipped with a sheaf $\mathcal{R}_X$ of functions with values in R, such that there exists a finite open cover $(U_i)_{i \in I}$ of X, with each $(U_i, \mathcal{R}_X|_{U_i})$ being an affine real algebraic variety. The sheaf $\mathcal{R}_X$ is called the* sheaf of regular functions *on X, and the topology of X is called the* Zariski topology. *If U is a Zariski open subset of X, we denote by $\mathcal{R}_X(U)$ (or $\mathcal{R}(U)$, by abuse of notation) the ring of continuous sections of $\mathcal{R}_X$ on U.*

If $(X, \mathcal{R}_X)$ and $(Y, \mathcal{R}_Y)$ are two real algebraic varieties, then a regular mapping *from X to Y is a continuous mapping $\varphi : X \to Y$, such that, if U is an open subset of Y and $f \in \mathcal{R}_Y(U)$, then $f \circ \varphi|_{\varphi^{-1}(U)} \in \mathcal{R}_X(\varphi^{-1}(U))$.*

Remark 3.2.12. According to the definitions in [293], we have defined only real pre-algebraic varieties and need a separation condition in order to obtain real algebraic varieties. We shall, however, not worry about this, since we shall be concerned almost exclusively with affine real algebraic varieties (for which the separation condition is automatically satisfied).

The only time we shall encounter real algebraic varieties that are not necessarily affine, is when we study the total space of an algebraic vector bundle (cf. Chap. 12, section 1). The general definition of real algebraic varieties is, nevertheless, useful, since grassmannians and projective spaces are naturally introduced as "abstract" algebraic varieties (over an arbitrary field). We note only afterwards that, in the real case, these are actually affine algebraic varieties.

Proposition 3.2.13. *If $(X, \mathcal{R}_X)$ is a real algebraic variety and U is a Zariski open subset of X, then $(U, \mathcal{R}_X|_U)$ is a real algebraic variety.*

Proof. Follows immediately from Proposition 3.2.10. □

Proposition 3.2.14. *Let X be the union of a finite family of subsets $(X_i)_{i \in I}$. Assume that each X_i is equipped with a structure of a real algebraic variety and that the following conditions are satisfied:*

(i) $X_i \cap X_j$ *is open in X_i, for every $i, j \in I$,*

(ii) *the structures induced by X_i and X_j on $X_i \cap X_j$ coincide for every $i, j \in I$.*

Then there exists exactly one structure of a real algebraic variety on X, for which the X_i are Zariski open, and which induces the given structure on each X_i.

Proof. Obvious. □

Remark 3.2.15. If two Zariski open subsets of algebraic sets are biregularly isomorphic, then they are, in particular, semi-algebraically homeomorphic in the euclidean topology. This shows two facts.

a) We can define the euclidean topology on a real algebraic variety X in a canonical way: a basis of open sets is given by the sets

$$\{x \in U \mid f_1(x) > 0, \ldots, f_m(x) > 0\},$$

where U is a Zariski open subset of X and $f_1, \ldots, f_m$ are regular functions on U.

b) We can define the notion of a semi-algebraic subset of X as a boolean combination of open sets belonging to the basis described above. If X is affine and $\varphi : X \to U$ is a biregular isomorphism onto a Zariski open subset of an algebraic set, then a subset S of X is semi-algebraic if and only if $\varphi(S)$ is semi-algebraic, in the sense of Definition 2.1.4. If X is an arbitrary real algebraic variety, then $S \subset X$ is semi-algebraic if and only if $S \cap Y$ is semi-algebraic for every affine Zariski open subset Y of X. Similarly it is possible to define the notion of a semi-algebraic function on a semi-algebraic subset of X.

It is worth mentioning that, when we speak of the *local ring $\mathcal{R}_{X,x}$ of germs of regular functions at a point x* of a real algebraic variety X, it is understood that the Zariski topology is used to define $\mathcal{R}_{X,x}$. Thus, $\mathcal{R}_{X,x} = \varinjlim \mathcal{R}_X(U)$, where the limit is taken over Zariski neighbourhoods U of x. If $V \subset R^n$ is an algebraic set and $x \in V$, then $\mathcal{R}_{V,x}$ is the localization $\mathcal{P}(V)_{\mathfrak{m}_x}$ at the maximal ideal $\mathfrak{m}_x$ of polynomial functions vanishing at x.

3.3 Nonsingular Points

We first recall an algebraic result valid over any field of characteristic zero.

Definition 3.3.1. *The* dimension *of an ideal I of $R[X_1, \ldots, X_n]$ is the dimension of the ring $R[X_1, \ldots, X_n]/I$.*

Proposition 3.3.2. *Let $I = (P_1, \ldots, P_k)$ be a prime ideal of dimension d of $R[X_1, \ldots, X_n]$. The rank of the image of the matrix $\left[\frac{\partial P_i}{\partial X_j}\right]$, $i = 1, \ldots, k$, $j = 1, \ldots, n$, in the field of fractions of $R[X_1, \ldots, X_n]/I$ is equal to $n - d$.*

References for the proof: See [163], Chap. 10, § 14, Theorem 1 or [286], Chap. 2, § 4.2, Lemma 2. □

In particular, if $x \in \mathcal{Z}(I)$, we have $\operatorname{rank}(\left[\frac{\partial P_i}{\partial X_j}(x)\right]) \leq n - d$ (this rank does not depend on the choice of generators of I).

Definition 3.3.3. *Let $V \subset R^n$ be an algebraic set and $\mathcal{I}(V) = (P_1, \ldots, P_k)$. Let $z \in V$. The* Zariski tangent space of V at z, *denoted $T_z^{\mathrm{Zar}}(V)$, is the linear subspace of R^n defined by*

$$T_z^{\mathrm{Zar}}(V) = \bigcap_{j=1}^{k} \left\{ x \in R^n \;\middle|\; \sum_{i=1}^{n} \frac{\partial P_j}{\partial X_i}(z) x_i = 0 \right\}.$$

The Zariski tangent space $T_z^{\text{Zar}}(V)$ does not depend on the choice of generators $P_1, \ldots, P_k$ of $\mathcal{I}(V)$. If V is irreducible, then Proposition 3.3.2 shows that the dimension of $T_z^{\text{Zar}}(V)$ is bigger than or equal to the dimension of V.

Definition 3.3.4. *Let $V \subset R^n$ be an irreducible algebraic set. A point z in V is said to be* nonsingular *if $\dim(T_z^{\text{Zar}}(V)) = \dim(V)$. In other words, if $\mathcal{I}(V) = (P_1, \ldots, P_k)$, then z is nonsingular if and only if the rank of the matrix $\left[\frac{\partial P_i}{\partial X_j}(z)\right]$ is equal to $n - \dim(V)$.*

The notion of Zariski tangent space is defined even at a singular point. However, when $V \subset \mathbb{R}^n$ is an irreducible algebraic set and $z \in V$ is a nonsingular point, a neighbourhood of z in V is a $\mathcal{C}^\infty$ submanifold of $\mathbb{R}^n$ and the tangent space of V at z (in the $\mathcal{C}^\infty$ sense) coincides with the Zariski tangent space. Since there is no risk of confusion, *we shall write $T_z(V)$ instead of $T_z^{\text{Zar}}(V)$ if z is a nonsingular point of $V \subset R^n$, and we shall speak of the tangent space instead of the Zariski tangent space.*

The definition of nonsingularity 3.3.4 is equivalent to a property whose formulation is more intrinsic, i.e. clearly invariant under a biregular isomorphism. Assume that z is the origin of R^n. Denote by $\mathfrak{m}_0$ the ideal of polynomials of $R[X_1, \ldots, X_n]$ vanishing at 0. There is a linear mapping $\theta : \mathfrak{m}_0 \to (R^n)^\vee$ (where $(R^n)^\vee$ denotes the dual of R^n), sending $P \in \mathfrak{m}_0$ to the linear form $x \mapsto \sum_{i=1}^n \frac{\partial P}{\partial X_i}(0)x_i$. The $\theta(X_i)$ form the canonical basis of $(R^n)^\vee$, and θ induces an isomorphism $\theta' : \mathfrak{m}_0/\mathfrak{m}_0^2 \to (R^n)^\vee$. Denote by $\mathfrak{m}_{V,0}$ the maximal ideal of the local ring $\mathcal{R}_{V,0}$. The dual of $T_0^{\text{Zar}}(V) \subset R^n$ can be identified with a quotient of $(R^n)^\vee$, and θ' induces an isomorphism from $\mathfrak{m}_0/(\mathfrak{m}_0^2 + \mathcal{I}(V)) \simeq \mathfrak{m}_{V,0}/\mathfrak{m}_{V,0}^2$ onto this quotient. Since V is irreducible, $\mathcal{R}_{V,0}$ and V have the same dimension.

Definition 3.3.5. *If A is a noetherian local ring, $\mathfrak{m}$ its maximal ideal and $k = A/\mathfrak{m}$ its residue field, then A is said to be* regular *if $\dim(A) = \dim_k \mathfrak{m}/\mathfrak{m}^2$, where $\mathfrak{m}/\mathfrak{m}^2$ is considered as a k-vector space.*

What we have just seen leads to the following characterization of nonsingularity.

Proposition 3.3.6. *Let $V \subset R^n$ be an irreducible algebraic set. A point z of V is nonsingular if and only if the local ring $\mathcal{R}_{V,z}$ of germs of regular functions at z is regular.*

Proposition 3.3.7. *Let A be a regular local ring of dimension d, $\mathfrak{m}$ its maximal ideal and $k = A/\mathfrak{m}$ its residue field. Then:*

(i) *The ring A is an integral domain and is integrally closed.*

(ii) *A system of d elements $f_1, \ldots, f_d$ generate $\mathfrak{m}$ if and only if their classes modulo $\mathfrak{m}^2$ form a basis of $\mathfrak{m}/\mathfrak{m}^2$ over k. In this case $f_1, \ldots, f_d$ are said to be a* regular system of parameters *of A.*

(iii) *If $\mathfrak{p}$ is an ideal of A, then $A/\mathfrak{p}$ is regular if and only if $\mathfrak{p}$ is generated by elements $f_1, \ldots, f_l$ of $\mathfrak{m}$ whose classes modulo $\mathfrak{m}^2$ are linearly independent over k (and then $l = \dim(A) - \dim(A/\mathfrak{p}))$.*

References for the proofs: [349], Chap. 8, § 11, Corollaries 1 and 2, Theorem 26.

Proposition 3.3.8. *Let $V \subset R^n$ be an irreducible algebraic set of dimension d and $z \in V$ a nonsingular point of V. Then there exist $n - d$ polynomials $P_1, \ldots, P_{n-d} \in \mathcal{I}(V)$ and a Zariski open neighbourhood U of z in R^n, such that*

(i) $\mathcal{Z}(P_1, \ldots, P_{n-d}) \cap U = V \cap U$,
(ii) *for every $x \in U$*, $\mathrm{rank}(\left[\frac{\partial P_j}{\partial X_i}(x)\right]) = n - d$.

Proof. The local ring $\mathcal{R}_{V,z} = \mathcal{R}_{R^n,z}/\mathcal{I}(V)\mathcal{R}_{R^n,z}$ is regular, hence, by Proposition 3.3.7, we can find a regular system $P_1, \ldots, P_n$ of parameters of $\mathcal{R}_{R^n,z}$, with $P_1, \ldots, P_{n-d}$ generating $\mathcal{I}(V)\mathcal{R}_{R^n,z}$. We can, of course, assume that $P_1, \ldots, P_{n-d}$ are polynomials. We take a Zariski open neighbourhood U of z small enough to have $(P_1, \ldots, P_{n-d})\mathcal{R}(U) = \mathcal{I}(V)\mathcal{R}(U)$ and $\left[\frac{\partial P_j}{\partial X_i}\right]$ of rank $n - d$ on U. □

We now define nonsingularity for a point of a real algebraic variety.

Definition 3.3.9. *Let $(X, \mathcal{R}_X)$ be a real algebraic variety. A point $x \in X$ is said to be* nonsingular in dimension d *if the local ring of germs of regular functions $\mathcal{R}_{X,x}$ is a regular local ring of dimension d. A real algebraic variety is* nonsingular *if all its points are nonsingular in the same dimension d.*

This definition does not conflict with the definition previously given for an irreducible algebraic set. If V is an irreducible algebraic set of dimension d, then a point of V is nonsingular (in the sense of Definition 3.3.4) if and only if it is nonsingular in dimension d. There are no nonsingular points in any other dimension.

It is useful to give the following characterization of a nonsingular point of an algebraic subset of R^n.

Proposition 3.3.10. *Let $V \subset R^n$ be an algebraic set, not necessarily irreducible, and x a point of V. The following properties are equivalent:*

(i) *The point x is nonsingular in dimension d.*
(ii) *There exists an irreducible component V' of V, with $\dim(V') = d$, such that V' is the only irreducible component of V containing x and x is a nonsingular point of V'.*
(iii) *There exist $n - d$ polynomials $P_1, \ldots, P_{n-d} \in \mathcal{I}(V)$ and an open neighbourhood U of x in R^n for the euclidean topology, such that $V \cap U = \mathcal{Z}(P_1, \ldots, P_{n-d}) \cap U$ and the rank of the jacobian matrix $\left[\frac{\partial P_j}{\partial X_i}(x)\right]$ is equal to $n - d$.*

3. Real Algebraic Varieties

Proof. (i) ⇒ (ii) The point x does not belong to the intersection of two irreducible components of V, since otherwise the ring $\mathcal{R}_{V,x}$ would not be an integral domain. Let V' be the only irreducible component of V such that $x \in V'$. Then $\mathcal{R}_{V,x} = \mathcal{R}_{V',x}$ and Proposition 3.3.6 implies (ii).

(ii) ⇒ (iii) Proposition 3.3.8 implies (iii), with V' instead of V. To get V, it is enough to multiply $P_1, \ldots, P_{n-d}$ by an equation of the union of all irreducible components of V other than V'.

(iii) ⇒ (i) We may assume that $\det\left(\left[\frac{\partial P_i}{\partial X_j}(x)\right]_{i=1,\ldots,n-d\,;\,j=d+1,\ldots,n}\right) \neq 0$. Applying the inverse function theorem 2.9.7 to the mapping $(X_1-x_1,\ldots,X_d-x_d,P_1,\ldots,P_{n-d})$, we get a Nash diffeomorphism φ from an open semi-algebraic neighbourhood Ω of 0 in R^n onto an open semi-algebraic neighbourhood Ω' of x in R^n such that $\varphi((R^d \times \{0\}) \cap \Omega) = V \cap \Omega'$. Thus, $\dim(V_x) = d$, and, hence, $\dim(\mathcal{R}_{V,x}) \geq d$. Moreover, $\mathcal{R}_{V,x}$ is a quotient of $\mathcal{R}_{R^n,x}/(P_1,\ldots,P_{n-d})$ and, by Proposition 3.3.7, the latter is a regular local ring (and hence an integral domain) of dimension d. Hence, we have $\mathcal{R}_{V,x} = \mathcal{R}_{R^n,x}/(P_1,\ldots,P_{n-d})$, which proves (i). □

In proving (iii) ⇒ (i) we have also proved the following result.

Proposition 3.3.11. *Let $V \subset R^n$ be an algebraic set and $x \in V$ a nonsingular point in dimension d. Then there is an open semi-algebraic neighbourhood of x in V which is a Nash submanifold of dimension d of R^n.*

The notion of a nonsingular point is quite subtle. The following examples show that intuition should not be followed blindly.

Example 3.3.12.

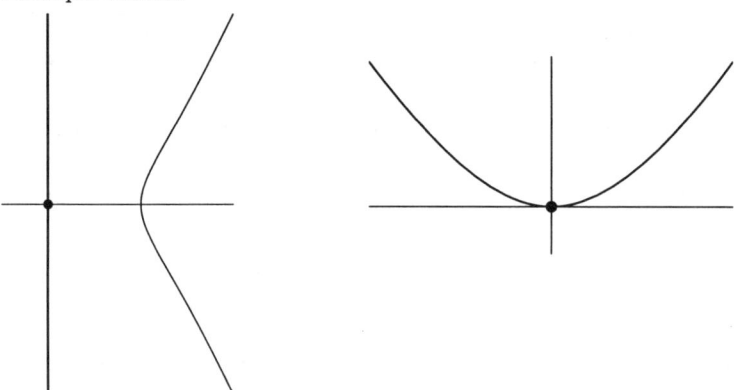

Figure 3.4. Examples a (left) and b (right)

a) Let $V = \mathcal{Z}(X(Y^2 + X^2 - X^3))$. Then V coincides with $\mathcal{Z}(X)$ in a neighbourhood of the origin 0 and the rank of the jacobian matrix $\left[\frac{\partial X}{\partial X}, \frac{\partial X}{\partial Y}\right]$ at 0 is, of course, equal to 1. However, the origin is not a nonsingular point in dimension 1 of V since it is the intersection of two irreducible components of V (the y-axis and the cubic curve given by the equation $y^2 = x^3 - x^2$, for which the origin is an isolated point). The condition $P_1, \ldots, P_{n-d} \in \mathcal{I}(V)$ is essential in Proposition 3.3.10.

b) Let $V \subset \mathbb{R}^2$ be the (irreducible) curve given by the equation $y^3 + 2x^2y - x^4 = 0$. We have $x^2 = y(1 + \sqrt{1+y})$ on V, from which it follows that y is a $\mathcal{C}^\infty$ function of x in a neighbourhood of 0 in $\mathbb{R}$. The set V is a $\mathcal{C}^\infty$ submanifold of $\mathbb{R}^2$, but the origin $0 \in \mathbb{R}^2$ is not a nonsingular point of V in the sense of Definition 3.3.4. Note that $T_0^{\text{Zar}}(V) = \mathbb{R}^2$ is different from the $\mathcal{C}^\infty$ tangent space of V.

Notation 3.3.13. *Let V be an algebraic set of dimension d. Denote by $\text{Reg}(V)$ the set of nonsingular points in dimension d of V, and denote by $\text{Sing}(V)$ the set $V \setminus \text{Reg}(V)$.*

Proposition 3.3.14. *If V is an algebraic set, then $\text{Sing}(V)$ is an algebraic subset of V of dimension smaller than the dimension of V. Hence $\text{Reg}(V)$ is a nonempty Zariski open subset of V of the same dimension as V.*

Proof. Let V_1 be an irreducible component of V. If $\dim(V_1) < \dim(V)$, then $V_1 \subset \text{Sing}(V)$. If $\dim(V_1) = \dim(V)$, then $\text{Sing}(V) \cap V_1$ is the union of $\text{Sing}(V_1)$ and of the intersection of V_1 with the union of the other irreducible components of V. This intersection is an algebraic set strictly contained in V_1. So we may assume that V is irreducible. Definition 3.3.4 clearly shows that $\text{Sing}(V)$ is algebraic and, by Proposition 3.3.2, $\mathcal{I}(\text{Sing}(V))$ strictly contains $\mathcal{I}(V)$, which proves that $\dim(\text{Sing}(V)) < \dim(V)$. □

Remark 3.3.15. $\text{Reg}(V)$ is not always dense in V in the euclidean topology, contrary to what happens (for an irreducible algebraic set) in the complex case: consider Examples 3.1.2 b), d), e) and f).

Proposition 3.3.16. *Let $I = (P_1, \ldots, P_k)$ be a prime ideal of dimension d of $R[X_1, \ldots, X_n]$. A point $x \in \mathcal{Z}(I)$ is said to be a nonsingular zero of I if $\text{rank}\left(\left[\frac{\partial P_i}{\partial X_j}(x)\right]\right) = n - d$. If I has a nonsingular zero, then $I = \mathcal{I}(\mathcal{Z}(I))$.*

Proof. If x is a nonsingular zero of I, then it is a nonsingular point in dimension d of $\mathcal{Z}(I)$, and, hence, $\dim(\mathcal{Z}(I)) = d$. Since $\mathcal{I}(\mathcal{Z}(I)) \supset I$ and the two ideals have the same dimension, we have $\mathcal{I}(\mathcal{Z}(I)) = I$. □

The next result will be useful in chapters 12 and 14.

Proposition 3.3.17. *Let $V \subset R^n$ be an algebraic set of dimension d and $W \subset V$ an algebraic subset such that, for every x in W, $\dim(W_x) = d$, and $W \subset \text{Reg}(V)$. Then $V \setminus W$ is an algebraic set, and $\text{Sing}(V \setminus W) = \text{Sing}(V)$, except when $W = \text{Reg}(V)$.*

Proof. Let $V_1, \ldots, V_p$ be the irreducible components of V and let $W_i = W \cap V_i$. The sets W_i are disjoint, since a nonsingular point of V cannot belong to two irreducible components of V. Each W_i is either empty or is an algebraic set of dimension d, and then $W_i = V_i$. The set W is thus the union of some irreducible components of V and is disjoint from the union of the other irreducible components, which proves the proposition. $\square$

Remark 3.3.18. The assumption "$W \subset \text{Reg}(V)$" cannot be replaced by "W nonsingular", as Example 3.3.12 a) shows, where W is the y-axis. The assumption "$\forall x \in W, \dim(W_x) = d$" cannot be weakened to "$\dim(W) = d$". To see this, take

$$V = \{(u,v) \mid u = 0\} \cup \{(u,v) \mid u = 1\}$$
$$W = \{(u,v) \mid u = 0\} \cup \{(1,0)\} \ .$$

We can, however, replace it with "$\dim(W) = d$ and W irreducible", or "$\dim(W) = d$ and W semi-algebraically connected".

3.4 Projective Spaces and Grassmannians

3.4.1 Projective Spaces as Real Algebraic Varieties

By definition, $\mathbb{P}_n(R)$ is the set of lines of R^{n+1} through the origin, which is equivalent to defining $\mathbb{P}_n(R)$ to be the quotient of $R^{n+1} \setminus \{0\}$ by the equivalence relation $x \sim y$ defined by "there exists $\lambda \in R$, $\lambda \neq 0$, $x = \lambda y$". Denote by $\Pi : R^{n+1} \setminus \{0\} \to \mathbb{P}_n(R)$ the canonical surjection. If $\Pi(x_0, \ldots, x_n) = t$, we denote $t = (x_0 : \ldots : x_n)$ and say that $(x_0, \ldots, x_n)$ are *homogeneous coordinates* of t. A *projective algebraic set* is a set of the form

$$\mathbb{P}\mathcal{Z}(P_1, \ldots, P_k) = \{\Pi(x) \in \mathbb{P}_n(R) \mid P_1(x) = \cdots = P_k(x) = 0\} \ ,$$

where $P_1, \ldots, P_k$ are homogeneous polynomials in $R[X_0, \ldots, X_n]$. The projective algebraic sets are the closed subsets of a topology on $\mathbb{P}_n(R)$: the Zariski topology.

Let $U_i = \{\Pi(x) \in \mathbb{P}_n(R) \mid x_i \neq 0\}$ for $i = 0, \ldots, n$. It is a Zariski open subset of $\mathbb{P}_n(R)$, and there is a bijection

$$\varphi_i : U_i \longrightarrow R^n$$
$$(x_0 : \ldots : x_n) \longmapsto (x_0/x_i, \ldots, x_{i-1}/x_i, x_{i+1}/x_i, \ldots, x_n/x_i) \ ,$$

which is a homeomorphism in the Zariski topology. Moreover, the mappings

$$\varphi_j \circ (\varphi_i)^{-1} : \varphi_i(U_i \cap U_j) \to \varphi_j(U_i \cap U_j)$$

are biregular isomorphisms. The structure of a real algebraic variety on $\mathbb{P}_n(R)$ is obtained by gluing the affine real algebraic varieties U_i together, as in

Proposition 3.2.14 (each U_i being here isomorphic to R^n). If W is a Zariski open subset of $\mathbb{P}_n(R)$, then a regular function on W is a function $f: W \to R$, such that, for $i = 0, \ldots, n$, $f \circ \varphi_i^{-1}|_{\varphi_i(U_i \cap W)}$ is regular on $\varphi_i(U_i \cap W)$.

A projective algebraic subset of $\mathbb{P}_n(R)$ has an obvious structure of a real algebraic variety. A real algebraic variety biregularly isomorphic to a projective algebraic set will be called a *projective real algebraic variety*.

3.4.2 Grassmannians as Real Algebraic Varieties

By definition, $\mathbb{G}_{n,k}(R)$ is the set of vector subspaces of dimension k of R^n. In particular, $\mathbb{G}_{n+1,1}(R) = \mathbb{P}_n(R)$. Denote by $e_1, \ldots, e_n$ the canonical basis of R^n. Let σ be a subset containing k elements of $\{1, \ldots, n\}$, V_σ the subspace of R^n generated by $\{e_i \mid i \in \sigma\}$ and W_σ the subspace generated by $\{e_i \mid i \notin \sigma\}$. Define $U_\sigma = \{V \in \mathbb{G}_{n,k}(R) \mid V \cap W_\sigma = \{0\}\}$. If $V \in U_\sigma$, then V has the form $\{x + \rho_V(x) \mid x \in V_\sigma\}$ for a uniquely determined linear mapping $\rho_V : V_\sigma \to W_\sigma$. There is a bijection $\varphi_\sigma : U_\sigma \to \mathbb{M}_{n-k,k}(R)$ from U_σ onto the set of matrices with $n - k$ rows and k columns with entries in R. We define $\varphi_\sigma(V) = [a_{i,j}]_{i \notin \sigma\, ;\, j \in \sigma}$ to be the matrix of ρ_V in the bases of V_σ and W_σ given above. If we define $a_{i,j} = \delta_{i,j}$ (of Kronecker), for $i \in \sigma$, $j \in \sigma$, then the matrix $A' = [a_{i,j}]_{i=1,\ldots,n\, ;\, j \in \sigma}$ is the matrix of the mapping $\mathrm{Id}_{V_\sigma} + \rho_V : V_\sigma \to R^n$, whose image is V. Now, let $\tau \neq \sigma$ be another subset containing k elements of $\{1, \ldots, n\}$. The subspace V belongs to U_τ if and only if the matrix $[a_{k,j}]_{k \in \tau\, ;\, j \in \sigma}$ extracted from A' has an inverse. In such a case, let B be this inverse and $C = A'B$. A linear mapping from V_τ to R^n corresponding to the matrix $C = [c_{i,k}]_{i=1,\ldots,n\, ;\, k \in \tau}$ has V as its image, and it can be written as $\mathrm{Id}_{V_\tau} + \mu_V$, with $\mu_V : V_\tau \to W_\tau$ linear. It follows that $\varphi_\tau(V)$ is the matrix $[c_{i,k}]_{i \notin \tau\, ;\, k \in \tau}$ extracted from C.

Clearly, the set $\mathbb{M}_{n-k,k}(R)$, identified with $R^{(n-k)k}$, is canonically equipped with a structure of an affine real algebraic variety. We transfer this structure from $\mathbb{M}_{n-k,k}(R)$ onto U_σ by using the bijection φ_σ. The observations made above show that $\varphi_\sigma(U_\sigma \cap U_\tau)$ is a Zariski open subset of $\mathbb{M}_{n-k,k}(R)$ and that $\varphi_\tau \circ (\varphi_\sigma)^{-1}|_{\varphi_\sigma(U_\sigma \cap U_\tau)}$ is a biregular isomorphism onto $\varphi_\tau(U_\sigma \cap U_\tau)$. By applying Proposition 3.2.14, we can glue together the structures of affine algebraic varieties constructed on each U_σ to obtain a structure of real algebraic variety on $\mathbb{G}_{n,k}(R)$.

Note that the construction of projective space is a particular case of the construction of the grassmannian. Note, also, that the fact that R is a real closed field has played no role until now.

We have seen that $\mathbb{P}_n(R)$ and $\mathbb{G}_{n,k}(R)$ are covered by open subsets biregularly isomorphic to R^n and $R^{(n-k)k}$, respectively. Hence:

Proposition 3.4.3. *The real algebraic varieties $\mathbb{P}_n(R)$ and $\mathbb{G}_{n,k}(R)$ are nonsingular.*

The next result shows dramatically the difference between the real and complex case.

Theorem 3.4.4. *The real algebraic varieties $\mathbb{P}_n(R)$ and $\mathbb{G}_{n,k}(R)$ are affine.*

Proof. It is, of course, enough to deal with $\mathbb{G}_{n,k}(R)$, since projective space is a particular case. We shall define an algebraic set $H_{n,k} \subset R^{n^2}$ and show that it is biregularly isomorphic to $\mathbb{G}_{n,k}(R)$. Consider R^n with its canonical basis and the scalar product: $(x,y) \mapsto x \cdot y = \sum_{i=1}^n x_i y_i$. A matrix $A \in \mathbb{M}_{n,n}(R)$ is the matrix of an orthogonal projection onto a subspace of dimension k if and only if A is symmetric, $A^2 = A$ and $\mathrm{trace}(A) = k$. Define

$$H_{n,k} = \{ A \in \mathbb{M}_{n,n}(R) \mid {}^t A = A, \; A^2 = A, \; \mathrm{trace}(A) = k \}.$$

The set $H_{n,k}$ is algebraic, and there is a bijection

$$\Psi : \mathbb{G}_{n,k}(R) \to H_{n,k}$$

sending $V \in \mathbb{G}_{n,k}(R)$ to the matrix of the orthogonal projection onto V. We use, now, the notations introduced earlier in this section. We need to show that $\Psi(U_\sigma)$ is a Zariski open subset of $H_{n,k}$ and that $\Psi \circ (\varphi_\sigma)^{-1}$ is a biregular isomorphism from $\mathbb{M}_{n-k,k}(R)$ onto $\Psi(U_\sigma)$. Let $V \in \mathbb{G}_{n,k}(R)$, $[h_{i,j}]_{i,j=1,\ldots,n} = \Psi(V)$. Denote $I = [h_{i,j}]_{i=1,\ldots,n;\, j \in \sigma}$ and $I' = [h_{i,j}]_{i \in \sigma;\, j \in \sigma}$. Then I' is the matrix of the orthogonal projection of V_σ onto V composed with the orthogonal projection of V onto V_σ. Hence $V \in U_\sigma$ if and only if I' is invertible, and $\Psi(U_\sigma)$ is a Zariski open subset of $H_{n,k}$. Moreover, if $V \in U_\sigma$, then $I \cdot (I')^{-1}$ is the matrix of $\mathrm{Id}_{V_\sigma} + \rho_V$, which shows that $\varphi_\sigma \circ (\Psi)^{-1}|_{\Psi(U_\sigma)}$ is a regular mapping.

It remains to show that $\Psi \circ (\varphi_\sigma)^{-1}$ is regular. If $V \in U_\sigma$, then a basis of V is given by the vectors

$$v_j = e_j + \rho_V(e_j) = e_j + \sum_{i \notin \sigma} a_{i,j} e_i \quad \text{for } j \in \sigma.$$

Let $\sum_{j' \in \sigma} \lambda_{j',k} v_{j'}$ be the orthogonal projection of e_k onto V, $k = 1, \ldots, n$. It is enough to show that $\lambda_{j',k}$ is a regular function of the $a_{i,j}$. We have $(\sum_{j' \in \sigma} \lambda_{j',k} v_{j'}) \cdot v_j = e_k \cdot v_j = a_{k,j}$ if $k \notin \sigma$, and $e_k \cdot v_j = \delta_{k,j}$ (of Kronecker) if $k \in \sigma$. It is then enough to note that the matrix $[v_{j'} \cdot v_j]_{j',j \in \sigma}$ is invertible since the vectors v_j are independent. □

An explicit algebraic embedding of $\mathbb{P}_n(R)$ into $\mathbb{M}_{n+1,n+1}(R) \simeq R^{(n+1)^2}$ is given by

$$(x_0 : \ldots : x_n) \longmapsto \left(\frac{x_i x_j}{\sum_{i=0}^n x_i^2} \right).$$

It is obvious from Theorem 3.4.4 that any projective real algebraic variety is affine.

Proposition 3.4.5. *The real algebraic varieties $\mathbb{G}_{n,k}(R)$ and $\mathbb{G}_{n,n-k}(R)$ are biregularly isomorphic.*

3.4 Projective Spaces and Grassmannians

Proof. The mapping $H_{n,k} \to H_{n,n-k}$ sending A to $\mathrm{Id}_{R^n} - A$ is a biregular isomorphism. It corresponds to the mapping $\mathbb{G}_{n,k}(R) \to \mathbb{G}_{n,n-k}(R)$ sending V to its orthogonal complement $V^\perp$ in R^n. □

We can study, in the same way, projective spaces and grassmannians over the field $C = R[i]$. The constructions analogous to those of 3.4.1 and 3.4.2 show that $\mathbb{G}_{n,k}(C)$, considered as a real algebraic variety over R, is biregularly isomorphic to

$$H'_{n,k} = \{A \in \mathbb{M}_{n,n}(C) \mid A = {}^t\overline{A},\ A^2 = A,\ \mathrm{trace}(A) = k\},$$

which is a real algebraic subset of $\mathbb{M}_{n,n}(C) \simeq R^{2n^2}$ (but not a "complex" algebraic subset of $\mathbb{M}_{n,n}(C) \simeq C^{n^2}$). Hence we have:

Proposition 3.4.6. *Let $C = R[i]$. Then $\mathbb{P}_n(C)$ and $\mathbb{G}_{n,k}(C)$, considered as varieties over R, are affine nonsingular real algebraic varieties.*

An explicit algebraic embedding of $\mathbb{P}_n(C)$ (considered as a real algebraic variety) into $\mathbb{M}_{n+1,n+1}(C) \simeq R^{2(n+1)^2}$ is given by

$$(x_0 : \ldots : x_n) \longmapsto \left(\frac{x_i \overline{x}_j}{\sum_{i=0}^n |x_i|^2} \right).$$

We shall use the following results in Chap. 12 and 14.

Proposition 3.4.7. *Let X be a real algebraic variety and $\varphi_i : X \to R^n$, $i = 1, \ldots, k$, $k < n$, regular mappings, such that, for every x in X, the vectors $\varphi_1(x), \ldots, \varphi_k(x)$ generate a vector subspace $\Phi(x)$ of dimension k of R^n. Then the mapping $\Phi : X \to \mathbb{G}_{n,k}(R)$ is regular.*

Proof. We identify $\mathbb{G}_{n,k}(R)$ with $H_{n,k}$, as in Theorem 3.4.4, and denote again by $\Phi(x)$ the matrix of the orthogonal projection of R^n onto $\Phi(x)$. Let $A(x)$ be the $k \times k$ matrix with entries $\varphi_i(x) \cdot \varphi_j(x)$. For every y in R^n, $(\Phi(x)) \cdot (y) = \sum_{i=1}^k a_i(x,y) \varphi_i(x)$, where

$$\begin{bmatrix} a_1(x,y) \\ \vdots \\ a_k(x,y) \end{bmatrix} = A(x)^{-1} \cdot \begin{bmatrix} y \cdot \varphi_1(x) \\ \vdots \\ y \cdot \varphi_k(x) \end{bmatrix}.$$

The entries of $\Phi(x)$ are regular functions of x, and, hence, Φ is regular. □

Corollary 3.4.8. *Let V be an algebraic subset of codimension k of R^n and X a Zariski open subset of $\mathrm{Reg}(V)$. Then the Gauss mappings*

$$\begin{array}{llll} \tau_X : X & \longrightarrow & \mathbb{G}_{n,n-k}(R) & \text{defined by} \quad \tau_X(x) = T_x(V) \\ \nu_X : X & \longrightarrow & \mathbb{G}_{n,k}(R) & \text{defined by} \quad \nu_X(x) = T_x(V)^\perp \end{array}$$

are regular (denoting by $T_x(V)^\perp$ the normal space of V at x, i.e. the orthogonal complement in R^n of the tangent space $T_x(V)$).

Proof. By Proposition 3.4.5, it is enough to prove that ν_X is regular. Let $\{q_1, \ldots, q_m\}$ be a set of generators of $\mathcal{I}(V)$. For each subset $I = \{i_1, \ldots, i_k\} \subset \{1, \ldots, m\}$ let U_I be the set of $x \in X$, such that $\mathrm{grad}(q_{i_1}(x)), \ldots, \mathrm{grad}(q_{i_k}(x))$ are linearly independent in R^n. The family of the U_I form an open cover of X in the Zariski topology. The mapping ν_X restricted to U_I sends x to the vector space generated by $\mathrm{grad}(q_{i_1}(x)), \ldots, \mathrm{grad}(q_{i_k}(x))$. The assumptions of Proposition 3.4.7 are satisfied, hence $\nu_X|_{U_I}$ is regular. The desired result now follows from Proposition 3.2.3. □

We now turn to another property of projective spaces.

Proposition 3.4.9. *Let $(X, \mathcal{R}_X)$ be an affine real algebraic variety and $F \subset X$ a semi-algebraic set. Then the following properties are equivalent:*

(i) For every biregular isomorphism $\varphi : X \to U$ onto a Zariski open subset of an algebraic subset of R^n, $\varphi(F)$ is bounded and closed in R^n.

(ii) There exists a biregular isomorphism $\varphi : X \to U$ onto a Zariski open subset of an algebraic subset of R^n such that $\varphi(F)$ is closed and bounded in R^n.

(iii) Every continuous semi-algebraic function on F is bounded.

(iv) Every regular function on a Zariski open subset of X containing F is bounded on F.

Proof. The implications (i) $\Rightarrow$ (ii) and (iii) $\Rightarrow$ (iv) are obvious.

(ii) $\Rightarrow$ (iii) because a continuous semi-algebraic function on a closed and bounded set is bounded (Theorem 2.5.8).

(iv) $\Rightarrow$ (i). Let $\varphi : X \to U$ be a biregular isomorphism onto a Zariski open subset of an algebraic subset of R^n. The fact that every function in $\mathcal{R}(U)$ is bounded on $\varphi(F)$ implies that $\varphi(F)$ is bounded. If $\varphi(F)$ is not closed in R^n, choose $x \in \mathrm{clos}(\varphi(F)) \setminus \varphi(F)$. Then $\left(\sum_{i=1}^n (X_i - x_i)^2\right)^{-1} \in \mathcal{R}(U \setminus \{x\})$ and is not bounded on $\varphi(F)$, contradicting (iv). □

Definition 3.4.10. *A semi-algebraic set with the properties described in Proposition 3.4.9 is said to be* complete.

Note that a semi-algebraic subset of R^n is complete if and only if it is closed and bounded.

Proposition 3.4.11. *The affine real algebraic varieties $\mathbb{P}_n(R)$, $\mathbb{G}_{n,k}(R)$, $\mathbb{P}^n(C)$ and $\mathbb{G}_{n,k}(C)$ are complete. Every projective real algebraic variety is complete.*

Proof. We consider the case of $\mathbb{G}_{n,k}(R)$, and we use the notations of Theorem 3.4.4. It is enough to verify that $H_{n,k}$ is bounded in R^{n^2}. If $[h_{i,j}] \in H_{n,k}$, then $|h_{i,j}| \leq 1$, since an orthogonal projection decreases the norm. The case of $\mathbb{G}_{n,k}(C)$ is similar. □

Note that a complete nonsingular affine real algebraic variety is projective. This can be seen using the theorem on resolution of singularities.

We conclude this section with the birational invariance of the number of semi-algebraically connected components.

Let X and Y be two irreducible affine real algebraic varieties. Recall that that X and Y are said to be *birationally equivalent* if the fields of fractions of $\mathcal{R}(X)$ and $\mathcal{R}(Y)$ are isomorphic over R. This is equivalent to the existence of a biregular isomorphism from a nonempty Zariski open subset of X onto a Zariski open subset of Y.

An example of birational equivalence is given by the normalization mapping. Let X be an irreducible affine real algebraic variety. We may assume that X is an algebraic subset of R^n. Let A be the integral closure of $\mathcal{P}(V)$ in its field of fractions. Since A is a finitely generated R-algebra, we may assume that $A = R[X_1, \ldots, X_p]/I$, for some ideal I of $R[X_1, \ldots, X_p]$. Define $X' = \mathcal{Z}(I) \subset R^p$. Then X' is called *the normalization of X* and the inclusion $\mathcal{P}(X) \hookrightarrow A = \mathcal{P}(X')$ induces a mapping $X' \to X$, which is a birational equivalence. The normalization of X does not depend on the choices made in its construction, up to a biregular isomorphism. Recall that, if X is a complete curve, then its normalization X' is complete and nonsingular.

Theorem 3.4.12. *Let X and Y be irreducible, complete, nonsingular real algebraic varieties. If X and Y are birationally equivalent, then they have the same number of semi-algebraically connected components.*

Proof. First observe that the theorem holds true for curves, because then X and Y are necessarily biregularly isomorphic (cf. [150] Chap. 1 Proposition 6.8).

Consider, thus, X and Y of arbitrary dimension. Let φ be a biregular isomorphism from a nonempty Zariski open subset U of X onto a Zariski open subset V of Y. Suppose that X has fewer semi-algebraically connected components than Y. Then there is a decomposition $Y = Y_1 \cup Y_2$ into disjoint open semi-algebraic subsets, and a semi-algebraically connected component A of X, such that $A_1 = \varphi^{-1}(Y_1) \cap A$ and $A_2 = \varphi^{-1}(Y_2) \cap A$ are both nonempty. We claim that there is a nonsingular semi-algebraically connected real algebraic curve C contained in A, such that $\dim(C \cap A_1) = \dim(C \cap A_2) = 1$. To find C, we proceed as follows. Since $U \cap A = A_1 \cup A_2$ is dense in A and A is semi-algebraically connected, there is a point x in $\text{clos}(A_1) \cap \text{clos}(A_2)$. We may assume that X is an algebraic subset of R^n, and then we obtain C by intersecting X with a sphere $S^{n-1}(x, r)$, for a sufficiently small r.

Let D be the Zariski closure of $\varphi(C \cap U)$ in Y. By construction, the curve D has at least two semi-algebraically connected components of dimension 1. Taking the normalization D' of D, we obtain a nonsingular complete real algebraic curve, birationally equivalent and, hence, biregularly isomorphic to C. Since the image of the normalization mapping contains a Zariski open subset of D, the curve D' has at least two semi-algebraically connected components, which is a contradiction. □

3.5 Some Useful Constructions

Proposition 3.5.1. Let
$$S^n = \{(x_1, \ldots, x_{n+1}) \in R^{n+1} \mid x_1^2 + \cdots + x_{n+1}^2 = 1\},$$
$P_N = (0, \ldots, 0, 1) \in S^n$, $P_S = (0, \ldots, 0, -1) \in S^n$. The stereographic projections

$$\Pi_N : S^n \setminus \{P_N\} \longrightarrow R^n, \quad \Pi_N(x) = \left(\frac{x_1}{1 - x_{n+1}}, \ldots, \frac{x_n}{1 - x_{n+1}}\right)$$

$$\Pi_S : S^n \setminus \{P_S\} \longrightarrow R^n, \quad \Pi_S(x) = \left(\frac{x_1}{1 + x_{n+1}}, \ldots, \frac{x_n}{1 + x_{n+1}}\right)$$

are biregular isomorphisms.

Proof. The inverse of the stereographic projection Π_N is

$$(t_1, \ldots, t_n) \mapsto \left(\frac{2t_1}{\|t\|^2 + 1}, \ldots, \frac{2t_n}{\|t\|^2 + 1}, \frac{\|t\|^2 - 1}{\|t\|^2 + 1}\right)$$

and the inverse of Π_S is

$$(t_1, \ldots, t_n) \mapsto \left(\frac{2t_1}{\|t\|^2 + 1}, \ldots, \frac{2t_n}{\|t\|^2 + 1}, \frac{-\|t\|^2 + 1}{\|t\|^2 + 1}\right).$$

$\square$

Note that, if $n = 1$, then $\Pi_N \circ \Pi_S^{-1}(x) = 1/x$ for $x \neq 0$. If $n = 2$, then $\Pi_N \circ \Pi_S^{-1}(x, y) = (x/(x^2 + y^2), y/(x^2 + y^2))$ for $(x, y) \neq (0, 0)$, which is equivalent to $\Pi_N \circ \Pi_S^{-1}(z) = 1/\bar{z}$ using the complex notation $z = x + iy$. This shows the following.

Proposition 3.5.2. *The real algebraic variety S^1 is biregularly isomorphic to $\mathbb{P}_1(R)$. The real algebraic variety S^2 is biregularly isomorphic to $\mathbb{P}_1(C)$ (where $C = R[i]$) considered as a real algebraic variety.*

There is also an algebraic Alexandrov compactification.

Proposition 3.5.3. *Let X be an affine real algebraic variety that is not complete. Then there exists a pair $(\dot{X}, i)$ such that*
  (i) *$\dot{X}$ is a complete affine real algebraic variety,*
  (ii) *$i : X \to \dot{X}$ is a biregular isomorphism from X onto $i(X)$,*
  (iii) *$\dot{X} \setminus i(X)$ consists of a single point.*

Proof. Let $V \subset R^n$ be an algebraic set biregularly isomorphic to X. We may assume that the origin 0 does not belong to V. By Proposition 3.4.9, V is not bounded. The inversion mapping $i : R^n \setminus \{0\} \to R^n \setminus \{0\}$, $i(x) = x/\|x\|^2$, is a biregular isomorphism. Thus, $i(V)$ is a Zariski closed subset of $R^n \setminus \{0\}$, and $\dot{X} = i(V) \cup \{0\}$, which is the closure of $i(V)$ in the euclidean topology, is a bounded algebraic subset of R^n. $\square$

3.5 Some Useful Constructions

Definition 3.5.4. *The pair $(\dot{X}, i)$ defined in Proposition 3.5.3 is called an algebraic Alexandrov compactification of X.*

Remark 3.5.5. An algebraic Alexandrov compactification of X is, in particular, a semi-algebraic Alexandrov compactification of X (Definition 2.5.11).

Note that the algebraic Alexandrov compactification is not unique up to a biregular isomorphism. Indeed, consider the curve C_1 given by the equation $y^2 = x^4 - x^6$. The Zariski open set $U = C_1 \setminus \{0\}$ is a real algebraic variety (which is not complete by Proposition 3.2.10), biregularly isomorphic to the Zariski open set $C_2 \setminus \{0\}$ of the curve C_2 given by the equation $t^2 = x^2 - x^4$. The isomorphism is given by the mapping sending (x, y) to $(x, y/x)$. The curves C_1 and C_2 are, thus, two algebraic Alexandrov compactifications of U. However, they are not biregularly isomorphic (consider the local rings at 0) (see Figure 3.5).

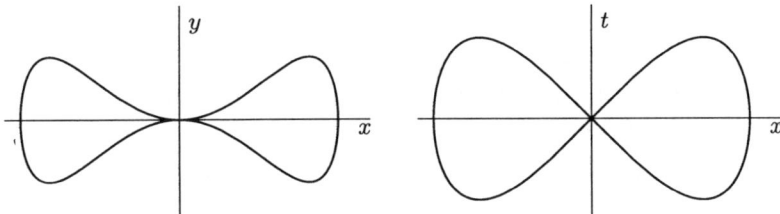

Figure 3.5. Curves C_1 (left) and C_2 (right)

It is possible to blow down an algebraic subset to a point.

Proposition 3.5.6. *Let X be an affine real algebraic variety and $Y \neq \emptyset$, a Zariski closed subset of X. There exist an affine real algebraic variety Z, a point $z \in Z$ and a regular mapping $\Phi: X \to Z$, such that*
  (i) $\Phi(Y) = \{z\}$,
  (ii) $\Phi|_{X \setminus Y}$ *is a biregular isomorphism onto $Z \setminus \{z\}$.*

Proof. It is enough to prove the proposition for an algebraic set $X \subset R^n$, with an algebraic subset Y. Let P be a polynomial such that $Y = P^{-1}(0)$. Let X' be the algebraic set obtained by sending Y to infinity, i.e.

$$X' = \{(x, t) \in X \times R \mid tP(x) = 1\}\,.$$

Let $Z \subset R^{n+1}$ be the union of the origin of R^{n+1} with the image of X' by the mapping $u \mapsto u/\|u\|^2$, $u \in R^{n+1} \setminus \{0\}$. Define Φ by:

$$\Phi(x) = \left(\frac{P^2(x)}{\|x\|^2 P^2(x) + 1} x,\ \frac{P(x)}{\|x\|^2 P^2(x) + 1} \right).$$

□

Remark 3.5.7. If $R = \mathbb{R}$ and X is compact, then the Z we have constructed is the topological quotient X/Y. This is not always true, if X is not compact. Let $X = \{(x,y) \in \mathbb{R}^2 \mid (xy-1)x = 0\}$ and $Y = \{(x,y) \in \mathbb{R}^2 \mid x = 0\}$. The Z constructed, as above, is connected (see Figure 3.6), while X/Y is not.

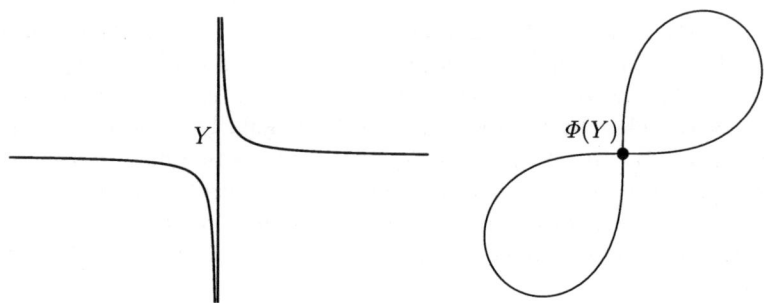

Figure 3.6. Curves X (left) and Z (right)

Finally, we describe a famous construction: blowing up. This construction can be performed over an arbitrary base field, but the result is an *affine* variety in the case of a real closed field.

Proposition 3.5.8. *Let X be an affine real algebraic variety and Y a Zariski closed subset of X with $\mathcal{I}_{\mathcal{R}(X)}(Y) = (f_1, \ldots, f_m)$. Define*
$$Z = \{(x, (f_1(x) : \ldots : f_m(x))) \in X \times \mathbb{P}_{m-1}(R) \mid x \in X \setminus Y\}.$$
Denote by $E(X,Y)$ the Zariski closure of Z in $X \times \mathbb{P}_{m-1}(R)$, ($E(X,Y)$ is thus an affine real algebraic variety) and $\sigma : E(X,Y) \to X$, the projection mapping. Then:

(i) The algebraic variety $E(X,Y)$ does not depend on the choice of generators of $\mathcal{I}_{\mathcal{R}(X)}(Y)$, up to a biregular isomorphism compatible with σ. The variety $E(X,Y)$ is called the blowing up *of X with centre Y.*

(ii) We have $Z = \sigma^{-1}(X \setminus Y)$ and $\sigma|_Z : Z \to X \setminus Y$ is a biregular isomorphism onto $X \setminus Y$.

(iii) If F is a complete semi-algebraic subset of X, then $\sigma^{-1}(F)$ is complete. If G is a closed semi-algebraic subset of $E(X,Y)$, then $\sigma(G)$ is closed.

Proof. (i) It is enough to see what happens if the ideal $(f_1, \ldots, f_m)$ is replaced by $(f_1, \ldots, f_{m+1})$, where $f_{m+1} = \sum_{i=1}^m \Lambda_i f_i$ with $\Lambda_i \in \mathcal{R}(X)$. The variety $X \times \mathbb{P}_{m-1}(R)$ is biregularly isomorphic to
$$\left\{(x, (y_1 : \ldots : y_{m+1})) \in X \times \mathbb{P}_m(R) \,\bigg|\, y_{m+1} = \sum_{i=1}^m \Lambda_i(x) y_i\right\},$$
which is a Zariski closed subset of $X \times \mathbb{P}_m(R)$. This induces the expected biregular isomorphism between the two constructions of the blowing up.

(ii) We have

$$Z = \{(x, (y_1 : \ldots : y_m)) \in (X \setminus Y) \times \mathbb{P}_{m-1}(R) \mid y_i f_j(x) = y_j f_i(x)\},$$

which shows that Z is a Zariski closed subset of $(X \setminus Y) \times \mathbb{P}_{m-1}(R)$. Hence, $E(X, Y) \cap ((X \setminus Y) \times \mathbb{P}_{m-1}(R)) = Z$. It is clear that $\sigma|_Z : Z \to X \setminus Y$ is a biregular isomorphism.

(iii) If F is a complete semi-algebraic subset of X, then $\sigma^{-1}(F)$ is closed in the euclidean topology in $F \times \mathbb{P}_{m-1}(R)$, which is complete. Hence $\sigma^{-1}(F)$ is complete. Let G be a closed semi-algebraic subset of $E(X, Y)$ and $x \in \text{clos}(\sigma(G))$. We may assume that X is embedded as an algebraic subset of some R^n, and then $\sigma^{-1}(\overline{B}_n(x, 1)) \cap G$ is complete. Hence, the set

$$\sigma(\sigma^{-1}(\overline{B}_n(x, 1)) \cap G) = \overline{B}_n(x, 1) \cap \sigma(G)$$

is closed, and $x \in \sigma(G)$. □

Proposition 3.5.9. *If X is irreducible, then $E(X, Y)$ is irreducible of the same dimension as X, and σ induces an isomorphism $\mathcal{K}(X) \to \mathcal{K}(E(X, Y))$ between the fields of rational functions (i.e. σ is a birational equivalence).*

Proof. If X is irreducible, then $X \setminus Y$ is Zariski-dense in X. On the other hand, Z is Zariski-dense in $E(X, Y)$ and $X \setminus Y$ and Z are biregularly isomorphic (Proposition 3.5.8 (ii)). The irreducibility of X implies the irreducibility of $E(X, Y)$. Moreover, $\mathcal{K}(X)$ (resp. $\mathcal{K}(E(X, Y))$) is the field of fractions of $\mathcal{R}(X \setminus Y)$ (resp. $\mathcal{R}(Z)$), and the two fields are, thus, isomorphic. □

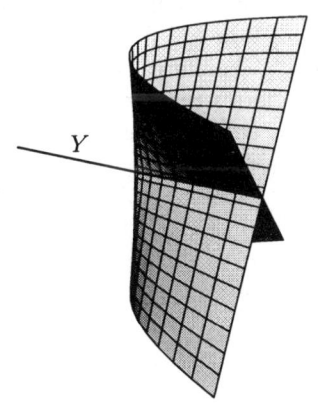 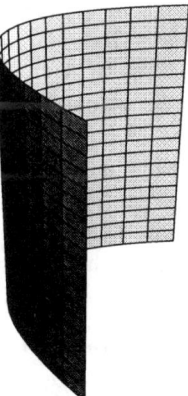

Figure 3.7. Surfaces X (left) and $E(X, Y)$ (right)

Example 3.5.10. Let X be the "umbrella" given by the equation $z^2 x = y^2$ and Y, the "stick" $z = y = 0$. In this case,

$$Z = \{(x, y, z, (y : z)) \in X \times \mathbb{P}_1(R) \mid y \neq 0 \text{ or } z \neq 0\}$$

and
$$E(X,Y) = \{(x,y,z,(u:v)) \in X \times \mathbb{P}_1(R) \mid uz = vy,\ v^2x = u^2\}$$
is biregularly isomorphic to $\{(x,t,z) \in R^3 \mid x = t^2\}$ (with $t = u/v$).

Note that $\sigma : E(X,Y) \to X$ is not surjective. The points of the stick where $x < 0$ do not belong to the image of σ.

The situation is simpler if everything is nonsingular.

Proposition 3.5.11. *Let X be a nonsingular affine real algebraic variety, and Y, a nonsingular Zariski closed subset of X, with $m = \dim(X) - \dim(Y)$. Let $\sigma : E(X,Y) \to X$ be the blowing up of X with centre Y. Then:*

(i) $E(X,Y)$ is nonsingular.

(ii) There exists a finite cover of Y by Zariski open subsets U_i, such that $\sigma^{-1}(U_i)$ is biregularly isomorphic to $U_i \times \mathbb{P}_{m-1}(R)$.

Proof. Let $y \in Y$. The local ring $\mathcal{R}_{X,y}$ is regular. We can find a regular system of parameters $f_1, \ldots, f_d$ of $\mathcal{R}_{X,y}$, such that $f_1, \ldots, f_m$ generate $\mathcal{I}(Y)\mathcal{R}_{X,y}$ (cf. Proposition 3.3.7). Let U be a Zariski open neighbourhood of y in X, such that $f_1, \ldots, f_m$ generate $\mathcal{I}(Y)\mathcal{R}(U)$.

We start by proving (ii). The set $E(U, Y \cap U)$ is the Zariski closure of
$$Z = \{(x, (f_1(x) : \ldots : f_m(x))) \in U \times \mathbb{P}_{m-1}(R) \mid x \in U \setminus Y\},$$
and $E(U, Y \cap U)$ is biregularly isomorphic to $\sigma^{-1}(U)$. Let $y' \in Y \cap U$. The classes of $f_1, \ldots, f_m$ modulo $(\mathfrak{m}_{X,y'})^2$ are linearly independent over R (Proposition 3.3.7), which means that the rank of the derivative $df_{y'} : T_{y'}(X) \to R^m$, where $f = (f_1, \ldots, f_m) : U \to R^m$, is equal to m. Using the implicit function theorem, we see that $f(\Omega)$ is a neighbourhood of 0 in R^m for every neighbourhood Ω of y' in X in the euclidean topology. Therefore, every point of $\{y'\} \times \mathbb{P}_{m-1}(R)$ belongs to the (euclidean and, a fortiori, Zariski) closure of Z, and, hence, $\sigma^{-1}(y') = \{y'\} \times \mathbb{P}_{m-1}(R)$. This is true for every point y' in $Y \cap U$, and, thus, $\sigma^{-1}(Y \cap U) = (Y \cap U) \times \mathbb{P}_{m-1}(R)$.

We now prove (i). Let $t = (y, (z_1 : \ldots : z_m))$ be a point of $U \times \mathbb{P}_{m-1}(R)$. We may assume that $z_m \neq 0$, and then we work in $U \times R^{m-1}$ with $t = (y, v_1, \ldots, v_{m-1})$, where $v_i = z_i/z_m$ for $i = 1, \ldots, m-1$. The equations of $E(U, Y \cap U)$ in $U \times R^{m-1}$ are
$$f_m(x)u_i = f_i(x) \quad \text{for } i = 1, \ldots, m-1.$$
The maximal ideal of $\mathcal{R}_{U \times R^{m-1}, t}$ is
$$(f_1, \ldots, f_d, u_1 - v_1, \ldots, u_{m-1} - v_{m-1}) =$$
$$(f_m u_1 - f_1, \ldots, f_m u_{m-1} - f_{m-1}, f_m, \ldots, f_d, u_1 - v_1, \ldots, u_{m-1} - v_{m-1}).$$
Hence, by Proposition 3.3.7 (iii),
$$\mathcal{R}_{E(U, Y \cap U), t} = \mathcal{R}_{U \times R^{m-1}, t}/(f_m u_1 - f_1, \ldots, f_m u_{m-1} - f_{m-1})$$
is a regular local ring and t is a nonsingular point of $E(U, Y \cap U)$. This proves that $E(X,Y)$ is nonsingular. □

Proposition 3.5.12. Let 0 be the origin of R^n. The blowing up $E(R^n, 0)$ is biregularly isomorphic to $\mathbb{P}_n(R)$ with a point removed.

Proof. The biregular isomorphism is constructed as follows: consider $R^n \subset R^{n+1}$, and take the radial projection ρ with centre $(0, 1) \in R^{n+1}$ of R^n into the sphere $S^{n+1}((0, \frac{1}{2}), \frac{1}{2})$. One has

$$\rho(x) = (1 + \|x\|^2)^{-1}(x, \|x\|^2) .$$

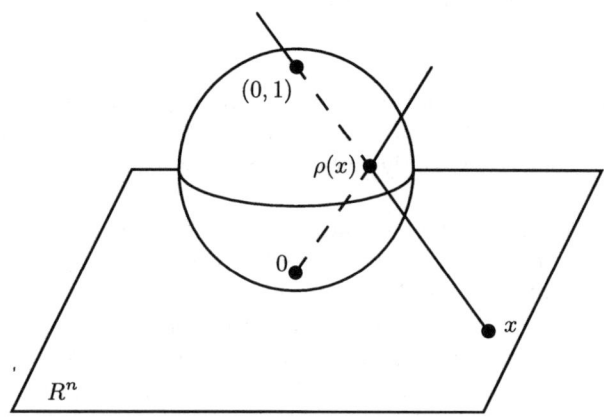

Figure 3.8.

Compose $\rho|_{R^n \setminus \{0\}}$ with the canonical surjection $\Pi : R^{n+1} \setminus \{0\} \to \mathbb{P}_n(R)$. Lift the composition mapping to the blowing up $E(R^n, 0)$, by sending the points of $\sigma^{-1}(\{0\})$ to the lines of R^{n+1} tangent to the sphere $S^{n+1}((0, \frac{1}{2}), \frac{1}{2})$ at the origin. Only the "vertical" line is not reached (see Figure 3.8). Explicitly, we have

$$E(R^n, 0) = \{(x, (t_1 : \ldots : t_n)) \in R^n \times \mathbb{P}_{n-1}(R) \mid x_i t_j = x_j t_i\} ,$$

and the biregular isomorphism

$$\varphi : E(R^n, 0) \to \mathbb{P}_n(R) \setminus \{(0 : \ldots : 0 : 1)\}$$

is defined by

$$\varphi(x, (t_1 : \ldots : t_n)) = \left(t_1 : \ldots : t_n : \sum_{i=1}^n t_i x_i \right) .$$

The inverse isomorphism φ^{-1} is defined by

$$\varphi^{-1}(u_1 : \ldots : u_{n+1}) = \left(\frac{u_1 u_{n+1}}{\sum_{i=1}^n u_i^2}, \ldots, \frac{u_n u_{n+1}}{\sum_{i=1}^n u_i^2}, (u_1 : \ldots : u_n) \right) .$$

□

Remark 3.5.13. ($R = \mathbb{R}$) Let X be a nonsingular affine real algebraic variety of dimension n and y, a point of X. In a neighbourhood of y, we can choose a regular system of parameters of $\mathcal{R}_{X,y}$ defining a Nash diffeomorphism onto an open neighbourhood of the origin of $\mathbb{R}^n$. The blowing up $E(X,y)$ is the *connected sum* $X \# \mathbb{P}_n(\mathbb{R})$ of X with $\mathbb{P}_n(\mathbb{R})$ ([160], p. 191). Let us describe the situation in dimension 2. By Proposition 3.5.12, we know that $E(\mathbb{R}^2, 0)$ is a Möbius band without boundary. Hence, the topology of $E(X, y)$ is as follows: remove a small disc with centre y from X and glue a Möbius band with boundary, by identifying this boundary with the boundary of the removed disc.

Bibliographic Notes. A large part of this chapter is valid for varieties over an arbitrary field of characteristic zero and can be found in the classical books and papers on algebraic geometry. We only give references for what is specific to the real case.

It is not easy to say who really is the originator of the umbrella in Example 3.1.2 d). It can be found in the papers [81] and [338] and in a note [74] that were published nearly simultaneously. Another umbrella (Example 3.5.10) can be found in a later paper [339] about stratification problems.

Several results have been taken from the paper [7]: property (iii) of Proposition 3.3.10 (the nonsingularity in dimension d), Proposition 3.3.17, the algebraic "Alexandrov compactification" (Definition 3.5.4) and the blowing down of an algebraic subset to a point (Proposition 3.5.6) (to be found also in [32]).

The example of a point analytically nonsingular but algebraically singular (Example 3.3.12 b)) is in [233].

Another proof of Theorem 3.4.12 is given in [108].

4. Real Algebra

Abstract. In the first section, we prove the Artin-Lang homomorphism theorem as a consequence of the Tarski-Seidenberg principle. We then present the real Nullstellensatz, which characterizes the ideals of polynomials vanishing on an algebraic set, when the ground field is real closed. The next two sections develop an "Artin-Schreier theory" for rings, which is an extension of the theory presented in Chap. 1 for fields. The notion of a prime cone of a ring, which is related to that of a ring homomorphism into a real closed field, plays an important role and we shall encounter it later on in the study of the real spectrum in Chap. 7. In Section 4 we use the Artin-Schreier theory for rings, together with the Artin-Lang homomorphism theorem, to establish various versions of the Positivstellensatz, which characterizes polynomials which are positive on certain semi-algebraic subsets of R^n. The last section presents the following criterion: an irreducible polynomial $f \in R[X_1, \ldots, X_n]$ generates the ideal of polynomials vanishing on $\mathcal{Z}(f)$ if and only if f changes sign on R^n.

Throughout this chapter R is a real closed field.

4.1 The Artin-Lang Homomorphism Theorem and the Real Nullstellensatz

We begin with a consequence of the Tarski-Seidenberg principle.

Proposition 4.1.1. *Let R_1 be a real closed extension of R. Let $\mathcal{B}(X)$ be a boolean combination of polynomial equations and inequalities in the variables $X = (X_1, \ldots, X_n)$, with coefficients in R. If $\mathcal{B}(y)$ holds true for some $y \in R_1^n$, then $\mathcal{B}(x)$ holds true for some $x \in R^n$.*

Proof. By induction on n, using Corollary 1.4.7. For $n = 0$, there is nothing to prove. Given $n \geq 1$, assume the result holds true for $n - 1$. By Corollary 1.4.7, there exists a boolean combination $\mathcal{C}(X')$ of polynomial equations and inequalities in the variables $X' = (X_1, \ldots, X_{n-1})$, with coefficients in R, such that, for every real closed field R_2 containing R and every $x' = (x_1, \ldots, x_{n-1}) \in R_2^{n-1}$, $\mathcal{B}(x', X_n)$ has a solution in R_2 if and only if $\mathcal{C}(x')$ holds true. Therefore, if $y = (y_1, \ldots, y_n)$ is a solution of $\mathcal{B}(X)$ in R_1^n, then $y' = (y_1, \ldots, y_{n-1})$ is a solution of $\mathcal{C}(X')$ in R_1^{n-1}. By induction, $\mathcal{C}(X')$ has a solution $x' = (x_1, \ldots, x_{n-1})$ in R^{n-1}. Hence, there exists $x_n \in R$, such that $x = (x', x_n)$ is a solution of $\mathcal{B}(X)$ in R^n. □

Theorem 4.1.2 (Artin-Lang Homomorphism Theorem). *Let R be a real closed field and A an R-algebra of finite type. If there exists an R-algebra homomorphism $\varphi : A \to R_1$ into a real closed extension R_1 of R, then there exists an R-algebra homomorphism $\psi : A \to R$.*

Proof. We may assume A to be of the form $R[X_1, \ldots, X_n]/I$, where I is the ideal of $R[X_1, \ldots, X_n]$ generated by $P_1, \ldots, P_m$. Let b_i be the image of the class of X_i by φ. Then $(b_1, \ldots, b_n)$ is a solution of the system of equations $P_1 = \cdots = P_m = 0$ in R_1^n. By Proposition 4.1.1, this system of equations also has a solution $(a_1, \ldots, a_n)$ in R^n. The homomorphism $\overline{\psi} : R[X_1, \ldots, X_n] \to R$ defined by $\overline{\psi}(X_i) = a_i$ obviously induces a homomorphism $\psi : A \to R$. □

The homomorphism theorem is used to prove the real Nullstellensatz, which characterizes the ideal of polynomials vanishing on an algebraic set.

Definition 4.1.3. *Let A be a commutative ring. An ideal I of A is said to be real if, for every sequence $a_1, \ldots, a_p$ of elements of A, we have*

$$a_1^2 + \cdots + a_p^2 \in I \implies a_i \in I, \text{ for } i = 1, \ldots, p.$$

Theorem 4.1.4 (Real Nullstellensatz). *Let R be a real closed field and I an ideal of $R[X_1, \ldots, X_n]$. Then $I = \mathcal{I}(\mathcal{Z}(I))$ if and only if I is real.*

Proof. Assume $I = \mathcal{I}(\mathcal{Z}(I))$. If $P_1, \ldots, P_s$ are polynomials such that $P_1^2 + \cdots + P_s^2 \in I$, then $P_i(x) = 0$, for all $x \in \mathcal{Z}(I)$ and $i = 1, \ldots, s$. Hence, $P_i \in I$, for $i = 1, \ldots, s$, which proves that the ideal I is real.

Before proving the converse we need the following lemmas.

Lemma 4.1.5. *Every real ideal I of a commutative ring A is radical. Moreover, if A is noetherian, then all minimal prime ideals containing I are real.*

Proof. If $a^n \in I$, $n > 1$, then $a^{n/2} \in I$ if n is even, and $a^{(n+1)/2} \in I$ if n is odd. In both cases the exponent has decreased, and we get $a \in I$ by iterating this process. Let $\mathfrak{p}_1, \ldots, \mathfrak{p}_q$ be the minimal prime ideals of A containing I. We can assume $q > 1$. If, for instance, $\mathfrak{p}_1$ is not real, then we can find $a_1, \ldots, a_p \in A \setminus \mathfrak{p}_1$, such that $a_1^2 + \cdots + a_p^2 \in \mathfrak{p}_1$. Choose $b_i \in \mathfrak{p}_i \setminus \mathfrak{p}_1$, for $i = 2, \ldots, q$, and set $b = \prod_{i=2}^q b_i$. Then $(a_1 b)^2 + \cdots + (a_p b)^2 \in \bigcap_{i=1}^q \mathfrak{p}_i = I$, but $a_1 b \notin \mathfrak{p}_1$, which is a contradiction. □

Lemma 4.1.6. *Let A be a commutative ring. A prime ideal I of A is real if and only if the field of fractions of A/I is real.*

Proof. Straightforward using Theorem 1.1.8 (iv). □

4.1 The Artin-Lang Homomorphism Theorem and the Real Nullstellensatz

Now we shall complete the proof of Theorem 4.1.4. By Lemma 4.1.5, we can assume that I is a real prime ideal. Denote by $\overline{P}$ the image of an element P of $R[X_1,\ldots,X_n]$ in $B = R[X_1,\ldots,X_n]/I$. Choose an ordering of the field of fractions of B, which is possible by Lemma 4.1.6. Let R_1 be the real closure of this ordered field. Let $P \notin I$, and let A be the ring of fractions $B_{\overline{P}}$. The ring A is contained in R_1. Hence, by the Artin-Lang homomorphism theorem, there is an R-algebra homomorphism $\psi : A \to R$. Define $x = (\psi(\overline{X_1}),\ldots,\psi(\overline{X_n}))$. Then $x \in \mathcal{Z}(I)$ and $P(x) \neq 0$, which shows that $P \notin \mathcal{I}(\mathcal{Z}(I))$. Hence, $I = \mathcal{I}(\mathcal{Z}(I))$. $\square$

Proposition 4.1.7. *Let A be a commutative ring and I an ideal of A. Then*

$$\sqrt[R]{I} = \{a \in A \mid \exists m \in \mathbb{N}\ \exists b_1,\ldots,b_p \in A\quad a^{2m} + b_1^2 + \cdots + b_p^2 \in I\}$$

is the smallest real ideal of A containing I. The ideal $\sqrt[R]{I}$, called the real radical of I, is the intersection of all real prime ideals containing I (or is A itself, if there is no real prime ideal containing I).

Proof. We first show that $\sqrt[R]{I}$ is an ideal. The delicate point is to check that $\sqrt[R]{I}$ is closed under addition. Suppose that

$$a^{2m} + b_1^2 + \cdots + b_p^2 \in I \quad \text{and} \quad (a')^{2m'} + (b_1')^2 + \cdots + (b_{p'}')^2 \in I\ .$$

We can write

$$(a+a')^{2(m+m')} + (a-a')^{2(m+m')} = a^{2m}c + a'^{2m'}c'\ ,$$

where c and c' are sums of squares of elements of A. Then

$$(a+a')^{2(m+m')} + (a-a')^{2(m+m')} + c(b_1^2 + \cdots + b_p^2) + c'(b_1'^2 + \cdots + b_{p'}'^2) \in I$$

and hence $a + a' \in \sqrt[R]{I}$.

It is clear that $\sqrt[R]{I}$ is real. By Definition 4.1.3 and the fact that J is radical (Lemma 4.1.5), every real ideal J containing I contains $\sqrt[R]{I}$. This shows that $\sqrt[R]{I}$ is the smallest real ideal of A containing I. Let $a \in A \setminus \sqrt[R]{I}$ and let J be maximal among the real ideals containing I but not a. We claim that J is prime. If J were not prime, there would be $b \notin J$ and $b' \notin J$ such that $bb' \in J$. Then $a \in \sqrt[R]{J + bA}$ and $a \in \sqrt[R]{J + b'A}$ and hence

$$a^{2m} + c_1^2 + \ldots + c_q^2 = j + bd \quad \text{and} \quad a^{2m'} + c_1'^2 + \cdots + c_{q'}'^2 = j' + b'd'\ ,$$

with $j, j' \in J$. From these two equalities we obtain

$$a^{2(m+m')} + \text{a sum of squares} = jj' + jb'd' + j'bd + bb'dd' \in J\ ,$$

and, thus, $a \in \sqrt[R]{J} = J$, which is impossible. This shows that $\sqrt[R]{I}$ is the intersection of all real prime ideals containing I. $\square$

Corollary 4.1.8. *Let $I \subset R[X_1,\ldots,X_n]$ be an ideal. Then $P \in \mathcal{I}(\mathcal{Z}(I))$ if and only if there exist finitely many polynomials $Q_1,\ldots,Q_p$ and an integer $m \in \mathbb{N}$, such that $P^{2m} + Q_1^2 + \cdots + Q_p^2 \in I$. In short, $\mathcal{I}(\mathcal{Z}(I)) = \sqrt[R]{I}$.*

Proof. The real Nullstellensatz says that $\mathcal{I}(\mathcal{Z}(I))$ is the smallest real ideal containing I, that is, by Proposition 4.1.7, the ideal $\sqrt[R]{I}$. □

Corollary 4.1.9. *Let $V \subset R^n$ be an algebraic set and $I \subset \mathcal{P}(V)$ an ideal. Then $P \in \mathcal{P}(V)$ vanishes on $\mathcal{Z}_V(I)$ if and only if there exist finitely many polynomial functions $Q_1,\ldots,Q_p \in \mathcal{P}(V)$ and an integer m, such that $P^{2m} + Q_1^2 + \cdots + Q_p^2 \in I$. In short, $\mathcal{I}_{\mathcal{P}(V)}(\mathcal{Z}_V(I)) = \sqrt[R]{I}$.*

Proof. Follows from Corollary 4.1.8. □

4.2 Cones, Convex Ideals

The notion of a cone, introduced for fields in Chap. 1, will now be extended to rings.

Definition 4.2.1. *Let A be a commutative ring. A cone P of A is a subset of A satisfying the following properties:*
  (i) $a \in P,\ b \in P \Rightarrow a + b \in P$,
  (ii) $a \in P,\ b \in P \Rightarrow ab \in P$,
  (iii) $a \in A \Rightarrow a^2 \in P$.
The cone P is said to be proper *if in addition*
  (iv) $-1 \notin P$.

Example 4.2.2.
  a) Let $A = R[X_1,\ldots,X_n]$, and let T be a nonempty subset of R^n. Then $\mathcal{W}(T) = \{P \in R[X] \mid \forall x \in T\ P(x) \geq 0\}$ is a proper cone of $R[X]$.
  b) The intersection of an arbitrary family of cones (resp. proper cones) of A is a cone (resp. proper cone) of A.
  c) We denote by $\sum A^2$ the set of sums of squares of elements of A. The set $\sum A^2$ is the smallest cone of A.
  d) Let P be a cone of A and $(a_i)_{i \in I}$ a family of elements of A. Denote by $M[(a_i)_{i \in I}]$ the multiplicative monoid generated by $(a_i)_{i \in I}$, i.e. the set of finite products of elements of $(a_i)_{i \in I}$. Denote by $P[(a_i)_{i \in I}]$ the smallest cone of A containing P and $(a_i)_{i \in I}$ (which exists, see Example b). We have

$$P[(a_i)_{i \in I}] = \left\{ p + \sum_{i=1}^{r} q_i b_i \mid p, q_1, \ldots, q_r \in P,\ b_1, \ldots, b_r \in M[(a_i)_{i \in I}] \right\}.$$

In particular, $\sum A^2[(a_i)_{i \in I}]$ is the *cone generated by* $(a_i)_{i \in I}$.

Definition 4.2.3. *Let P be a cone of a commutative ring A. An ideal I of A is said to be P-convex (resp. P-radical) if*

$$p_1, p_2 \in P, \ p_1 + p_2 \in I \Rightarrow p_1 \in I \text{ (and hence } p_2 \in I)$$
$$(resp. \quad a \in A, \ p \in P, \ a^2 + p \in I \Rightarrow a \in I \text{)}.$$

Example 4.2.4.
  a) An ideal is $\sum A^2$-radical if and only if it is real.
  b) Let $A = R[X_1, \ldots, X_n]$, $T \subset R^n$ a nonempty set and $P = \mathcal{W}(T)$ (cf. Example 4.2.2 a)). If V is an algebraic subset of R^n, such that $T \cap V$ is Zariski-dense in V, then $\mathcal{I}(V)$ is P-convex.

Proposition 4.2.5. *An ideal I is P-radical if and only if it is radical and P-convex.*

Proof. Assume that I is P-radical. Let $p_1, p_2 \in P$, with $p_1 + p_2 \in I$. Then $p_1^2 + p_1 p_2 \in I$, and, thus, $p_1 \in I$. Moreover, the same proof as for Lemma 4.1.5 shows that I is radical. Conversely, if I is radical and P-convex, and if $a^2 + p \in I$ with $p \in P$, then $a^2 \in I$ since $a^2 \in P$, and, hence, $a \in I$. $\square$

Proposition 4.2.6. *Let I be an ideal of a commutative ring A and P a cone of A. Then*

$$\sqrt[P]{I} = \{a \in A \mid \exists m \in \mathbb{N} \ \exists p \in P \quad a^{2m} + p \in I\}$$

is the smallest P-radical ideal of A containing I. The ideal $\sqrt[P]{I}$, called the P-radical of I, is the intersection of all P-convex prime ideals containing I (or is A itself, if there is no P-convex prime ideal containing I).

Proof. Repeat the proof of Proposition 4.1.7, with appropriate changes. $\square$

Proposition 4.2.7. *Let P be a cone of A. If there exists a proper P-convex ideal of A, then P is proper.*

Proof. Let I be a proper P-convex ideal. If P is not proper, then $-1 \in P$ and $1 + (-1) \in I$. Hence $1 \in I$, which is impossible. $\square$

Proposition 4.2.8.
  (i) *Let $P \subset Q$ be two cones of A. If I is a Q-convex ideal, then I is P-convex.*
  (ii) *Let P_1 and P_2 be two cones of A and I a $(P_1 \cap P_2)$-convex prime ideal. Then I is P_1-convex or P_2-convex.*

Proof. (i) is obvious.
  (ii) If I is neither P_1-convex nor P_2-convex, then there exist $a \in P_1$, $p \in P_1$, with $a + p \in I$ and $a \notin I$, and there also exist $b \in P_2$, $q \in P_2$, with $b + q \in I$ and $b \notin I$. Then $(a+p)^2 b^2 + (b+q)^2 a^2 \in I$. Set $r = (a+p)^2 b^2 + (b+q)^2 a^2 - a^2 b^2$. We have $r = b^2(p^2 + 2ap) + (b+q)^2 a^2 \in P_1$

and $r = (a+p)^2b^2 + (q^2+2bq)a^2 \in P_2$. Since $a^2b^2 \in P_1 \cap P_2$ and I is $(P_1 \cap P_2)$-convex, then $a^2b^2 \in I$, so that $a \in I$ or $b \in I$, which is impossible. □

We shall use the next result in Chap. 7.

Proposition 4.2.9. *Let K be a field, A a subring of K, P a proper cone of K and I a $(P \cap A)$-convex prime ideal of A. Then there exists an ordering of K, whose positive cone Q contains P, and such that I is $(Q \cap A)$-convex.*

Proof. Let Q be a cone, which is a maximal element of the family of proper cones P' of K containing P and such that I is $(P' \cap A)$-convex. The cone Q is the positive cone of an ordering of K. Otherwise, we could find $x \in K$, such that $x \notin Q$ and $-x \notin Q$. Then $Q[x]$ and $Q[-x]$ are proper cones of K (Lemma 1.1.7) and Proposition 1.1.10 shows that $Q[x] \cap Q[-x] = Q$. By Proposition 4.2.8 (ii), I is $(Q[x] \cap A)$-convex or $(Q[-x] \cap A)$-convex, which contradicts the fact that Q is maximal. □

4.3 Prime Cones

Definition 4.3.1. *Let A be a commutative ring. A* prime cone P *of A is a proper cone of A satisfying:*

$$ab \in P \Rightarrow a \in P \text{ or } -b \in P.$$

Proposition 4.3.2. *Let P be a prime cone of A, $-P = \{a \in A \mid -a \in P\}$. Then:*

(i) $P \cup -P = A$.

(ii) $P \cap -P$ *is a prime ideal of A, called the* support *of P and denoted by* $\mathrm{supp}(P)$.

Proof. (i) We have $a^2 \in P$, for every a in A, and, hence, $a \in P$ or $-a \in P$.

(ii) It is clear that $P \cap -P$ is an additive subgroup of A. Let $a \in P \cap -P$, $b \in A$. If $b \in P$, then ba and $b(-a)$ also belong to P. If $(-b) \in P$, then $(-b)a$ and $(-b)(-a)$ belong to P. So $ab \in P \cap -P$ and $P \cap -P$ is an ideal. If $ab \in P \cap -P$ and $a \notin P \cap -P$, then there are two cases:

– either $a \notin P$, and then $-b \in P$ since $ab \in P$, and $b \in P$ since $a(-b) \in P$,
– or $a \notin -P$, and then $b \in P$ since $ba \in P$, and $-b \in P$ since $(-b)a \in P$.

In both cases, $b \in P \cap -P$. □

Example 4.3.3.

a) If F is a field, then the prime cones of F are the positive cones of orderings of F.

b) If $f : A \to B$ is a ring homomorphism and P is a prime cone of B, then $f^{-1}(P)$ is a prime cone of A.

c) Let $A = R[X]$ be the ring of polynomials in one variable over a real closed field R. The set $\{f \in R[X] \mid f(0) \geq 0\}$ is a prime cone whose support is (X). The set $\{f \in R[X] \mid \exists \epsilon \in R, \, \epsilon > 0, \, f(]0, \epsilon[) \subset [0, +\infty[\,\}$ is a prime cone whose support is (0). The first of these two prime cones is the inverse image of the positive cone of R by the mapping $f \mapsto f(0)$, while the latter is the intersection of $R[X]$ with the positive cone of the ordering of $R(X)$ given by the cut 0_+ (Example 1.1.2).

Proposition 4.3.4. *Let A be a commutative ring. A subset $P \subset A$ is a prime cone of A if and only if there exists an ordered field $(F, \leq)$ and a homomorphism $\varphi : A \to F$, such that $P = \{a \in A \mid \varphi(a) \geq 0\}$.*

Proof. Given $(F, \leq)$ and $\varphi : A \to F$ as in the proposition, $\{a \in A \mid \varphi(a) \geq 0\}$ is a prime cone of A (cf. Example 4.3.3 a) and b)).

The converse follows from the next lemma.

Lemma 4.3.5. *Let A be a commutative ring and P a prime cone of A. Let $k(\mathrm{supp}(P))$ denote the field of fractions of $A/\mathrm{supp}(P)$. Then*

$$\overline{P} = \{\overline{a}/\overline{b} \in k(\mathrm{supp}(P)) \mid ab \in P\}$$

is the positive cone of an ordering of $k(\mathrm{supp}(P))$, and P is the inverse image of $\overline{P}$ under the canonical homomorphism $A \to k(\mathrm{supp}(P))$.

Proof. It is easy to check that $\overline{P}$ is a cone and that $\overline{P} \cup -\overline{P} = k(\mathrm{supp}(P))$. Let us show that P is the inverse image of $\overline{P}$, which implies that $\overline{P}$ is proper. If $\overline{a} \in \overline{P}$, then $\overline{a} = \overline{b}/\overline{c}$, with $c \notin \mathrm{supp}(P)$ and $bc \in P$. We can assume that $c \in P$, and then $b \in P$, since $c \notin -P$. We have $ca = b + d$ for some $d \in \mathrm{supp}(P)$, and, hence $ca \in P$. Since $c \notin -P$, then $a \in P$. Thus both the lemma and Proposition 4.3.4 are proved. □ □

Corollary 4.3.6. *The support of a prime cone is always a real prime ideal.*

Proof. Obvious from Lemma 4.3.5. □

Theorem 4.3.7. *Let A be a commutative ring. The following conditions are equivalent:*
  (i) *The ring A has a proper cone.*
  (ii) *The ring A has a prime cone.*
  (iii) *There is a homomorphism $\varphi : A \to K$, where K is a real closed field.*
  (iv) *The ring A has a real prime ideal.*
  (v) *The element -1 is not a sum of squares in A, i.e. $-1 \notin \sum A^2$.*

Proof. (i) $\Rightarrow$ (ii) Let P be a maximal proper cone of A. If P is not prime, then there are $a \notin P$, $b \notin -P$, with $ab \in P$. Hence $P[a]$ and $P[-b]$ are not proper, and $-1 = p_1 + q_1 a = p_2 - q_2 b$, with $p_1, q_1, p_2, q_2 \in P$. It follows that $1 + p_1 = -q_1 a$ and $1 + p_2 = q_2 b$. Hence, $1 + p_1 + p_2 + p_1 p_2 = -q_1 q_2 ab$, and $-1 \in P$, which is impossible.

(ii) ⇒ (iii) Proposition 4.3.4 gives a homomorphism from A into an ordered field and hence into its real closure.

(iii) ⇒ (iv) The kernel of φ is a real prime ideal.

(iv) ⇒ (v) If I is a real prime ideal of A and $-1 \in \sum A^2$, then there is $s \in \sum A^2$ such that $1 + s = 0$. Hence $1 \in I$, which is impossible.

(v) ⇒ (i) If $-1 \notin \sum A^2$, then $\sum A^2$ is a proper cone. □

Proposition 4.3.8. *Let A be a commutative ring, P a cone of A and I a P-convex prime ideal. Then there exists a prime cone Q containing P and such that $\operatorname{supp}(Q) = I$.*

Proof. Let Q be a cone, which is a maximal element of the family of cones P' containing P and such that I is P'-convex. The cone Q is proper (Proposition 4.2.7). We first show that $I = Q \cap -Q$. If $a \in Q \cap -Q$, then $a \in I$, since $a + (-a) \in I$ and I is Q-convex. Conversely, if $a \in I$ and $b^2 + p + qa \in I$, with $p, q \in Q$, then $b^2 + p \in I$ and $b \in I$, since I is Q-convex. This shows that I is $Q[a]$-convex, and, hence, $a \in Q$ since Q is maximal. We proceed in the same way for $-a$.

We now show that Q is prime. Let $a, b \in A$, with $ab \in Q$ and $a \notin Q$. Then $Q[a]$ strictly contains Q, and I is not $Q[a]$-convex; a fortiori, it is not $Q[a]$-radical. We can find $p, q \in Q$, $c \notin I$, such that $c^2 + p + qa \in I$. We verify that I is $Q[-b]$-convex. Let $p', q' \in Q$, $c' \in A$, such that $c'^2 + p' - q'b \in I$. Then

$$d = (c^2 + p + qa)(c'^2 + p' + q'b) + (c^2 + p - qa)(c'^2 + p' - q'b) \in I$$

and

$$d - c^2 c'^2 = c^2 c'^2 + 2(pc'^2 + p'c^2 + pp' + qq'ab) \in Q \ .$$

Hence $c^2 c'^2 \in I$ and $c' \in I$, since $c \notin I$. Since I is $Q[-b]$-convex, $-b \in Q$. □

Proposition 4.3.9. *Let A be a commutative ring and P a prime cone of A. The mapping $Q \mapsto \operatorname{supp}(Q)$ is a bijection from the set of prime cones containing P onto the set of P-convex prime ideals.*

Proof. The mapping is surjective by Proposition 4.3.8. If Q is a prime cone containing P, then $x \notin -Q \Rightarrow x \notin -P \Rightarrow x \in P$, hence $Q = \operatorname{supp}(Q) \cup P$, which proves that the mapping is injective. □

4.4 The Positivstellensatz

The Positivstellensatz gives an algebraic characterization of functions positive on certain semi-algebraic subsets of R^n. We first state a formal theorem for an arbitrary ring in terms of prime cones (or homomorphisms into a real closed field). We next obtain a geometric statement using the Artin-Lang homomorphism theorem.

4.4 The Positivstellensatz

Proposition 4.4.1 (Formal Positivstellensatz). *Let A be a commutative ring. Let $(a_j)_{j \in J}$, $(b_k)_{k \in K}$, $(c_\ell)_{\ell \in L}$ be arbitrary families of elements of A. Denote by P the cone generated by $(a_j)_{j \in J}$, M the multiplicative monoid generated by $(b_k)_{k \in K}$ and I the ideal generated by $(c_\ell)_{\ell \in L}$. Then the following properties are equivalent:*

(i) *There is no prime cone Q of A, such that*

$$\forall j \in J \; a_j \in Q, \quad \forall k \in K \; b_k \notin \mathrm{supp}(Q), \quad \forall \ell \in L \; c_\ell \in \mathrm{supp}(Q).$$

(ii) *There is no homomorphism $\varphi : A \to F$ into a real closed field F, such that*

$$\forall j \in J \; \varphi(a_j) \geq 0, \quad \forall k \in K \; \varphi(b_k) \neq 0, \quad \forall \ell \in L \; \varphi(c_\ell) = 0.$$

(iii) *There exist $p \in P$, $b \in M$, $c \in I$, such that:*

$$p + b^2 + c = 0.$$

Proof. (i) $\Leftrightarrow$ (ii) by Proposition 4.3.4.

(ii) $\Rightarrow$ (iii) Without loss of generality, we can assume that $M \cap I = \emptyset$. Define $A_1 = A/I$, $A_2 = (\overline{M})^{-1} A_1$ (where the bar denotes the image in A_1) and $A_3 = A_2[(T_j)_{j \in J}]/\mathfrak{a}$, where $(T_j)_{j \in J}$ are indeterminates and $\mathfrak{a}$ is the ideal generated by the family $(T_j^2 - \overline{a}_j)_{j \in J}$. Condition (ii) implies that there is no homomorphism $\Psi : A_3 \to F$ into a real closed field F. Hence, by Theorem 4.3.7, we can find $\delta_1, \ldots, \delta_n \in A_3$ such that $-1 = \delta_1^2 + \cdots + \delta_n^2$. Each δ_λ can be written as

$$\delta_\lambda = \sum_{r=1}^{s_\lambda} \gamma_{\lambda,r} T_{J_r} + \mathfrak{a},$$

where $\gamma_{\lambda,r} \in A_2$, J_r is a (possibly empty) finite subset of J and $T_{J_r} = \prod_{j \in J_r} T_j$ with $T_\emptyset = 1$. The $T_{J'} + \mathfrak{a}$, for J' a finite subset of J, form a basis of A_3 over A_2. The equality $-1 = \delta_1^2 + \cdots + \delta_n^2$ implies

$$-1 = \sum_{\lambda=1}^{n} \sum_{r=1}^{s_\lambda} \gamma_{\lambda,r}^2 \prod_{j \in J_r} \overline{a}_j,$$

where $\gamma_{\lambda,r} = \overline{x}_{\lambda,r}/\overline{\beta}_{\lambda,r}$, with $x_{\lambda,r} \in A$ and $\beta_{\lambda,r} \in M$. Clearing denominators we obtain

$$\overline{m}^2 (\prod_{\lambda,r} \overline{\beta}_{\lambda,r}^2 + \overline{q}) = 0$$

in A_1, for some $m \in M$ and $q \in P$. Set $b = m \prod_{\lambda,r} \beta_{\lambda,r} \in M$, $p = m^2 q \in P$. We have $p + b^2 \in I$, and there exists $c \in I$ such that $p + b^2 + c = 0$.

(iii) $\Rightarrow$ (ii) If there is a homomorphism $\varphi : A \to F$ into a real closed field F, satisfying the conditions of (ii), then φ satisfies $\varphi(p) \geq 0$ since $p \in P$, $\varphi(b) \neq 0$ since $b \in M$, and $\varphi(c) = 0$ since $c \in I$. Hence $\varphi(p + b^2 + c) > 0$, which is impossible. $\square$

Theorem 4.4.2. *Let R be a real closed field. Let $(f_j)_{j=1,\ldots,s}$, $(g_k)_{k=1,\ldots,t}$ and $(h_\ell)_{\ell=1,\ldots,u}$ be finite families of polynomials in $R[X_1,\ldots,X_n]$. Denote by P the cone generated by $(f_j)_{j=1,\ldots,s}$, M the multiplicative monoid generated by $(g_k)_{k=1,\ldots,t}$ and I the ideal generated by $(h_\ell)_{\ell=1,\ldots,u}$. Then the following properties are equivalent:*
  (i) *The set*
$$\{x \in R^n \mid f_j(x) \geq 0,\ j=1,\ldots,s,\quad g_k(x) \neq 0,\ k=1,\ldots,t,$$
$$h_\ell(x) = 0,\ \ell = 1,\ldots,u\}$$
is empty.
  (ii) *There exist $f \in P$, $g \in M$ and $h \in I$ such that $f + g^2 + h = 0$.*

Proof. By Proposition 4.1.1, the negation of the property (i) is equivalent to the existence of a real closed extension F of R and an element y of F^n such that
$$f_j(y) \geq 0 \text{ for } j=1,\ldots,s, \quad g_k(y) \neq 0 \text{ for } k=1,\ldots,t,$$
$$h_\ell(y) = 0 \text{ for } \ell = 1,\ldots,u\ .$$
Hence the property (i) is equivalent to the fact that there is no homomorphism $\varphi : R[X_1,\ldots,X_n] \to F$ into a real closed field F, such that
$$\varphi(f_j) \geq 0 \text{ for } j=1,\ldots,s, \quad \varphi(g_k) \neq 0 \text{ for } k=1,\ldots,t,$$
$$\varphi(h_\ell) = 0 \text{ for } \ell = 1,\ldots,u\ .$$
The equivalence (i) $\Leftrightarrow$ (ii) now follows from Proposition 4.4.1. $\square$

Corollary 4.4.3 (Positivstellensatz). *Let $V \subset R^n$ be an algebraic set, $g_1,\ldots,g_s \in \mathcal{P}(V)$ and*
$$W = \{x \in V \mid g_1(x) \geq 0,\ldots,\ g_s(x) \geq 0\}\ .$$
Let P be the cone of $\mathcal{P}(V)$ generated by $g_1,\ldots,g_s$, and let $f \in \mathcal{P}(V)$. Then:
  (i) $\forall x \in W\ f(x) \geq 0 \Leftrightarrow \exists m \in \mathbb{N}\ \exists g, h \in P\ fg = f^{2m} + h$.
  (ii) $\forall x \in W\ f(x) > 0 \Leftrightarrow \exists g, h \in P\ fg = 1 + h$.
  (iii) $\forall x \in W\ f(x) = 0 \Leftrightarrow \exists m \in \mathbb{N}\ \exists g \in P\ f^{2m} + g = 0$.

Proof. Let $u_1,\ldots,u_k$ generate $\mathcal{I}(V)$. We denote by the same symbol polynomials in $R[X_1,\ldots,X_n]$ and their restrictions to V.

For (i), we apply Theorem 4.4.2 to the set
$$\{x \in R^n \mid g_1(x) \geq 0,\ldots,\ g_s(x) \geq 0,\ -f(x) \geq 0,\ f(x) \neq 0,$$
$$u_1(x) = \ldots = u_k(x) = 0\}\ ,$$
obtaining g and h in P, m in $\mathbb{N}$, such that $h - fg + f^{2m} = 0$.

For (ii), we apply Theorem 4.4.2 to the set

$$\{x \in R^n \mid g_1(x) \geq 0, \ldots, g_s(x) \geq 0, -f(x) \geq 0,$$
$$u_1(x) = \ldots = u_k(x) = 0\},$$

obtaining g and h in P, such that $h - fg + 1 = 0$.

For (iii), we apply Theorem 4.4.2 to the set

$$\{x \in R^n \mid g_1(x) \geq 0, \ldots, g_s(x) \geq 0, f(x) \neq 0,$$
$$u_1(x) = \ldots = u_k(x) = 0\},$$

obtaining g in P and m in $\mathbb{N}$ such that $g + f^{2m} = 0$. □

Corollary 4.4.4. *Let $V, g_1, \ldots, g_s, W, P$ be as in Corollary 4.4.3. Then an ideal I of $\mathcal{P}(V)$ is P-radical if and only if $I = \mathcal{I}_{\mathcal{P}(V)}(W \cap \mathcal{Z}_V(I))$. For every ideal I of $\mathcal{P}(V)$, one has $\mathcal{I}_{\mathcal{P}(V)}(W \cap \mathcal{Z}_V(I)) = \sqrt[P]{I}$.*

Proof. Let $I = (h_1, \ldots, h_t)$ and $\mathcal{I}(V) = (u_1, \ldots, u_k)$, and let $f \in \mathcal{P}(V)$. Theorem 4.4.2 applied to the set

$$\{x \in R^n \mid g_1(x) \geq 0, \ldots, g_s(x) \geq 0, f(x) \neq 0,$$
$$h_1(x) = \ldots = h_t(x) = u_1(x) = \ldots = u_k(x) = 0\}$$

implies

$$f \in \mathcal{I}_{\mathcal{P}(V)}(W \cap \mathcal{Z}_V(I)) \iff \exists g \in P \quad f^{2m} + g \in I \iff f \in \sqrt[P]{I}.$$
□

Corollary 4.4.5. *Let $V \subset R^n$ be an algebraic set. Denote*

$$\Sigma_1 = \{1 + f_1^2 + \cdots + f_p^2 \mid f_1, \ldots, f_p \in \mathcal{P}(V), p \in \mathbb{N}\}.$$

Then $\mathcal{R}(V) = \Sigma_1^{-1}\mathcal{P}(V)$.

Proof. It is enough to prove that, if $g \in \mathcal{P}(V)$ has no zero in V, then g divides an element of Σ_1. Since $g^2 > 0$ on V, by applying Corollary 4.4.3, with $P = \sum \mathcal{P}(V)^2$, we deduce that g^2 divides an element of Σ_1. □

Finally we shall prove the real Nullstellensatz and the Positivstellensatz for regular functions.

Proposition 4.4.6. *Let $V \subset R^n$ be an algebraic set and I an ideal of $\mathcal{R}(V)$. Then $f \in \mathcal{R}(V)$ vanishes on $\mathcal{Z}_V(I)$ if and only if there exist finitely many regular functions $g_1, \ldots, g_p \in \mathcal{R}(V)$ and $m \in \mathbb{N}$, such that*

$$f^{2m} + g_1^2 + \cdots + g_p^2 \in I.$$

In short, $\mathcal{I}_{\mathcal{R}(V)}(\mathcal{Z}_V(I)) = \sqrt[R]{I}$.

Proof. Let $I = (H_1,\ldots,H_q)\mathcal{R}(V)$, with $H_1,\ldots,H_q \in \mathcal{P}(V)$. If $f = P/D$, with $P, D \in \mathcal{P}(V)$ and D has no zero in V, then Corollary 4.1.9 implies that $P^{2m} + Q_1^2 + \cdots + Q_p^2 \in (H_1,\ldots,H_q)\mathcal{P}(V)$, for some $Q_i \in \mathcal{P}(V)$. The result follows by dividing the last relation by D^{2m}. □

Proposition 4.4.7. *Let $V \subset R^n$ be an algebraic set, $g_1,\ldots,g_s \in \mathcal{R}(V)$ and $W = \{x \in V \mid g_1(x) \geq 0, \ldots, g_s(x) \geq 0\}$. Let P be the cone of $\mathcal{R}(V)$ generated by $g_1,\ldots,g_s$ and let $f \in \mathcal{R}(V)$. Then:*
  (i) $\forall x \in W \ \ f(x) \geq 0 \Leftrightarrow \exists m \in \mathbb{N} \ \exists g, h \in P \ \ fg = f^{2m} + h$.
  (ii) $\forall x \in W \ \ f(x) > 0 \Leftrightarrow \exists g, h \in P \ \ fg = 1 + h$.
  (iii) $\forall x \in W \ \ f(x) = 0 \Leftrightarrow \exists m \in \mathbb{N} \ \exists g \in P \ \ f^{2m} + g = 0$.

Proof. We may assume that all g_j belong to $\mathcal{P}(V)$, by replacing $g_j = Q_j/E_j$ with $Q_j E_j = g_j E_j^2 \in \mathcal{P}(V)$. If $f = P/D$, then we apply Corollary 4.4.3 to $PD \in \mathcal{P}(V)$, and we get the desired results for f, by dividing by an appropriate power of the polynomial D. □

Example 4.4.8. We conclude this section with an illustration of the Positivstellensatz (Corollary 4.4.3 (ii)). Let $V = R^2$, $W = \{(x,y) \mid x \geq 0, \ y \geq 0\}$. A polynomial $f \in R[X,Y]$ is positive on W if and only if it can be written in the form
$$f = \frac{1 + p + qX + rY + sXY}{p' + q'X + r'Y + s'XY},$$
where p,q,r,s,p',q',r',s' are sums of squares of polynomials in $R[X,Y]$.

4.5 Real Principal Ideals

The following theorem gives a criterion to decide whether a principal prime ideal of $R[X_1,\ldots,X_n]$ is real.

Theorem 4.5.1. *Let R be a real closed field and f an irreducible polynomial in $R[X_1,\ldots,X_n]$. Then the following properties are equivalent:*
  (i) *The ideal (f) is real.*
  (ii) $(f) = \mathcal{I}(\mathcal{Z}(f))$.
  (iii) *The polynomial f has a nonsingular zero in R^n (i.e. there is an $x \in R^n$ such that $f(x) = 0$ and $\frac{\partial f}{\partial X_i}(x) \neq 0$ for some $i \in \{1,\ldots,n\}$).*
  (iv) *The sign of the polynomial f changes on R^n (i.e. $f(x)f(y) < 0$ for some x, y in R^n).*
  (v) $\dim(\mathcal{Z}(f)) = n - 1$.

Proof. (i) $\Leftrightarrow$ (ii) follows from the real Nullstellensatz 4.1.4.
  (ii) $\Rightarrow$ (iii) By Proposition 3.3.14, $\text{Reg}(\mathcal{Z}(f))$ is nonempty, and (ii) implies that $\text{Reg}(\mathcal{Z}(f))$ coincides with the set of nonsingular zeros of f.
  (iii) $\Rightarrow$ (iv) Let x^0 be a nonsingular zero of f in R^n. Then, for some i, one has $\frac{\partial f}{\partial X_i}(x^0) \neq 0$. The function

$$x_i \mapsto f(x_1^0, \ldots, x_{i-1}^0, x_i, x_{i+1}^0, \ldots, x_n^0)$$

is strictly monotone on an open interval containing x_i^0, and hence the sign of f changes.

(iv) $\Rightarrow$ (v) follows from the next lemma applied to

$$B = R^n, \ U_1 = \{x \in R^n \mid f(x) > 0\} \text{ and } U_2 = \{x \in R^n \mid f(x) < 0\}.$$

Lemma 4.5.2. *Let B be an open ball of R^n (or $B = R^n$) and U_1 and U_2 two disjoint nonempty semi-algebraic open subsets of B. Then*

$$\dim(B \setminus (U_1 \cup U_2)) \geq n - 1.$$

Proof. Let $F = B \setminus (U_1 \cup U_2)$ and $x \in U_1$, $y \in U_2$. Let H be a hyperplane of R^n containing y but not x, H' the hyperplane parallel to H containing x, and $\Pi : R^n \setminus H' \to H$ the projection with centre x onto H. The set $\Pi(F \cap (R^n \setminus H'))$ contains $H \cap U_2$. If $z \in H \cap U_2$, then the intersection of the line $\overline{xz}$ with B is an open interval containing at least one point of F (otherwise it would not be semi-algebraically connected). Since $H \cap U_2$ is a neighbourhood of y in H, we have $\dim(H \cap U_2) = n - 1$ (Proposition 2.8.4). Then $\dim(F) \geq n - 1$ by Theorem 2.8.8. □

(v) $\Rightarrow$ (ii) If $\dim(\mathcal{Z}(f)) = n - 1$, then the ideal $\mathcal{I}(\mathcal{Z}(f))$ has height 1 in $R[X_1, \ldots, X_n]$, and it contains (f) which is prime of height 1. Hence, $(f) = \mathcal{I}(\mathcal{Z}(f))$. □

Bibliographic Notes. We deduce the Artin-Lang homomorphism theorem [15, 209] from the Tarski-Seidenberg principle. The direct proof of this result was a major step in Artin's solution of Hilbert's 17th problem. A weak form of the real Nullstellensatz (giving a necessary and sufficient condition for a family of polynomials in $R[X_1, \ldots, X_n]$ to have no common zero in R^n) first appeared in [194], but this article remained unnoticed by geometers. Dubois [121] gave a version of the Nullstellensatz that involves rational fractions, and the final version (Theorem 4.1.4) was proved by Risler [267].

A version of the Positivstellensatz is given in a paper of Stengle [310] containing Corollary 4.4.3 (i); the notions of a cone and of the P-radical of an ideal can also be found in this paper. The use of prime cones in the proof of the Positivstellensatz goes back to Prestel [258]. Our presentation follows [89]. A constructive proof of the Positivstellensatz has been given recently in [217]. The paper [288] contains a form of the Positivstellensatz valid only for basic compact semi-algebraic subsets of $\mathbb{R}^n$: if polynomials $f, g_1, \ldots, g_s$ are such that $K = \{x \in \mathbb{R}^n \mid g_1(x) \geq 0, \ldots, g_s(x) \geq 0\}$ is compact and $f > 0$ on K, then f belongs to the cone generated by $g_1, \ldots, g_s$.

Theorem 4.5.1 can be found in [123] and the equivalence of (ii) and (iii) in 4.5.1 appears in [233].

Much of the content of this chapter can be found in the paper [206] and also in the book [191].

5. The Tarski-Seidenberg Principle as a Transfer Tool

Abstract. We show in the first section that a semi-algebraic set defined over a real closed field R can be naturally extended to any real closed field K containing R. We then explore further the Tarski-Seidenberg principle. It has already been used in two different ways: to prove that the projection of a semi-algebraic set is semi-algebraic (in Chap. 2), and to obtain the Artin-Lang homomorphism theorem (in Chap. 4). Here we study some additional applications of the Tarski-Seidenberg principle. We use its full strength in the last section in order to establish certain properties of the extension of semi-algebraic sets to a larger real closed field, and to show that semi-algebraic functions can also be extended. The possibility of transfer given by the Tarski-Seidenberg principle will be used in subsequent chapters.

5.1 Extension of Semi-algebraic Sets

Throughout this section R denotes a real closed field and K a real closed extension of R.

Proposition 5.1.1. *Let $S \subset R^n$ be a semi-algebraic set given by a boolean combination $\mathcal{B}(X)$ of sign conditions on polynomials in $R[X]$, where $X = (X_1, \ldots, X_n)$. The subset $\{x \in K^n \mid \mathcal{B}(x)\}$ of K^n, denoted S_K, is semi-algebraic and depends only on the set S and not on the boolean combination chosen to describe it.*

Proof. Let $\mathcal{B}'(X)$ be another boolean combination of sign conditions on polynomials in $R[X]$ describing S:
$$S = \{x \in R^n \mid \mathcal{B}(X)\} = \{x \in R^n \mid \mathcal{B}'(X)\}\,.$$

This is equivalent to the fact that the boolean combination: "($\mathcal{B}$ and not $\mathcal{B}'$) or ($\mathcal{B}'$ and not $\mathcal{B}$)" has no solution in R^n. But then, by Proposition 4.1.1 (which is equivalent to the Artin-Lang homomorphism theorem), this boolean combination has no solution in K^n either. This means that
$$\{x \in K^n \mid \mathcal{B}(x)\} = \{x \in K^n \mid \mathcal{B}'(x)\}\,.$$

Hence S_K is well defined, independently of the choice of $\mathcal{B}$. □

Definition 5.1.2. *The set S_K defined in the preceding proposition is called the extension of S to K.*

Proposition 5.1.3. *The mapping $S \mapsto S_K$ preserves the boolean operations (finite intersection, finite union and taking complement).* □

Proposition 5.1.4. *If $S \subset T$, then $S_K \subset T_K$. Hence $S = \emptyset$ implies $S_K = \emptyset$, and $S = T$ implies $S_K = T_K$.*

Proof. Suppose S and T are given by the boolean combinations $\mathcal{B}(X)$ and $\mathcal{C}(X)$, respectively. The inclusion $S \subset T$ means that the combination "$\mathcal{B}$ and not $\mathcal{C}$" has no solution in R, and, hence, by Proposition 4.1.1, no solution in K either. This shows that $S_K \subset T_K$. □

Remark 5.1.5. Of course, $S_K \cap R^n = S$. But S_K is not the only semi-algebraic subset of K^n with this property: if $S = [0,4] \subset \mathbb{R}_{\mathrm{alg}}$ (the real algebraic numbers), then $S_\mathbb{R} = [0,4] \subset \mathbb{R}$, but also $([0, \pi[\cup]\pi, 4]) \cap \mathbb{R}_{\mathrm{alg}} = S$, where $\pi = 3.14\ldots$.

5.2 The Full Strength of the Tarski-Seidenberg Principle

Until now, we have used the Tarski-Seidenberg principle (Theorem 1.4.2) in two different ways:

- In Chap. 2, with a fixed real closed field R, to prove that the projection of a semi-algebraic set is semi-algebraic.
- In Chap. 4, to prove the Artin-Lang homomorphism theorem, that is, to prove that a system of polynomial equations and inequalities over a real closed field R has a solution in R if and only if it has a solution in a real closed extension of R.

We have just seen how to extend a semi-algebraic set, by using the Artin-Lang homomorphism theorem. But this tool alone is not sufficient to extend semi-algebraic functions. In order to prove that, if $G \subset R^{n+1}$ is the graph of a semi-algebraic function, then G_K is still the graph of a function, we need a more general transfer principle than the Artin-Lang homomorphism theorem. This transfer principle will also be obtained as a consequence of the Tarski-Seidenberg principle.

We reformulate the Tarski-Seidenberg principle geometrically (compare with Theorem 2.2.1).

Proposition 5.2.1. *Let R be a real closed field and $S \subset R^{n+1}$ a semi-algebraic set. Denote by $\Pi : R^{n+1} \to R^n$ the projection onto the space of the first n coordinates. Then $\Pi(S)$ is semi-algebraic. Moreover, if K is an arbitrary real closed extension of R and $\Pi_K : K^{n+1} \to K^n$ is the projection onto the space of the first n coordinates, then $\Pi_K(S_K) = (\Pi(S))_K$.*

5.2 The Full Strength of the Tarski-Seidenberg Principle   99

Proof. We use Corollary 1.4.7 again. The set S is given by a boolean combination $\mathcal{B}(X,Y)$ of sign conditions on polynomials in $R[X,Y]$, where $X = (X_1, \ldots, X_n)$. By Corollary 1.4.7, there exists a boolean combination $\mathcal{C}(X)$ of sign conditions on polynomials in $R[X]$, such that, for every real closed extension K of R and for every x in K^n, $\mathcal{B}(x,Y)$ has a solution in K if and only if $\mathcal{C}(x)$ holds true in K. So
$$\Pi(S) = \{x \in R^n \mid \mathcal{C}(x)\}$$
and
$$\Pi_K(S_K) = \{x \in K^n \mid \mathcal{C}(x)\} = (\Pi(S))_K \ .$$
□

We return to the logical terminology introduced in Definition 2.2.3. Denote by $\mathcal{L}(R)$ the first-order language of ordered fields with parameters in R. A quantifier-free formula of $\mathcal{L}(R)$ is, of course, a formula in which no variable is quantified either by $\forall$ or by $\exists$; it is a boolean combination of sign conditions on polynomials with coefficients in R.

For every semi-algebraic subset A of R^k, there exists a quantifier-free formula θ of $\mathcal{L}(R)$ such that $A = \{x \in R^k \mid \theta(x)\}$. By Proposition 5.1.1, $A_K = \{x \in K^k \mid \theta(x)\}$ for every real closed extension K of R.

Proposition 5.2.2 (Quantifier Elimination). *Let Φ be a formula of $\mathcal{L}(R)$. Then there exists a quantifier-free formula Ψ of $\mathcal{L}(R)$, with the same free variables $x_1, \ldots, x_n$ as Φ, such that for every real closed extension K of R and every x in K^n, $\Phi(x) \Leftrightarrow \Psi(x)$.*

Proof. We proceed by induction on the construction of the formula Φ, exactly as in Proposition 2.2.4. The cases of conjunction, disjunction and negation are obvious. If $\Phi(x)$ has the form $\exists y \ \theta(x,y)$, then
$$\{x \in K^n \mid \exists y \ \theta(x,y)\} = \Pi_K(\{(x,y) \in K^{n+1} \mid \theta(x,y)\}) \ ,$$
and we can apply Proposition 5.2.1. The universal quantification can be reduced to the existential one, since "$\forall y \ldots$" is equivalent to "not $\exists y$ not $\ldots$".
□

Proposition 5.2.3 (Transfer Principle). *Let Φ be a formula of $\mathcal{L}(R)$ without a free variable, and K, a real closed extension of R. Then Φ holds true in R if and only if it holds true in K.*

Proof. Proposition 5.2.2 gives a quantifier-free formula without a free variable (thus, only with parameters in R), and then the result is clear.
□

Propositions 5.2.2 and 5.2.3 are important properties from the point of view of model theory. Proposition 5.2.2 says that the theory of real closed fields admits quantifier elimination in the language of ordered fields, and Proposition 5.2.3 says that the theory of real closed fields is model-complete. We shall use them in the following form.

Corollary 5.2.4. *Let R be a real closed field and K a real closed extension of R. Let (P) be a property whose formulation is constructed from expressions "$x \in A_i$", where A_i is a semi-algebraic subset of some R^{q_i}, using a finite number of conjunctions, disjunctions and negations and a finite number of universal and existential quantifications over variables ranging over semi-algebraic subsets B_j of R^{q_j}. Let (P_K) be the property obtained by replacing the semi-algebraic sets A_i and B_j occurring in (P) with their extensions $(A_i)_K$ and $(B_j)_K$ to K.*

Then the property (P) holds true in R if and only if (P_K) holds true in K.

Proof. Let θ_{A_i} and θ_{B_j} be formulas of $\mathcal{L}(R)$ such that

$$A_i = \{x \in R^{q_i} \mid \theta_{A_i}(x)\}$$
$$B_j = \{x \in R^{q_j} \mid \theta_{B_j}(x)\}.$$

Replace in (P)
   each expression $x \in A_i$ with $\theta_{A_i}(x)$,
   each quantification $\exists x \in B_j \ldots$ with $\exists x \, (\theta_{B_j}(x) \text{ and } \ldots)$,
   each quantification $\forall x \in B_j \ldots$ with $\forall x \, (\theta_{B_j}(x) \Rightarrow \ldots)$.
This defines a formula Φ of $\mathcal{L}(R)$.

The property (P) holds true in R if and only if Φ holds true in R, and, hence, by Proposition 5.2.3, if and only if Φ holds true in K, that is, if and only if (P_K) holds true in K. □

5.3 Further Results on Extension of Semi-algebraic Sets and Mappings

Throughout this section, R is a real closed field, $A \subset R^m$ and $B \subset R^n$ are semi-algebraic sets, and $f : A \to B$ is a semi-algebraic mapping whose graph is $G \subset A \times B$.

Proposition 5.3.1. *Let K be a real closed extension of R. Then G_K is the graph of a semi-algebraic mapping $f_K : A_K \to B_K$.*

Proof. The fact that G is the graph of a mapping from A to B can be expressed by the following property (P):

$$\forall x \in R^m \; \Big[\; (x \in A \Leftrightarrow \exists y \in R^n \; (x,y) \in G) \text{ and}$$
$$(\forall y \in R^n \; (x,y) \in G \Rightarrow y \in B) \text{ and}$$
$$(\forall y \in R^n \; \forall y' \in R^n \; (x,y) \in G \text{ and } (x,y') \in G \Rightarrow y = y') \; \Big].$$

We can apply Corollary 5.2.4 to the property (P_K) expressing the fact that G_K is the graph of a mapping from A_K to B_K. □

5.3 Further Results on Extension of Semi-algebraic Sets and Mappings

Definition 5.3.2. *The semi-algebraic mapping f_K of the previous proposition is called the* extension of f to K.

Proposition 5.3.3. *The mapping f is injective (resp. surjective, resp. bijective) if and only if f_K is injective (resp. surjective, resp. bijective).*

Proof. We note that the mapping f is injective if and only if G satisfies the property
$$\forall x \in A \ \forall x' \in A \ \forall y \in B \ \big((x,y) \in G \text{ and } (x',y) \in G\big) \Rightarrow x = x',$$
and we apply Corollary 5.2.4. The other cases can be treated in the same way. $\square$

Proposition 5.3.4. *Let C (resp. D) be a semi-algebraic subset of A (resp. B). Then*
$$(f(C))_K = f_K(C_K) \quad \text{and} \quad (f^{-1}(D))_K = f_K^{-1}(D_K) \, .$$

Proof. We observe that $f(C)$ is the projection of $G \cap (C \times R^n)$ onto R^n and use Proposition 5.2.1. We prove the statement about $f^{-1}(D)$ similarly. $\square$

The extension behaves well with respect to topology.

Proposition 5.3.5.
(i) *The semi-algebraic set A is open (resp. closed) in R^n if and only if A_K is open (resp. closed) in K^n. More generally, $\operatorname{clos}(A_K) = (\operatorname{clos}(A))_K$.*
(ii) *The semi-algebraic mapping f is continuous if and only if f_K is continuous.*

Proof. (i) We can describe $\operatorname{clos}(A)$ by means of projections (Proposition 2.2.2), or formulas of $\mathcal{L}(R)$, and then use Proposition 5.2.1 or Proposition 5.2.2.
(ii) The continuity of f is equivalent to the following property of its graph G:
$$\forall x \in A \ \forall \epsilon > 0 \ \exists \eta > 0 \ \forall x' \in A \ \ \|x - x'\|^2 < \eta \Rightarrow$$
$$\big(\forall y \in B \ \forall y' \in B \ (x,y) \in G \text{ and } (x',y') \in G \Rightarrow \|y - y'\|^2 < \epsilon\big) \, .$$
We obtain (ii) by applying Corollary 5.2.4. $\square$

Proposition 5.3.6.
(i) *The semi-algebraic set A is closed and bounded if and only if A_K is closed and bounded.*
(ii) *The semi-algebraic set A is semi-algebraically connected if and only if A_K is semi-algebraically connected. More generally, if $C_1, \ldots, C_l$ are the semi-algebraically connected components of A, then $(C_1)_K, \ldots, (C_l)_K$ are the semi-algebraically connected components of A_K.*

102   5. The Tarski-Seidenberg Principle as a Transfer Tool

Proof. (i) It remains to verify that A_K is bounded if and only if A bounded. This is clear, using Corollary 5.2.4, since the fact that A is bounded can be expressed by
$$\exists m \in R \quad \forall x \in A \quad \|x\|^2 \leq m \ .$$

(ii) Theorem 2.3.6 gives a decomposition $A = \bigcup_{i=1}^m A_i$, such that, for each i, there is a semi-algebraic homeomorphism $\varphi_i :]0,1[^{d_i} \to A_i$. By extending this to K, we get a decomposition $A_K = \bigcup_{i=1}^m (A_i)_K$, and semi-algebraic homeomorphisms $(\varphi_i)_K : (]0,1[_K)^{d_i} \to (A_i)_K$. We used such a decomposition in the proof of Theorem 2.4.4 to characterize semi-algebraically connected components of A_K. This characterization implies (ii). □

Remark 5.3.7. In this section we have given several properties of extension. The general idea can be summarized as follows (Proposition 5.2.3): "Every property that can be expressed in the first-order language of ordered fields with parameters in R can be transferred to any real closed extension of R".

Bibliographic Notes. The reader is invited to consult the references given in the bibliographic notes of Chapter 1 or 6 for the Tarski-Seidenberg principle and the model-theoretic notions related to it (quantifier elimination, model-completeness). See [22, 138, 151, 266] for information on recent works on the algorithmic complexity of quantifier elimination for real closed fields and related problems.

6. Hilbert's 17th Problem.
Quadratic Forms in Real Algebraic Geometry

Abstract. Hilbert's 17th problem has played a major role in the development of real algebraic geometry. We first prove the following result: a nonnegative polynomial over a real closed field is a sum of squares of rational functions. We also prove a few generalizations. The second section deals with the equivariant version of Hilbert's 17th problem. In the third section, we address the problem of representing nonnegative polynomials as sum of squares of polynomials. A theorem of Hilbert describes exactly the couples (n, m), for which every nonnegative form of degree m in n variables is a sum of squares of forms. The fourth section studies the quantitative aspects of the problem, namely, the number of squares needed. The theory of quadratic forms, especially Pfister forms, plays an important role in the solution of this question. The use of Pfister forms is also crucial in the proof, given in Section 5, of a striking result of Bröcker and Scheiderer which says that a basic open semi-algebraic subset of an algebraic set of dimension $d > 0$ can always be defined by d inequalities.

We conclude with more extensive bibliographic and historical notes than in the other chapters, due to the importance of Hilbert's 17th problem in the development of real algebraic geometry.

6.1 Solution of Hilbert's 17th Problem

The 17th problem stated by Hilbert is the question whether every polynomial nonnegative on $\mathbb{R}^n$ is a sum of squares of rational functions. This problem was solved by Emil Artin.

Theorem 6.1.1. *Let R be a real closed field and $f \in R[X_1, \ldots, X_n]$. If f is nonnegative on R^n, then f is a sum of squares in the field of rational functions $R(X_1, \ldots, X_n)$.*

Proof. If f is not a sum of squares in $R(X_1, \ldots, X_n)$, then, by Corollary 1.1.11, there exists an ordering on $R(X_1, \ldots, X_n)$ for which f is negative. Let K be a real closure of $R(X_1, \ldots, X_n)$ equipped with this ordering. Then $-f$ has a nonzero square root in K, and hence there is an R-algebra homomorphism $R[X_1, \ldots, X_n][T]/(fT^2 + 1) \to K$. By the Artin-Lang homomorphism theorem 4.1.2, there is an R-algebra homomorphism

$R[X_1,\ldots,X_n][T]/(fT^2+1) \to R$. This implies the existence of a point x in R^n such that $f(x) < 0$. □

Remark 6.1.2. The Positivstellensatz (Corollary 4.4.3 (i)) gives more precise information: if f is nonnegative on R^n, then $fg = f^{2m} + h$, where h and g are sums of squares of polynomials ($h, g \in \sum R[X_1,\ldots,X_n]^2$). This implies that $f(f^{2m}+h) = f^2 g \in \sum R[X_1,\ldots,X_n]^2$, and, thus, $f = g_1/(f^{2m}+h)$ with g_1 and h belonging to $\sum R[X_1,\ldots,X_n]^2$. The fraction $f = g_1(f^{2m}+h)/(f^{2m}+h)^2$ is a sum of squares in $R(X_1,\ldots,X_n)$. Hence, we know in addition that f can be represented as a sum of squares of rational functions which are defined outside the set of zeros of f.

Theorem 6.1.3. *Let $(F, \leq)$ be an ordered field and R its real closure. Let $f \in F[X_1,\ldots,X_n]$. If f is nonnegative on R^n, then there exist positive elements $a_1,\ldots,a_p$ in F and rational funtions $g_1,\ldots,g_p$ in $F(X_1,\ldots,X_n)$, such that $f = a_1 g_1^2 + \cdots + a_p g_p^2$ in $F(X_1,\ldots,X_n)$.*

Proof. We need the following lemma.

Lemma 6.1.4. *Let $(F, \leq)$ be an ordered field and F_1 a field extension of F. The intersection of the positive cones of orderings of F_1 extending the ordering of F is the set Λ of elements of F_1 of the form $a_1 g_1^2 + \cdots + a_p g_p^2$, where $a_1,\ldots,a_p$ are positive elements of F and $g_1,\ldots,g_p$ are arbitrary elements of F_1. If F_1 has no ordering extending the ordering of F, then $\Lambda = F_1$.*

Proof. The set Λ is the cone of F_1 generated by the positive cone of F. An ordering of F_1 extends the ordering of F if and only if its positive cone contains Λ. The conclusion follows by applying Proposition 1.1.10. □

We now return to the proof of Theorem 6.1.3: if f is not of the form $a_1 g_1^2 + \cdots + a_p g_p^2$ as in the statement, then, by Lemma 6.1.4, there exists an ordering of $F(X_1,\ldots,X_n)$, extending the ordering of F, for which f is negative. Let K be the real closure of $F(X_1,\ldots,X_n)$ with respect to this ordering. The field K is an extension of R, and there is an R-algebra homomorphism $R[X_1,\ldots,X_n][T]/(fT^2+1) \to K$. This implies that there exists a point $x \in R^n$ such that $f(x) < 0$ (as in the proof of Theorem 6.1.1). □

Remark 6.1.5. Lemma 6.1.4 gives the following criterion, sometimes called *Serre's criterion*: if F is an ordered field and F_1 is a field extension of F, then F_1 has an ordering extending the ordering of F if and only if -1 cannot be represented as $a_1 g_1^2 + \cdots + a_p g_p^2$, with $a_1,\ldots,a_p$ positive elements of F and $g_1,\ldots,g_p$ elements of F_1.

Corollary 6.1.6. *Let F be a subfield of $\mathbb{R}$, ordered by the restriction of the ordering of $\mathbb{R}$, and $f \in F[X_1,\ldots,X_n]$. If f is nonnegative on F^n, then f can be represented as $f = a_1 g_1^2 + \cdots + a_p g_p^2$, where $a_1,\ldots,a_p$ are positive elements of F and $g_1,\ldots,g_p$ are elements of $F(X_1,\ldots,X_n)$.*

6.1 Solution of Hilbert's 17th Problem

Proof. Since F contains $\mathbb{Q}$, F^n is dense in R^n with respect to the euclidean topology, and, thus, if f is nonnegative on F^n, it is nonnegative on R^n and, a fortiori, on R^n, where R is the real closure of F. We use Theorem 6.1.3 to finish the proof. □

Corollary 6.1.6 uses the fact that F is dense in R. The following examples show that the property of being dense in the real closure is necessary.

Example 6.1.7. a) Let $F = \mathbb{R}(t)$, with the ordering 0_+ (cf. Example 1.1.2). The real closure of F is the field $R = \mathbb{R}(t)^\wedge_{\text{alg}}$ of real algebraic Puiseux series (Example 1.3.6). Let $f(X) = (X^2 - t)^2 - t^3 \in F[X]$; f is positive at every point z in F since the first term of the expansion of $f(z) \in \mathbb{R}(t)$ as a formal series in t is a positive constant or t^2. However, f is not a sum of squares, even in the field $R(X)$. Indeed, f is negative on the union of the intervals I and $-I$, where

$$I = \left]\sqrt{t(1-\sqrt{t})}, \sqrt{t(1+\sqrt{t})}\right[.$$

Note that I contains no point of F.

b) Now, let F be the smallest subfield of $R = \mathbb{R}(t)^\wedge_{\text{alg}}$ containing $\mathbb{R}(t)$ and closed under the operation of taking square roots of positive elements. The field F has only one ordering, whose positive cone is the set of squares, and its real closure is R. Take $f(X) = (X^3 - t)^2 - t^3 \in F[X]$. Since $f(t^{1/3}) < 0$, f cannot be a sum of squares in $R(X)$. However, $f(z)$ is positive for every z in F. We have $f(0) > 0$. If $z \neq 0$ and if the first term of its Puiseux series expansion is $at^{p/2^n}$, with $p \in \mathbb{Z}$, then the first term of $f(z)$ is t^2 if $p > 2^n/3$, and $a^2 t^{3p/2^{n-1}}$ if $p < 2^n/3$; in both cases $f(z) > 0$.

E. Artin also considered Hilbert's 17th problem for an irreducible algebraic subset V of R^n, instead of R^n. This is different from the case of affine space: a polynomial that is a sum of squares in $R(X_1, \ldots, X_n)$ is clearly nonnegative on R^n, but an element of $\mathcal{P}(V)$ that is a sum of squares in $\mathcal{K}(V)$ (the field of fractions of $\mathcal{P}(V)$) is not necessarily nonnegative everywhere on V.

Example 6.1.8. Let V be the Cartan umbrella (Example 3.1.2 d)) in R^3, given by the equation $x^3 = z(x^2 + y^2)$. Then $f = x^2 + y^2 - z^2 \in \mathcal{P}(V)$ is negative on the stick $x = y = 0$ outside the origin. Nevertheless, f is a sum of squares in $\mathcal{K}(V)$:

$$f = x^2 + y^2 - \frac{x^6}{(x^2+y^2)^2} = \frac{3x^4y^2 + 3x^2y^4 + y^6}{(x^2+y^2)^2}.$$

Theorem 6.1.9. *Let R be a real closed field, $V \subset R^n$ an irreducible algebraic set of dimension d and $f \in \mathcal{P}(V)$. Then the following properties are equivalent:*

(i) *f is a sum of squares in $\mathcal{K}(V)$.*
(ii) *f is nonnegative on $V^{(d)} = \{x \in V \mid \dim(V_x) = d\}$.*
(iii) *f is nonnegative on $\text{Reg}(V)$ (the set of nonsingular points of V).*
(iv) *f is nonnegative on some nonempty Zariski open subset of V.*

Proof. (i) ⇒ (ii) Suppose that $f = (g_1/h_1)^2 + \cdots + (g_p/h_p)^2$, with $g_1,\ldots,g_p$, $h_1,\ldots,h_p$ in $\mathcal{P}(V)$ and $h_1,\ldots,h_p$ different from zero. Then $Z = \mathcal{Z}_V(h_1,\ldots,h_p)$ is a proper algebraic subset of V, and, hence, the dimension of Z is less than d. It is clear that, if $x \in \text{clos}(V \setminus Z)$, then $f(x) \geq 0$. On the other hand, if $\dim(V_x) = d$, then no euclidean neighbourhood of x in V can be contained in Z, and, hence, $x \in \text{clos}(V \setminus Z)$. We have proved (ii).

(ii) ⇒ (iii) We have $\text{Reg}(V) \subset V^{(d)}$ by Proposition 3.3.11.

(iii) ⇒ (iv) The set $\text{Reg}(V)$ is a nonempty Zariski open subset of V, by Proposition 3.3.14.

(iv) ⇒ (i) Suppose that $f \geq 0$ on the Zariski open subset $U \neq \emptyset$, and let $Z = V \setminus U$. Let $h \in \mathcal{P}(V)$ be such that $Z = h^{-1}(0)$. Clearly h is different from zero in $\mathcal{P}(V)$. Suppose f is not a sum of squares in $\mathcal{K}(V)$. Then there exists an ordering of $\mathcal{K}(V)$ for which f is negative. Let K be the real closure of $\mathcal{K}(V)$ for this ordering. There is a homomorphism of R-algebras

$$(\mathcal{P}(V)_h)[T]/(fT^2+1) \to K$$

and, hence, by the Artin-Lang homomorphism theorem, there is a point x in V, such that $h(x) \neq 0$ and $f(x) < 0$. This is a contradiction. □

Remark 6.1.10. The statements of Theorems 6.1.1, 6.1.3 and 6.1.9 are still valid, if f is a rational function and if the condition "f is nonnegative on a set S" is replaced with "$f(x) \geq 0$ at every point x in S where f is defined". Indeed, if $f = g/h$, where g and h are polynomials, then $f = gh/h^2$, and $gh(x) < 0$ if and only if f is defined at x and $f(x) < 0$.

6.2 The Equivariant Version of Hilbert's 17$^\text{th}$ Problem

We shall see, in this section, that a symmetric polynomial nonnegative on R^n can be represented as $\sum_{i=1}^r s_i \delta_i$, where the s_i are sums of squares of symmetric rational functions, and δ_i belong to a finite list of (explicitly given) symmetric polynomials nonnegative on R^n. The key point is the fact that the set of $(a_1,\ldots,a_n)$ in R^n, such that the polynomial $X^n - a_1 X^{n-1} + \cdots + (-1)^n a_n$ has its n roots in R, is generically basic closed ("generically" will shortly be defined). First we give a representation for polynomials nonnegative on a generically basic closed semi-algebraic set.

The Positivstellensatz (Corollary 4.4.3) implies, in particular, that, if V is an irreducible algebraic set and W is a basic closed semi-algebraic subset of V,

$$W = \{x \in V \mid g_1(x) \geq 0, \ldots, g_k(x) \geq 0\},$$

with $g_1,\ldots,g_k \in \mathcal{P}(V)$, then there exist $f_1,\ldots,f_r \in \mathcal{P}(V)$, all nonnegative on W, such that every polynomial $f \in \mathcal{P}(V)$ nonnegative on W can be represented as $f = \sum_{i=1}^r s_i f_i$, with $s_i \in \sum \mathcal{K}(V)^2$. Explicitly, we can take for f_i the products $\prod_{j=1}^k g_j^{\epsilon_j}$, where $\epsilon_j = 0$ or 1.

6.2 The Equivariant Version of Hilbert's 17th Problem

The conclusion is no longer true, if we replace W with an arbitrary closed semi-algebraic set, as we shall soon see. The property of being basic – at least generically – is crucial here.

Definition 6.2.1. *Two semi-algebraic subset S and T of an irreducible algebraic set V are* generically equal *in V if*

$$\dim\left((S \setminus T) \cup (T \setminus S)\right) < \dim(V).$$

A semi-algebraic subset of V is generically basic *in V if it is generically equal to a basic closed semi-algebraic set.*

Remark 6.2.2. The generic equality is an equivalence relation on semi-algebraic subsets of V. Observe that every semi-algebraic subset of V, of dimension less than $\dim(V)$, is generically equal to the empty set. In the definition of "generically basic" we could have replaced "basic closed" with "basic open", since a basic open semi-algebraic subset of V is generically equal to the basic closed semi-algebraic subset obtained by relaxing the strict inequalities (if there is no inequality $0 > 0$, which we may always assume).

Theorem 6.2.3. *Let V be an irreducible algebraic set and T a semi-algebraic subset of V. Then the following properties are equivalent:*
  (i) *The set T is generically basic in V.*
  (ii) *There exist a finite number of polynomials $f_1, \ldots, f_r \in \mathcal{P}(V)$, all non-negative on T, such that every polynomial $f \in \mathcal{P}(V)$ nonnegative on T can be represented as $f = \sum_{i=1}^{r} s_i f_i$, with $s_i \in \sum \mathcal{K}(V)^2$.*

Proof. (i) $\Rightarrow$ (ii) We assume that T is generically equal to the basic closed semi-algebraic subset W of V. Choose $h \in \mathcal{P}(V)$, $h \neq 0$, such that $W \setminus T \subset \mathcal{Z}_V(h)$; it is possible to find such an h, since $\dim(W \setminus T) < \dim(V)$. Let $f \in \mathcal{P}(V)$ be nonnegative on T. Then fh^2 is nonnegative on W. The conclusion concerning f follows from the observation made earlier in this section.

(ii) $\Rightarrow$ (i) Let $W = \{x \in V \mid f_1(x) \geq 0, \ldots, f_r(x) \geq 0\}$. Of course, $W \supset T$. By Proposition 2.8.13, in order to prove that $\dim(W \setminus T) < \dim(V)$, it is enough to prove that $\dim(W \setminus \operatorname{clos}(T)) < \dim(V)$. Let $y \in W \setminus \operatorname{clos}(T)$. There exists $\epsilon > 0$ such that $f(x) = \|x - y\|^2 - \epsilon^2$ is nonnegative on T, and, by assumption (ii), we have $f = \sum_{i=1}^{r} s_i f_i$, with $s_i \in \sum \mathcal{K}(V)^2$. Since f is negative on $B = \{x \in W \mid \|x - y\| < \epsilon\}$, the product of the denominators of the s_i's is identically zero on B, and, hence, $\dim(W_y) < \dim(V)$. $\square$

The previous theorem is useful in studying Hilbert's 17th problem for subalgebras of $\mathcal{P}(V)$.

Corollary 6.2.4. *Let V be an irreducible algebraic set, A an R-subalgebra of $\mathcal{P}(V)$ and $\operatorname{Fr}(A)$ its field of fractions. Assume that A has a finite number of generators $q_1, \ldots, q_k$ (as an R-algebra). Define $q = (q_1, \ldots, q_k) : V \to R^k$*

and Z the Zariski-closure of $q(V)$ in R^k. Then the following properties are equivalent:
  (i) The set $q(V)$ is generically basic in Z.
  (ii) There exist a finite number of elements $f_1, \ldots, f_r$ of A, all nonnegative on V, such that every f in A that is nonnegative on V can be represented as $f = \sum_{i=1}^r s_i f_i$, where $s_i \in \sum \mathrm{Fr}(A)^2$.

Proof. Observe that Z is irreducible and the mapping $\varphi : \mathcal{P}(Z) \to A$, $\varphi(g) = g \circ q$, is an R-algebra isomorphism. Moreover, an element $g \in \mathcal{P}(Z)$ is nonnegative on $T = q(V)$ if and only if $g \circ q$ is nonnegative on V. The corollary follows, by applying Theorem 6.2.3 to the couple (Z, T). □

Example 6.2.5. The condition (i) of Corollary 6.2.4 is not always satisfied. For instance, if $q : R^2 \to R^2$ is defined by

$$q(x, y) = (x^6 - 3x^2 y^4, 3x^4 y^2 - y^6)$$

(or $q(x + iy) = (x^2 + iy^2)^3$ in complex notation), then

$$q(R^2) = \{(x, y) \in R^2 \mid x \leq 0 \text{ or } y \geq 0\},$$

and this set is not generically basic in R^2. Indeed, since $q(R^2)$ lies on one side of the positive x-axis, if $q(R^2)$ were generically equal to

$$\{x \in R^2 \mid f_1(x) \geq 0, \ldots, f_r(x) \geq 0\},$$

then one of the f_i should be $x^{2l+1} g$ with g not divisible by x. This is impossible since $q(R^2)$ lies on both sides of the negative x-axis.

Now we shall study the symmetric version of Hilbert's 17th problem. We consider the subalgebra A of symmetric polynomials in $R[X_1, \ldots, X_n]$. It is well known that $A = R[\sigma_1, \ldots, \sigma_n]$, where the σ_i are the elementary symmetric polynomials in $X_1, \ldots, X_n$, and that the σ_i, $i = 1, \ldots, n$, are algebraically independent over R. Define $\sigma = (\sigma_1, \ldots, \sigma_n) : R^n \to R^n$. Then $\sigma(R^n)$ is the set of $(a_1, \ldots, a_n)$ in R^n such that the polynomial

$$X^n - a_1 X^{n-1} + \cdots + (-1)^n a_n$$

has its n roots in R (i.e., it decomposes into a product of factors of degree 1 over R). We want to show that this set $\sigma(R^n)$ is generically basic. Actually, we shall show that it is even basic closed.

Using Remark 1.2.13 it is not difficult to see that $\sigma(R^n)$ is generically basic. We compute the Sturm sequence of

$$f = X^n - a_1 X^{n-1} + \cdots + (-1)^n a_n$$

and f' considered as polynomials with coefficients in the field of rational functions in the indeterminates $a_1, \ldots, a_n$. The leading coefficients of the polynomials in this sequence are rational functions in $a_1, \ldots, a_n$. Multiplying them by the squares of their denominators, we get polynomials in $a_1,\ldots,a_n$. The set of points where these polynomials are positive is generically equal to $\sigma(R^n)$.

We can give more explicitly a system of inequalities generically defining $\sigma(R^n)$. We first describe a method for counting real roots, which is different from Sturm's sequence, and is due to Hermite [152] and Sylvester [313].

Let $f = X^n - a_1 X^{n-1} + \cdots + (-1)^n a_n$ be a monic polynomial in $R[X]$, and let $x_1,\ldots,x_n$ be the n roots of f in an algebraic closure of R. For every $i \in \mathbb{N}$, define $N_i = \sum_{j=1}^n x_j^i$ (these N_i are usually called the Newton sums of the roots of f). Since N_i is a symmetric polynomial in $x_1,\ldots,x_n$ with coefficients in $\mathbb{Z}$, it can be expressed as a polynomial in $a_1,\ldots,a_n$ with coefficients in $\mathbb{Z}$.

Denote by $\mathcal{H}(a_1,\ldots,a_n) = (b_{i,j})$ the symmetric $n \times n$ matrix whose entries $b_{i,j} = N_{i+j-2}$, $i,j = 1,\ldots,n$.

Proposition 6.2.6. *Let $f = X^n - a_1 X^{n-1} + \cdots + (-1)^n a_n \in R[X]$. Then:*

(i) The signature of the quadratic form with matrix $\mathcal{H}(a_1,\ldots,a_n)$ is equal to the number of distinct roots of f in R.

(ii) The rank of $\mathcal{H}(a_1,\ldots,a_n)$ is equal to the total number of distinct roots of f (in an algebraic closure of R).

Proof. Let Q be the quadratic form with matrix $\mathcal{H}(a_1,\ldots,a_n)$. An easy computation shows that

$$Q(Y_1,\ldots,Y_n) = (Y_1 + x_1 Y_2 + \cdots + x_1^{n-1} Y_n)^2 + \cdots + (Y_1 + x_n Y_2 + \cdots + x_n^{n-1} Y_n)^2.$$

Moreover, if x_i is a root of f in $R[\sqrt{-1}] \setminus R$, then its conjugate $\bar{x}_i$ is also a root of f, and

$$\Big(\sum_{j=1}^n x_i^{j-1} Y_j\Big)^2 + \Big(\sum_{j=1}^n \bar{x}_i^{j-1} Y_j\Big)^2 = 2\Big(\sum_{j=1}^n \operatorname{Re}(x_i^{j-1}) Y_j\Big)^2 - 2\Big(\sum_{j=1}^n \operatorname{Im}(x_i^{j-1}) Y_j\Big)^2.$$

The rank of Q is the rank of the Vandermonde matrix of the x_i's, i.e. the number of distinct roots of f in $R[\sqrt{-1}]$. Since the pairs of conjugate roots in $R[\sqrt{-1}] \setminus R$ contribute 0 to the signature, this signature is equal to the number of distinct roots of f in R. $\square$

Corollary 6.2.7. *The set $\sigma(R^n)$ consists of all $(a_1,\ldots,a_n)$ in R^n such that the quadratic form with matrix $\mathcal{H}(a_1,\ldots,a_n)$ is positive semi-definite on R^n (i.e., it takes only nonnegative values).*

Let $\Delta_1,\ldots,\Delta_n$ be the principal minors of the matrix $\mathcal{H}(a_1,\ldots,a_n)$ (i.e., Δ_k is the determinant of the submatrix obtained by taking the first k rows and the first k columns of $\mathcal{H}(a_1,\ldots,a_n)$).

Lemma 6.2.8. *The set $\sigma(R^n)$ is generically equal to the basic closed semi-algebraic subset of R^n defined by the $n-1$ inequalities $\Delta_2 \geq 0,\ldots,\Delta_n \geq 0$.*

6. Hilbert's 17th Problem. Quadratic Forms

Proof. If $\mathcal{H}(a_1, \ldots, a_n)$ is positive semi-definite, then $\Delta_1 \geq 0, \ldots, \Delta_n \geq 0$. Moreover, if $\Delta_1 > 0, \ldots, \Delta_n > 0$, then $\mathcal{H}(a_1, \ldots, a_n)$ is positive definite (cf. [133], Chap. 10, §4, Theorems 3 and 4). Hence, the difference

$$\{(a_1, \ldots, a_n) \in R^n \mid \Delta_1 \geq 0, \ldots, \Delta_n \geq 0\} \setminus \sigma(R^n)$$

is contained in the set of zeros of the product $\Delta_1 \cdots \Delta_n$. Since $\Delta_1 = N_0 = n$, we can remove the inequality $\Delta_1 \geq 0$. □

Remark 6.2.9. It is obvious from the preceding argument that the subset of $(a_1, \ldots, a_n) \in R^n$, such that the polynomial $X^n - a_1 X^{n-1} + \cdots + (-1)^n a_n$ has n distinct roots in R, is the basic open semi-algebraic subset of R^n defined by the $n-1$ inequalities $\Delta_2 > 0, \ldots, \Delta_n > 0$.

The question arises whether $\sigma(R^n)$ is defined by the inequalities $\Delta_2 \geq 0, \ldots, \Delta_n \geq 0$. This is, however, not the case. The polynomial $X^4 + 1$ has no real root, but $\Delta_2 = \Delta_3 = 0$, $\Delta_4 = 256$.

Theorem 6.2.10 (Symmetric Hilbert's 17th Problem). *Let f be a symmetric polynomial in $R[X_1, \ldots, X_n]$. If f is nonnegative on R^n, then it can be represented as*

$$f = \sum_{i=1}^{r} s_i \delta_i,$$

where the s_i are sums of squares of symmetric rational functions and the δ_i are products $\prod_{j=2}^{n} (\Delta_j(\sigma_1, \ldots, \sigma_n))^{\epsilon_j}$, with $\epsilon_j = 0$ or 1.

Proof. Follows from Corollary 6.2.4 and Lemma 6.2.8. □

Let us now describe another representation of symmetric nonnegative polynomials. First, we show that $\sigma(R^n)$ is a basic closed semi-algebraic set (although it is not defined by $\Delta_2 \geq 0, \ldots, \Delta_n \geq 0$). Let $D_1, \ldots, D_{2^n-1}$ be the $2^n - 1$ symmetric minors of $\mathcal{H}(a_1, \ldots, a_n)$: these are the determinants of the submatrices obtained by taking the rows and the columns with indices in the same nonempty subset of $\{1, \ldots, n\}$. The $\Delta_1, \ldots, \Delta_n$ are among the D_j.

Lemma 6.2.11. *One has*

$$\sigma(R^n) = \{(a_1, \ldots, a_n) \in R^n \mid D_j(a_1, \ldots, a_n) \geq 0, \ j = 1, \ldots, 2^n - 1\}.$$

Proof. $\mathcal{H}(a_1, \ldots, a_n)$ is positive semi-definite if and only if all D_j are nonnegative (cf. [133], Chap. 10, §4, Theorem 4). □

Theorem 6.2.12. *Let $f \in R[X_1, \ldots, X_n]$ be a symmetric polynomial nonnegative on R^n. Then there exist a nonnegative integer m and polynomials g and h, both of the form $\sum_j s_j d_j$, where the s_j are sums of squares of symmetric polynomials and the d_j are products $\prod_{i=1}^{2^n-1} (D_i(\sigma_1, \ldots, \sigma_n))^{\epsilon_i}$, with $\epsilon_i = 0$ or 1, such that $fg = f^{2m} + h$.*

Proof. The theorem follows from the Positivstellensatz (Corollary 4.4.3 (i)) and Lemma 6.2.11. □

Let G be a compact Lie group, $GL(n,\mathbb{R})$ the linear group of $n \times n$ invertible matrices with entries in $\mathbb{R}$ and $\varphi : G \to GL(n,\mathbb{R})$ a linear representation of G. This representation φ induces an action of G on $\mathbb{R}^n$. Let $\mathbb{R}[X_1,\ldots,X_n]^G$ be the $\mathbb{R}$-algebra of polynomials invariant under this action. By the classical Hilbert-Nagata theorem, $\mathbb{R}[X_1,\ldots,X_n]^G$ is generated as an $\mathbb{R}$-algebra by a finite number of polynomials $q_1,\ldots,q_k$ (cf. [116], p. 42). Define $q = (q_1,\ldots,q_k) : \mathbb{R}^n \to \mathbb{R}^k$.

Theorem 6.2.13. *The semi-algebraic set $q(\mathbb{R}^n)$ is basic closed in $\mathbb{R}^k$.*

Proof. See [260]. □

Corollary 6.2.14. *Let G be a compact Lie group. There exists a finite family $f_1,\ldots,f_m$ of G-invariant polynomials nonnegative on $\mathbb{R}^n$, such that every G-invariant polynomial f that is nonnegative on $\mathbb{R}^n$ can be represented as $f = \sum_{i=1}^m s_i f_i$, where the s_i are sums of squares in the field of fractions of $\mathbb{R}[X_1,\ldots,X_n]^G$.*

Proof. Follows from Corollary 6.2.4 and Theorem 6.2.13. □

Theorem 6.2.13 is in general not valid if G is not compact [260].

6.3 Hilbert's Theorem about Positive Forms

We have seen that a polynomial nonnegative on $\mathbb{R}^n$ is a sum of squares of rational functions. A theorem of Hilbert describes the couples (n,m) such that every polynomial of degree m nonnegative on $\mathbb{R}^n$ is a sum of squares of *polynomials*. Actually, we shall consider only homogeneous polynomials, or *forms*. The problem about general polynomials can be reduced to the problem about forms by homogenization.

Notation 6.3.1. *Throughout this section, we fix a real closed field R. Let $P_{n,m}$ denote the set of nonzero forms in n variables of degree m, with coefficients in R, that are nonnegative on R^n, and let $\Sigma_{n,m}$ denote the subset of $P_{n,m}$ of those forms which are sums of squares of polynomials.*

Remark 6.3.2. The comparative study of $P_{n,m}$ and $\Sigma_{n,m}$ is only interesting for even m, since otherwise $P_{n,m} = \emptyset$. We shall always assume that m is even. If $f \in \Sigma_{n,m}$, then f is, in fact, a sum of squares of forms of degree $m/2$ (by decomposing polynomials into sums of their homogeneous components).

Proposition 6.3.3. *If $n \leq 2$, or if $m = 2$, then $P_{n,m} = \Sigma_{n,m}$.*

Proof. The case $n = 1$ is obvious. The case $n = 2$ follows from the factorization of forms in two variables over a real closed field (given by the factorization of polynomials in one variable): a form in $P_{2,m}$ is a square or a sum of two squares of forms. The case $m = 2$ follows from the diagonalization of quadratic forms: a form in $P_{n,2}$ is a sum of squares of n linear forms. $\square$

Proposition 6.3.4. $P_{3,4} = \Sigma_{3,4}$.

Proof. First we need the following lemma:

Lemma 6.3.5. *If $s \in P_{3,4}$, then there exists a nonzero quadratic form q such that $s \geq q^2$ on R^3.*

Proof. Let $\mathbb{P}\mathcal{Z}(s) \subset \mathbb{P}_2(R)$ be the set of points $(x:y:z)$ in $\mathbb{P}_2(R)$ such that $s(x,y,z) = 0$. We shall distinguish between three cases:

(i) The set $\mathbb{P}\mathcal{Z}(s)$ is empty. Then $s(X,Y,Z)/(X^2 + Y^2 + Z^2)^2$ does not vanish on the unit-sphere S^2, hence there exists $\epsilon \in R$, $\epsilon > 0$ such that $s \geq \epsilon(X^2 + Y^2 + Z^2)^2$ on S^2. This inequality also holds on R^3, and we take $q = \sqrt{\epsilon}(X^2 + Y^2 + Z^2)$.

(ii) The set $\mathbb{P}\mathcal{Z}(s)$ has exactly one element. We may assume that this point is $(1:0:0)$, by changing coordinates, if necessary. Then the degree of s with respect to X is less than 4, and we may assume that it is equal to 2 (otherwise, X does not occur in s, and we are reduced to the case of two variables). Thus we have

$$s(X,Y,Z) = X^2 f(Y,Z) + 2X g(Y,Z) + h(Y,Z),$$

and

$$fs = (Xf + g)^2 + (fh - g^2),$$

with f, h and $fh - g^2$ all nonnegative on R^2, since $s \geq 0$ on R^3.

If the quadratic form f is nondegenerate, then $f > 0$ on $R^2 \setminus \{(0,0)\}$. The discriminant $fh - g^2$ has to be positive on $R^2 \setminus \{(0,0)\}$, since, if $(fh - g^2)(b,c) = 0$ with $(b,c) \neq (0,0)$, then $(-g(b,c)/f(b,c):b:c) \in \mathbb{P}\mathcal{Z}(s)$, contradicting the assumption. Hence, there exists $\epsilon \in R$, $\epsilon > 0$, such that $(fh - g^2)/f^3 \geq \epsilon$ on S^1, and, thus, $fh - g^2 \geq \epsilon f^3$ on R^2. Then $fs \geq fh - g^2 \geq \epsilon f^3$ on R^3, from which we deduce $s \geq \epsilon f^2$ on R^3. We take $q = \sqrt{\epsilon} f$.

If f is degenerate, then it is the square of a linear form f_1, and, since $f_1^2 h - g^2 \geq 0$, f_1 divides g. There is a quadratic form $g_1(Y,Z)$ such that $g = g_1 f_1$. Then $fs \geq (Xf + g)^2 = f(Xf_1 + g_1)^2$, and, thus, $s \geq (Xf_1 + g_1)^2$ on R^3. We take $q = Xf_1 + g_1$.

(iii) The set $\mathbb{P}\mathcal{Z}(s)$ has at least two elements. We may assume that $(1:0:0)$ and $(0:1:0)$ belong to $\mathbb{P}\mathcal{Z}(s)$, by changing coordinates if necessary. Then we may assume that the degrees of s with respect to X and with respect to Y are both equal to 2 (otherwise, one of the variables X or Y does not occur, and we are reduced to the case of two variables). We have

$$s(X,Y,Z) = X^2 f(Y,Z) + 2XZg(Y,Z) + Z^2 h(Y,Z),$$

and

$$fs = (Xf + Zg)^2 + Z^2(fh - g^2),$$

where f, g and h are quadratic forms in Y, Z and f, h and $fh - g^2$ are all nonnegative on R^2.

If f is degenerate, then we proceed as in case (ii). If h is degenerate, we proceed in the same way, using $hs = (Zh + Xg)^2 + X^2(fh - g^2)$.

It remains the case where neither f nor h are degenerate. In this case $f > 0$ and $h > 0$ on $R^2 \setminus \{(0,0)\}$. There are two subcases:

a) The form $fh - g^2$ has a nontrivial zero (b, c). Define $\alpha = -g(b,c)/f(b,c)$ and

$$s_1(X,Y,Z) = s(X + \alpha Z, Y, Z) = X^2 f + 2XZ(g + \alpha f) + Z^2(h + 2\alpha g + \alpha^2 f).$$

Then (b, c) is a zero of $h + 2\alpha g + \alpha^2 f$, and the problem is reduced to the case when h is degenerate.

b) The form $fh - g^2 > 0$ is positive on $R^2 \setminus \{(0,0)\}$. Then there exists $\epsilon > 0$, such that $(fh - g^2)/(Y^2 + Z^2)f \geq \epsilon$ on S^1. Hence $fh - g^2 \geq \epsilon(Y^2 + Z^2)f$ on R^2, and $s \geq \epsilon Z^2(Y^2 + Z^2) \geq \epsilon Z^4$ on R^3. We take $q = \sqrt{\epsilon}Z^2$. □

Now we return to the proof of Proposition 6.3.4. A form $f \in P_{n,m}$ is said to be *extremal* if, for every g, h in $P_{n,m}$ such that $f = g + h$, there exists $a \in R$, $0 < a < 1$, such that $g = af$ and $h = (1-a)f$. Every form in $P_{n,m}$ is the sum of finitely many extremal forms of $P_{n,m}$; this is a consequence of the theory of convex sets applied to the closed convex cone $P_{n,m} \cup \{0\}$ ([278] p. 167). In order to prove that $P_{3,4} = \Sigma_{3,4}$, it is enough to prove that every extremal form of $P_{3,4}$ belongs to $\Sigma_{3,4}$. Let s be such an extremal form. By Lemma 6.3.5, we have $s = q^2 + t$, with $q \neq 0$ and $t = 0$ or $t \in P_{3,4}$. Since s is extremal, $q^2 = as$ for some a, $0 < a \leq 1$, and, hence, $s \in \Sigma_{3,4}$. □

Proposition 6.3.6. *If $n \geq 3$, $m \geq 4$ and $(n,m) \neq (3,4)$, then $\Sigma_{n,m} \neq P_{n,m}$.*

Proof. We give two counter-examples:

(i) For $(n,m) = (4,4)$, the form

$$q(X,Y,Z,W) = W^4 + X^2Y^2 + Y^2Z^2 + Z^2X^2 - 4XYZW$$

is nonnegative on R^4, since $XYZW$ is the geometric mean of W^4, X^2Y^2, Y^2Z^2 and Z^2X^2. If q were a sum of squares of quadratic forms, $q = \sum_{i=1}^{k} q_i^2$, then no q_i could contain X^2, Y^2, Z^2, since q does not contain X^4, Y^4, Z^4. But then no q_i could contain WX, WY, WZ, since q does not contain W^2X^2, W^2Y^2 and W^2Z^2 either. Hence, each q_i would be a linear combination of the monomials XY, YZ, ZX and W^2, but then there would be no $XYZW$ in $\sum_{i=1}^{k} q_i^2$. This shows that $q \in P_{4,4} \setminus \Sigma_{4,4}$.

If $n \geq 4$ and $m > 4$, then the form $X^{m-4}q$ belongs to $P_{n,m} \setminus \Sigma_{n,m}$.

(ii) For $(n, m) = (3, 6)$, the form

$$s(X, Y, Z) = Z^6 + X^4Y^2 + X^2Y^4 - 3X^2Y^2Z^2$$

is nonnegative on R^3, since $X^2Y^2Z^2$ is the geometric mean of Z^6, X^4Y^2 and X^2Y^4. If s were a sum of squares of cubic forms, $s = \sum_{i=1}^{k} s_i^2$, then no s_i could contain X^3, Y^3, hence no s_i could contain X^2Z, Y^2Z (since s does not contain X^4Z^2, Y^4Z^2), and, finally, no s_i could contain XZ^2, YZ^2 (since s does not contain X^2Z^4, Y^2Z^4). Then each s_i would be a linear combination of XY^2, X^2Y, XYZ, Z^3, and $X^2Y^2Z^2$ would necessarily have a nonnegative coefficient in $\sum_{i=1}^{k} s_i^2$. Hence, $s \in P_{3,6} \setminus \Sigma_{3,6}$.

If $n \geq 3$ and $m > 6$, then the form $X^{m-6}s$ belongs to $P_{n,m} \setminus \Sigma_{n,m}$. $\square$

We can sum up the results of this section in the following theorem.

Theorem 6.3.7. *The equality $P_{n,m} = \Sigma_{n,m}$ holds if and only if $n \leq 2$ or $m = 2$ or $(n, m) = (3, 4)$.*

6.4 Quantitative Aspects of Hilbert's 17th Problem

We know that a polynomial nonnegative on R^n can be represented as a sum of squares in $R(X_1, \ldots, X_n)$. The quantitative Hilbert's 17th problem concerns the number of squares needed.

Definition 6.4.1. *Let A be a commutative ring. Suppose there is a positive integer r, such that every sum of squares of elements of A is a sum of r squares of elements of A. The smallest such r is called* the Pythagoras number *of A and is denoted $p(A)$. If such an r does not exist, set $p(A) = \infty$.*

Example 6.4.2. If R is real closed, then $p(R) = 1$, $p(R(X)) = p(R[X]) = 2$ (where X denotes a single variable). It is known that $p(\mathbb{Z}) = p(\mathbb{Q}) = 4$ (Lagrange's theorem [294], Chap. 4, Appendix, Corollary 1).

The study of the quantitative aspect of Hilbert's 17th problem is based on the theory of quadratic forms. Throughout this section, F will denote a field of characteristic $\neq 2$, and $F^* = F \setminus \{0\}$, its multiplicative group. A *quadratic form φ of dimension n over F* is a homogeneous polynomial of degree 2 in n variables: $\varphi(X) = \sum_{1 \leq i \leq j \leq n} a_{ij} X_i X_j$, $a_{ij} \in F$. Two quadratic forms φ and ψ of dimension n are *equivalent* (we use the notation $\varphi \simeq \psi$) if there exists $B \in GL(n, F)$, such that $\varphi(X) = \psi(B \cdot X)$. Every quadratic form is equivalent to a diagonal form, i.e. to a form $\varphi(X) = \sum_{i=1}^{n} a_i X_i^2$. Such a diagonal form will be denoted $\varphi = \langle a_1, \ldots, a_n \rangle$.

If φ is a quadratic form of dimension n over F, then there exists a unique symmetric bilinear form B_φ over F^n, such that $\varphi(x) = B_\varphi(x, x)$, defined by

6.4 Quantitative Aspects of Hilbert's 17th Problem

$$B_\varphi(x, y) = \varphi\left(\frac{x+y}{2}\right) - \varphi\left(\frac{x-y}{2}\right).$$

The quadratic form φ is *nondegenerate* if B_φ is nondegenerate, i.e. if the mapping $x \mapsto B_\varphi(x, -)$ is an isomorphism from F^n onto its dual $(F^n)^\vee$. The form $\langle a_1, \ldots, a_n \rangle$ is nondegenerate if and only if all $a_1, \ldots, a_n$ belong to F^*. *All quadratic forms considered in this section are nondegenerate.*

We define the operations of orthogonal sum and tensor product for quadratic forms as follows: the *orthogonal sum* $\varphi \perp \psi$ is defined by

$$(\varphi \perp \psi)(x \oplus y) = \varphi(x) + \psi(y),$$

and the *tensor product* $\varphi \otimes \psi$ by

$$B_{\varphi \otimes \psi}(x \otimes y, x' \otimes y') = B_\varphi(x, x') B_\psi(y, y').$$

Definition 6.4.3. *Let φ be a nondegenerate quadratic form of dimension n over F. The form φ represents $b \in A$ over A (where A is an F-algebra) if there exists $x \in A^n$ such that $\varphi(x) = b$. If $A = F$, we simply say that φ represents b. The form φ is isotropic (over F) if there exists $x \neq 0$ in F^n such that $\varphi(x) = 0$, and anisotropic otherwise.*

Theorem 6.4.4. *Let φ be a nondegenerate quadratic form of dimension n over F.*

(i) *If φ is isotropic, then there exist $a_3, \ldots, a_n \in F^*$ such that*

$$\varphi \simeq \langle 1, -1, a_3, \ldots, a_n \rangle.$$

(ii) *If φ is isotropic, then φ represents every $b \in F^*$ (φ is universal).*
(iii) *The form φ represents $b \in F^*$ if and only if $\varphi \perp \langle -b \rangle$ is isotropic.*

Proof. See [204], Chap. 1, Theorem 3.4(2) for (i), Theorem 3.4(3) for (ii) and Theorem 3.5 for (iii) □

Theorem 6.4.5. *Let $f \in F[X]$ be a polynomial in a single variable X. Let φ be a nondegenerate quadratic form over F. If φ represents f over the field $F(X)$, then φ represents f over the ring $F[X]$.*

Proof. If φ is isotropic, then, applying Theorem 6.4.4 (i) and the equality $f = ((f+1)/2)^2 - ((f-1)/2)^2$, we deduce that φ represents every element of $F[X]$ over $F[X]$. Hence, we may assume that φ is anisotropic, and that $f \neq 0$.

Consider a representation of f by φ over $F(X)$:

(∗) $$f = \varphi(g_1/g_0, \ldots, g_n/g_0),$$

where $g_0, g_1, \ldots, g_n \in F[X]$ and $g_0 \neq 0$. We introduce the quadratic form $\psi = \langle -f \rangle \perp \varphi$ of dimension $n+1$ over $F(X)$ i.e. $\psi(u) = -fu_0^2 + \varphi(u_1, \ldots, u_n)$. Define $g = (g_0, g_1, \ldots, g_n) \in F[X]^{n+1}$. From the equality (∗), we deduce

$\psi(g) = 0$. In order to prove the theorem, it is enough to construct from g a zero of ψ of the form $(1, p_1, \ldots, p_n)$, with $p_1, \ldots, p_n$ in $F[X]$. Then $f = \varphi(p_1, \ldots, p_n)$ will be a representation of f over $F[X]$.

For $i = 1, \ldots, n$, we make the euclidean division $g_i = q_i g_0 + r_i$, with $\deg r_i < \deg g_0$ or $r_i = 0$. If $q = (1, q_1, \ldots, q_n)$ is a zero of ψ, we are done. So, we may assume $\psi(q) \neq 0$. Then g and q are linearly independent over $F(X)$, and, hence, $h = \psi(q)g - 2B_\psi(g,q)q$ is not the zero vector of $F(X)^{n+1}$. By construction, $h \in F[X]^{n+1}$, and h is a zero of ψ:

$$\psi(h) = (\psi(q))^2 \psi(g) - 4\psi(q)(B_\psi(g,q))^2 + 4(B_\psi(g,q))^2 \psi(q) = 0.$$

The first component h_0 of h is

$$h_0 = \psi(q)g_0 - 2B_\psi(g,q) = \frac{1}{g_0}\varphi(r_1, \ldots, r_n),$$

and $r_1, \ldots, r_n$ are not all zero, since, otherwise, we would have $g = g_0 q$. Since φ is anisotropic over F, $\varphi(r_1, \ldots, r_n) \neq 0$, and

$$\begin{aligned}\deg(h_0 g_0) &= \deg(\varphi(r_1, \ldots, r_n)) \\ &= 2\max(\{\deg(r_i) \mid i = 1, \ldots, n\}) < 2\deg(g_0).\end{aligned}$$

Hence we have constructed a zero h of ψ in $F[X]^{n+1}$, with $h_0 \neq 0$ and $\deg h_0 < \deg g_0$. Since the degree has decreased, we can reiterate this construction and finally obtain a zero of ψ in $F[X]^{n+1}$ of the form $(1, p_1, \ldots, p_n)$. This completes the proof. $\square$

The next lemma will be useful in Section 6.5.

Lemma 6.4.6. *Let φ be a quadratic form over F, and $f \in F[X_1, \ldots, X_m]$. If φ represents f over $F(X_1, \ldots, X_m)$, then, for every $a_1, \ldots, a_m \in F$, φ represents $f(a_1, \ldots, a_m)$ over F.*

Proof. If φ represents f over $F(X_1, \ldots, X_m) = F(X_1, \ldots, X_{m-1})(X_m)$, then, by Theorem 6.4.5, $f = \varphi(A_1(X_m), \ldots, A_n(X_m))$, for some polynomials $A_1, \ldots, A_n$ with coefficients in $F(X_1, \ldots, X_{m-1})$. For every a_m in F, we obtain $f(X_1, \ldots, X_{m-1}, a_m) = \varphi(A_1(a_m), \ldots, A_n(a_m))$; hence, φ represents $f(X_1, \ldots, X_{m-1}, a_m)$ over $F(X_1, \ldots, X_{m-1})$. The result is obtained by induction on m. $\square$

Proposition 6.4.7. *Let $a \in F$, $b_1, \ldots, b_n \in F^*$, $n > 1$ and $\varphi = \langle b_1, \ldots, b_n \rangle$. Then the following properties are equivalent:*
  (i) *φ represents $b_1 X^2 + a$ over $F(X)$.*
  (ii) *$\varphi' = \langle b_2, \ldots, b_n \rangle$ represents a, or φ is isotropic.*

Proof. (i) $\Rightarrow$ (ii) By Theorem 6.4.5, φ represents $b_1 X^2 + a$ over $F[X]$:

$$b_1 X^2 + a = \sum_{j=1}^n b_j f_j^2, \quad f_j \in F[X].$$

If φ is anisotropic over F, we have

$$\deg\left(\sum_{j=1}^{n} b_j f_j^2\right) = 2\max(\{\deg(f_j) \mid j = 1,\ldots,n\}),$$

and, thus, $f_j = c_j X + d_j$, with $c_j, d_j \in F$ for $j = 1,\ldots,n$. There exists at least one $e \in F$ such that $e^2 = (c_1 e + d_1)^2$, and, for such an e, one has

$$b_1 e^2 + a = b_1 e^2 + \sum_{j=2}^{n} b_j (c_j e + d_j)^2.$$

This shows that φ' represents a over F.

(ii) $\Rightarrow$ (i) If φ' represents a over F, the implication is obvious. If φ is isotropic, we use Theorem 6.4.4 (ii). □

If $b_1 = \cdots = b_n = 1$, Proposition 6.4.7 means that $X^2 + a$ is a sum of $n > 1$ squares in $F(X)$ if and only if -1 is a sum of $n-1$ squares in F or a is a sum of $n-1$ squares in F. A proof by induction on n gives the following results:

Corollary 6.4.8. *Let F be a real field.*

(i) The polynomial $1 + X_1^2 + \cdots + X_n^2$ is not a sum of n squares in $F(X_1,\ldots,X_n)$, and, hence, $X_1^2 + \cdots + X_n^2$ is not a sum of $n-1$ squares in $F(X_1,\ldots,X_n)$.

(ii) One has $p(F(X_1,\ldots,X_n)) \geq p(F) + n$.

Corollary 6.4.8 (ii) gives us a lower bound on the Pythagoras number. We use the theory of Pfister multiplicative forms in order to obtain an upper bound. We shall only explain the part of this theory needed for this purpose.

Lemma 6.4.9. *If the form $\langle a, b\rangle$ represents $c \in F^*$, then $\langle a, b\rangle \simeq c\langle 1, ab\rangle$.*

Proof. There exists $(u, v) \in F^2$ such that $c = au^2 + bv^2 \neq 0$. The equality

$$\begin{bmatrix} u & v \\ -bv & au \end{bmatrix} \begin{bmatrix} a & 0 \\ 0 & b \end{bmatrix} \begin{bmatrix} u & -bv \\ v & au \end{bmatrix} = c\begin{bmatrix} 1 & 0 \\ 0 & ab \end{bmatrix}$$

proves the result. □

Definition 6.4.10. *A Pfister form φ over F is a quadratic form of dimension 2^n over F, such that there exist elements $a_1,\ldots,a_n$ in F^*, with*

$$\varphi = \langle 1, a_1\rangle \otimes \langle 1, a_2\rangle \otimes \cdots \otimes \langle 1, a_n\rangle.$$

We abbreviate it to $\varphi = \langle\!\langle a_1,\ldots,a_n\rangle\!\rangle$. The form of dimension $2^n - 1$

$$\varphi' = \langle a_1,\ldots,a_n, a_1 a_2, \ldots, a_1 a_2 \cdots a_n\rangle$$

*is called the **pure subform** of the Pfister form φ. We have $\varphi = \langle 1\rangle \perp \varphi'$.*

A quadratic form φ of dimension d over F is said to be **multiplicative** if it has the following property

$$\forall x \in F^d \ \varphi(x) \neq 0 \Rightarrow \varphi(x)\varphi \simeq \varphi \ .$$

Theorem 6.4.11. *Every Pfister form over F is multiplicative.*

Proof. Let φ be a Pfister form of dimension 2^n. The proof is by induction on n. For $n = 0$, φ is the form of dimension 1, $\varphi(X) = X^2$, and it is obviously multiplicative. Hence, it is enough to prove that, if φ is multiplicative and $a \in F^*$, then $\varphi \perp a\varphi = \langle 1, a \rangle \otimes \varphi$ is multiplicative. Let $x, x' \in F^{2^n}$ be such that $\varphi(x) + a\varphi(x') \neq 0$.

If $\varphi(x') = 0$, then $\varphi(x)(\varphi \perp a\varphi) = (\varphi(x)\varphi) \perp (a\varphi(x)\varphi) \simeq \varphi \perp a\varphi$.

If $\varphi(x) = 0$, then

$$a\varphi(x')(\varphi \perp a\varphi) = (a\varphi(x')\varphi) \perp (a^2\varphi(x')\varphi) \simeq a\varphi \perp a^2\varphi \simeq \varphi \perp a\varphi \ .$$

The case $\varphi(x)\varphi(x') \neq 0$ remains. In this case:

$$\begin{aligned}(\varphi(x) + a\varphi(x'))(\varphi \perp a\varphi) &\simeq (\varphi(x) + a\varphi(x'))(\varphi \perp a\varphi(x)\varphi(x')\varphi) \\ &\simeq (\varphi(x) + a\varphi(x'))\langle 1, a\varphi(x)\varphi(x') \rangle \otimes \varphi \\ &\simeq \langle \varphi(x), a\varphi(x') \rangle \otimes \varphi \quad \text{(by Lemma 6.4.9)} \\ &\simeq \varphi(x)\varphi \perp a\varphi(x')\varphi \simeq \varphi \perp a\varphi \ .\end{aligned}$$

□

Corollary 6.4.12. *Let φ be a Pfister form, and let G_φ denote the set of elements of F^* represented by φ over F. Then G_φ is a subgroup of the multiplicative group F^*.*

Proof. The set G_φ contains 1. Theorem 6.4.11 shows that G_φ is closed under multiplication. It is also closed under taking inverses since $(\varphi(x))^{-1} = \varphi(x/\varphi(x))$. □

Corollary 6.4.13. *Let n be a positive integer. The product of two sums of 2^n squares in F is again a sum of 2^n squares in F.*

Proof. Apply Corollary 6.4.12 to $\varphi = \langle\!\langle 1, \ldots, 1 \rangle\!\rangle$ (n times). □

The next lemma gives more information than we need for our present purpose, but this extra information will be useful in Section 5.

Lemma 6.4.14. *Let $a_1, \ldots, a_n \in F^*$. For $1 \leq i \leq n$, denote by φ_i the Pfister form $\langle\!\langle a_1, \ldots, a_i \rangle\!\rangle$. Let u_1 be a square in F, and, for $1 < i \leq n$, let u_i be an element of F represented by φ_{i-1}. Define $b_i = \sum_{j=i}^{n} a_j u_j$, for $1 \leq i \leq n$. If $b_1, \ldots, b_n$ all belong to F^*, then*

$$\varphi_n \simeq \langle\!\langle b_1, a_1 b_2, \ldots, a_{n-1} b_n \rangle\!\rangle \ .$$

6.4 Quantitative Aspects of Hilbert's 17th Problem

Proof. We use the following two facts:

(i) If a, b and $a+b$ belong to F^*, then $\langle\!\langle a, b\rangle\!\rangle \simeq \langle\!\langle a+b, ab\rangle\!\rangle$. This follows from Lemma 6.4.9, with $c = a+b$.

(ii) If $b \in F^*$ is represented by the Pfister form ψ, then $\psi \otimes \langle\!\langle a\rangle\!\rangle \simeq \psi \otimes \langle\!\langle ab\rangle\!\rangle$. This follows from the fact that ψ is multiplicative.

We prove, by decreasing induction on k, that

$$\varphi_n \simeq \varphi_k \otimes \langle\!\langle b_{k+1}, a_{k+1}b_{k+2}, \ldots, a_{n-1}b_n\rangle\!\rangle \,,$$

and we get the desired result, for $k = 0$ (with $\varphi_0 = \langle 1\rangle$). First, by (ii), we have

$$\varphi_n = \varphi_{n-1} \otimes \langle\!\langle a_n\rangle\!\rangle \simeq \varphi_{n-1} \otimes \langle\!\langle a_n u_n\rangle\!\rangle = \varphi_{n-1} \otimes \langle\!\langle b_n\rangle\!\rangle$$

(note that $u_n \neq 0$, since $b_n \neq 0$). Then it is enough to show that

$$\varphi_{k-1} \otimes \langle\!\langle a_k, b_{k+1}\rangle\!\rangle \simeq \varphi_{k-1} \otimes \langle\!\langle b_k, a_k b_{k+1}\rangle\!\rangle \,,$$

for $0 < k < n$. If $u_k \neq 0$, we have, by (i) and (ii),

$$\begin{aligned}
\varphi_{k-1} \otimes \langle\!\langle a_k, b_{k+1}\rangle\!\rangle &\simeq \varphi_{k-1} \otimes \langle\!\langle a_k u_k, b_{k+1}\rangle\!\rangle \\
&\simeq \varphi_{k-1} \otimes \langle\!\langle a_k u_k + b_{k+1}, a_k u_k b_{k+1}\rangle\!\rangle \\
&\simeq \varphi_{k-1} \otimes \langle\!\langle b_k, a_k b_{k+1}\rangle\!\rangle \,.
\end{aligned}$$

If $u_k = 0$, then $b_k = b_{k+1}$, and we have by (ii) that

$$\langle\!\langle a_k, b_{k+1}\rangle\!\rangle \simeq \langle\!\langle b_k, a_k\rangle\!\rangle \simeq \langle\!\langle b_k, a_k b_{k+1}\rangle\!\rangle \,.$$

□

Lemma 6.4.15. *Let $\varphi = \langle\!\langle a_1, \ldots, a_n\rangle\!\rangle$, with $a_1, \ldots, a_n \in F^*$. Let $b_1 \in F^*$ be an element represented by the pure subform φ'. Then there exist $c_2, \ldots, c_n \in F^*$ such that $\varphi \simeq \langle\!\langle b_1, c_2, \ldots, c_n\rangle\!\rangle$.*

Proof. We use the notation φ_i of Lemma 6.4.14. Note that, for $1 \leq r \leq n$, the pure subform $(\varphi_r)'$ is equal to

$$\langle a_1\rangle \perp a_2\varphi_1 \perp \cdots \perp a_r\varphi_{r-1}{}' \,.$$

Hence, we have $b_1 = a_1 u_1 + a_2 u_2 + \cdots + a_n u_n$, where u_1 is a square and u_i is represented by φ_{i-1} for $1 < i \leq n$. Without loss of generality, we may assume that n is the smallest integer $r \geq 1$ such that b_1 is represented by $(\varphi_r)'$. Then, with the notation of Lemma 6.4.14, $b_1, \ldots, b_n$ all belong to F^*, and, hence, $\varphi \simeq \langle\!\langle b_1, a_1 b_2, \ldots, a_{n-1} b_n\rangle\!\rangle$, by this lemma. □

The other ingredient we need, in addition to Pfister forms, is the Tsen-Lang theorem.

Theorem 6.4.16. *Let K be an algebraically closed field, and F, a field of transcendence degree n over K. Then every homogeneous polynomial φ of degree d in $m > d^n$ variables, with coefficients in F, has a nontrivial zero in F^m. In particular, every quadratic form over F of dimension greater than 2^n is isotropic.*

Proof. Without loss of generality, we may assume that F is generated by the coefficients of φ. Then F is a finite extension of a pure transcendental extension $G = K(t_1, \ldots, t_n)$.

Given a basis B of F over G and an element γ of F, we call the coordinates of γ with respect to B, the *B-coordinates of γ*. Let us choose a basis $B = (\alpha_i)_{i=1,\ldots,k}$ of F over G, such that the B-coordinates of the products $\alpha_i \alpha_j$, for $i, j = 1, \ldots, k$, and the B-coordinates of the coefficients of φ are polynomials in $K[t_1, \ldots, t_n]$. Let r be the maximum of the degrees of these polynomials.

Given a positive integer s, let us look for a nontrivial zero $(q_1, \ldots, q_m)$ of φ in F^m, such that the B-coordinates of the q_i are polynomials in $K[t_1, \ldots, t_n]$ of degree less than s, with respect to each of the variables $t_1, \ldots, t_n$. Consider the coefficients of these polynomials as indeterminates τ_j; their number is thus mks^n. The equation $\varphi(q_1, \ldots, q_m) = 0$ is equivalent to a system S of at most $k(d(s-1) + (d+1)r)^n$ homogeneous equations of degree d in the variables $\tau_1, \ldots, \tau_{mks^n}$. Since $m > d^n$, we have $e = mks^n - k(d(s-1) + (d+1)r)^n > 0$ for s large enough. Since F is algebraically closed, the set of solutions of S is an algebraic set of dimension at least e. Hence, there is a nontrivial zero of φ in F^m. □

Theorem 6.4.17. *Let R be a real closed field, and F, an extension of transcendence degree n over R. Let $\varphi = \langle\!\langle a_1, \ldots, a_n \rangle\!\rangle$, with $a_1, \ldots, a_n \in F^*$. Let b be a sum of squares of elements of F. Then φ represents b over F.*

Proof. If φ is isotropic over F, or, if b is a square, then it is obvious that φ represents b. So we may assume that φ is anisotropic, and that b is not a square.

We first consider the case $b = b_1^2 + b_2^2$, $b_1 b_2 \neq 0$. The Tsen-Lang theorem 6.4.16 and Theorem 6.4.4 (iii) imply that φ represents every element of $F(i)$ over $F(i)$ (where $i = \sqrt{-1}$). If $F = F(i)$, then φ represents b over F; therefore, we assume that $F \neq F(i)$. The form φ represents $\beta = b_1 + ib_2$ over $F(i)$, and $F(i) = F(\beta)$. Hence, there exist $u, v \in F^{2^n}$, such that $\beta = \varphi(u + \beta v)$, and $v \neq 0$. We have

$$\beta = \varphi(u) + 2\beta B_\Phi(u, v) + \beta^2 \varphi(v) .$$

Comparing with the minimal polynomial of β over F

$$\beta^2 - 2b_1 \beta + b = 0 ,$$

we get $b\varphi(v) = \varphi(u)$, and, hence, by Corollary 6.4.12, φ represents b over F.

Let k be an integer, $k \geq 2$, and suppose that each Pfister form over F represents every sum of k squares in F. It is enough to show that φ represents every element of the form $c = 1 + b$, $c \neq 0$, where b is a sum of k squares in F. Let φ' be the pure subform of φ. By induction, φ represents b, and, hence, $b = b_1^2 + b_2$, where b_2 is represented by φ'. We may assume $b_2 \neq 0$ (otherwise, c is a sum of two squares). Define another Pfister form $\psi = \varphi \otimes \langle 1, -c \rangle$. Its pure subform $\psi' = \varphi' \perp (-c\varphi)$ represents $b_2 - c = -1 - b_1^2$. By Lemma 6.4.15, there exist $c_1, \ldots, c_n \in F^*$, such that

$$\psi \simeq \langle\!\langle -1 - b_1^2, c_1, \ldots, c_n \rangle\!\rangle .$$

The theorem having been proved for sums of two squares, it follows that the Pfister form $\langle\!\langle c_1, \ldots, c_n \rangle\!\rangle$ represents $1 + b_1^2$. Hence, ψ is isotropic, and there exist u, v in F^{2^n}, such that $\psi(u, v) = \varphi(u) - c\varphi(v) = 0$ and $(u, v) \neq (0, 0)$. Since φ is anisotropic and $c \neq 0$, both $\varphi(u)$ and $\varphi(v)$ are different from 0. By Corollary 6.4.12, we deduce that φ represents c. □

The following theorem is due to A. Pfister.

Theorem 6.4.18. *Let R be a real closed field, and $V \subset R^m$, an irreducible algebraic set of dimension n. Every polynomial function in $\mathcal{P}(V)$, that is nonnegative on a nonempty Zariski open subset of V, is a sum of 2^n squares in the field $\mathcal{K}(V)$.*

Proof. Apply Theorem 6.1.9 and Theorem 6.4.17 for $a_1 = \cdots = a_n = 1$. □

Theorem 6.4.18 and Corollary 6.4.8 (ii) give the following bounds.

Corollary 6.4.19. *Let R be a real closed field. Then the Pythagoras number of $R(X_1, \ldots, X_n)$ satisfies*

$$n + 1 \leq p(R(X_1, \ldots, X_n)) \leq 2^n .$$

Remark 6.4.20. The lower bound can be improved to $n + 2$ for $n \geq 2$. In the proof of Proposition 6.3.6 (ii), it is shown that the polynomial $s(X, Y, 1) = 1 + X^4 Y^2 + X^2 Y^4 - 3 X^2 Y^2$ is not a sum of squares of polynomials. The paper [83] contains the proof that this polynomial is not a sum of three squares of rational functions. The proof is based on the theory of elliptic curves over $R(X)$. This result implies that the Pythagoras number $p(R(X, Y))$ is equal to 4, and, hence, by Corollary 6.4.8 (ii), that

$$n + 2 \leq p(R(X_1, \ldots, X_n)) \leq 2^n .$$

for $n \geq 2$.

6.5 A Bound on the Number of Inequalities Needed to Define a Basic Open Semi-algebraic Set

We recall that a basic open semi-algebraic subset of an algebraic set $V \subset R^n$ is a subset of the form

$$\mathcal{U}(g_1, \ldots, g_k) = \{x \in V \mid g_1(x) > 0, \ldots, g_k(x) > 0\},$$

where $g_1, \ldots, g_k$ belong to $\mathcal{P}(V)$. We shall always assume that $g_1, \ldots, g_k$ are nonzero.

The aim of this section is to prove the following result of Bröcker and Scheiderer.

Theorem 6.5.1. *Let V be an algebraic subset of R^n of dimension $d > 0$. Then every basic open semi-algebraic subset U of V can be defined by d simultaneous strict polynomial inequalities: there exist $f_1, \ldots, f_d$ in $\mathcal{P}(V)$ such that*

$$U = \mathcal{U}(f_1 \ldots, f_d).$$

In other words, on a real algebraic set of dimension $d > 0$, any finite system of strict polynomial inequalities can be reduced to a system of d inequalities.

The proof relies heavily on the theory of quadratic forms, especially Pfister forms, and we shall use the results of Section 6.4. The fact that Pfister forms are related to basic open semi-algebraic subsets may seem surprising. It is better understood with the following lemma.

Lemma 6.5.2. *Let $V \subset R^n$ be an irreducible algebraic subset, and let $g_1, \ldots, g_\ell$ and $h_1, \ldots, h_\ell$ be nonzero polynomial functions in $\mathcal{P}(V)$. Assume that*

$$\langle\!\langle g_1, \ldots, g_\ell \rangle\!\rangle \simeq \langle\!\langle h_1, \ldots, h_\ell \rangle\!\rangle,$$

as Pfister forms over the field of rational functions $\mathcal{K}(V)$. Then the basic open semi-algebraic subsets $\mathcal{U}(g_1, \ldots, g_\ell)$ and $\mathcal{U}(h_1, \ldots, h_\ell)$ are generically equal in V (cf. Definition 6.2.1).

This enlightening lemma actually plays no role in the proof of Theorem 6.5.1 given in this section. We postpone its proof to Section 7.6.

The first step in the proof of Theorem 6.5.1 is a result on Pfister forms (Theorem 6.5.5), whose geometric translation is a generic reduction of inequalities (Corollary 6.5.6): if V is irreducible, a basic open semi-algebraic subset of V can be generically defined by d strict inequalities.

Lemma 6.5.3. *Let F be a field and φ a quadratic form anisotropic over F. Then φ is anisotropic over $F(X)$.*

Proof. Suppose that φ is of dimension n and isotropic over $F(X)$. There are $f_1, \ldots, f_n \in F(X)$, not all zero, such that $\varphi(f_1, \ldots, f_n) = 0$. Multiplying, if necessary, the f_i by an appropriate power of X, we may assume that all f_i are defined at 0, and that not all $f_1(0), \ldots, f_n(0)$ are zero. Then $\varphi(f_1(0), \ldots, f_n(0)) = 0$, which shows that φ is isotropic over F. □

Proposition 6.5.4. *Let $\varphi = \langle a_1, \ldots, a_m \rangle$, $\psi = \langle b_1, \ldots, b_n \rangle$ be quadratic forms over a field F, where $a_1, \ldots, a_m, b_1, \ldots, b_n \in F^*$. Assume that φ is anisotropic over F, and that φ represents $\psi(X_1, \ldots, X_n) = \sum_{i=1}^n b_i X_i^2$ over the field $F(X_1, \ldots, X_n)$. Then φ contains ψ, i.e. there exists a quadratic form θ over F, such that φ is equivalent to $\psi \perp \theta$.*

Proof. We proceed by induction on the dimension m of φ. There is nothing to prove for $m = 0$. Given $m > 0$, assume that the result holds true for $m - 1$. By Lemma 6.4.6, the form φ represents b_1 over F; hence, $\varphi \simeq \langle b_1 \rangle \perp \rho$. The form φ is anisotropic over $F(X_2, \ldots, X_n)$, by Lemma 6.5.3, and it represents $b_1 X_1^2 + (b_2 X_2^2 + \cdots + b_n X_n^2)$ over the field $F(X_2, \ldots, X_n)(X_1)$. It follows, by Proposition 6.4.7, that ρ represents $b_2 X_2^2 + \cdots + b_n X_n^2$ over $F(X_2, \ldots, X_n)$. The form ρ is clearly anisotropic over F. Hence, by the inductive assumption, we have $\rho \simeq \langle b_2, \ldots, b_n \rangle \perp \theta$ and $\varphi \simeq \psi \perp \theta$. □

Theorem 6.5.5. *Let F be a field of transcendence degree d over the real closed field R. Let $\varphi = \langle\!\langle a_1, \ldots, a_k \rangle\!\rangle$ be a Pfister form over F, where $k > \max(d, 1)$. Let φ' be the pure subform of φ (recall that $\varphi = \langle 1 \rangle \perp \varphi'$). Then φ' represents 1 over F.*

Proof. If φ is isotropic, then φ contains the form $\langle 1, -1 \rangle$ (Theorem 6.4.4(i)). Hence φ' represents -1 and, thus, by Lemma 6.4.15, there exist $b_2, \ldots, b_k$ in F^* such that $\varphi \simeq \langle\!\langle -1, b_2, \ldots, b_k \rangle\!\rangle$. By Witt's cancellation theorem ([210], Corollary 4, p. 362) and the inequality $2 \le k$, the form φ' contains $\langle b_2, -b_2 \rangle$, which represents 1. So we may assume that φ is anisotropic over F.

The form φ is of dimension 2^k and, by the Tsen-Lang theorem 6.4.16, it is isotropic over the field $F(\sqrt{-1})$. This implies that $\sqrt{-1} \notin F$. We have also that φ is universal over $F(\sqrt{-1})(X, Y)$, and, in particular, it represents $X + \sqrt{-1} Y$ over this field. Hence, there are $f, g \in F(X, Y)^{2^k}$ such that

$$\varphi(f + (X + \sqrt{-1} Y)g) = X + \sqrt{-1} Y,$$

so one has

$$\varphi(g)(X + \sqrt{-1} Y)^2 + (2B_\varphi(f, g) - 1)(X + \sqrt{-1} Y) + \varphi(f) = 0,$$

where B_φ is the symmetric bilinear form associated to φ. It is clear that $g \ne 0$, and, since, by Lemma 6.5.3, φ is anisotropic over $F(X, Y)$, we have $\varphi(g) \ne 0$. By comparing the equality above with the minimal polynomial of $X + \sqrt{-1} Y$ over $F(X, Y)$, which is

$$T^2 - 2XT + X^2 + Y^2,$$

it follows that
$$\varphi(g)(X^2 + Y^2) = \varphi(f) .$$
Since φ is multiplicative (Theorem 6.4.11) and $\varphi(g) \neq 0$, the form φ represents $X^2 + Y^2$ over $F(X, Y)$. By Proposition 6.5.4, we have $\varphi \simeq \langle 1, 1 \rangle \perp \theta$. Hence, by Witt's cancellation theorem, φ' represents 1 over F. □

The next corollary of Theorem 6.5.5 is a weak form of Theorem 6.5.1, that we want to prove. It is not used in the proof. Nevertheless, we give the statement here, because it helps in understanding the geometric content of Theorem 6.5.5.

Corollary 6.5.6. *Let $V \subset R^n$ be an irreducible algebraic set of dimension $d > 0$, and $U = \mathcal{U}(g_1, \ldots, g_k)$, a basic open semi-algebraic subset of V. Then there exist $f_1, \ldots, f_d \in \mathcal{P}(V)$ such that U is generically equal to $\mathcal{U}(f_1, \ldots, f_d)$ in V.*

Proof. Clearly, it is enough to prove the corollary for $k = d + 1$. Set $\varphi = \langle\!\langle g_1, \ldots, g_{d+1} \rangle\!\rangle$. By Theorem 6.5.5, the pure subform φ' represents 1 over $\mathcal{K}(V)$, hence, by Lemma 6.4.15, we have $\varphi \simeq \langle\!\langle 1, f_1, \ldots, f_d \rangle\!\rangle$, for some $f_1, \ldots, f_d$ which can be assumed to be in $\mathcal{P}(V)$. We conclude that U and $\mathcal{U}(1, f_1, \ldots, f_d) = \mathcal{U}(f_1, \ldots, f_d)$ are generically equal in V, by applying Lemma 6.5.2. □

Until now, we have been working with quadratic forms over fields, and we obtained a weak version of Theorem 6.5.1. The proof of this theorem will follow more or less the same lines, but we shall now consider quadratic forms over commutative rings. It suffices, for our purpose, to consider only diagonal forms. If A is any commutative ring, and $a_1, \ldots, a_m$ are elements of A, then $\langle a_1, \ldots, a_m \rangle$ denotes the quadratic form $a_1 x_1^2 + \cdots + a_m x_m^2$. A Pfister form $\langle\!\langle a_1, \ldots, a_m \rangle\!\rangle$ over a ring A is defined in the same way as for fields.

Definition 6.5.7. *Let A be a ring, B an A-algebra. A quadratic form $\varphi = \langle a_1, \ldots, a_m \rangle$ with coefficients in A is nondegenerate over B if the image of $\prod_{i=1}^m a_i$ is invertible in B. An element $b \in B$ is weakly represented by φ over B if there are elements $u_1, \ldots, u_k$ of B^m, such that $b = \varphi(u_1) + \cdots + \varphi(u_k)$.*

Note that the definition of a nondegenerate form agrees with the definition we gave in the case of fields. Note, also, that an element is weakly represented by φ over a real closed field F if and only if it is represented by φ over F.

Let $V \subset R^n$ be an algebraic set and $x \in V$. If $\varphi = \langle a_1, \ldots, a_m \rangle$ is a quadratic form with coefficients in $\mathcal{P}(V)$, we denote by φ_x the quadratic form $\langle a_1(x), \ldots, a_m(x) \rangle$ over R.

Lemma 6.5.8. *Let $g_1, \ldots, g_m$ be polynomial functions in $\mathcal{P}(V)$. Denote by φ_i the Pfister form $\langle\!\langle g_1, \ldots, g_i \rangle\!\rangle$, for $i = 1, \ldots, m$. Let v_1 be a sum of squares in $\mathcal{P}(V)$, and v_{i+1}, an element weakly represented by φ_i over $\mathcal{P}(V)$, for $i = 1, \ldots, m - 1$. Define $w_i = \sum_{j=i+1}^m g_j v_j$, for $i = 0, \ldots, m - 1$, and*

$\psi = \langle\!\langle w_0, g_1 w_1, \ldots, g_{m-1} w_{m-1} \rangle\!\rangle$. Let $x \in V$ be such that both forms $(\varphi_m)_x$ and ψ_x are nondegenerate. Then $(\varphi_m)_x \simeq \psi_x$, and, hence, both forms have the same signature.

Proof. Note that $v_1(x)$ is a square in R, and that $v_{i+1}(x)$ is represented by $(\varphi_i)_x$ over R for $i = 1, \ldots, m-1$. Hence, we can apply Lemma 6.4.14 for the field R. □

We now introduce a special class of R-algebras, which includes the rings of regular functions on subsets of R^n.

Definition 6.5.9. *Let A be a ring of fractions of an R-algebra of finite type. Then A is a RFR (for Regular Functions Ring) over R if it satisfies the following equivalent conditions:*
  (i) *Every maximal ideal of A is real.*
  (ii) *All elements of the form $1 + \sum_{i=1}^{k} a_i^2$ are invertible in A.*

It is clear that, if A is a RFR over R, then any quotient of A is also a RFR over R. The *transcendence degree of A over R* is, by definition, the maximum of the transcendence degrees over R of the residue fields of A at its prime ideals. Theorem 6.5.5 has a weak version for RFR.

Theorem 6.5.10. *Let A be a RFR of transcendence degree d over R. Let $\varphi = \langle\!\langle a_1, \ldots, a_n \rangle\!\rangle$ be a nondegenerate Pfister form, with coefficients in A, where $n > \max(d, 1)$. Then 1 is weakly represented by the pure subform φ' over A.*

Proof. Let N be the nilradical of A. It is clear that A/N is a RFR over R, with the same transcendence degree. If 1 is weakly represented by φ' over A/N, then we have $1 + b = \varphi'(a_1) + \cdots + \varphi'(a_k)$, where $b, a_1, \ldots, a_k$ belong to A and b is nilpotent. Then $1 + b$ is invertible and has a square root s in A. We get a weak representation $1 = \varphi'(a_1/s) + \cdots + \varphi'(a_k/s)$ over A. Hence, we may assume that A is radical. We now proceed by induction on d.

If $d = 0$, then A is a finite product of copies of R. Since $n \geq 2$, it is clear that in each copy of R the image of at least one of the coefficients of φ' is positive. Hence φ' represents 1 over A.

Given $d > 0$, assume that the result holds true for all RFR of transcendence degree less than d over R. Since A is radical, we can embed it into its total ring of fractions K, which is obtained from A by inverting all elements which are not zero divisors. It is well known that K is the product of the residue fields of A at its minimal prime ideals. Each of these fields is of transcendence degree at most d over R, so, by Theorem 6.5.5, we get a representation of 1 by φ' over K. Clearing denominators, we obtain an identity $f^2 = \varphi'(u)$, where f and u belong to A, and f is not a zero divisor. If f is invertible in A, then $1 = \varphi'(u/f)$. Otherwise, A/f is a RFR of transcendence degree less than d over A. By the inductive assumption, 1 is weakly represented by φ' over A/f. Hence we have

$$1 - \varphi'(a_1) - \cdots - \varphi'(a_k) = bf,$$

for some $a_1, \ldots, a_k, b$ in A. Taking squares, we get

$$(1 - \varphi'(a_1) - \cdots - \varphi'(a_k))^2 = b^2 f^2 = \varphi'(bu).$$

Expanding the expression on the left side, we obtain

$$1 + s^2 = \varphi'(y_1) + \varphi'(y_2) + \cdots + \varphi'(y_{k+1}),$$

where $s = \varphi'(a_1) + \cdots + \varphi'(a_k)$, $y_1 = bu$ and $y_{i+1} = \sqrt{2}\, a_i$ for $i = 1, \ldots, k$. Since $1 + s^2$ is invertible in A, we obtain

$$1 = \sum_{i=1}^{k+1} \varphi'\left(\frac{y_i}{1+s^2}\right) + \sum_{i=1}^{k+1} \varphi'\left(\frac{sy_i}{1+s^2}\right).$$

This shows that 1 is weakly represented by φ' over A. □

Proof of Theorem 6.5.1. It is enough to prove the theorem in the case where $U = \mathcal{U}(g_1, \ldots, g_{d+1})$. Let ΣU be the set of polynomial functions h in $\mathcal{P}(V)$ such that $h(x) > 0$ for all $x \in U$. Let B be the ring of fractions $(\Sigma U)^{-1}\mathcal{P}(V)$. Then the Pfister form $\varphi = \langle\!\langle g_1, \ldots, g_{d+1} \rangle\!\rangle$ is nondegenerate over B, and B is a RFR of transcendence degree at most d over R. By Theorem 6.5.10, the pure subform φ' of φ weakly represents 1 over B, and, hence, it also weakly represents the square $(1 - g_{d+1} + g_{d+1}^2)^2$. We have

$$\varphi' = g_1\langle 1\rangle \perp g_2\varphi_1 \perp \cdots \perp g_{d+1}\varphi_d,$$

where $\varphi_i = \langle\!\langle g_1, \ldots, g_i \rangle\!\rangle$. The weak representation of $(1 - g_{d+1} + g_{d+1}^2)^2$ gives an identity

$$(1 + g_{d+1}^2)^2 = g_1 u_1 + g_2 u_2 + \cdots + g_d u_d + g_{d+1}(u_{d+1} + 2(1 + g_{d+1}^2)),$$

where u_1 is a sum of squares of B, and u_{i+1} is weakly represented by φ_i over B, for $i = 1, \ldots, d$. Note that $u_{d+1} + 2(1 + g_{d+1}^2)$ is positive on U. Clearing denominators, we get an identity in $\mathcal{P}(V)$:

$$w_0 = g_1 v_1 + g_2 v_2 + \cdots + g_d v_d + g_{d+1} v_{d+1},$$

where w_0 is a square in $\mathcal{P}(V)$, v_1 is a sum of squares in $\mathcal{P}(V)$, v_{i+1} is weakly represented by φ_i over $\mathcal{P}(V)$, for $i = 1, \ldots, d$, and v_{d+1} is positive on U. Set $w_i = \sum_{j=i+1}^{d+1} g_j v_j$ and $\psi = \langle\!\langle w_0, g_1 w_1, \ldots, g_d w_d \rangle\!\rangle$. By Lemma 6.5.8, for every $x \in V$ such that both forms φ_x and ψ_x are nondegenerate, we have $\varphi_x \simeq \psi_x$. Set $f_i = g_i w_i$, for $i = 1, \ldots, d$. We now prove that $U = \mathcal{U}(f_1, \ldots, f_d)$.

Since v_{d+1} is positive on U, w_d and, hence, all w_i are positive on U. It follows that $f_1, \ldots, f_d$ are positive on U, and $U \subset \mathcal{U}(f_1, \ldots, f_d)$.

Let x be a point of $\mathcal{U}(f_1, \ldots, f_d)$. Since $g_1, \ldots, g_{d+1}$ all divide $\prod_{i=1}^{d} f_i$ in $\mathcal{P}(V)$, the form φ_x is nondegenerate. Since $g_1 w_0 = g_1^2 v_1 + f_1$ is positive on $\mathcal{U}(f_1, \ldots, f_d)$, we have $w_0(x) \neq 0$ and $w_0(x) > 0$, since w_0 is a square.

Hence, the form ψ_x is also nondegenerate, and its signature is positive. The two forms φ_x and ψ_x are equivalent, since both are nondegenerate. It follows that the signature of φ_x is positive, which implies that all $g_i(x)$ are positive, and, hence, $x \in U$. □

The reduction to d inequalities is optimal. We shall see in Chap. 10 that every algebraic subset of R^n of dimension $d > 0$ has a basic open semi-algebraic subset which cannot be described by less than d inequalities. Here, we prove this fact in the case of affine space.

Theorem 6.5.11. *There do not exist polynomials* $f_1, \ldots, f_{n-1}$ *in* $R[X_1, \ldots, X_n]$ *such that*

$$\mathcal{U}(f_1, \ldots, f_{n-1}) = \{x \in R^n \mid f_1(x) > 0, \ldots, f_{n-1}(x) > 0\}\,.$$

is generically equal to $A_n = \{x \in R^n \mid x_1 > 0, \ldots, x_n > 0\}$.

Proof. We proceed by induction on n. The theorem is clearly true for $n = 2$, otherwise the degree of f_1 would be at the same time odd and even.

Assume that, for some $n > 2$ and some $f_1, \ldots, f_{n-1}$ in $R[X_1, \ldots, X_n]$, the sets A_n and $\mathcal{U}(f_1, \ldots, f_{n-1})$ are generically equal, and that n is the smallest number having this property. Let $X_n^{e_i}$ be the largest power of X_n dividing f_i. We have $f_i = X_n^{e_i} g_i$, where $g_i(X', 0) \neq 0$ in $R[X'] = R[X_1, \ldots, X_{n-1}]$. All e_i cannot be even, since at least one of the f_i has to change sign along $x_n = 0$. Hence, we may assume that e_i is odd (resp. even), for $i = 1, \ldots, r-1$ (resp. $i = r, \ldots, n-1$), where r is some integer satisfying $1 < r \leq n-1$. Define

$$B = \{x' \in R^{n-1} \mid g_1(x', 0)g_i(x', 0) > 0 \text{ for } 1 < i \leq r-1,$$
$$g_j(x', 0) > 0 \text{ for } j = r, \ldots, n-1\}\,.$$

Observe that g_1 does not play a role in the definition of B if $r = 2$. Clearly B is defined by $n-2$ inequalities. Hence, to complete the proof, it suffices to show that our assumptions imply that B is generically equal to A_{n-1}.

Since f_ℓ is nonnegative on A_n, it follows that g_ℓ is nonnegative on A_n, and, therefore,

$$A_{n-1} \setminus \{x' \in R^{n-1} \mid \prod_{i=1}^n g_i(x', 0) = 0\} \subset B\,.$$

Consider now $b' \in B$ such that $g_1(b', 0) \neq 0$. Then there exists an open ball Γ in R^n, centred at $(b', 0)$, such that $g_i(x)$ has the same sign as $g_1(b', 0)$ on Γ, for $i = 1, \ldots, r-1$, and that $g_j(x) > 0$ on Γ, for $j = r, \ldots, n-1$. Hence, there must be a half-ball cut by $x_n = 0$, on which all f_ℓ are positive, for $\ell = 1, \ldots, n-1$. This half-ball is contained in A_n, and, therefore, b' belongs to A_{n-1}. We have

$$B \setminus \{x' \in R^{n-1} \mid g_1(x', 0) = 0\} \subset A_{n-1}\,.$$

This completes the proof. □

6.6 Bibliographic and Historical Notes

6.6.1 Representation of nonnegative polynomials as sums of squares of polynomials

The interest of Hilbert in the representation of nonnegative polynomials as sums of squares goes back to Minkowski's thesis defence in 1885. On this occasion, Minkowski made the following assertion: "It is not likely that every nonnegative form can be represented by a sum of squares of forms" ([157], p. 342). Hilbert was at first sceptical, but later worked on this question (which turned out to play a role in showing the possibility of certain geometrical constructions by elementary means [156]). He published a paper [153] that gave a complete answer: it is Theorem 6.3.7 in the present chapter. The proof that we give here is taken from [86]. Hilbert's proof uses nontrivial facts from the theory of complex projective algebraic curves, and Hilbert gave no explicit example of a nonnegative polynomial that is not the sum of squares of polynomials. Such an example was given for the first time in 1967 by Motzkin [240]: it is the form s in the proof of Proposition 6.3.6. Other examples, for the two cases $(3,6)$ and $(4,4)$, are constructed in [276]. The paper [86] contains examples that are simpler and more symmetric (namely, the form q of Proposition 6.3.6 and $X^4Y^2 + Y^4Z^2 + Z^4X^2 - 3X^2Y^2Z^2$, which are also examples of extremal forms).

Several results about the Pythagoras number $p(\mathcal{P}(V))$ of rings of polynomial functions are proved in [85]. In particular, it is shown that $p(\mathbb{R}[X_1,\ldots,X_n]) = \infty$ for $n \geq 2$, and also that, for each V of dimension ≥ 3, one has $p(\mathcal{P}(V)) = \infty$. On the other hand, for every affine real algebraic curve C, one has $p(\mathcal{P}(C)) < \infty$. The rings $\mathbb{R}[X,Y]/(Y^n - X^{2n-1})$ have unbounded Pythagoras number (as $n \to \infty$).

A few questions remain open:

(i) Is $p(\mathcal{P}(C))$ bounded by a universal constant, if C is an irreducible nonsingular affine real algebraic curve?

(ii) Does there exist a real algebraic surface S with $p(\mathcal{P}(S)) < \infty$?

6.6.2 Representation of nonnegative polynomials as sums of squares of rational functions

In his famous address at the 1900 International Congress of Mathematicians in Paris, David Hilbert proposed the question whether every nonnegative polynomial in $\mathbb{R}[X_1,\ldots,X_n]$ is a sum of squares in $\mathbb{R}(X_1,\ldots,X_n)$. This question became known as Hilbert's 17th problem. Hilbert himself proved that every nonnegative polynomial in $\mathbb{R}[X_1,X_2]$ is a sum of four squares in $\mathbb{R}(X_1,X_2)$, i.e. $p(\mathbb{R}(X_1,X_2)) \leq 4$ [155]. The problem was solved by E. Artin in 1927 [15]. Artin's own proof used a specialization argument, Sturm's theorem on counting the real zeros of polynomials, and, of course, the Artin-Schreier theory (cf. Chap. 1). There is no doubt that this theory was developed precisely to solve Hilbert's 17th problem. Later Lang [209] replaced the specialization argument with the use of real places.

A. Robinson's model-theoretic solution of Hilbert's 17th problem basically applied the model-completeness of the theory of real closed fields deduced from the Tarski-Seidenberg principle [274]. It is essentially Robinson's proof that we give in Section 6.1, since the Artin-Lang homomorphism theorem used in the proof is obtained as a consequence of the Tarski-Seidenberg principle (Theorem 4.1.2). However, we keep this intermediary step in order to emphasize that the solution of Hilbert's 17th problem does not require the full strength of the Tarski-Seidenberg principle (cf. Chap. 5).

The representation given by Hilbert in the case of two variables was constructive. The next attempt to find a constructive solution to Hilbert's 17th problem is contained in Habicht's paper [143]. He proved that, if f is a homogeneous polynomial such that $f(x) > 0$ for all $x \neq 0$, then it is possible to *construct* a representation $f = \sum_{i=1}^{\ell}(g_i/g_0)^2$, where $g_0 \neq 0$ and g_i are homogeneous polynomials. Later Kreisel gave a constructive representation in the general case [193], and Delzell found a representation that is both constructive and continuous [111]. Delzell also proved that the denominators of the rational functions of the representation can be chosen in such a way that the codimension of their set of zeros is at least 3 (in the case of two variables it means that a nonnegative polynomial is a sum of squares of *regular* functions).

Example 6.1.7 (a) (resp. 6.1.7 (b)) is given in [182] (resp. [120]).

The equivariant versions of Hilbert's 17th problem are studied in [259, 260]. A few results of Section 2, concerning the subalgebras of $R[X_1, \ldots, X_n]$, are in [43].

6.6.3 Quantitative aspects of Hilbert's 17th problem

The computation of the Pythagoras number $p_n = p(\mathbb{R}(X_1, \ldots, X_n))$, for $n \geq 2$, is a complicated matter. After the initial result of 1893 of Hilbert, giving the estimate $2 \leq p_2 \leq 4$, nothing more was known for over 70 years. Substantial progress was made only in the sixties. The lower bound $n + 1 \leq p_n$ was obtained in the influential paper by Cassels [82] and improved later to $n+2 \leq p_n$, once it was known that the representation of the Motzkin polynomial needs four squares [83]. Another proof that $p_2 = 4$ was given recently in [90].

Around 1966 J. Ax proved that $p_3 \leq 8$. His idea was to find a link between this question and the Tsen-Lang theorem. He also used methods from cohomology theory, which were sufficient for the case $n \leq 3$, but did not extend to the case $n \geq 4$. In 1967 A. Pfister published the paper [255] containing the proof that $p_n \leq 2^n$. He realized that the cohomological approach can be successfully replaced with his theory of multiplicative forms developed in [254].

A result of Mahé allows one to represent a nonnegative polynomial $f \in \mathbb{R}[X_1, \ldots, X_n]$ as a sum of 2^n squares of rational functions having their poles contained in $f^{-1}(0)$ [221]. This additional fact is useful in studying Hilbert's 17th problem for real analytic functions.

The exact values of p_n are not known for $n \geq 3$ and the estimates $n+2 \leq p_n \leq 2^n$ remain the best, at present. Even less is known in the case of other real fields. The case of $\mathbb{Q}(X)$ was studied in [208], where the bound $p(\mathbb{Q}(X)) \leq 8$ is proved. In fact, one has $p(\mathbb{Q}(X)) = 5$. More generally, if $K \subset \mathbb{R}$ is a number field then $p(K(X)) = p(K) + 1 \leq 5$ [257]. The question of computing $p(K(X_1, \ldots, X_n))$ is far from solved, but the estimates $p(K(X_1, X_2)) \leq 8$ and $p(K(X_1, X_2, X_3)) \leq 16$ are shown in [91].

It is interesting to mention another invariant of a ring A, its *level* $s(A)$. Recall that
$$s(A) = \min\{n \mid -1 = a_1^2 + \cdots + a_n^2, \ a_i \in A\}$$
if -1 is a sum of squares in A. Otherwise, we set $s(A) = \infty$. Of course, $s(\mathcal{P}(V)) = \infty$, for every real algebraic set V. On the other hand, if the set of zeros of an ideal I of $\mathbb{R}[X_1, \ldots, X_n]$ is empty, the level of the quotient ring $\mathbb{R}[X_1, \ldots, X_n]/I$ is finite and, in fact, bounded by a number depending only on the dimension of I [221]. The study of $s(A)$ for finitely generated $\mathbb{R}$-algebras was stimulated by a surprising discovery of Dai and Lam, that $s(\mathbb{R}[X_1, \ldots, X_n]/(1 + \sum_{i=1}^n X_i^2)) = n$. For further results about the level of finitely generated $\mathbb{R}$-algebras, see [105, 221]. An excellent book by Pfister [256] contains a wealth of results concerning the relationship between quadratic forms, number theory, algebraic geometry and topology, including multiplicative forms, Hilbert's 17th problem, and the level and Pythagoras number of rings.

6.6.4 Hilbert's 17th problem for other rings of functions

There exists a nonnegative $\mathcal{C}^\infty$ function from $\mathbb{R}$ into $\mathbb{R}$ that is not a sum of squares of $\mathcal{C}^\infty$ functions (examples due to P. Cohen and D. Epstein). On the other hand, it is easy to see that each nonnegative $\mathcal{C}^\infty$ function is the square of the quotient of two $\mathcal{C}^\infty$ functions.

Each germ of a nonnegative real analytic function is the sum of squares of germs of meromorphic functions [269]. No bound is known for the number of squares, except in one or two variables. In the global case, let M be a connected real analytic manifold, and $\mathcal{O}(M)$, the ring of real analytic functions on M. If M is compact, then each $f \in \mathcal{O}(M)$ is a sum of squares in the field of fractions of $\mathcal{O}(M)$ [282]. If M is a noncompact (resp. compact) surface, then each $f \in \mathcal{O}(M)$ nonnegative on M is a sum of two (resp. three) squares in $\mathcal{O}(M)$ [60].

The case of Nash functions is considered in Chap. 8.

6.6.5 The theorem of Bröcker and Scheiderer

The first bound for the number of inequalities needed to define a basic open semi-algebraic subset can be found in [70]. The improved optimal statement 6.5.1 was obtained by Scheiderer [287] and Bröcker (unpublished). The proof presented here is taken from [78]. The proof of Theorem 6.5.11 is due to Risler.

There is also a precise statement concerning the number of inequalities needed to define a basic *closed* semi-algebraic set, which we shall give in Chap. 10.

Finally there are bounds, given in [71], for the number t (resp. $\bar{t}$), such that every open (resp. closed) semi-algebraic subset of $\mathbb{R}^n$ is the union of t (resp. $\bar{t}$) basic open (resp. closed) semi-algebraic sets. The statements concerning these upper bounds are quantitative versions of the finiteness theorem 2.7.2.

7. Real Spectrum

Abstract. In Chap. 4, we introduced the notion of a prime cone of a ring. Now we define a topology on the set of prime cones of a ring. The topological space thus defined is the real spectrum of the ring. In the first section, we give some general properties of the real spectrum of a ring. Some of these properties are analogous to the properties of the Zariski spectrum (compactness, for instance), while others are more specific (for instance, the fact that the closed points form a compact Hausdorff space). The second section deals with the ring of polynomial functions on an algebraic set V: the real spectrum of this ring contains V, equipped with its euclidean topology, as a subspace, and there is a bijection (the tilde operation) between open semi-algebraic subsets of V and compact open subsets of the real spectrum. This tilde operation gives a dictionary between geometric properties of V and algebraic properties of the real spectrum of the ring of polynomial functions on V. In Section 3, we define the value of a semi-algebraic function at a point of the real spectrum; we also show that the continuous semi-algebraic functions are the sections of a sheaf on the real spectrum. Section 4 deals with semi-algebraic families of sets or mappings. In Section 5, the tilde operation is used to show, on the one hand, that the semi-algebraically connected components correspond to the connected components in the real spectrum, and, on the other hand, that the dimension of a semi-algebraic set S is related to the lengths of specialization chains in $\widetilde{S}$. In Section 6, we study the central points of an irreducible algebraic set V, that is, the limits of nonsingular points of V for the euclidean topology; these central points are strongly related to the orderings of the field of fractions $\mathcal{K}(V)$ in the real spectrum.

7.1 Definition and General Properties of the Real Spectrum

Throughout this section A is a unitary commutative ring.

The notion of a prime cone of the ring A was introduced in Chap. 4. We recall that a prime cone α of A is a subset of A that satisfies (i) $\alpha + \alpha \subset \alpha$, (ii) $\alpha \cdot \alpha \subset \alpha$, (iii) $a^2 \in \alpha$ for every a in A, (iv) $-1 \notin \alpha$ and (v) $ab \in \alpha \Rightarrow (a \in \alpha$ or $-b \in \alpha)$, for every a, b in A (Definition 4.3.1).

The set $\operatorname{supp}(\alpha) = \alpha \cap -\alpha$ is a real prime ideal of A (Corollary 4.3.6). Let $k(\operatorname{supp}(\alpha))$ denote *the residue field of A at* $\operatorname{supp}(\alpha)$, i.e. the field of fractions of $A/\operatorname{supp}(\alpha)$. The prime cone α induces an ordering $\leq_\alpha$ of the residue field $k(\operatorname{supp}(\alpha))$. This ordering is defined by $0 \leq_\alpha \bar{a} \Leftrightarrow a \in \alpha$, for every $a \in A$ (where $\bar{a}$ denotes the class of a in $k(\operatorname{supp}(\alpha))$).

Notation 7.1.1. *If α is a prime cone of A, we denote by $k(\alpha)$ the real closure of the ordered field $(k(\operatorname{supp}(\alpha)), \leq_\alpha)$. If $a \in A$, we denote by $a(\alpha)$ the image of a by the canonical homomorphism from A into $k(\alpha)$.*

Note that
$$a(\alpha) \geq 0 \Leftrightarrow a \in \alpha$$
$$a(\alpha) > 0 \Leftrightarrow a \notin -\alpha$$
$$a(\alpha) = 0 \Leftrightarrow a \in \operatorname{supp}(\alpha).$$

Proposition 7.1.2. *The following data are equivalent:*
(i) A prime cone α of A.
(ii) A couple $(\mathfrak{p}, \leq)$, where $\mathfrak{p}$ is a prime ideal of A, and $\leq$ is an ordering of the residue field $k(\mathfrak{p})$. (The ideal $\mathfrak{p}$ is then necessarily real, by Lemma 4.1.6).
(iii) An equivalence class of homomorphisms $\varphi : A \to R$ with values in a real closed field, for the smallest equivalence relation, such that φ and φ' are equivalent if there exists a commutative diagram of homomorphisms:

$$\begin{array}{ccc} A & \xrightarrow{\varphi} & R \\ & \searrow{\varphi'} & \downarrow \\ & & R' \end{array}$$

More precisely, one goes from (i) to (ii) by taking $(\mathfrak{p}, \leq) = (\operatorname{supp}(\alpha), \leq_\alpha)$, from (ii) to (iii) by taking $\varphi : A \to k(\mathfrak{p}) \to R$, where R is the real closure of $k(\mathfrak{p})$ for $\leq$, and from (iii) to (i) by taking $\alpha = \{a \in A \mid \varphi(a) \geq 0\}$.

Proof. See Proposition 4.3.4 and Lemma 4.3.5. □

Definition 7.1.3. *The* real spectrum of A, *denoted* $\operatorname{Spec}_r(A)$, *is the topological space whose points are the prime cones of A, and whose topology is given by the basis of open subsets*

$$\widetilde{\mathcal{U}}(a_1, \ldots, a_n) = \{\alpha \in \operatorname{Spec}_r(A) \mid a_1(\alpha) > 0, \ldots, a_n(\alpha) > 0\},$$

where $a_1, \ldots, a_n$ is any finite family of elements of A. This topology is called the spectral topology.

A prime cone of A is now considered as a point in a topological space, and the elements of A, as "functions" on this space. The viewpoint has changed from that of Chap. 4, where the prime cones were only considered as subsets of A. This change is reflected in the change of notation. Now the points of the real spectrum will usually be denoted by $\alpha, \beta, \ldots$

7.1 Definition and General Properties of the Real Spectrum

Example 7.1.4.

a) If F is a field, then the prime cones of F are the positive cones of orderings of F and $\mathrm{Spec}_r(F)$ is (homeomorphic to) the space of orderings of F equipped with the *Harrison topology*, often denoted $X(F)$. In this case, $\mathrm{Spec}_r(F)$ is a compact, totally disconnected Hausdorff space ([235], Chap. 3, §2, Lemma 2.8). The set $\mathrm{Spec}_r(F)$ is nonempty if and only if F is real.

b) If α is a prime cone of $\mathbb{R}[X]$, then either $\mathrm{supp}(\alpha)$ is a maximal ideal whose residue field is $\mathbb{R}$ (since $k(\mathrm{supp}(\alpha))$ is real), or $\mathrm{supp}(\alpha) = (0)$. In the first case, $\leq_\alpha$ is the ordering of $\mathbb{R}$, while in the second case it is one of the orderings of $\mathbb{R}(X)$ given by the cuts of $\mathbb{R}$ ($-\infty$, $+\infty$, x_- and x_+ for $x \in \mathbb{R}$, cf. Example 1.1.2). Finally, the points of $\mathrm{Spec}_r(\mathbb{R}[X])$ are (up to the natural identifications):

- the points x of $\mathbb{R}$,
- the points $-\infty, +\infty, x_-$ and x_+ for $x \in \mathbb{R}$.

Let us specify some of the corresponding prime cones:

$$P_x = \{f \in \mathbb{R}[X] \mid f(x) \geq 0\}$$
$$P_{-\infty} = \{f \in \mathbb{R}[X] \mid \exists m \in \mathbb{R} \quad \forall x < m \quad f(x) \geq 0\}$$
$$P_{x+} = \{f \in \mathbb{R}[X] \mid \exists \epsilon > 0 \quad \forall y \in]x, x + \epsilon[\quad f(y) \geq 0\}.$$

We easily see, by factoring polynomials, that the topology of $\mathrm{Spec}_r(\mathbb{R}[X])$ has a basis of open subsets consisting of the following "intervals":

$$\widetilde{\mathcal{U}}(X - x, -X + y) = [x_+, y_-]$$
$$= \{z \in \mathbb{R} \mid x < z < y\} \cup \{z_- \mid z \in \mathbb{R}, \ x < z \leq y\} \cup \{z_+ \mid z \in \mathbb{R}, \ x \leq z < y\},$$

where x and y are two elements of $\mathbb{R}$, such that $x < y$,

$$\widetilde{\mathcal{U}}(X - x) = [x_+, +\infty]$$
$$= \{z \in \mathbb{R} \mid z > x\} \cup \{z_- \mid z \in \mathbb{R}, \ z > x\} \cup \{z_+ \mid z \in \mathbb{R}, \ z \geq x\} \cup \{+\infty\},$$

and

$$\widetilde{\mathcal{U}}(-X + y) = [-\infty, y_-]$$
$$= \{z \in \mathbb{R} \mid z < y\} \cup \{z_- \mid z \in \mathbb{R}, \ z \leq y\} \cup \{z_+ \mid z \in \mathbb{R}, \ z < y\} \cup \{-\infty\},$$

where x and y belong to $\mathbb{R}$.

Note that $\mathrm{Spec}_r(\mathbb{R}[X])$ is not a Hausdorff space: x belongs to the closure of both $\{x_+\}$ and $\{x_-\}$. It is clear from the description of the topology that $\mathrm{Spec}_r(\mathbb{R}[X])$ is compact. Finally, $\mathbb{R}$ with its euclidean topology is a subspace of $\mathrm{Spec}_r(\mathbb{R}[X])$. This is a general fact, as we shall see in the next proposition.

Proposition 7.1.5. *Let R be a real closed field, and V, an algebraic subset of R^n. Then the mapping $V \to \mathrm{Spec}_r(\mathcal{P}(V))$, that sends $x \in V$ to the prime cone $P_x = \{f \in \mathcal{P}(V) \mid f(x) \geq 0\}$, is injective and induces a homeomorphism from V, with its euclidean topology, onto its image in $\mathrm{Spec}_r(\mathcal{P}(V))$.*

From now on, we shall always identify V with its image in $\operatorname{Spec}_r(\mathcal{P}(V))$ by this mapping.

Proof. It is clear that P_x is a prime cone of $\mathcal{P}(V)$, and that $P_x \neq P_y$ if $x \neq y$. The inverse image of the open set $\widetilde{\mathcal{U}}(f_1,\ldots,f_p)$ by the mapping $V \to \operatorname{Spec}_r(\mathcal{P}(V))$ is the open set

$$\mathcal{U}(f_1,\ldots,f_p) = \{x \in V \mid f_1(x) > 0,\ldots, f_p(x) > 0\},$$

and the $\mathcal{U}(f_1,\ldots,f_p)$ form a basis of the euclidean topology of V. □

Remark 7.1.6. Note that the identification of the point x with the prime cone P_x is consistent with Notation 7.1.1: if $f \in \mathcal{P}(V)$, then $f(P_x)$ is indeed equal to $f(x)$. Moreover, note that, if α is any point of $\operatorname{Spec}_r(\mathcal{P}(V))$, then $f(\alpha) = \overline{f}(X_1(\alpha),\ldots,X_n(\alpha))$, where $\overline{f}$ is the image of f in $k(\alpha)[X_1,\ldots,X_n]$.

Proposition 7.1.7. *Let $\varphi : A \to B$ be a ring homomorphism. If β is a prime cone of B, then $\varphi^{-1}(\beta)$ is a prime cone of A and the mapping*

$$\operatorname{Spec}_r(\varphi) : \operatorname{Spec}_r(B) \longrightarrow \operatorname{Spec}_r(A)$$

defined by $\operatorname{Spec}_r(\varphi)(\beta) = \varphi^{-1}(\beta)$ is a continuous mapping. In other words, Spec_r is a contravariant functor from the category of rings to the category of topological spaces.

Proof. The fact that $\varphi^{-1}(\beta)$ is a prime cone of A follows from the definition of prime cones, or by using Proposition 7.1.2 (iii). The continuity follows from the equality $(\operatorname{Spec}_r(\varphi))^{-1}(\widetilde{\mathcal{U}}(a_1,\ldots,a_n)) = \widetilde{\mathcal{U}}(\varphi(a_1),\ldots,\varphi(a_n))$. □

Proposition 7.1.8. *The mapping* $\operatorname{supp} : \operatorname{Spec}_r(A) \to \operatorname{Spec}(A)$, *that sends a prime cone to its support, is a continuous mapping, whose image is the set of real prime ideals of A.*

Proof. Since $\mathfrak{p} \in \operatorname{Spec}(A)$ is real if and only if its residue field $k(\mathfrak{p})$ is real, it is clear that the image of supp is the set of real prime ideals. Moreover, if $D_a = \{\mathfrak{p} \in \operatorname{Spec}(A) \mid a \notin \mathfrak{p}\}$ is a basic open subset of $\operatorname{Spec}(A)$, then

$$\operatorname{supp}^{-1}(D_a) = \{\alpha \mid a(\alpha) > 0 \text{ or } a(\alpha) < 0\} = \widetilde{\mathcal{U}}(a) \cup \widetilde{\mathcal{U}}(-a).$$

□

Remark 7.1.9. In the terminology of the theory of categories, supp is a natural transformation from the functor Spec_r to the functor Spec.

It is useful to introduce another, finer, topology in order to establish some general topological properties of the real spectrum.

7.1 Definition and General Properties of the Real Spectrum

Definition 7.1.10.
(i) A *constructible subset* of $\text{Spec}_r(A)$ *is a boolean combination of basic open subsets* $\widetilde{\mathcal{U}}(a_1,\ldots,a_n)$ *(i.e. obtained from the* $\widetilde{\mathcal{U}}(a_1,\ldots,a_n)$, *by taking finite unions, finite intersections and complements).*
(ii) *The* constructible topology *on* $\text{Spec}_r(A)$ *is the topology defined by the basis of open subsets consisting of the constructible subsets of* $\text{Spec}_r(A)$.

Remark 7.1.11.
a) If $V \subset R^n$ is an algebraic set over a real closed field R, then the intersections of constructible subsets of $\text{Spec}_r(\mathcal{P}(V))$ with V are precisely the semi-algebraic subsets of V (cf. Proposition 7.1.5 and its proof). We shall say more on this topic in the second section.

b) If A is a field, then the complement of $\widetilde{\mathcal{U}}(a_1,\ldots,a_n)$ is $\widetilde{\mathcal{U}}(-a_1) \cup \ldots \cup \widetilde{\mathcal{U}}(-a_n)$, and the constructible topology coincides with the spectral topology of the real spectrum. As previously remarked, this topology is already known as the Harrison topology, and $\text{Spec}_r(A)$ equipped with this topology is a compact, totally disconnected Hausdorff space (Example 7.1.4 a)). This is still the case for the real spectrum of any ring equipped with its constructible topology. Before showing this, we recall the construction the Stone space of a boolean algebra.

Let E be a set, and Λ, a boolean subalgebra of the boolean algebra of subsets of E. This means that Λ contains $\emptyset$ and E and is closed under taking finite unions, finite intersections and complements in E. A subset $\mathcal{F}$ of Λ is called a *filter* of Λ if it satisfies:
(i) $E \in \mathcal{F}$,
(ii) $D \cap D' \in \mathcal{F} \Leftrightarrow (D \in \mathcal{F} \text{ and } D' \in \mathcal{F})$, for $D, D' \in \Lambda$.
A subset $\mathcal{F}$ is called an *ultrafilter* of Λ if it satisfies in addition:
(iii) $\emptyset \notin \mathcal{F}$,
(iv) $D \in \Lambda \Rightarrow (D \in \mathcal{F} \text{ or } E \setminus D \in \mathcal{F})$.

The *Stone space* of Λ is the set of ultrafilters of Λ equipped with the topology defined by the basis of open sets consisting of the subsets

$$\widehat{D} = \{\mathcal{F} \mid \mathcal{F} \text{ is an ultrafilter of } \Lambda, D \in \mathcal{F}\},$$

for $D \in \Lambda$. The Stone space of Λ is a compact, totally disconnected Hausdorff space. The mapping $D \mapsto \widehat{D}$ is an isomorphism from the boolean algebra Λ onto the boolean algebra of closed and open subsets of the Stone space of Λ. For all of this, we refer the reader to [28].

Proposition 7.1.12. *Let φ be the mapping from $\text{Spec}_r(A)$ to the Stone space of the boolean algebra of constructible subsets of $\text{Spec}_r(A)$, defined by*

$$\varphi(\alpha) = \{C \mid C \text{ is a constructible subset of } \text{Spec}_r(A), \ \alpha \in C\}.$$

The mapping φ is a homeomorphism from $\text{Spec}_r(A)$ equipped with its constructible topology onto this Stone space. Hence, $\text{Spec}_r(A)$ with its constructible topology is a compact, totally disconnected Hausdorff space, whose closed and open subsets are precisely the constructible subsets.

Proof. We define the inverse mapping of φ. We use the following notation: if $a \in A$, then $\widetilde{\mathcal{W}}(a) = \{\alpha \in \mathrm{Spec}_r(A) \mid a(\alpha) \geq 0\}$. If $\mathcal{F}$ is an ultrafilter of the boolean algebra of constructible subsets of $\mathrm{Spec}_r(A)$, we define

$$\psi(\mathcal{F}) = \{a \in A \mid \widetilde{\mathcal{W}}(a) \in \mathcal{F}\}.$$

This is a prime cone of A, since

$$\widetilde{\mathcal{W}}(x+y) \supset \widetilde{\mathcal{W}}(x) \cap \widetilde{\mathcal{W}}(y), \qquad \widetilde{\mathcal{W}}(xy) \supset \widetilde{\mathcal{W}}(x) \cap \widetilde{\mathcal{W}}(y),$$
$$\widetilde{\mathcal{W}}(x^2) = \mathrm{Spec}_r(A), \qquad \widetilde{\mathcal{W}}(-1) = \emptyset,$$
$$\widetilde{\mathcal{W}}(xy) \subset \widetilde{\mathcal{W}}(x) \cup \widetilde{\mathcal{W}}(-y).$$

The composition $\psi \circ \varphi$ is the identity mapping, since $\widetilde{\mathcal{W}}(a) \in \varphi(\alpha)$ if and only if $a \in \alpha$. On the other hand, the composition $\varphi \circ \psi$ is the identity mapping, since every constructible subset is a boolean combination of constructible subsets of the form $\widetilde{\mathcal{W}}(a)$, and $\widetilde{\mathcal{W}}(a) \in \mathcal{F}$ if and only if $\widetilde{\mathcal{W}}(a) \in \varphi \circ \psi(\mathcal{F})$. To conclude, we note that the images by φ of the constructible subsets of $\mathrm{Spec}_r(A)$ are precisely the closed and open subsets of the Stone space, by the very definition of φ. □

Corollary 7.1.13. *Every constructible subset of $\mathrm{Spec}_r(A)$ is compact (with respect to the spectral topology of $\mathrm{Spec}_r(A)$). In particular, the basic open subsets $\widetilde{\mathcal{U}}(a_1, \ldots, a_n)$, and $\mathrm{Spec}_r(A)$ itself, are compact. An open subset of $\mathrm{Spec}_r(A)$ is constructible if and only if it is compact.*

Proof. By Proposition 7.1.12, the constructible subsets are compact for the constructible topology, which is finer than the spectral topology. □

There is a result similar to Proposition 7.1.12 for $\mathrm{Spec}_r(A)$ equipped with its spectral topology and the family of constructible open subsets of $\mathrm{Spec}_r(A)$. Of course, this family is not in general a boolean algebra, so we need an extension of the classical notion of Stone space.

Definition 7.1.14. *Let E be a set and Λ a family of subsets of E containing E and the empty set $\emptyset$ and closed under taking finite unions and intersections. A subset $\mathcal{F}$ of Λ is called a* prime filter *of Λ if it satisfies:*
   (i) $E \in \mathcal{F}$,
   (ii) $D \cap D' \in \mathcal{F} \Leftrightarrow (D \in \mathcal{F}$ and $D' \in \mathcal{F})$, for $D, D' \in \Lambda$,
   (iii) $\emptyset \notin \mathcal{F}$,
   (iv) $D \cup D' \in \mathcal{F} \Leftrightarrow (D \in \mathcal{F}$ or $D' \in \mathcal{F})$, for $D, D' \in \Lambda$.

The Stone space *of Λ is the set of prime filters of Λ equipped with the topology which has a basis of open subsets consisting of all subsets*

$$\widehat{D} = \{\mathcal{F} \mid \mathcal{F} \text{ is a prime filter of } \Lambda, D \in \mathcal{F}\},$$

for $D \in \Lambda$.

Note that the notions of ultrafilter and prime filter coincide if Λ is a boolean algebra. Hence, in the case of a boolean algebra, Definition 7.1.14 is the usual definition of a Stone space. But the Stone space of Λ is, in general, neither Hausdorff nor totally disconnected, if Λ is not a boolean algebra.

Proposition 7.1.15. *Let θ be the mapping from $\mathrm{Spec}_r(A)$ to the Stone space of the family of constructible open subsets of $\mathrm{Spec}_r(A)$, defined by*

$$\theta(\alpha) = \{C \mid C \text{ is a constructible open subset of } \mathrm{Spec}_r(A),\ \alpha \in C\}.$$

The mapping θ is a homeomorphism from $\mathrm{Spec}_r(A)$ onto this Stone space.

Proof. We repeat the proof of Proposition 7.1.12, with the following changes. The inverse mapping ρ of θ is defined by

$$\rho(\mathcal{F}) = \{a \in A \mid \widetilde{\mathcal{U}}(-a) \notin \mathcal{F}\}.$$

We can verify that $\rho(\mathcal{F})$ is a prime cone. The composition $\rho \circ \theta$ is the identity mapping, since $a \in \alpha$ if and only if $\widetilde{\mathcal{U}}(-a) \notin \theta(\alpha)$. The other composition $\theta \circ \rho$ is also the identity mapping, since a constructible open set is a finite union of finite intersections of open sets of the form $\widetilde{\mathcal{U}}(b)$. The mapping θ is a homeomorphism, by the definition of the topology of the Stone space.
□

Corollary 7.1.16. *An irreducible closed subset of $\mathrm{Spec}_r(A)$ is the closure of a unique point.*

Proof. Let F be an irreducible closed subset of $\mathrm{Spec}_r(A)$. We associate to F the prime filter $\mathcal{F}$ of the family of constructible open subsets defined by

$$\mathcal{F} = \{C \subset \mathrm{Spec}_r(A) \mid C \text{ is a constructible open subset},\ C \cap F \neq \emptyset\}.$$

It is easy to verify that this defines a bijection from the set of irreducible closed subsets to the Stone space of the family of constructible open subsets, and the corollary now follows from Proposition 7.1.15.
□

Remark 7.1.17. The space $\mathrm{Spec}_r(A)$ has the following properties:

(i) it has a basis of compact open subsets, closed under taking finite intersections,

(ii) each irreducible closed subset is the closure of a unique point.

Following the terminology of [162], a space with these properties is a *spectral space*. Hochster shows that the spectral spaces are precisely the spaces homeomorphic to the prime spectrum of a ring. The duality established in Proposition 7.1.15 is a particular case of the duality between spectral spaces and distributive lattices [181].

Corollary 7.1.16 leads to the notion of specialization.

Proposition 7.1.18. *Let α, β be two points of $\text{Spec}_r(A)$. Then the following conditions are equivalent:*
  (i) $\alpha \subset \beta$.
  (ii) $\forall a \in A \quad a(\alpha) \geq 0 \Rightarrow a(\beta) \geq 0$.
  (iii) $\forall a \in A \quad a(\beta) > 0 \Rightarrow a(\alpha) > 0$.
  (iv) $\beta \in \text{clos}(\{\alpha\})$.

Proof. (i) $\Leftrightarrow$ (ii) $\Leftrightarrow$ (iii) is obvious. The equivalence (iii) $\Leftrightarrow$ (iv) comes from the fact that both conditions are equivalent to

$$\beta \in \widetilde{\mathcal{U}}(a_1, \ldots, a_n) \Rightarrow \alpha \in \widetilde{\mathcal{U}}(a_1, \ldots, a_n).$$

$\square$

Definition 7.1.19. *If the equivalent conditions of Proposition 7.1.18 are satisfied, we say that β is a specialization of α, or that α is a generization of β.*

Example 7.1.20. In $\text{Spec}_r(\mathbb{R}[X])$ (cf. Example 7.1.4 b)), the point x in $\mathbb{R}$ is a specialization of x_+ and of x_-.

The constructible topology, together with the partial order on $\text{Spec}_r(A)$ given by the relation of specialization, completely determine the spectral topology of the real spectrum. More precisely:

Proposition 7.1.21. *Let C be a constructible subset of $\text{Spec}_r(A)$. Then $\alpha \in \text{Spec}_r(A)$ belongs to the closure $\text{clos}(C)$ for the spectral topology if and only if there exists $\beta \in C$, such that α is a specialization of β.*

Proof. By Proposition 7.1.18 (iv), it is clear that the closure of C contains the set of specializations of points of C. Let $\alpha \in \text{clos}(C)$. For every basic open subset $\widetilde{\mathcal{U}}(a_1, \ldots, a_n)$ containing α, $\widetilde{\mathcal{U}}(a_1, \ldots, a_n) \cap C$ is nonempty. Since the sets $\widetilde{\mathcal{U}}(a_1, \ldots, a_n)$ and C are compact in the constructible topology, there exists a point β in C belonging to all $\widetilde{\mathcal{U}}(a_1, \ldots, a_n)$ which contain α. Hence α is a specialization of β. $\square$

In what follows, the subspaces of $\text{Spec}_r(A)$ we consider will be equipped with the topology induced by the spectral topology of $\text{Spec}_r(A)$, unless otherwise stated.

Corollary 7.1.22. *Let $C \subset D$ be two constructible sets of $\text{Spec}_r(A)$. Then C is closed (resp. open) in D if and only if it is stable under specialization (resp. generization) in D.*

The notion of specialization, the results of Proposition 7.1.21 and Corollary 7.1.22 are valid for any spectral space (Remark 7.1.17). The following properties are more specific to the real spectrum.

7.1 Definition and General Properties of the Real Spectrum

Proposition 7.1.23. *Let α be a point of $\mathrm{Spec}_r(A)$. The specializations of α are totally ordered with respect to inclusion: if $\alpha \subset \beta$ and $\alpha \subset \gamma$, then $\beta \subset \gamma$ or $\gamma \subset \beta$.*

Proof. If the conclusion does not hold, we can find $b \in \beta \setminus \gamma$ and $c \in \gamma \setminus \beta$. We know that $b - c \in \alpha$, or $c - b \in \alpha$ (Proposition 4.3.2 (i)). In the first case, $b = c + (b - c) \in \gamma$, while in the latter $c = b + (c - b) \in \beta$. We get a contradiction in both cases. □

Now we consider the subspace of closed points of a constructible set. We recall that a point α of a constructible set C is closed in C if and only if there is no proper specialization of α in C (i.e., a specialization different from α).

Proposition 7.1.24. *Let C be a constructible subset of $\mathrm{Spec}_r(A)$ and $\alpha \in C$. There exists a unique closed point in C, denoted $\rho_C(\alpha)$, such that $\rho_C(\alpha)$ is a specialization of α.*

Proof. The uniqueness is an obvious consequence of Proposition 7.1.23. For the existence, we use the fact (proved as in Proposition 2.1.8) that C is a finite union of constructible sets C_i, $1 \leq i \leq p$, of the form

$$C_i = \{\gamma \in \mathrm{Spec}_r(A) \mid f_1(\gamma) > 0, \ldots, f_p(\gamma) > 0, \ g(\gamma) = 0\},$$

where $f_1, \ldots, f_p, g \in A$. If α has a specialization in C_i, we define β_i to be the union of all $\beta \in C_i$ such that β is a specialization of α. Then β_i is a prime cone of A (this is easily checked, using Proposition 7.1.23). It belongs to C_i, since, otherwise, we would have $f_j(\beta_i) = 0$ for some j, and, hence, there would be a specialization β of α in C_i such that $f_j(\beta) = 0$, which is impossible. Let $\rho_C(\alpha)$ be the largest of these β_i. Then $\rho_C(\alpha)$ belongs to C, is a specialization of α and has no proper specialization in C. □

Proposition 7.1.25. *Let C be a constructible subset of $\mathrm{Spec}_r(A)$.*

(i) Let F and G be two disjoint closed subsets of C. Then there exist two disjoint constructible open subsets U and V of C, such that $F \subset U$ and $G \subset V$.

(ii) The subspace $\mathrm{Max}(C)$ of closed points of C is compact Hausdorff.

(iii) The mapping $\rho_C : C \to \mathrm{Max}(C)$ is a continuous and closed retraction (with the notations of Proposition 7.1.24).

Proof. (i) Suppose that for every two constructible open subsets U, V of C such that $F \subset U$ and $G \subset V$, we have $U \cap V \neq \emptyset$. Then, by compactness of the constructible topology, there exists a point α contained in every constructible open subset of C containing F and in every constructible open subset of C containing G. Since F and G are compact in the constructible topology, both have a basis of constructible open neighbourhoods in C. Hence, $\mathrm{clos}(\{\alpha\})$ has nonempty intersection with both F and G. Then $\rho_C(\alpha) \in F \cap G$, which contradicts the hypothesis.

(ii) By (i), two closed points of C are separated by disjoint open subsets of C; hence, a fortiori, Max(C) is Hausdorff. The compactness of Max(C) follows from the compactness of C and the fact that an open subset of C containing Max(C) is equal to C (Corollary 7.1.22).

(iii) Let U be an open subset of C, $\alpha \in \rho_C^{-1}(U \cap \text{Max}(C))$. By (i), we can find an open subset V and a closed subset F of C, both constructible, such that $\rho_C(\alpha) \in V \subset F \subset U$. Then $\alpha \in V$, and $V \subset F \subset \rho_C^{-1}(U \cap \text{Max}(C))$ by Corollary 7.1.22. This proves the continuity of ρ_C. The mapping ρ_C is closed, because C is compact, and Max(C) is compact Hausdorff, by (ii) (it is even true that the image of every constructible subset of C by ρ_C is closed). $\square$

Example 7.1.26. The subset of closed points of $\text{Spec}_r(\mathbb{R}[X])$ is $\mathbb{R} \cup \{-\infty, +\infty\}$ (cf. Example 7.1.4 b)). This space of closed points has a basis of open sets consisting of the intervals $[-\infty, s[, \,]r, s[, \,]r, +\infty]$. So it is a compact Hausdorff space, homeomorphic to a segment.

7.2 Real Spectrum of a Ring of Polynomial Functions

For the remainder of this chapter, R will be a real closed field.

Let $V \subset R^n$ be an algebraic set. We are going to explain the relations existing between V and the real spectrum of the ring $\mathcal{P}(V)$.

Proposition 7.2.1. *Let I be an ideal of $R[X_1, \ldots, X_n]$, and $V = \mathcal{Z}(I)$, its set of zeros in R^n. Denote by $\varphi : R[X_1, \ldots, X_n]/I \to \mathcal{P}(V)$ the canonical surjection and $i : \mathcal{P}(V) \hookrightarrow \mathcal{R}(V)$ the canonical injection. Then $\text{Spec}_r(\varphi)$ and $\text{Spec}_r(i)$ are homeomorphisms.*

Proof. We have $\mathcal{P}(V) = R[X_1, \ldots, X_n]/\sqrt[R]{I}$, and a real prime ideal of $R[X_1, \ldots, X_n]$ contains I if and only if it contains $\sqrt[R]{I}$ (Proposition 4.1.7). On the other hand, $\mathcal{R}(V) = \Sigma_1^{-1} \mathcal{P}(V)$, where Σ_1 is the set of elements of the form "1 + a sum of squares" (Corollary 4.4.5), and a real prime ideal of $\mathcal{P}(V)$ has empty intersection with Σ_1 (otherwise, it would contain 1). So, the homomorphisms φ and i induce bijections between the sets of real prime ideals of the rings $R[X_1, \ldots, X_n]/I$, $\mathcal{P}(V)$, and $\mathcal{R}(V)$. Moreover, if $\mathfrak{p}$ is a real prime ideal of $\mathcal{R}(V)$, then $i^{-1}(\mathfrak{p})$ and $\varphi^{-1}(i^{-1}(\mathfrak{p}))$ have the same residue field as $\mathfrak{p}$. It is then clear, by the description of the points of the real spectrum in Proposition 7.1.2 (ii), that $\text{Spec}_r(\varphi)$ and $\text{Spec}_r(i)$ are bijective. It is easy to check that they are homeomorphisms. Note that, if $V = \emptyset$, then $\mathcal{P}(V)$ and $\mathcal{R}(V)$ are the null ring, and the three real spectra are empty. $\square$

In Proposition 7.1.5, we have seen that the identification of the point x of V with the prime cone $P_x = \{f \in \mathcal{P}(V) \mid f(x) \geq 0\}$ of $\mathcal{P}(V)$ allows one to consider V as a subspace of $\text{Spec}_r(\mathcal{P}(V))$. With this identification, we can state the following result.

7.2 Real Spectrum of a Ring of Polynomial Functions

Proposition 7.2.2. *Let S be a semi-algebraic subset of V.*

(i) There exists a unique constructible subset of $\mathrm{Spec}_r(\mathcal{P}(V))$, denoted by $\widetilde{S}$, such that $\widetilde{S} \cap V = S$.

(ii) If S is a boolean combination of
$$\mathcal{U}(f_i) = \{x \in V \mid f_i(x) > 0\},$$
where $f_i \in \mathcal{P}(V)$, then $\widetilde{S}$ is the same boolean combination of
$$\widetilde{\mathcal{U}}(f_i) = \{\alpha \in \mathrm{Spec}_r(\mathcal{P}(V)) \mid f_i(\alpha) > 0\}.$$

(iii) If Φ is a formula of the first-order language of ordered fields with parameters in R and n free variables $x_1, \ldots, x_n$, such that
$$S = \{x = (x_1, \ldots, x_n) \in V \mid \Phi(x_1, \ldots, x_n)\},$$
then
$$\widetilde{S} = \{\alpha \in \mathrm{Spec}_r(\mathcal{P}(V)) \mid \Phi(X_1(\alpha), \ldots, X_n(\alpha))\ \text{holds true in}\ k(\alpha)\}.$$

Proof. (i) The mapping, which takes a constructible subset C of $\mathrm{Spec}_r(\mathcal{P}(V))$ to $C \cap V$, is a surjective homomorphism from the boolean algebra of constructible subsets of $\mathrm{Spec}_r(\mathcal{P}(V))$ onto the boolean algebra of semi-algebraic subsets of V. It remains to prove that this homomorphism is injective. It is sufficient to show that, if C is a constructible subset of $\mathrm{Spec}_r(\mathcal{P}(V))$ such that $C \cap V = \emptyset$, then $C = \emptyset$. Since C is a finite union of constructible subsets of the form
$$\{\alpha \in \mathrm{Spec}_r(\mathcal{P}(V)) \mid f_1(\alpha) > 0, \ldots, f_k(\alpha) > 0, g(\alpha) = 0\},$$
we may assume that C is of this form. Then, since $C \cap V$ is empty, there is no homomorphism from the R-algebra of finite type
$$\mathcal{P}(V)[Y_1, \ldots, Y_k]/(f_1 Y_1^2 - 1, \ldots, f_k Y_k^2 - 1, g)$$
into R. Hence, by the Artin-Lang homomorphism theorem 4.1.2, there is no homomorphism from this R-algebra into a real closed field. This implies that C is empty.

(ii) It is clear that $\widetilde{\mathcal{U}}(f_i) \cap V = \mathcal{U}(f_i)$ (which justifies the notation $\widetilde{\mathcal{U}}$).

(iii) Using the Tarski-Seidenberg principle (as in Proposition 5.2.2), we replace Φ with an equivalent quantifier-free formula. Then we apply (ii). $\square$

In the following, we shall use the alternative notation $(S)\tilde{\ }$ for $\widetilde{S}$, in order to avoid writing long tildes.

Theorem 7.2.3. *Let S be a semi-algebraic subset of V.*

(i) The mapping $S \mapsto \widetilde{S}$ is an isomorphism from the boolean algebra of semi-algebraic subsets of V onto the boolean algebra of constructible subsets of $\mathrm{Spec}_r(\mathcal{P}(V))$.

(ii) S is open (resp. closed) in V if and only if $\widetilde{S}$ is open (resp. closed) in $\mathrm{Spec}_r(\mathcal{P}(V))$. Hence, the isomorphism $S \mapsto \widetilde{S}$ induces a bijection from the family of open semi-algebraic subsets of V onto the family of compact open subsets of $\mathrm{Spec}_r(\mathcal{P}(V))$.

(iii) The tilde operation commutes with the closure and the interior:

$$\mathrm{clos}(\widetilde{S}) = (\mathrm{clos}(S))\tilde{}\,, \qquad \mathrm{int}(\widetilde{S}) = (\mathrm{int}(S))\tilde{}\,.$$

Proof. (i) has already been proved in the previous proposition.

(ii) It is clear that, if $\widetilde{S}$ is open, then $S = \widetilde{S} \cap V$ is open in V. If S is an open semi-algebraic subset of V, then S is a finite union of basic open semi-algebraic subsets

$$\mathcal{U}(f_1, \ldots, f_k) = \{x \in V \mid f_1(x) > 0, \ldots, f_k(x) > 0\}, \ f_i \in \mathcal{P}(V),$$

by the finiteness theorem 2.7.2. Then, by Proposition 7.2.2 (ii), $\widetilde{S}$ is a finite union of $\widetilde{\mathcal{U}}(f_1, \ldots, f_k)$, and, hence, it is a compact open subset of $\mathrm{Spec}_r(\mathcal{P}(V))$.

(iii) Since the compact open subsets form a basis of the topology of $\mathrm{Spec}_r(\mathcal{P}(V))$, we deduce from (ii) that $\mathrm{int}(\widetilde{S})$ is the union of all $\widetilde{U}$ such that U is an open semi-algebraic subset of V contained in S. Since $\mathrm{int}(S)$ is semi-algebraic, $\mathrm{int}(\widetilde{S}) = (\mathrm{int}(S))\tilde{}$. The analogous result for closures follows by taking complements. $\square$

Corollary 7.2.4. *Let S be a semi-algebraic subset of V.*

(i) The mapping that sends a point $\alpha \in \widetilde{S}$ to the ultrafilter of semi-algebraic subsets T of S, such that $\alpha \in \widetilde{T}$, is a homeomorphism from $\widetilde{S}$ equipped with the constructible topology onto the Stone space of the boolean algebra of semi-algebraic subsets of S.

(ii) The mapping that sends a point $\alpha \in \widetilde{S}$ to the prime filter of open semi-algebraic subsets U of S, such that $\alpha \in \widetilde{U}$, is a homeomorphism from $\widetilde{S}$ onto the Stone space of the family of open semi-algebraic subsets of S.

Proof. If $S = V$, then (i) (resp. (ii)) is an immediate consequence of Theorem 7.2.3 and Proposition 7.1.12 (resp. Proposition 7.1.15). In the general case, it is easily checked that we can replace in Propositions 7.1.12 and 7.1.15 $\mathrm{Spec}_r(A)$ with any constructible subset of $\mathrm{Spec}_r(A)$. We also use the fact that an open semi-algebraic subset of S (resp. a compact open subset of $\widetilde{S}$) is the intersection of an open semi-algebraic subset of V with S (resp. of a compact open subset of $\mathrm{Spec}_r(\mathcal{P}(V))$ with $\widetilde{S}$). $\square$

7.2 Real Spectrum of a Ring of Polynomial Functions 145

Remark 7.2.5. Corollary 7.2.4 shows that $\widetilde{S}$ is defined intrinsically by S (up to a homeomorphism) and does not depend on the choice of the algebraic set V. We can, thus, take the tilde of a semi-algebraic set, without specifying the ring, in the real spectrum of which $\widetilde{S}$ is constructible. We shall also use $\widetilde{V}$ instead of $\mathrm{Spec}_r(\mathcal{P}(V))$, when it is more convenient.

Example 7.2.6. We now describe the tilde operation in the case of the real spectrum of $\mathbb{R}[X]$. A semi-algebraic subset of $\mathbb{R}$ is a finite union of points and open intervals. The tilde of a point is the point itself (again using the identification of $\mathbb{R}$ with a subspace of $\mathrm{Spec}_r(\mathbb{R}[X])$). For open intervals:

$$(]a,b[)^\sim = [a_+, b_-]\,, \quad (]a, +\infty[)^\sim = [a_+, +\infty]\,, \quad (]-\infty, b[)^\sim = [-\infty, b_-]\,,$$

with the notations of Example 7.1.4. Note that two distinct open subsets of $\mathrm{Spec}_r(\mathbb{R}[X])$ can have the same intersection with $\mathbb{R}$. The intersection of the open set $]a_+, b_-[= \bigcup_{a<c<d<b}[c_+, d_-]$ with $\mathbb{R}$ is also the interval $]a, b[$. But $]a_+, b_-[$ is not compact. In the general case, if U is an open semi-algebraic subset of V, we can characterize, in the following way, $\widetilde{U}$ among the open subsets of $\mathrm{Spec}_r(\mathcal{P}(V))$ whose intersection with V is U.

Proposition 7.2.7.
(i) *Let U be an open semi-algebraic subset of V. Then $\widetilde{U}$ is the largest open subset of $\mathrm{Spec}_r(\mathcal{P}(V))$ whose intersection with V is U.*
(ii) *Let F be a closed semi-algebraic subset of V. Then $\widetilde{F}$ is the smallest closed subset of $\mathrm{Spec}_r(\mathcal{P}(V))$ whose intersection with V is F.*

Proof. (i) Let Ω be an open subset of $\mathrm{Spec}_r(\mathcal{P}(V))$, such that $\Omega \cap V = U$. Since the constructible open subsets form a basis of the topology of $\mathrm{Spec}_r(\mathcal{P}(V))$, it follows that Ω is the union of $\widetilde{T}$, for all semi-algebraic subsets T of V such that $\widetilde{T} \subset \Omega$. If $\widetilde{T} \subset \Omega$, then $T = \widetilde{T} \cap V \subset U$, and, hence, $\widetilde{T} \subset \widetilde{U}$. This shows that $\Omega \subset \widetilde{U}$.
(ii) is obtained from (i) by taking complements. □

In this section, we have studied the tilde operation for semi-algebraic sets. Now we explain how this tilde operation can be defined for semi-algebraic mappings. Of particular interest is the case of semi-algebraic homeomorphisms, which is needed later.

Proposition 7.2.8. *Let S and T be two semi-algebraic sets, and $f: S \to T$, a semi-algebraic mapping. Then there exists a unique mapping $\widetilde{f}: \widetilde{S} \to \widetilde{T}$, such that $\widetilde{f}^{-1}(\widetilde{T'}) = (f^{-1}(T'))^\sim$ for every semi-algebraic subset T' of T. If f is a bijection, then $\widetilde{f}$ is a bijection. If f is a homeomorphism, then $\widetilde{f}$ is a homeomorphism.*

Proof. Let $\alpha \in \widetilde{S}$. The family of semi-algebraic subsets T' of T, such that $\alpha \in (f^{-1}(T'))^\sim$, is an ultrafilter of semi-algebraic subsets of T. Hence, it determines a point $\widetilde{f}(\alpha)$ of $\widetilde{T}$, by Corollary 7.2.4 (i). We have $\widetilde{f}(\alpha) \in \widetilde{T'} \Leftrightarrow$

$\alpha \in (f^{-1}(T'))\tilde{}$, and this property characterizes completely the mapping $\tilde{f}: \tilde{S} \to \tilde{T}$. If f is bijective, the mapping $T' \mapsto f^{-1}(T')$ is an isomorphism from the boolean algebra of semi-algebraic subsets of T onto the boolean algebra of semi-algebraic subsets of S, and, hence, by Corollary 7.2.4 (i) again, the mapping $\tilde{f}$ is bijective. Finally, if f is a homeomorphism, the equivalences:

$$\tilde{f}^{-1}(\tilde{U}) \text{ open } \Leftrightarrow (f^{-1}(U))\tilde{} \text{ open } \Leftrightarrow f^{-1}(U) \text{ open } \Leftrightarrow U \text{ open } \Leftrightarrow \tilde{U} \text{ open}$$

show that $\tilde{f}$ is a homeomorphism as well. □

7.3 Semi-algebraic Functions on the Real Spectrum

Let $T \subset R^n$ be a semi-algebraic set. We know how to evaluate a polynomial f at a point $\alpha \in \tilde{T}$. We can also evaluate semi-algebraic functions:

Proposition 7.3.1. *Let $f : T \to R$ be a semi-algebraic function, and $\alpha \in \tilde{T}$. Denote by $f(\alpha)$ the element $f_{k(\alpha)}(X(\alpha))$ of $k(\alpha)$, where $f_{k(\alpha)} : T_{k(\alpha)} \to k(\alpha)$ is the extension of f to $k(\alpha)$ (cf. Definition 5.3.2), and $X(\alpha) = (X_1(\alpha), \ldots, X_n(\alpha))$ is the evaluation of the coordinates at α. The mapping*

$$\mathrm{ev}_\alpha : \mathcal{S}^0(T) \to k(\alpha)$$

defined by $\mathrm{ev}_\alpha(f) = f(\alpha)$ is a ring homomorphism. Moreover, if $\mathcal{U}(f) = \{x \in T \mid f(x) > 0\}$, then $(\mathcal{U}(f))\tilde{} = \{\alpha \in \tilde{T} \mid f(\alpha) > 0\}$.

Proof. The fact that ev_α is a ring homomorphism follows from the fact that the mapping from $\mathcal{S}^0(T)$ into $\mathcal{S}^0(T_{k(\alpha)})$, that sends f to its extension $f_{k(\alpha)}$, is a ring homomorphism. Let $\Phi(x,t)$ be a formula of the first-order language of ordered fields with parameters in R, such that $t = f(x) \Leftrightarrow \Phi(x,t)$. The element $f(\alpha)$ is the unique element t of $k(\alpha)$ satisfying $\Phi(X(\alpha),t)$. We have

$$\mathcal{U}(f) = \{x \in T \mid \forall t \in R \quad \Phi(x,t) \Rightarrow t > 0\},$$

and hence

$$(\mathcal{U}(f))\tilde{} = \{\alpha \in \tilde{T} \mid \forall t \in k(\alpha) \quad \Phi(X(\alpha),t) \Rightarrow t > 0\} = \{\alpha \in \tilde{T} \mid f(\alpha) > 0\}.$$

□

A continuous semi-algebraic function on T can, thus, be viewed as a function on $\tilde{T}$ taking its values in a field $k(\alpha)$ that depends on the point $\alpha \in \tilde{T}$. It is in fact a section of a sheaf of local rings over $\tilde{T}$, that we now define.

Proposition 7.3.2. *There exists a unique sheaf $\widetilde{\mathcal{S}^0}_{\tilde{T}}$ of rings over $\tilde{T}$, such that $\widetilde{\mathcal{S}^0}_{\tilde{T}}(\tilde{U}) = \mathcal{S}^0(U)$ for every open semi-algebraic subset U of T. This sheaf is denoted simply $\widetilde{\mathcal{S}^0}$, if no confusion is possible.*

Proof. The $\widetilde{U}$ form a basis of open subsets of $\widetilde{T}$. Hence, in order to define a sheaf over $\widetilde{T}$, it is enough to describe its sections over the $\widetilde{U}$. It remains to check that $\widetilde{\mathcal{S}^0}$, as defined in the proposition, is indeed a sheaf. Let $\widetilde{U} = \bigcup_i \widetilde{U}_i$ be an open cover of $\widetilde{U}$, $f_i \in \mathcal{S}^0(U_i)$, with $f_i|_{U_i \cap U_j} = f_j|_{U_i \cap U_j}$. Since $\widetilde{U}$ is compact (Theorem 7.2.3), we may assume that the cover is finite. The functions f_i can be glued together to give a continuous function $f : U = \bigcup_i U_i \to R$, and f is semi-algebraic, since its graph is the finite union of the graphs of the f_i. □

Remark 7.3.3. The compactness of $\widetilde{T}$ is essential. The continuous semi-algebraic functions do not form a sheaf on T. Take, for instance, $T = \mathbb{R}$ and $f : \mathbb{R} \to \mathbb{R}$ defined by $f(x) = |x - \frac{1}{2} - \lfloor x \rfloor|$, where $\lfloor x \rfloor$ denotes the greatest integer function. The function f is continuous semi-algebraic on each open interval of the cover $\mathbb{R} = \bigcup_{n \in \mathbb{Z}}]n, n+2[$, but f is not globally semi-algebraic on $\mathbb{R}$.

Proposition 7.3.4. *Let $\alpha \in \widetilde{T}$. The stalk*

$$\widetilde{\mathcal{S}^0}_{\widetilde{T}, \alpha} = \varinjlim_{\alpha \in \widetilde{U}} \mathcal{S}^0(U) \quad \text{(for U an open semi-algebraic subset of T)}$$

is a local ring whose residue field is $k(\alpha)$.

Proof. The evaluation homomorphisms of Proposition 7.3.1 induce a homomorphism, also denoted $\mathrm{ev}_\alpha : \widetilde{\mathcal{S}^0}_\alpha \to k(\alpha)$, that sends f to $f(\alpha)$. We show that, if $f(\alpha) \neq 0$, then f is invertible in $\widetilde{\mathcal{S}^0}_\alpha$. The function f is defined on some open semi-algebraic subset U of T, with $\alpha \in \widetilde{U}$. Since $f(\alpha) \neq 0$, one has

$$\alpha \in \{\beta \in \widetilde{U} \mid f(\beta) \neq 0\} = (\{x \in U \mid f(x) \neq 0\})\widetilde{}$$

(by Proposition 7.3.1). Since f is invertible on $\{x \in U \mid f(x) \neq 0\}$ and its inverse is continuous semi-algebraic, it follows that f is invertible in $\widetilde{\mathcal{S}^0}_\alpha$. This already shows that $\widetilde{\mathcal{S}^0}_\alpha$ is local and that its maximal ideal is the set of f such that $f(\alpha) = 0$. It remains to prove that the homomorphism $\mathrm{ev}_\alpha : \widetilde{\mathcal{S}^0}_\alpha \to k(\alpha)$ is surjective. Let $a \in k(\alpha)$. Since a is algebraic over $k(\mathrm{supp}(\alpha))$, it has a minimal polynomial $b_p Y^p + \cdots + b_0$ over $k(\mathrm{supp}(\alpha))$. We may assume that $b_p, \ldots, b_0$ are classes in $k(\mathrm{supp}(\alpha))$ of polynomials $g_p, \ldots, g_0$ in $R[X_1, \ldots, X_n]$. We set $h(X, Y) = g_p(X) Y^p + \cdots + g_0(X)$ and denote by $h', \ldots, h^{(p)}$ the derivatives of h with respect to Y. Let $\epsilon_1, \ldots, \epsilon_p$ be the signs of $h'(\alpha, a), \ldots, h^{(p)}(\alpha, a)$. None of these signs is equal to zero, since $h(\alpha, Y) = b_p Y^p + \cdots + b_0$ is the minimal polynomial of a over $k(\mathrm{supp}(\alpha))$. We define

$$\Omega = \{x \in T \mid \exists y \in R \quad h(x, y) = 0 \text{ and}$$
$$\mathrm{sign}(h^{(k)}(x, y)) = \epsilon_k \text{ for } k = 1, \ldots, p\}.$$

The set Ω is semi-algebraic, and $\alpha \in \widetilde{\Omega}$. For every x in Ω, there exists a unique $f(x) \in R$ such that $h(x, f(x)) = 0$ and $\mathrm{sign}(h^{(k)}(x, f(x))) = \epsilon_k$ for

$k = 1, \ldots, p$, by Thom's lemma (Proposition 2.5.4). It is clear that f is semi-algebraic. Moreover, Ω is open and f is continuous by the implicit function theorem 2.9.8. Finally, $f(\alpha) = a$. □

Corollary 7.3.5. *Let $\alpha \in \widetilde{T}$ and $a \in k(\alpha)$. Then there exists an open semi-algebraic subset U of T, such that $\alpha \in \widetilde{U}$, and a continuous semi-algebraic function $f : U \to R$, such that $f(\alpha) = a$.*

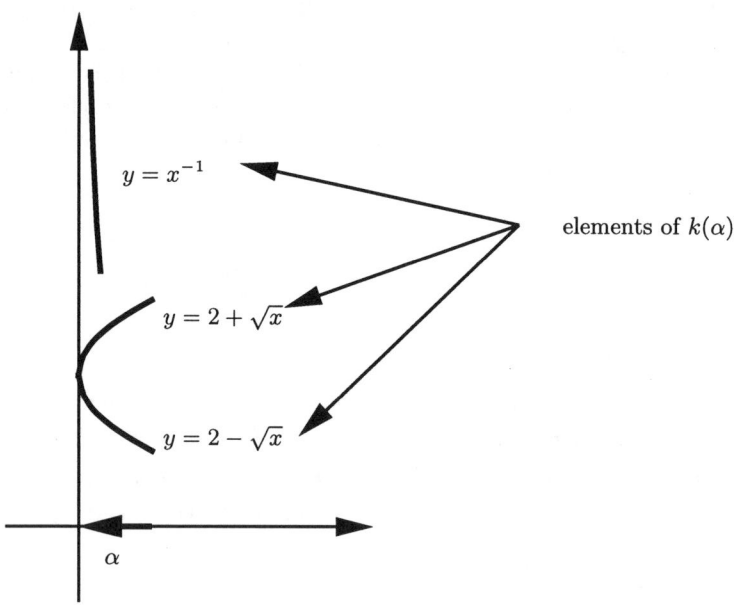

Figure 7.1.

Example 7.3.6. Consider $\alpha = 0_+ \in \widetilde{\mathbb{R}}$ (Example 7.1.4 b)). The field $k(0_+)$ is the field of algebraic Puiseux series $\mathbb{R}(X)^\wedge_{\text{alg}}$ (Example 1.3.6 b)). We claim that $\widetilde{S^0}_{\widetilde{\mathbb{R}},0_+} = k(0_+)$. It is enough to prove that, if $f :]0, \epsilon[\to \mathbb{R}$ is a continuous semi-algebraic function such that $f(0_+) = 0$, then its image in $\widetilde{S^0}_{\widetilde{\mathbb{R}},0_+}$ is null. This is clear, since, by Proposition 7.2.2 (iii), $f(0_+) = 0$ means that f is identically zero on some interval $]0, \eta[$. An algebraic Puiseux series can, thus, be identified with the germ of a continuous semi-algebraic function f at 0_+. It is equivalent to the assertion that, on some small interval $]0, \epsilon[$, $x \mapsto (x, f(x))$ is a parametrization of a piece of a plane algebraic curve (cf. Fig. 7.1).

All that has been said in this example still holds for an arbitrary real closed field R. The field of algebraic Puiseux series $R(X)^\wedge_{\text{alg}}$ is defined as for $\mathbb{R}$. It is the real closure of the field $R(X)$ equipped with the ordering 0_+, for which X is positive and smaller than any positive element of R.

7.4 Semi-algebraic Families of Sets and Mappings

It is natural to consider a semi-algebraic subset X of $R^n \times R^p$ as a *semi-algebraic family of subsets of R^n parametrized by R^p*. The *fibre of the family* X at a point $t \in R^p$ is

$$X_t = \{x \in R^n \mid (x,t) \in X\}.$$

The *restriction of the family* X to a semi-algebraic subset S of R^p is

$$X|_S = X \cap (R^n \times S).$$

If $X \subset R^n \times R^p$ and $Y \subset R^m \times R^p$ are two semi-algebraic families of sets, then a *semi-algebraic family of mappings* from X to Y is a semi-algebraic mapping $f : X \to Y$ such that the following diagram commutes :

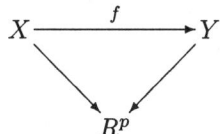

If $t \in R^p$, the *fibre of the semi-algebraic family f at t* is the semi-algebraic mapping $f_t : X_t \to Y_t$ defined by $f_t(x) = y \Leftrightarrow f(x,t) = (y,t)$.

In this section, we study the fibres of semi-algebraic families of sets and mappings at a point $\alpha \in \widetilde{R^p}$. The general idea is that *a property of the fibre at α still holds true over some semi-algebraic subset $S \subset R^p$, such that $\alpha \in \widetilde{S}$*. We shall use the results of this section in Chap. 9.

Denote by $T = (T_1, \ldots, T_p)$ the coordinate functions of R^p. If $\alpha \in \widetilde{R^p}$, denote by $T(\alpha)$ the point $(T_1(\alpha), \ldots, T_p(\alpha)) \in k(\alpha)^p$ (cf. Notation 7.1.1). We recall that $\mathcal{L}(R)$ denotes the first-order language of ordered fields with parameters in R.

Proposition 7.4.1. *Let $X \subset R^n \times R^p$ be a semi-algebraic family of subsets of R^n, defined by a formula $\Phi(x,t)$ of $\mathcal{L}(R)$, with free variables $x = (x_1, \ldots, x_n)$ and $t = (t_1, \ldots, t_p)$. Let $\alpha \in \widetilde{R^p}$.*

(i) *The semi-algebraic set*

$$X_\alpha = \{x \in k(\alpha)^n \mid \Phi(x, T(\alpha))\} \subset k(\alpha)^n$$

depends only on X, and not on the choice of Φ. This set X_α is called the fibre of the family X at α.

(ii) *Let $Y \subset R^n \times R^p$ be another semi-algebraic family of subsets of R^n. Then $X_\alpha = Y_\alpha$ if and only if there exists a semi-algebraic subset S of R^p, such that $\alpha \in \widetilde{S}$ and $X|_S = Y|_S$.*

Proof. Suppose that Y is defined by the formula $\Psi(x,t)$ of $\mathcal{L}(R)$, and let
$$Y_\alpha = \{x \in k(\alpha)^n \mid \Psi(x, T(\alpha))\} \ .$$
Then $X_\alpha = Y_\alpha$ if and only if
$$\forall x \in k(\alpha)^n \quad (\Phi(x, T(\alpha)) \Leftrightarrow \Psi(x, T(\alpha)))$$
holds true in $k(\alpha)$. Define
$$S = \{t \in R^p \mid \forall x \in R^n \ (\Phi(x,t) \Leftrightarrow \Psi(x,t))\} = \{t \in R^p \mid X_t = Y_t\} \ .$$
Then, by Proposition 7.2.2 (iii), $X_\alpha = Y_\alpha$ if and only if $\alpha \in \widetilde{S}$. This implies (ii), and also (i), if $X = Y$. □

Example 7.4.2. Consider two semi-algebraic sets $S \subset R^p$ and
$$F = \{x \in R^n \mid \Phi(x)\} \subset R^n \ ,$$
where Φ is a formula of $\mathcal{L}(R)$. Let X be the constant semi-algebraic family $F \times S \subset R^n \times R^p$, and $\alpha \in \widetilde{S}$. Then the fibre
$$X_\alpha = \{x \in k(\alpha)^n \mid \Phi(x)\}$$
is simply the extension $F_{k(\alpha)}$ of the semi-algebraic set F to $k(\alpha)$ (cf. Definition 5.1.2).

Proposition 7.4.3. *Let $\alpha \in \widetilde{R^p}$. For every semi-algebraic subset Λ of $k(\alpha)^n$, there exists a semi-algebraic family $X \subset R^n \times R^p$ of subsets of R^n such that $X_\alpha = \Lambda$.*

Proof. The semi-algebraic set Λ is defined by a formula $\Phi(x, a)$ of $\mathcal{L}(k(\alpha))$, where $x = (x_1, \ldots, x_n)$ are free variables, and $a = (a_1, \ldots, a_q)$ are parameters in $k(\alpha)$. By Corollary 7.3.5, there is a semi-algebraic set $S \subset R^p$, with $\alpha \in \widetilde{S}$, and a semi-algebraic mapping $f : S \to R^q$, such that $f(\alpha) = (f_1(\alpha), \ldots, f_q(\alpha)) = a$. Define
$$X = \{(x,t) \in R^n \times R^p \mid t \in S \text{ and } \Phi(x, f(t))\} \ .$$
It is clear that $X_\alpha = \Lambda$. □

Proposition 7.4.4. *Let $X \subset R^n \times R^p$ and $Y \subset R^m \times R^p$ be two semi-algebraic families of sets parametrized by R^p. Let $\alpha \in \widetilde{R^p}$.*

  (i) *Let $f : X \to Y$ be a semi-algebraic family of mappings. Define*
$$\Gamma f = \{(x, y, t) \in R^n \times R^m \times R^p \mid (x,t) \in X \text{ and } f(x,t) = (y,t)\} \ .$$
Then the fibre $(\Gamma f)_\alpha \subset k(\alpha)^n \times k(\alpha)^m$ is the graph of a semi-algebraic mapping $f_\alpha : X_\alpha \to Y_\alpha$, called the fibre of the family f at α.

  (ii) *Let $\varphi : X_\alpha \to Y_\alpha$ be a semi-algebraic mapping. Then there exists a semi-algebraic set $S \subset R^p$ such that $\alpha \in \widetilde{S}$, and a semi-algebraic family of mappings $f : X|_S \to Y|_S$ such that $f_\alpha = \varphi$.*

7.4 Semi-algebraic Families of Sets and Mappings   151

Proof. (i) Let $\Phi(x,y,t)$ be a formula of $\mathcal{L}(R)$ describing Γf, and let $\Psi(y,t)$ be a formula describing Y. Define

$$X_\alpha = \{x \in k(\alpha)^n \mid \exists y \in k(\alpha)^m \ \Phi(x,y,T(\alpha))\}$$

and

$$Y_\alpha = \{y \in k(\alpha)^m \mid \Psi(y,T(\alpha))\} \ .$$

Then, by Proposition 7.2.2 (iii), the formula, which says that

$$(\Gamma f)_\alpha = \{(x,y) \in k(\alpha)^n \times k(\alpha)^m \mid \Phi(x,y,T(\alpha))\}$$

is the graph of a semi-algebraic mapping from X_α to Y_α, holds true in $k(\alpha)$.

(ii) Let $G \subset X_\alpha \times Y_\alpha$ be the graph of φ. By Proposition 7.4.3, there exists a semi-algebraic family of sets $\Gamma \subset R^n \times R^m \times R^p$ parametrized by R^p, such that $\Gamma_\alpha = G$. Define S to be the set of $t \in R^p$ such that Γ_t is the graph of a mapping from X_t to Y_t. By Proposition 7.2.2 (iii), $\alpha \in \widetilde{S}$. Let $f : X|_S \to Y|_S$ be the semi-algebraic family of mappings defined by

$$f(x,t) = (y,t) \iff t \in S \text{ and } (x,y,t) \in \Gamma \ .$$

It is clear that $f_\alpha = \varphi$. □

Remark 7.4.5. a) The fibre $f_\alpha : X_\alpha \to Y_\alpha$ of a semi-algebraic family of mappings $f : X \to Y$ parametrized by R^p is injective (resp. surjective, resp. bijective) if and only if there exists a semi-algebraic subset S of R^p, such that $\alpha \in \widetilde{S}$ and $f|_{X|_S}$ is injective (resp. surjective, resp. bijective). This is an easy application of Proposition 7.2.2 (iii).

b) If two semi-algebraic families of mappings $f : X \to Y$ and $g : X \to Y$ parametrized by R^p satisfy $f_\alpha = g_\alpha$, then there exists a semi-algebraic subset S of R^p, such that $\alpha \in \widetilde{S}$ and $f|_{X|_S} = g|_{X|_S}$. This follows again from 7.2.2 (iii).

c) If $f : X \to Y$ and $g : Y \to Z$ are two semi-algebraic families of mappings parametrized by R^p, then $(g \circ f)_\alpha = g_\alpha \circ f_\alpha$.

d) Let $g : F \to G$ be a semi-algebraic mapping. Then the fibre at $\alpha \in \widetilde{R^p}$ of the constant family $g \times \mathrm{Id}_{R^p} : F \times R^p \to G \times R^p$ is the extension $g_{k(\alpha)} : F_{k(\alpha)} \to G_{k(\alpha)}$.

Proposition 7.4.6. *Let $X \subset Y \subset R^n \times R^p$ be two semi-algebraic families of subsets of R^n. Let $\alpha \in \widetilde{R^p}$. Then the fibre X_α is open (resp. closed) in Y_α if and only if there exists a semi-algebraic set $S \subset R^p$, such that $\alpha \in \widetilde{S}$ and $X|_S$ is open (resp. closed) in $Y|_S$.*

Proof. It is enough to prove the proposition for $Y = R^n \times R^p$. If $X|_S$ is open in $R^n \times S$, then by the finiteness theorem 2.7.2:

$$X|_S = \bigcup_{i=1}^m \{(x,t) \in R^n \times S \mid f_{i,1}(x,t) > 0 ,\ldots, f_{i,\ell_i}(x,t) > 0\} \ ,$$

where $f_{i,j} \in \mathcal{P}(R^n \times R^p)$, and, hence,

$$X_\alpha = \bigcup_{i=1}^m \{x \in k(\alpha)^n \mid f_{i,1}(x, T(\alpha)) > 0, \ldots, f_{i,\ell_i}(x, T(\alpha)) > 0\}$$

is open in $k(\alpha)^n$. Conversely, if X_α is open in $k(\alpha)^n$, one has, again by the finiteness theorem:

$$X_\alpha = \bigcup_{i=1}^m \{x \in k(\alpha)^n \mid f_{i,1}(x, a) > 0, \ldots, f_{i,\ell_i}(x, a) > 0\},$$

with $f_{i,j} \in \mathbb{Z}[X_1, \ldots, X_n, U_1, \ldots, U_q]$ and $a = (a_1, \ldots, a_q) \in k(\alpha)^q$. By Corollary 7.3.5, we can find a semi-algebraic subset S of R^p such that $\alpha \in \widetilde{S}$, and a *continuous* semi-algebraic mapping $g : S \to R^q$ such that $a = g(\alpha)$. By Proposition 7.4.1 (ii), we may assume that

$$X|_S = \bigcup_{i=1}^m \{(x,t) \in R^n \times S \mid f_{i,1}(x, g(t)) > 0, \ldots, f_{i,\ell_i}(x, g(t)) > 0\},$$

and this set is clearly open in $R^n \times S$. The proposition for closed subsets follows by taking complements. $\square$

Lemma 7.4.7. *Let $F \subset R^n$ and $G \subset R^m$ be two semi-algebraic sets and $f : F \to G$ a semi-algebraic mapping.*

(i) If f is continuous, there exist a finite open semi-algebraic cover $F = \bigcup_{i=1}^m U_i$ and, for each $i = 1, \ldots, m$, a regular function $g_i : U_i \to R$, such that $\|f\| \le g_i$ on U_i.

(ii) If the graph of f is closed in $F \times R^m$, and if there exists a continuous semi-algebraic function $g : F \to R$ such that $\|f\| \le g$ on F, then f is continuous.

Proof. Let $\alpha \in \widetilde{F}$, and set $a = \|f(\alpha)\| = \sqrt{f_1^2(\alpha) + \cdots + f_m^2(\alpha)} \in k(\alpha)$. We know that a is algebraic over $k(\text{supp}(\alpha))$. Lemma 1.2.11 implies that a root of a monic polynomial is bounded by a polynomial function of its coefficients, and, hence, there exists a regular function $g_\alpha = N_\alpha/D_\alpha$, with $N_\alpha, D_\alpha \in R[X_1, \ldots, X_n]$ and $D_\alpha(\alpha) \ne 0$, such that $a < g_\alpha(\alpha)$. Let U_α be the open semi-algebraic subset of F defined by

$$U_\alpha = \{x \in F \mid D_\alpha(x) \ne 0 \text{ and } \|f(x)\| < g_\alpha(x)\}.$$

We have $\alpha \in \widetilde{U}_\alpha$, and the $\widetilde{U}_\alpha$ cover $\widetilde{F}$. By compactness of $\widetilde{F}$ (Corollary 7.1.13), we can extract a finite subcover $\widetilde{U}_1, \ldots, \widetilde{U}_m$, which proves (i).

(ii) Suppose that f is not continuous at a point $x^0 \in F$. Then we can find $\epsilon \in R$, $\epsilon > 0$, such that, for every $\eta \in R$, $\eta > 0$, there exists $x \in F$ with $\|x - x^0\| < \eta$ and $\|f(x) - f(x^0)\| \ge \epsilon$. The point x^0 belongs to the closure of the semi-algebraic set

$$L = \{x \in F \mid \|f(x) - f(x^0)\| \geq \epsilon\}.$$

By the curve-selection lemma (Theorem 2.5.5), there exists a continuous semi-algebraic mapping $\varphi : [0,1] \to R^n$, such that $\varphi(0) = x^0$ and $\varphi(]0,1]) \subset L$. Since $\|f\|$ is bounded from above by a continuous semi-algebraic function, it is clear that $\|f \circ \varphi\|$ is bounded from above on $]0,1]$ by some $M \in R$. Hence, by Proposition 2.5.3, $f \circ \varphi(t)$ has a limit $y^0 \in R^m$ as $t \to 0$. Since the graph of f is closed in $F \times R^m$, one has $y^0 = f(x^0)$. Since $\|f(\varphi(t)) - f(x^0)\| \geq \epsilon$ for every $t \in]0,1]$, we get $\|f(x^0) - f(x^0)\| \geq \epsilon$. □

Proposition 7.4.8. *Let $X \subset R^n \times R^p$ and $Y \subset R^m \times R^p$ be two semi-algebraic families of sets parametrized by R^p, and $f : X \to Y$, a semi-algebraic family of mappings. Let $\alpha \in \tilde{R^p}$. Then the semi-algebraic mapping $f_\alpha : X_\alpha \to Y_\alpha$ is continuous if and only if there exists a semi-algebraic subset S of R^p, such that $\alpha \in \tilde{S}$ and $f|_{X|_S}$ is continuous.*

Proof. Suppose that f_α is continuous. By Lemma 7.4.7 (i), we have a finite open semi-algebraic cover $X_\alpha = \bigcup_{i=1}^r \Omega_i$, and regular functions $g_i : \Omega_i \to k(\alpha)$ (defined over $k(\alpha)$), such that $\|f_\alpha\| \leq g_i$ on Ω_i. We have $g_i(x) = N_i(x,a)/D_i(x,a)$, where $N_i, D_i \in \mathbb{Z}[X_1, \ldots, X_n, V_1, \ldots, V_q]$ and $a \in k(\alpha)^q$. It follows from Corollary 7.3.5 and from the previous results of this section that we can find:

- a semi-algebraic subset S of R^p, such that $\alpha \in \tilde{S}$ and the set $\Gamma f|_S$ is closed in $\{(x,y,t) \in R^n \times R^m \times R^p \mid (x,t) \in X|_S\}$ (this implies that the graph of $f|_{X|_S}$ is closed in $(X|_S) \times R^m \times R^p$),
- an open semi-algebraic cover $X|_S = \bigcup_{i=1}^r U_i$, such that $(U_i)_\alpha = \Omega_i$ for $i = 1, \ldots, r$,
- a continuous semi-algebraic mapping $h : S \to R^q$, such that $h(\alpha) = a$, $D_i(x, h(t)) \neq 0$ and $\|f(x,t)\| \leq N_i(x, h(t))/D_i(x, h(t))$, for every $(x,t) \in U_i$ and for $i = 1, \ldots, r$.

Since f has a closed graph and $\|f\|$ is bounded by a continuous semi-algebraic function on each U_i, it follows by Lemma 7.4.7 (ii) that $f|_{X|_S}$ is continuous.

Suppose, now, that there exists a semi-algebraic subset S of R^p, such that $\alpha \in \tilde{S}$ and $f|_{X|_S}$ is continuous. Let Ω be an open semi-algebraic subset of Y_α. Shrinking S, if necessary, we may assume, by Proposition 7.4.6, that $\Omega = U_\alpha$ where U is an open semi-algebraic subset of $Y|_S$. Then $f_\alpha^{-1}(\Omega) = (f^{-1}(U))_\alpha$ is an open subset of X_α, since $f^{-1}(U)$ is an open semi-algebraic subset of $X|_S$. Hence, f_α is continuous. □

Remark 7.4.9. It is easy to prove (using Proposition 7.2.2 (iii)) that f_α is continuous if and only if there exists a semi-algebraic subset S of R^p, such that $\alpha \in \tilde{S}$ and $f_t : X_t \to Y_t$ is continuous for every $t \in S$. This last property is, of course, weaker than the continuity of $f|_{X|_S}$. If the semi-algebraic family of mappings $f : X \to Y$ is such that f_t is continuous for every t in the semi-algebraic subset S of R^p, then Proposition 7.4.8 shows that, for every

$\alpha \in \widetilde{S}$, there exists a semi-algebraic set S_α, such that $\alpha \in \widetilde{S}_\alpha$ and $f|_{X|S_\alpha}$ is continuous. By compactness of $\widetilde{S}$ with respect to the constructible topology, we get a finite partition $S = \bigcup_{i=1}^q S_i$ into semi-algebraic subsets, such that $f|_{X|S_i}$ is continuous for $i = 1, \ldots, q$.

Corollary 7.4.10. *Let $X \subset R^n \times R^p$ and $Y \subset R^m \times R^p$ be semi-algebraic families of sets parametrized by R^p, and let $\alpha \in \widetilde{R^p}$. If $\varphi : X_\alpha \to Y_\alpha$ is a semi-algebraic homeomorphism, there exist a semi-algebraic subset S of R^p such that $\alpha \in \widetilde{S}$ and a semi-algebraic homeomorphism $f : X|_S \to Y|_S$, such that the diagram*

$$X|_S \xrightarrow{f} Y|_S$$
$$\searrow \quad \swarrow$$
$$R^p$$

commutes and $f_\alpha = \varphi$.

Proof. Apply Propositions 7.4.4 and 7.4.8 to φ and to the inverse homeomorphism φ^{-1}. $\square$

7.5 Semi-algebraically Connected Components. Dimension

Proposition 7.5.1. *Let S be a semi-algebraic set.*
  (i) *The set S is semi-algebraically connected if and only if $\widetilde{S}$ is connected.*
  (ii) *If $S_1, \ldots, S_k$ are the semi-algebraically connected components of S, then $\widetilde{S}_1, \ldots, \widetilde{S}_k$ are the connected components of $\widetilde{S}$.*

Proof. It is enough to prove (i). A closed and open subset of $\widetilde{S}$ is necessarily compact, and, thus, by Theorem 7.2.3, it has the form $\widetilde{T}$, where T is a closed and open semi-algebraic subset of S. The result follows from the definition of semi-algebraic connectedness (Definition 2.4.2). $\square$

Example 7.5.2. Proposition 7.5.1 shows, in particular, that $\widetilde{R} = \mathrm{Spec}_r(R[X])$ is always connected, while R is connected only if $R = \mathbb{R}$ (see Example 2.4.1). It is interesting to see how the real spectrum "fills the holes", for example in the case of $R = \mathbb{R}_{\mathrm{alg}}$, the field of real algebraic numbers. The set $\mathrm{Spec}_r(\mathbb{R}_{\mathrm{alg}}[X])$ contains, on the one hand, the real algebraic numbers, and on the other hand the orderings of the field $\mathbb{R}_{\mathrm{alg}}(X)$ (identified with the prime cones of $\mathbb{R}_{\mathrm{alg}}[X]$ whose support is (0)). As in the case of $\mathbb{R}$, an ordering determines a cut (I, J) where $I = \{x \in \mathbb{R}_{\mathrm{alg}} \mid x < X\}$ and $J = \{x \in \mathbb{R}_{\mathrm{alg}} \mid X < x\}$, and this gives a bijection between the set of orderings of $\mathbb{R}_{\mathrm{alg}}(X)$ and the set of cuts in $\mathbb{R}_{\mathrm{alg}}$. These cuts are $-\infty$, $+\infty$, x_- and x_+ for $x \in \mathbb{R}_{\mathrm{alg}}$ (with the same notations as in Example 1.1.2), and also the cuts given by transcendental real numbers in $\mathbb{R} \setminus \mathbb{R}_{\text{alg}}$. So the number $\pi = 3.14\ldots$ belongs to $\text{Spec}_r(\mathbb{R}_{\text{alg}}[X])$, identified with the prime cone $P_\pi = \{f \in \mathbb{R}_{\text{alg}}[X] \mid f(\pi) \geq 0\}$. A constructible subset $\widetilde{S}$ of $\text{Spec}_r(\mathbb{R}_{\text{alg}}[X])$ contains π if and only if S contains an interval $]a, b[$ of $\mathbb{R}_{\text{alg}}$, with $a < \pi < b$. Note, also, that the subspace of closed points of $\text{Spec}_r(\mathbb{R}_{\text{alg}}[X])$ coincides with the subspace of closed points of $\text{Spec}_r(\mathbb{R}[X])$.

We now explain the relationship between the dimension of semi-algebraic sets and the lengths of specialization chains in the real spectrum.

Definition 7.5.3. *Let A be a commutative ring. A specialization chain of length n in $\text{Spec}_r(A)$ is a chain of prime cones*

$$\alpha_n \subsetneq \alpha_{n-1} \subsetneq \cdots \subsetneq \alpha_0.$$

Let C be a constructible subset of $\text{Spec}_r(A)$. If the lengths of specialization chains consisting of elements of C are bounded, then the dimension of C (denoted $\dim(C)$) is the maximal length of such chains. Otherwise, the dimension of C is infinite.

First, we construct a specialization chain of length n in the real spectrum of $R[X_1, \ldots, X_n]$.

Proposition 7.5.4. *There exists a specialization chain*

$$\alpha_n \subsetneq \alpha_{n-1} \subsetneq \alpha_{n-2} \subsetneq \cdots \subsetneq \alpha_0 = 0 \quad (\text{the origin of } R^n)$$

in $\text{Spec}_r(R[X_1, \ldots, X_n])$.

Proof. The prime cones α_i that we construct satisfy $\text{supp}(\alpha_n) = (0)$ and $\text{supp}(\alpha_i) = (X_{i+1}, \ldots, X_n)$ for $0 \leq i < n$. We construct, by induction, the orderings $\leq_{\alpha_i}$ of the residue fields. Of course, $\leq_{\alpha_0}$ is the ordering of R. Given $i > 0$, assume that the ordering $\leq_{\alpha_{i-1}}$ has been constructed. Consider a polynomial

$$f = g_p X_i^p + g_{p-1} X_i^{p-1} + \cdots + g_m X_i^m, \quad g_m \neq 0,$$

where $g_p, \ldots, g_m \in R(X_1, \ldots, X_{i-1}) = k(\text{supp}(\alpha_{i-1}))$ (if $i = 1$, then f has its coefficients in R). Set

$$0 <_{\alpha_i} f \iff 0 <_{\alpha_{i-1}} g_m.$$

This defines an ordering $\leq_{\alpha_i}$ of the residue field $k(\text{supp}(\alpha_i)) = R(X_1, \ldots, X_i)$ (this ordering makes X_i positive and smaller than every positive element of $R(X_1, \ldots, X_{i-1})$). It is clear from the construction that α_i is a generization of α_{i-1}. □

156   7. Real Spectrum

Remark 7.5.5. It is interesting to see the prime cones of the specialization chain constructed above as ultrafilters of semi-algebraic subsets of R^n (Corollary 7.2.4). Let $\mathcal{F}_i$ be the ultrafilter of semi-algebraic subsets S of R^n such that $\alpha_i \in \widetilde{S}$. The ultrafilter $\mathcal{F}_0$ is, of course, the ultrafilter of semi-algebraic subsets containing the origin. By the construction of α_1, for every $f \in R[X_1, \ldots, X_n]$, we have

$$f(\alpha_1) > 0 \iff \exists \epsilon > 0 \quad \forall x \in]0, \epsilon[\quad f(x, 0, \ldots, 0) > 0 .$$

Hence, $S \in \mathcal{F}_1$ if and only if S contains some interval $]0, \epsilon[$ of the X_1 axis. In general, $\mathcal{F}_i$ is generated by semi-algebraic subsets of R^i identified with the subspace of coordinates $X_1, \ldots, X_i$ of R^n. The ultrafilter $\mathcal{F}_{i+1}$ can be obtained from $\mathcal{F}_i$ in the following way: $S \in \mathcal{F}_{i+1}$ if and only if S contains a semi-algebraic subset of R^{i+1} of the form $\{(t, y) \in R^i \times R \mid t \in T \text{ and } 0 < y < f(t)\}$, where $T \subset R^i$ is an element of $\mathcal{F}_i$, and $f : T \to R$ is a continuous semi-algebraic function which is strictly positive on T.

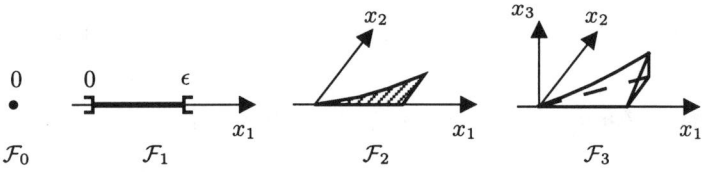

Figure 7.2. Typical generators of $\mathcal{F}_i$

Proposition 7.5.6. *Let S be a semi-algebraic subset of R^n. Then*

$$\dim(S) = \dim(\widetilde{S}) .$$

Proof. If $\alpha_k \subsetneq \cdots \subsetneq \alpha_0$ is a specialization chain in $\widetilde{S}$, then, by Proposition 4.3.9, the inclusions $\mathrm{supp}(\alpha_i) \subset \mathrm{supp}(\alpha_{i-1})$ for $i = 1, \ldots, k$ are strict. If $f \in \mathcal{I}(S)$, then $f(\alpha_k) = 0$, and, thus, $\mathcal{I}(S) \subset \mathrm{supp}(\alpha_k)$. This shows that $\dim(\widetilde{S})$ is not greater than the dimension of $\mathcal{I}(S)$, which is, by definition, the dimension of S. It remains to prove that $\dim(S) \leq \dim(\widetilde{S})$. If $\dim(S) = d$, then S contains a semi-algebraic subset T, semi-algebraically homeomorphic to an open hypercube $]-1, 1[^d$ (cf. Corollary 2.8.9 – one takes here $]-1, 1[$ instead of $]0, 1[$, so that the origin belongs to the open hypercube). The semi-algebraic homeomorphism between T and $]-1, 1[^d$ gives a homeomorphism φ between $\widetilde{T}$ and $(]-1, 1[^d)\widetilde{}$ through the tilde operation (Proposition 7.2.8). We have constructed a specialization chain of length d in $\mathrm{Spec}_r(R[X_1, \ldots, X_d])$, whose last element is the origin (Proposition 7.5.4). Since $]-1, 1[^d$ is open, $(]-1, 1[^d)\widetilde{}$ is stable under generization, and this specialization chain is entirely contained in $(]-1, 1[^d)\widetilde{}$. Using the homeomorphism φ, we obtain a specialization chain of length d in $\widetilde{T}$, and, hence, in $\widetilde{S}$. This shows that $\dim(\widetilde{S}) \geq \dim(S)$. □

Definition 7.5.7. *Let A be a commutative ring, and α, a prime cone of A. The dimension of α (denoted $\dim(\alpha)$) is the dimension of the prime ideal $\operatorname{supp}(\alpha)$ of A, i.e. the dimension of the ring $A/\operatorname{supp}(\alpha)$.*

Proposition 7.5.8.
(i) *Let S be a semi-algebraic subset of R^n. Then*
$$\dim(S) = \max\{\dim(\alpha) \mid \alpha \in \widetilde{S}\}.$$
(ii) *Let $\alpha \in \operatorname{Spec}_r(R[X_1,\ldots,X_n])$. Then*
$$\dim(\alpha) = \min\{\dim(S) \mid S \text{ semi-algebraic subset of } R^n,\ \alpha \in \widetilde{S}\}.$$

Proof. First note that, if $\alpha \in \widetilde{S}$, then $\mathcal{I}(S) \subset \operatorname{supp}(\alpha)$, and, hence, $\dim(S) \geq \dim(\alpha)$. Assertion (i) follows from Proposition 7.5.6, which gives a specialization chain $\alpha_d \subsetneq \cdots \subsetneq \alpha_0$ of length $d = \dim(S)$ in $\widetilde{S}$; then $\dim(\alpha_d) \geq d$. Assertion (ii) is proved by taking $S = \mathcal{Z}(\operatorname{supp}(\alpha))$, so that $\alpha \in \widetilde{S}$, and $\dim(\alpha) = \dim(S)$. □

Remark 7.5.9. It may happen that $\dim(\alpha) = p$, and that there is no specialization chain $\alpha = \alpha_p \subsetneq \alpha_{p-1} \subsetneq \cdots \subsetneq \alpha_0$ of length p. Consider, for instance, the prime cone α of $\mathbb{R}[X,Y]$ defined by
$$\alpha = \{f \in \mathbb{R}[X,Y] \mid \exists \epsilon > 0 \quad \forall x \in]0,\epsilon[\quad f(x,e^x) \geq 0\}.$$
It is clear that $\operatorname{supp}(\alpha) = (0)$, and, thus, $\dim(\alpha) = 2$. However, we shall show in Chap. 10 that α has no specialization of dimension 1, and that its only proper specialization is the point $(0,1)$.

7.6 Orderings of the Field of Rational Functions and Central Points

Throughout this section $V \subset R^n$ is an irreducible algebraic set of dimension d. The canonical injection $\operatorname{Spec}_r(\mathcal{K}(V)) \hookrightarrow \operatorname{Spec}_r(\mathcal{P}(V))$ identifies $\operatorname{Spec}_r(\mathcal{K}(V))$ with the subspace of $\operatorname{Spec}_r(\mathcal{P}(V))$ of prime cones whose support is (0) (i.e. those of dimension d). In this section, we shall always make this identification.

Proposition 7.6.1. *Let S and T be two semi-algebraic subsets of V. Then S is generically equal to T in V if and only if*
$$\widetilde{S} \cap \operatorname{Spec}_r(\mathcal{K}(V)) = \widetilde{T} \cap \operatorname{Spec}_r(\mathcal{K}(V)).$$

Proof. Let $X = (S \setminus T) \cup (T \setminus S)$ be the symmetric difference of S and T. By definition, S and T are generically equal if and only if $\dim(X) < d$. On the other hand, $\widetilde{S} \cap \operatorname{Spec}_r(\mathcal{K}(V)) = \widetilde{T} \cap \operatorname{Spec}_r(\mathcal{K}(V))$ if and only if $\widetilde{X} \cap \operatorname{Spec}_r(\mathcal{K}(V)) = \emptyset$, i.e. if and only if there is no $\alpha \in \widetilde{X}$ such that $\dim(\alpha) = d$. Proposition 7.5.8 (i) gives the equivalence between this last property and the property $\dim(X) < d$. □

Proposition 7.6.1 allows one to give an easy *proof of Lemma 6.5.2*. Let $g_1,\ldots,g_\ell$ and $h_1,\ldots,h_\ell$ be nonzero polynomial functions in $\mathcal{P}(V)$. Assume that
$$\langle\!\langle g_1,\ldots,g_\ell\rangle\!\rangle \simeq \langle\!\langle h_1,\ldots,h_\ell\rangle\!\rangle\,,$$
as Pfister forms over the field of rational functions $\mathcal{K}(V)$. Let α be any ordering of $\mathcal{K}(V)$. Then the signature of $\langle\!\langle g_1,\ldots,g_\ell\rangle\!\rangle$ at α is equal to 2^ℓ if and only if $\alpha \in \widetilde{\mathcal{U}}(g_1,\ldots,g_\ell)$. Similarly for $\langle\!\langle h_1,\ldots,h_\ell\rangle\!\rangle$. Since the signatures of both Pfister forms are equal, Proposition 7.6.1 implies that $\mathcal{U}(g_1,\ldots,g_\ell)$ is generically equal to $\mathcal{U}(h_1,\ldots,h_\ell)$. □

Proposition 7.6.2. *Let x be a point of V. Then the following properties are equivalent:*
  (i) *x belongs to the closure (with respect to the euclidean topology) of the set $\operatorname{Reg}(V)$ of nonsingular points of V.*
  (ii) *$\dim(V_x) = d$ (i.e. $x \in V^{(d)}$).*
  (iii) *x is a specialization in $\operatorname{Spec}_r(\mathcal{P}(V))$ of a prime cone of dimension d (i.e. x is a specialization of an ordering of $\mathcal{K}(V)$).*

Proof. (iii) ⇒ (ii) If U is a semi-algebraic neighbourhood of x in V, then $\widetilde{U}$ contains a prime cone of dimension d in $\mathcal{P}(V)$, and, hence, $\dim(U) = d$, by Proposition 7.5.8 (i). This proves (ii).

(ii) ⇒ (i) If $x \notin \operatorname{clos}(\operatorname{Reg}(V))$, there exists an open semi-algebraic subset U of V containing x, such that $U \subset \operatorname{Sing}(V)$, and, hence, $\dim(U) < d$ (Proposition 3.3.14). Thus, $\dim(V_x) < d$.

(i) ⇒ (iii) If x is a nonsingular point of V, then, by Proposition 3.3.11, there exists a semi-algebraic homeomorphism from a semi-algebraic neighbourhood U of x in V onto a semi-algebraic neighbourhood U' of the origin in R^d (mapping x to the origin). We obtain a specialization chain $\alpha_d \subsetneq \cdots \subsetneq \alpha_0 = x$, such that $\operatorname{supp}(\alpha_d) = (0)$, using homeomorphism between $\widetilde{U}$ and $\widetilde{U}'$ and applying Proposition 7.5.4.

Now suppose $x \in \operatorname{clos}(\operatorname{Reg}(V))$. As observed above, for every semi-algebraic neighbourhood U of x in V there exists a prime cone in $\widetilde{U}$ whose support is (0). Hence $\widetilde{U} \cap \operatorname{Spec}_r(\mathcal{K}(V)) \neq \emptyset$. The sets $\operatorname{Spec}_r(\mathcal{K}(V))$ and $\widetilde{U}$ are compact with respect to the constructible topology of $\operatorname{Spec}_r(\mathcal{P}(V))$. By compactness we get $\alpha \in \operatorname{Spec}_r(\mathcal{K}(V))$ such that x is a specialization of α. □

Definition 7.6.3. *The set $\operatorname{Cent}(V)$ of central points of V is the closure of the set of nonsingular points $\operatorname{Reg}(V)$.*

Proposition 7.6.4.
  (i) $\operatorname{Spec}_r(\mathcal{K}(V)) \subset (\operatorname{Reg}(V))\widetilde{}$.
  (ii) *The set $(\operatorname{Cent}(V))\widetilde{}$ is the closure of $\operatorname{Spec}_r(\mathcal{K}(V))$ in $\operatorname{Spec}_r(\mathcal{P}(V))$.*

7.6 Orderings and Central Points 159

Proof. (i) If there were an ordering of $\mathcal{K}(V)$ in $(\mathrm{Sing}(V))\tilde{} = \widetilde{V} \setminus (\mathrm{Reg}(V))\tilde{}$, then one would have $\dim(\mathrm{Sing}(V)) = \dim((\mathrm{Sing}(V))\tilde{}) = d$ (Propositions 7.5.6 and 7.5.8(i)). This is impossible, since, by Proposition 3.3.14, the dimension of $\mathrm{Sing}(V)$ is less than d.

(ii) By (i), $\mathrm{clos}(\mathrm{Spec}_r(\mathcal{K}(V))) \subset \mathrm{clos}((\mathrm{Reg}(V))\tilde{})$. By Theorem 7.2.3(iii), $\mathrm{clos}((\mathrm{Reg}(V))\tilde{}) = (\mathrm{Cent}(V))\tilde{}$. It follows from Proposition 7.6.2 (iii) that $\mathrm{Cent}(V)$ is contained in $\mathrm{clos}(\mathrm{Spec}_r(\mathcal{K}(V)))$. Since, by Proposition 7.2.7(ii), $(\mathrm{Cent}(V))\tilde{}$ is the smallest closed subset of $\mathrm{Spec}_r(\mathcal{P}(V))$ containing $\mathrm{Cent}(V)$, we have $(\mathrm{Cent}(V))\tilde{} \subset \mathrm{clos}(\mathrm{Spec}_r(\mathcal{K}(V)))$. □

Theorem 7.6.5. *Let I be an ideal of $\mathcal{P}(V)$ and U a basic open semi-algebraic subset defined by*

$$U = \{x \in V \mid f_1(x) > 0, \ldots, f_\ell(x) > 0\},$$

where $f_1, \ldots, f_\ell$ are nonzero polynomial functions in $\mathcal{P}(V)$. Let P be the cone $\sum \mathcal{K}(V)^2[f_1, \ldots, f_\ell] \cap \mathcal{P}(V)$ of $\mathcal{P}(V)$. A polynomial function f in $\mathcal{P}(V)$ vanishes on $\mathcal{Z}_V(I) \cap \mathrm{clos}(U \cap \mathrm{Reg}(V))$ if and only if it belongs to the P-radical of I.

Proof. Suppose that $f^{2m} + p \in I$, with $p \in P$. We know that p is of the form $p = q/s^2$, where $s \in \mathcal{P}(V)$, $s \neq 0$, and q belongs to the cone of $\mathcal{P}(V)$ generated by $f_1, \ldots, f_\ell$. If $x \in U \cap \mathrm{Reg}(V)$, then $\dim(U_x) = \dim(V_x) = d$. Since $\dim(\mathcal{Z}_V(s)) < d$, the point x belongs to the closure of $U \setminus (\mathcal{Z}_V(s) \cap U)$, and, thus, $p(x) \geq 0$. Hence, p is nonnegative on $\mathrm{clos}(U \cap \mathrm{Reg}(V))$, so that $\mathcal{Z}_V(I) \cap \mathrm{clos}(U \cap \mathrm{Reg}(V)) \subset \mathcal{Z}_V(f)$.

Conversely, suppose that $\mathcal{Z}_V(I) \cap \mathrm{clos}(U \cap \mathrm{Reg}(V)) \subset \mathcal{Z}_V(f)$. If P is not proper, then the P-radical of I is $\mathcal{P}(V)$, and, hence, it contains f. So, we may assume that P is proper. By Proposition 4.2.6, it is enough to prove that, if J is a P-convex prime ideal containing I, then $f \in J$. By Proposition 4.2.9, there exists an ordering of $\mathcal{K}(V)$, whose positive cone Q contains P, such that J is α-convex, where $\alpha = Q \cap \mathcal{P}(V) \in \mathrm{Spec}_r(\mathcal{P}(V))$. Since $Q \supset P$, one has $\alpha \in \widetilde{U}$, and, therefore, by Proposition 7.6.4(i), $\alpha \in (\mathrm{Reg}(V))\tilde{}$. By Proposition 4.3.9, there exists a unique $\beta \in \mathrm{Spec}_r(\mathcal{P}(V))$, such that $\alpha \subset \beta$ and $\mathrm{supp}(\beta) = J$. By Proposition 7.1.21 and Theorem 7.2.3(iii), we have $\beta \in \mathrm{clos}(\widetilde{U} \cap (\mathrm{Reg}(V))\tilde{}) = (\mathrm{clos}(U \cap \mathrm{Reg}(V)))\tilde{}$. On the other hand, since $I \subset \mathrm{supp}(\beta)$, we have $\beta \in (\mathcal{Z}_V(I))\tilde{}$. By assumption, it follows that $\beta \in (\mathcal{Z}_V(f))\tilde{}$, and, hence, $f \in \mathrm{supp}(\beta) = J$. □

Corollary 7.6.6 (Central Nullstellensatz). *Let I be an ideal of $\mathcal{P}(V)$. A polynomial function f in $\mathcal{P}(V)$ vanishes on $\mathcal{Z}_V(I) \cap \mathrm{Cent}(V)$ if and only if there exist rational functions $g_1, \ldots, g_k \in \mathcal{K}(V)$ and an integer $m \in \mathbb{N}$, such that $f^{2m} + g_1^2 + \cdots + g_k^2 \in I$.*

Proof. We apply Theorem 7.6.5 to the special case $U = V$ and $P = \sum \mathcal{K}(V)^2 \cap \mathcal{P}(V)$. □

Corollary 7.6.7. *Let I be an ideal of $\mathcal{P}(V)$. Then*
$$I = \mathcal{I}_{\mathcal{P}(V)}(\mathcal{Z}_V(I) \cap \operatorname{Cent}(V))$$
if and only if I is $(\sum \mathcal{K}(V)^2 \cap \mathcal{P}(V))$-radical.

Example 7.6.8. Consider the two umbrellas of Example 3.1.2 d) and e). For each of them, consider the ideal $I = (x, y)$ of the stick. In the case of the first umbrella (given by the equation $z(x^2 + y^2) - x^3 = 0$), the ideal I does not satisfy the condition of Corollary 7.6.7. The intersection of the stick with the set of central points (the cloth) consists only of the origin. In the case of the second umbrella (given by the equation $x^3 + zx^2 - y^2 = 0$), the ideal I satisfies the condition. The intersection of the stick with the set of central points is, in this case, the subset $[0, +\infty[$ of the z-axis.

Remark 7.6.9. It is clear that a $(\sum \mathcal{K}(V)^2 \cap \mathcal{P}(V))$-radical ideal is real. Example 3.1.2 d) shows that the converse is, in general, false.

Bibliographic Notes. The real spectrum of a ring was initially introduced by Coste and Roy in order to give a real analogue of the étale topos, and it originated from considerations of categorical logic [99]. The notion of a prime cone had already been used in [258], and the real spectrum equipped with its constructible topology can be found in an earlier paper of model theory [117]. The theory of the real spectrum is presented in several articles [26, 100, 114, 189, 206]. The real spectrum shares the general properties of all spectral spaces [162]. On the other hand, it has more specific properties, like the Hausdorff property of the set of closed points, which can be found in [69] (cf. also [80, 112]). The tilde operation was defined in [100]. The presentation of $\widetilde{T}$ as the Stone space of the boolean algebra of semi-algebraic subsets of T is contained in the "ultrafilter theorem" of [69]. The ringed spaces $(\widetilde{T}, \widetilde{\mathcal{S}^0})$ are essentially equivalent to the affine semi-algebraic spaces in [108]. These spaces are the starting point of the theory of locally semi-algebraic and weakly semi-algebraic spaces [107, 110, 190]. The sheaf of continuous semi-algebraic functions on the real spectrum is also studied in the "abstract" setting (i.e. on the real spectrum of an arbitrary ring) by several authors ([77, 106] and, above all, [289]). The notion of a central point can be found in [122]; the central Nullstellensatz, which is a generalization of a version of the real Nullstellensatz given by Dubois [121], can be found in [284].

8. Nash Functions

Abstract. In the first section, we define an isomorphism between the ring of germs of Nash functions at the origin of R^n, the ring of power series which are algebraic over the polynomials, and (for $R = \mathbb{R}$) the ring of germs of analytic algebraic functions. From the study of the ring of algebraic power series in the second section, we deduce the local properties of Nash functions. In the third section, we state a theorem of approximation of formal solutions of a system of Nash equations by Nash solutions. The study of global properties of Nash functions is greatly simplified by the Artin-Mazur description of Nash functions given in Section 4. We use this description in Section 5 to obtain the substitution theorem, from which we deduce a Nullstellensatz and a Positivstellensatz for Nash functions. In Section 6, we study the sets of zeros of Nash functions, and we prove that the germs of such sets are Nash equivalent to germs of algebraic sets. In Section 7, we prove that the ring of germs of Nash functions at the origin of R^n is the henselization of the local ring of germs of regular functions, and we use this fact to prove that the ring of global Nash functions is noetherian. In Section 8, we consider Nash functions on the real spectrum, and we give a proof of Efroymson's theorem of approximation of continuous semi-algebraic functions by Nash functions. Efroymson's extension theorem is given in Section 9, where we also study the tubular neighbourhoods of Nash submanifolds. The last section is devoted to families of Nash functions, and we show that there is $r = r(n,d) \in \mathbb{N}$, such that every semi-algebraic function of class C^r on R^n, satisfying a polynomial equation of total degree $\leq d$, is Nash.

Throughout this chapter, R is a real closed field.

8.1 Germs of Nash Functions and Algebraic Power Series

We have defined Nash functions in Section 2.9 as C^∞ semi-algebraic functions. Usually, a Nash function on an open semi-algebraic subset of $\mathbb{R}^n$ is defined as an *analytic algebraic function*, i.e. an analytic function φ that is a solution of an equation $a_d(x)(\varphi(x))^d + \cdots + a_0(x) = 0$, where $a_d, \ldots, a_0$ are polynomials, $a_d \neq 0$ [19]. Our first aim in this section is to show that the two notions

162   8. Nash Functions

coincide. We shall also identify germs of Nash functions at the origin of R^n with algebraic power series (i.e. power series which are algebraic over the field of rational functions). This identification is useful for the study of local properties of Nash functions.

We denote by $R[[X]]$ the ring of power series in the variables $X = (X_1, \ldots, X_n)$ and by $R((X))$ its field of fractions.

First we introduce the notion of an étale regular mapping. We shall be interested only in the nonsingular case.

Definition 8.1.1. *Let $f : V \to W$ be a regular mapping between algebraic sets and t a nonsingular point of V such that $f(t)$ is a nonsingular point of W. The regular mapping f is said to be étale at t if $df_t : T_t(V) \to T_{f(t)}(W)$ is an isomorphism.*

Proposition 8.1.2. *Let $f : V \to W$ be a regular mapping between algebraic sets and t a nonsingular point of V such that $f(t)$ is nonsingular in W. If f is étale at t, there exist open semi-algebraic neighbourhoods U (resp. U') of t (resp. $f(t)$) in V (resp. W), such that $f|_U$ is a Nash diffeomorphism from U onto U'.*

Proof. By Proposition 3.3.11, we can find neighbourhoods of t in V and $f(t)$ in W which are Nash submanifolds. The fact that f is a Nash diffeomorphism from a neighbourhood of t onto a neighbourhood of $f(t)$ follows from the inverse function theorem 2.9.7. □

We now recall a few facts about the Taylor expansion of elements of a regular local ring (cf. [295], Chap. 2, §2). Let V be an irreducible algebraic set of dimension d and t a nonsingular point of V. Let $(v_1, \ldots, v_d)$ be a regular system of parameters of the regular local ring $\mathcal{R}_{V,t}$. For every $h \in \mathcal{R}_{V,t}$, there exists a unique series $\tau(h) \in R[[X_1, \ldots, X_d]]$ such that, for every $i \in \mathbb{N}$, $h - \tau_i(h)(v_1, \ldots, v_d)$ belongs to the $(i+1)^{\text{th}}$ power of the maximal ideal of $\mathcal{R}_{V,t}$, where $\tau_i(h)$ is the polynomial obtained by deleting all terms of degree $> i$ in $\tau(h)$. The mapping $\tau : \mathcal{R}_{V,t} \to R[[X_1, \ldots, X_d]]$ is a local homomorphism, and it induces an isomorphism $\widehat{\tau}$ between the completion $\widehat{\mathcal{R}}_{V,t}$ and $R[[X_1, \ldots, X_d]]$. We call $\tau(h)$ *the Taylor expansion of h with respect to* $(v_1, \ldots, v_d)$.

Proposition 8.1.3. *Let $f : V \to W$ be a regular mapping between two irreducible algebraic sets of dimension d. Let t be a point of V such that $f(t)$ is a nonsingular point of W. Then the following properties are equivalent:*

(i) *The point t is nonsingular in V, and f is étale at t.*

(ii) *The homomorphism $\widehat{\mathcal{R}}_{W,f(t)} \to \widehat{\mathcal{R}}_{V,t}$, induced by f, between the completions of the local rings is an isomorphism.*

Proof. (i) $\Rightarrow$ (ii) Since f is étale at t, the local homomorphism

$$f^* : \mathcal{R}_{W,f(t)} \to \mathcal{R}_{V,t}$$

8.1 Germs of Nash Functions and Algebraic Power Series

induced by f maps a regular system of parameters $(w_1, \ldots, w_d)$ of $\mathcal{R}_{W,f(t)}$ to a regular system of parameters of $\mathcal{R}_{V,t}$. If $\tau : \mathcal{R}_{V,t} \to R[[X_1, \ldots, X_d]]$ is the Taylor expansion with respect to $(f^*(w_1), \ldots, f^*(w_d))$, then

$$\tau \circ f^* : \mathcal{R}_{W,f(t)} \to R[[X_1, \ldots, X_d]]$$

is the Taylor expansion with respect to $(w_1, \ldots, w_d)$. By taking the completions, we get a commutative diagram

$$\begin{array}{ccc}
\widehat{\mathcal{R}}_{W,f(t)} & \xrightarrow{\widehat{f^*}} & \widehat{\mathcal{R}}_{V,t} \\
\scriptstyle{\widehat{\tau \circ f^*}} \searrow & & \swarrow \scriptstyle{\widehat{\tau}} \\
& R[[X_1, \ldots, X_d]] &
\end{array}$$

where $\widehat{\tau}$ and $\widehat{\tau \circ f^*}$ are isomorphisms. Hence, $\widehat{f^*}$ is an isomorphism.

(ii) $\Rightarrow$ (i) We recall that, if $\mathfrak{m}_{V,t}$ is the maximal ideal of $\mathcal{R}_{V,t}$, then $T_t^{\mathrm{Zar}}(V)$ is canonically isomorphic to the dual of $\mathfrak{m}_{V,t}/\mathfrak{m}_{V,t}^2$. If $\widehat{\mathcal{R}}_{W,f(t)}$ is isomorphic to $\widehat{\mathcal{R}}_{V,t}$, then their vector spaces $\mathfrak{m}/\mathfrak{m}^2$ are isomorphic, and, therefore, $\mathfrak{m}_{V,t}/\mathfrak{m}_{V,t}^2$ and $\mathfrak{m}_{W,f(t)}/\mathfrak{m}_{W,f(t)}^2$ are isomorphic. Hence, we have the following equalities between dimensions of R-vector spaces:

$$\begin{aligned}
\dim(T_t^{\mathrm{Zar}}(V)) &= \dim(\mathfrak{m}_{V,t}/\mathfrak{m}_{V,t}^2) = \dim(\mathfrak{m}_{W,f(t)}/\mathfrak{m}_{W,f(t)}^2) \\
&= \dim(T_{f(t)}(W)) = d \,.
\end{aligned}$$

This shows that t is a nonsingular point of V. Furthermore,

$$df_t : T_t(V) \to T_{f(t)}(W)$$

is the dual of the isomorphism $\mathfrak{m}_{W,f(t)}/\mathfrak{m}_{W,f(t)}^2 \to \mathfrak{m}_{V,t}/\mathfrak{m}_{V,t}^2$. Hence df_t is an isomorphism. $\square$

Recall that the *integral closure* of an integral domain A is the subring of the field of fractions of A consisting of those elements which are integral over A. If A is an algebra of finite type over a field, then the integral closure of A is an A-module of finite type. An *integrally closed* ring is an integral domain which is equal to its integral closure. A regular local ring is integrally closed. If $\overline{A}$ is the integral closure of the integral domain A, and if $f : A \to B$ is an injective homomorphism into an integrally closed ring B, then there exists a unique homomorphism $\overline{f} : \overline{A} \to B$ such that $\overline{f}|_A = f$. We refer the reader to [349] for these results.

Theorem 8.1.4. *Let $\sigma \in R[[X_1, \ldots, X_n]]$ be a power series algebraic over $R[X_1, \ldots, X_n]$. There exist an irreducible algebraic set $V \subset R^{n+p}$ of dimension n, a nonsingular point t of V and a polynomial function $f \in \mathcal{P}(V)$, such that:*

(i) If $\Pi : R^{n+p} \to R^n$ is the projection onto the space of the first n coordinates, $\Pi|_V : V \to R^n$ is étale at the point t (and, hence, by Proposition 8.1.2, there is a Nash diffeomorphism s from an open neighbourhood U' of 0 in R^n onto an open neighbourhood U of t in V, which is the inverse of $\Pi|_U$).

(ii) The series σ is the Taylor series of the composition $f \circ s$ at $0 \in R^n$:

$$\begin{array}{ccc} t \in U \subset V & \xrightarrow{f} & R \\ {\scriptstyle s}\uparrow & {\scriptstyle \Pi|_V}\downarrow & \\ 0 \in U' \subset R^n & & \end{array}$$

Proof. Let $\varphi : R[X,T] \to R[[X]]$ be the R-algebra homomorphism mapping every X_i to X_i and T to σ. Since σ is algebraic over $R[X]$, the kernel $\mathrm{Ker}(\varphi)$ of φ is different from zero. It is obvious that $\mathrm{Ker}(\varphi)$ is a real prime ideal. In the quotient $R[X,T]/\mathrm{Ker}(\varphi)$, the images of $X_1,\ldots,X_n$ are algebraically independent over R, and hence the transcendence degree of this quotient over R is equal to n. Therefore the height of $\mathrm{Ker}(\varphi)$ is 1. Define $W = \mathcal{Z}(\mathrm{Ker}(\varphi)) \subset R^{n+1}$. The algebraic set W is irreducible of dimension n, and $\mathcal{P}(W) = R[X,T]/\mathrm{Ker}(\varphi)$ (Theorem 4.1.4). Now we construct the normalization of W. Let A be the integral closure of $\mathcal{P}(W)$ in its field of fractions $\mathcal{K}(W)$. The ring A is a finite $\mathcal{P}(W)$-algebra of the form

$$A = \mathcal{P}(W)[Z_1,\ldots,Z_{p-1}]/J = R[X,T,Z]/I ,$$

where I is a real prime ideal. Take $V = \mathcal{Z}(I) \subset R^{n+p}$. The algebraic set V is irreducible of dimension n, and $A = \mathcal{P}(V)$ (Theorem 4.1.4).

Since the ring $R[[X]]$ is integrally closed, the injective homomorphism $\mathcal{P}(W) \to R[[X]]$ induces a unique homomorphism $\eta : A = \mathcal{P}(V) \to R[[X]]$ such that the following diagram commutes:

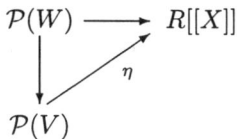

Composing η with the homomorphism $R[[X]] \to R$ which sends a series to its constant term, we get a homomorphism $\theta : \mathcal{P}(V) \to R$. Let $t = (0,y) \in R^n \times R^p$ be the point of V such that $\theta(f) = f(t)$. The homomorphism η factors as $\mathcal{P}(V) \to \widehat{\mathcal{R}}_{V,t} \xrightarrow{\rho} R[[X]]$. Consider the following commutative diagram of local homomorphisms

$$\begin{array}{ccc} \widehat{\mathcal{R}}_{V,t} & \xrightarrow{\rho} & R[[X]] \\ {\scriptstyle \Pi^*}\nwarrow & & \nearrow{\scriptstyle \mathrm{Id}} \\ & \widehat{\mathcal{R}}_{R^n,0} = R[[X]] & \end{array} ,$$

where Π^* is induced by the projection mapping $\Pi|_V$ of V onto R^n.

The local ring $\mathcal{R}_{V,t}$ is integrally closed of dimension n. By Zariski's "main theorem" ([350], Chap. 8, §13, Theorem 32), $\widehat{\mathcal{R}}_{V,t}$ is integrally closed as well, and, hence, it is an integral domain of dimension n ([20], Corollary 11.19, p.122). The homomorphism $\rho : \widehat{\mathcal{R}}_{V,t} \to R[[X]]$ is obviously surjective. If it were not injective, its kernel would be a nonzero prime ideal. Since $\widehat{\mathcal{R}}_{V,t}$ is an integral domain, its dimension would be at least $n+1$, which is a contradiction. Hence ρ is an isomorphism with inverse $\Pi^* : \widehat{\mathcal{R}}_{R^n,0} \to \widehat{\mathcal{R}}_{V,t}$. By Proposition 8.1.3, this implies that $\Pi|_V : V \to R^n$ is étale at t, and Proposition 8.1.2 gives a Nash section $s : U' \to {}^tU$ of $\Pi|_V$ defined on a neighbourhood of the origin. Let f be the image of the coordinate T in $\mathcal{P}(V)$. The homomorphism $\eta : \mathcal{P}(V) \to R[[X]]$ maps f to σ. To show (ii), we remark that $\eta : \mathcal{P}(V) \to \widehat{\mathcal{R}}_{V,t} \to R[[X]]$ maps every polynomial function h of $\mathcal{P}(V)$ to the Taylor series of $h \circ s$ at the origin. $\square$

Notation 8.1.5. *Let $X = (X_1, \ldots, X_n)$. We denote by $R[[X]]_{\mathrm{alg}}$ the ring of power series which are algebraic over $R[X]$. A power series in $R[[X]]_{\mathrm{alg}}$ is called an* algebraic power series.

Corollary 8.1.6. *The homomorphism sending the germ of a Nash function at the origin of R^n to its Taylor series at the origin is an R-algebra isomorphism from $\mathcal{N}_{R^n,0}$ onto $R[[X]]_{\mathrm{alg}}$.*

Proof. This homomorphism takes its values in $R[[X]]_{\mathrm{alg}}$, since a semi-algebraic function satisfies a polynomial equation (Lemma 2.6.3). We have already shown that it is injective (Proposition 2.9.5). Theorem 8.1.4 shows that it is surjective. $\square$

Corollary 8.1.7. *($R = \mathbb{R}$) Every algebraic power series in n variables is the Taylor series of an analytic algebraic function defined in a neighbourhood of the origin in $\mathbb{R}^n$.*

Proof. The implicit function theorem for analytic functions implies that the section s of $\Pi|_V$ in the statement of Theorem 8.1.4 is analytic, and, hence, $f \circ s$ is analytic. $\square$

Proposition 8.1.8. *($R = \mathbb{R}$) Let U be an open semi-algebraic subset of $\mathbb{R}^n$. A function $f : U \to \mathbb{R}$ is Nash if and only if it is analytic algebraic on U.*

Proof. According to Corollaries 8.1.6 and 8.1.7, it suffices to show that an analytic algebraic function f on U is semi-algebraic. The function f satisfies an equation $g(x, f(x)) = 0$ on U, where $g \in \mathbb{R}[X, Y]$ is a nonzero polynomial. Consider a slicing $(A_i, (\xi_{i,j})_{j=1,\ldots,\ell_i})_{i=1,\ldots,m}$ of the polynomial g (Definition 2.3.4). We recall that the semi-algebraic sets A_i form a partition of $\mathbb{R}^n$ and that $g^{-1}(0) \cap (A_i \times \mathbb{R})$ is either $A_i \times \mathbb{R}$ or the union of the graphs of the continuous semi-algebraic functions $\xi_{i,1} < \cdots < \xi_{i,\ell_i} : A_i \to \mathbb{R}$. Furthermore, we may assume that the A_i are connected and that U is the union of some of the A_i. Let U' be the union of those $A_i \subset U$ such that g is not identically zero on $A_i \times \mathbb{R}$. If $A_i \subset U'$, then $f|_{A_i}$ coincides with one of the $\xi_{i,j}$, and, hence, $f|_{U'}$ is semi-algebraic. Since U' is dense in U, the graph of f is the closure of the graph of $f|_{U'}$ in $U \times \mathbb{R}$. This proves that f is semi-algebraic. □

The identity theorem holds for Nash functions over any real closed field.

Proposition 8.1.9. *Let $M \subset R^n$ be a semi-algebraically connected Nash submanifold, U a nonempty open subset of M and $f : M \to R$ a Nash function. If $f|_U = 0$, then $f = 0$.*

Proof. It suffices to prove that the set S of points of M where the germ of f is zero is an open and closed semi-algebraic subset of M. The set S is semi-algebraic, since the fact that the germ of f at a point x of M is equal to zero may be expressed by a formula of the first-order language of ordered fields with parameters in R (Proposition 2.2.4). The set S is open by definition. It is also closed, because the partial derivatives of a Nash function are continuous, and the germ of a Nash function is zero if its Taylor series is zero (Corollary 8.1.6). □

Proposition 8.1.10. *Let $M \subset R^n$ be a semi-algebraically connected Nash submanifold and $f \in \mathcal{N}(M)$. If f is not identically zero on M, then $\dim(f^{-1}(0)) < \dim(M)$.*

Proof. By Proposition 2.8.13, we have

$$\dim\left(\operatorname{clos}(M \setminus f^{-1}(0)) \setminus (M \setminus f^{-1}(0))\right) < \dim(M) .$$

Hence, if $\dim(f^{-1}(0)) = \dim(M)$, the interior of $f^{-1}(0)$ in M is nonempty. By Proposition 8.1.9, f is then identically zero on M. □

We conclude this section with a comparison between germs of Nash functions at $0 \in R$ and germs of continuous semi-algebraic functions defined on intervals of type $]0, \delta[$. We set $S = \varinjlim_{\delta > 0} \mathcal{S}^0(]0, \delta[)$. With the notations of Proposition 7.3.4, S is the stalk of the sheaf $\widetilde{\mathcal{S}^0}_{\widetilde{R}}$ at the point $0_+ \in \widetilde{R}$. Let $\rho : \mathcal{N}_{R,0} \to S$ be the homomorphism which sends a germ f defined on $]-\delta, \delta[$ to its restriction on $]0, \delta[$. We have seen, in Example 7.3.6, that there is a $R[T]$-algebra isomorphism $j : S \to R(T)^\wedge_{\text{alg}}$ (where T is the variable, and $R(T)^\wedge_{\text{alg}}$ is the field of algebraic Puiseux series). On the other hand, by Corollary 8.1.6, there is a $R[T]$-algebra isomorphism $\tau : \mathcal{N}_{R,0} \to R[[T]]_{\text{alg}}$ given by the Taylor expansion.

Lemma 8.1.11. *The composition*

$$j \circ \rho \circ \tau^{-1} : R[[T]]_{\text{alg}} \to \mathcal{N}_{R,0} \to S \to R(T)^\wedge_{\text{alg}}$$

is the canonical embedding $R[[T]]_{\text{alg}} \hookrightarrow R(T)^\wedge_{\text{alg}}$.

Proof. The homomorphism $j \circ \rho \circ \tau^{-1}$ is an injective $R[T]$-algebra homomorphism. It induces a field homomorphism $\varphi : R((T))_{\text{alg}} \to R(T)^{\wedge}_{\text{alg}}$ such that $\varphi|_{R(T)}$ is the identity mapping. We know that $R(T)^{\wedge}_{\text{alg}}$ is the real closure of $R(T)$ for the ordering 0_+. In order to prove that φ coincides with the canonical embedding $i : R((T))_{\text{alg}} \to R(T)^{\wedge}_{\text{alg}}$, it suffices, by Theorem 1.3.2, to prove that φ and i induce the same ordering of $R((T))_{\text{alg}}$. Every nonzero element $f \in R((T))_{\text{alg}}$ has a representation $f = T^v g$, where $v \in \mathbb{Z}$ and $g \in R[[T]]_{\text{alg}}$, $g(0) \neq 0$. Moreover, either g or $-g$ is a square in $R[[T]]_{\text{alg}}$. This shows that $\varphi(f)$ and $i(f)$ have the same sign. $\square$

Proposition 8.1.12. *Let $g : [0, \delta[\to R$ be a continuous semi-algebraic function. There exist a positive integer p, an element $\epsilon \in R$, $0 < \epsilon \le \delta^{1/p}$ and a Nash function $f :]-\epsilon, \epsilon[\to R$, such that $f(t) = g(t^p)$ for every $t \in [0, \epsilon[$.*

Proof. We denote also by g the image of g in S. Then, $j(g) \in R(T)^{\wedge}_{\text{alg}}$ is a series in $T^{1/p}$ for some $p > 0$. Since g is continuous at 0, there is no term with a negative exponent in $j(g)$. By Lemma 8.1.11, it suffices to prove that $j(g \circ \alpha_p) \in R[[T]]_{\text{alg}}$, where α_p is defined by $\alpha_p(t) = t^p$. The endomorphism $\varphi_p : R[T] \to R[T]$ defined by $\varphi_p(T) = T^p$ can be extended in a unique way to an endomorphism, also denoted by $\varphi_p : R(T)^{\wedge}_{\text{alg}} \to R(T)^{\wedge}_{\text{alg}}$. The uniqueness of the extension follows from Theorem 1.3.2 and implies that $j(s \circ \alpha_p) = \varphi_p(j(s))$, for every $s \in S$. In particular, $j(g \circ \alpha_p) = \varphi_p(j(g)) \in R[[T]]_{\text{alg}}$. $\square$

Proposition 8.1.13 (Nash Curve Selection Lemma). *Let $A \subset R^n$ be a semi-algebraic set and $x \in \text{clos}(A)$. There exists a Nash mapping $\varphi :]-1, 1[\to R^n$ such that $\varphi(0) = x$ and $\varphi(]0, 1[) \subset A$.*

Proof. The curve selection lemma 2.5.5 gives a continuous semi-algebraic mapping $\gamma : [0, 1] \to R^n$ such that $\gamma(0) = x$ and $\gamma(]0, 1]) \subset A$. Applying Proposition 8.1.12 to each coordinate of γ, we get a positive integer p and a Nash mapping $\psi :]-\epsilon, \epsilon[\to R^n$ such that $\psi(t) = \gamma(t^p)$. Take φ to be defined by $\varphi(t) = \psi(\epsilon t)$. $\square$

8.2 Local Properties of Nash Functions

By Corollary 8.1.6, if M is a Nash submanifold of dimension n of an affine space over R and x is a point of M, the local ring $\mathcal{N}_{M,x}$ of germs of Nash functions at x is isomorphic to $R[[X_1, \ldots, X_n]]_{\text{alg}}$. Hence, the study of local properties of Nash functions is reduced to the study of algebraic power series.

We begin with a division theorem. First we recall some results about power series. Set $X = (X_1, \ldots, X_n)$ and $X' = (X_1, \ldots, X_{n-1})$.

Definition 8.2.1. *A power series $f(X', X_n) \in R[[X]]$ is regular of order k with respect to X_n if $f(0, X_n) = X_n^k g(X_n)$, with $g \in R[[X_n]]$ and $g(0) \neq 0$. A polynomial*

$$a_k X_n^k + a_{k-1} X_n^{k-1} + \cdots + a_0 \in R[[X']][X_n]$$

is distinguished if $a_k = 1$ and $a_0(0) = \cdots = a_{k-1}(0) = 0$.

Theorem 8.2.2 (Formal Division Theorem). *Let $f \in R[[X]]$ be a regular series of order k with respect to X_n. Every series $g \in R[[X]]$ has a representation*

$$g(X) = f(X)h(X) + r_{k-1}(X')X_n^{k-1} + \ldots + r_0(X'),$$

where $h \in R[[X]]$ and $r_i \in R[[X']]$, for $i = 0, \ldots, k-1$. Moreover, this representation is unique.

Reference for the proof: [350], Chap. 7, §1, Theorem 5.

Corollary 8.2.3 (Formal Preparation Theorem). *Let $f \in R[[X]]$ be a regular series of order k with respect to X_n. There exists a unique distinguished polynomial $P_f \in R[[X']][X_n]$ of degree k with respect to X_n, such that $f = u P_f$, where $u \in R[[X]]$ and $u(0) \neq 0$.*

Proof. Apply the division theorem 8.2.2 to $g = X_n^k$. □

We want to prove a division theorem for algebraic power series. We shall use a substitution technique. Let $f \in R[[X']][X_n]$ be a distinguished polynomial of degree k with respect to X_n. Denote by K the algebraic closure of $R((X'))$, and let y be a root of f in K: $f(X', y) = 0$. Let $g \in R[[X]]$. By Theorem 8.2.2, we have a unique representation

$$g(X) = f(X)h(X) + r_{k-1}(X')X_n^{k-1} + \cdots + r_0(X').$$

Set

$$g(X', y) = r_{k-1}(X')y^{k-1} + \cdots + r_0(X') \in K.$$

Lemma 8.2.4. *The mapping $g(X) \mapsto g(X', y)$ is an R-algebra homomorphism from $R[[X]]$ to K.*

Proof. The only nontrivial fact to prove is

$$(g_1 g_2)(X', y) = g_1(X', y) g_2(X', y).$$

This equality holds if g_1 and g_2 are polynomials in $R[[X']][X_n]$. The general case can be reduced to this case, replacing g_1 and g_2 with their remainders in the division by f. □

Lemma 8.2.5. *If there is $g \in R[[X]]_{\mathrm{alg}}$, $g \neq 0$, such that $g(X', y) = 0$, then y is algebraic over $R(X')$.*

Proof. If $g \in R[[X]]_{\mathrm{alg}}$, then there is an equation of algebraic dependence $a_p(X)(g(X))^p + \cdots + a_0(X) = 0$, where $a_i \in R[X]$ and $a_0 \neq 0$. By Lemma 8.2.4, we can substitute y for X_n, and we get $a_0(X', y) = 0$. □

8.2 Local Properties of Nash Functions

Proposition 8.2.6. *Let $f \in R[[X']][X_n]$ be the distinguished polynomial*

$$X_n^k + f_{k-1}(X')X_n^{k-1} + \cdots + f_0(X') .$$

Assume that there exists $h \in R[[X]]$, $h \neq 0$, such that $fh \in R[[X]]_{\mathrm{alg}}$. Then $f_i \in R[[X']]_{\mathrm{alg}}$, for $i = 0, \ldots, k-1$, and $h \in R[[X]]_{\mathrm{alg}}$.

Proof. Let y be a root of f in K (with the same notations as above). Set $g = fh$. We have $g(X', y) = 0$ and, thus, by Lemma 8.2.5, y is algebraic over $R(X')$. Since $(-1)^{k-i}f_i$ is the i-th elementary symmetric polynomial of the roots of f in K, f_i is algebraic over $R(X')$. Hence, all f_i belong to $R[[X']]_{\mathrm{alg}}$. Since f and g belong to $R[[X]]_{\mathrm{alg}}$, h belongs to $R[[X]]_{\mathrm{alg}}$ as well. □

Corollary 8.2.7 (Preparation Theorem for Nash Functions). *Let f in $R[[X]]_{\mathrm{alg}}$ be a regular series of order k with respect to X_n. There exists a unique distinguished polynomial $P_f \in R[[X']]_{\mathrm{alg}}[X_n]$ of degree k with respect to X_n, such that $f = gP_f$, where $g \in R[[X]]_{\mathrm{alg}}$ and $g(0) \neq 0$.*

Corollary 8.2.8. *The ring $R[[X]]_{\mathrm{alg}}$ is factorial. A decomposition of $f \in R[[X]]_{\mathrm{alg}}$ into a product of irreducible factors in $R[[X]]_{\mathrm{alg}}$ is a decomposition of f into a product of irreducible factors in $R[[X]]$.*

Proof. We know that $R[[X]]$ is factorial ([350], Chap. 7, §1, Theorem 6). Let $f \in R[[X]]_{\mathrm{alg}}$. We may assume that f is regular with respect to X_n, performing, if needed, a linear change of variables. In $R[[X]]$, f has a decomposition of the form $f = hg_1 \cdots g_p$, where h is invertible and $g_1, \ldots, g_p$ are irreducible distinguished polynomials with respect to X_n. By Proposition 8.2.6, $h, g_1, \ldots, g_p$ belong to $R[[X]]_{\mathrm{alg}}$, from which the corollary follows. □

Theorem 8.2.9 (Division Theorem for Nash Functions). *Let f in $R[[X]]_{\mathrm{alg}}$ be a regular series of order k with respect to X_n. Every series g in $R[[X]]_{\mathrm{alg}}$ has a representation*

$$g(X) = f(X)h(X) + r_{k-1}(X')X_n^{k-1} + \cdots + r_0(X') ,$$

where $h \in R[[X]]_{\mathrm{alg}}$ and $r_i \in R[[X']]_{\mathrm{alg}}$, for $i = 0, \ldots, k-1$. Moreover, this representation is unique.

Proof. It suffices to verify that the $r_i(X')$ of Theorem 8.2.2 are algebraic under the assumptions of the present theorem. By Corollaries 8.2.7 and 8.2.8, we may assume that f is a distinguished polynomial of degree k with respect to X_n. We may also assume that f is irreducible, since, if there is a division by f_1 and by f_2, then there is a division by the product f_1f_2. Let $y_1, \ldots, y_k$ be the roots of f in K. These roots are all distinct. For $j = 1, \ldots, k$,

$$g(X', y_j) = r_{k-1}(X')y_j^{k-1} + \cdots + r_0(X') .$$

This system of linear equations gives the $r_i(X')$ as rational functions of the y_j and the $g(X', y_j)$ (the determinant of the system is a Vandermonde determinant which is different from zero, because the y_j are distinct). We know that the y_j are algebraic over $R(X')$. The proof will be complete if we show that the $g(X', y_j)$ are also algebraic over $R(X')$. The series $g(X)$ satisfies an equation of algebraic dependence $a_p(X)(g(X))^p + \cdots + a_0(X) = 0$, where $a_0, \ldots, a_p \in R[X]$. We may assume that $a_0, \ldots, a_p$ have no nonconstant common factor. For every $j = 1, \ldots, k$, at least one of the $a_i(X', y_j)$, $i = 0, \ldots, p$, is nonzero, since, otherwise, the minimal polynomial of y_j over $R(X')$ would be a common factor of the a_i. Hence, $g(X', y_j)$ is algebraic over $R(X')(y_j)$ and therefore over $R(X')$ as well. □

Now we can obtain the local properties of Nash functions, using standard arguments based on the division theorem.

Proposition 8.2.10. *The ring $R[[X]]_{\mathrm{alg}}$ is local, and its maximal ideal is generated by $X_1, \ldots, X_n$.*

Proof. We have to prove that, if $f \in R[[X]]_{\mathrm{alg}}$ and $f(0) = 0$, then $f = \sum_{i=1}^n X_i f_i$, with $f_i \in R[[X]]_{\mathrm{alg}}$. This is obvious for $n = 1$. If $n > 1$, we divide $f = X_n g + r(X')$. Since $r(0) = 0$, the result follows by induction. □

Proposition 8.2.11. *The ring $R[[X]]_{\mathrm{alg}}$ is noetherian.*

Proof. The theorem for $n = 1$ follows from the fact that every ideal of $R[[X_1]]_{\mathrm{alg}}$ is generated by a power of X_1. Let $n > 1$, and assume the result holds true for $R[[X']]_{\mathrm{alg}}$. Let I be a nonzero ideal of $R[[X]]_{\mathrm{alg}}$. We may assume that I contains some f which is regular with respect to X_n. Let I' be the set of elements of I which are regular with respect to X_n. Then I' generates I: if $g \in I \setminus I'$, $h = g - f \in I'$ and $g = f + h$. Hence, by Corollary 8.2.7, I is generated by $I \cap R[[X']]_{\mathrm{alg}}[X_n]$. Since $R[[X']]_{\mathrm{alg}}[X_n]$ is noetherian, I is finitely generated. □

Proposition 8.2.12. *Let A (resp. B) be a noetherian local ring whose maximal ideal is $\mathfrak{m}$ (resp. $\mathfrak{n}$). Assume that $A \subset B \subset \widehat{A}$ and $\mathfrak{n} = \mathfrak{m}B$. Then,*
  (i) *B is a faithfully flat A-module.*
  (ii) *For every $k \geq 1$, $\mathfrak{n}^k = \mathfrak{m}^k B = \widehat{\mathfrak{m}}^k \cap B$.*
  (iii) *The inclusion $A \subset B$ induces an isomorphism $\widehat{A} \to \widehat{B}$.*

Reference for the proof: [66], Chap. 3, §3, Proposition 11.

Proposition 8.2.13. *The local ring $R[[X_1, \ldots, X_n]]_{\mathrm{alg}}$ is regular of dimension n. Its completion is $R[[X_1, \ldots, X_n]]$.*

Proof. By Propositions 8.2.10, 8.2.11 and 8.2.12 applied to
$$A = R[X_1, \ldots, X_n]_{(X_1, \ldots, X_n)} \text{ and } B = R[[X_1, \ldots, X_n]]_{\mathrm{alg}},$$

the completion of $R[[X]]_{\text{alg}}$ is $R[[X]]$. Hence, $\dim(R[[X_1,\ldots,X_n]]_{\text{alg}}) = \dim(R[[X_1,\ldots,X_n]]) = n$ ([20], Corollary 11.19). This implies that $R[[X]]_{\text{alg}}$ is regular. □

Corollary 8.2.14. *Let $M \subset R^p$ be a Nash submanifold of dimension n and $x \in M$. The local ring $\mathcal{N}_{M,x}$ of germs of Nash functions at x is a regular local ring of dimension n.*

Remark 8.2.15. We could also have proved the regularity (and hence the noetherian property and factoriality) of $R[[X_1,\ldots,X_n]]_{\text{alg}} = \mathcal{N}_{R^n,0}$ using the fact that $\mathcal{N}_{R^n,0}$ is the henselization of $\mathcal{R}_{R^n,0}$ (Corollary 8.7.11). However, we preferred to obtain these properties from a preparation theorem which is of independent interest.

8.3 Approximation of Formal Solutions of a System of Nash Equations

Throughout this section, $X = (X_1,\ldots,X_n)$.

The main result of this section is an approximation theorem of M. Artin [17]. We begin with some notations. Let $s_1(X), s_2(X) \in R[[X]]^p$ and $\nu \in \mathbb{N}$. We write $s_1(X) \simeq_\nu s_2(X)$ if all the derivatives of $s_1(X) - s_2(X)$ of order $\leq \nu$ vanish at the origin, or, equivalently, if all the components of $s_1(X) - s_2(X)$ belong to the $(\nu+1)^{\text{th}}$ power of the maximal ideal of $R[[X]]$.

Theorem 8.3.1. *Let $f(X,Y) = (f_1(X,Y),\ldots,f_q(X,Y)) \in (R[[X,Y]]_{\text{alg}})^q$, where $Y = (Y_1,\ldots,Y_p)$, such that $f(0,0) = 0$. Let $y(X) \in R[[X]]^p$ such that $y(0) = 0$ and $f(X,y(X)) = 0$. Then for every integer $\nu \in \mathbb{N}$, there exists $y^\nu(X)$ in $(R[[X]]_{\text{alg}})^p$, such that $f(X,y^\nu(X)) = 0$ and $y(X) \simeq_\nu y^\nu(X)$.*

References for the proof: [327], Chap. 3, Theorem 4.2. Actually, the theorem proved in this excellent book is a theorem of approximation of formal solutions of analytic equations by analytic solutions. But the proof uses only the implicit function theorem and the local properties (division theorem, etc...) that we have just established for Nash functions. Thus it can be adapted to the present context.

Alternatively, Theorem 8.3.1 may be proved using the fact that $R[[X]]_{\text{alg}}$ is the henselization of the localization of $R[X]$ at the ideal $(X_1,\ldots,X_n)$ and applying Theorem 1.10 in [18]. □

We denote by $\mathbb{R}\{X\}$ the ring of germs of analytic functions at the origin of $\mathbb{R}^n$.

Corollary 8.3.2. *Let I be a prime (resp. primary, resp. real) ideal of $R[[X]]_{\text{alg}}$. Then $IR[[X]]$ is a prime (resp. primary, resp. real) ideal of $R[[X]]$. If, in addition, $R = \mathbb{R}$, $I\mathbb{R}\{X\}$ is a prime (resp. primary, resp. real) ideal of $\mathbb{R}\{X\}$.*

Proof. We consider only the case where I is prime. The other cases are similar. Let $fg \in IR[[X]]$. By Theorem 8.3.1, there exist, for every $\nu \in \mathbb{N}$, $f^\nu, g^\nu \in R[[X]]_{\text{alg}}$ such that $f \simeq_\nu f^\nu$, $g \simeq_\nu g^\nu$ and $f^\nu g^\nu \in I$. Hence, there exists a sequence $\nu_i \to \infty$ such that $f^{\nu_i} \in I$ (or $g^\nu \in I$, for all sufficiently large ν). Then $f = \lim_{i \to \infty} f^{\nu_i} \in IR[[X]]$ (or $g \in IR[[X]]$), which shows that $IR[[X]]$ is a prime ideal.

In the case $R = \mathbb{R}$, the ring $\mathbb{R}[[X]]$ is faithfully flat over $\mathbb{R}\{X\}$. Hence, $I\mathbb{R}\{X\} = (I\mathbb{R}[[X]]) \cap \mathbb{R}\{X\}$, and $I\mathbb{R}\{X\}$ is prime. □

Corollary 8.3.3. *Let I be an ideal of $R[[X]]_{\text{alg}}$ and $I = \bigcap_{i=1}^q I_i$, a primary decomposition. Then $IR[[X]] = \bigcap_{i=1}^q I_i R[[X]]$ is a primary decomposition of $IR[[X]]$. If $R = \mathbb{R}$, then $I\mathbb{R}\{X\} = \bigcap_{i=1}^q I_i \mathbb{R}\{X\}$ is a primary decomposition of $I\mathbb{R}\{X\}$.*

Proof. Each $I_i R[[X]]$ is primary by Corollary 8.3.2. Since $R[[X]]$ is faithfully flat over $R[[X]]_{\text{alg}}$, we have $IR[[X]] = \bigcap_{i=1}^q I_i R[[X]]$ and, for every i, $IR[[X]] \neq \bigcap_{j \neq i} I_j R[[X]]$. The same proof applies to $\mathbb{R}\{X\}$. □

We shall return to the geometric meaning of these corollaries when we study the germs of Nash sets in Section 6.

8.4 The Artin-Mazur Description of Nash Functions

Proposition 8.4.1. *Let $M \subset R^p$ be a semi-algebraically connected Nash submanifold of dimension n. Let $Z = \text{clos}_{\text{Zar}}(M)$ be the Zariski closure of M in R^p and x a point of M. Then:*

(i) *The algebraic set Z is irreducible of dimension n.*

(ii) *The completion $\widehat{\mathcal{N}}_{M,x}$ of the local ring of germs of Nash functions at x is the quotient of the completion $\widehat{\mathcal{R}}_{Z,x}$ of the local ring of germs of regular functions by a minimal prime ideal.*

Proof. (i) The ideal $\mathcal{I}(M)$ of polynomials vanishing on M is the kernel of the canonical homomorphism $R[X_1, \ldots, X_p] \to \mathcal{N}(M)$. By the identity theorem 8.1.9, the ring $\mathcal{N}(M)$ is an integral domain. Therefore $\mathcal{I}(M)$ is prime. The algebraic set Z is irreducible and of dimension n by Proposition 2.8.2.

(ii) By the definition of Nash submanifolds and Proposition 8.2.13, there exists in $\mathcal{N}_{R^p,x}$ a regular system of parameters $f_1, \ldots, f_p$ such that $\mathcal{N}_{M,x} \simeq \mathcal{N}_{R^p,x}/(f_{n+1}, \ldots, f_p)$. We consider the following commutative diagram of local homomorphisms:

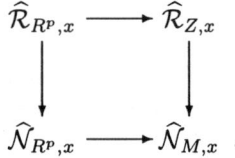

8.4 The Artin-Mazur Description of Nash Functions 173

The homomorphism $\widehat{\mathcal{R}}_{R^p,x} \to \widehat{\mathcal{N}}_{R^p,x}$ is an isomorphism (Proposition 8.2.13), and $\widehat{\mathcal{N}}_{R^p,x} \to \widehat{\mathcal{N}}_{M,x}$ is surjective. Hence, $\widehat{\mathcal{R}}_{Z,x} \to \widehat{\mathcal{N}}_{M,x}$ is surjective. Since $\widehat{\mathcal{N}}_{M,x}$ is an integral domain and both $\widehat{\mathcal{N}}_{M,x}$ and $\widehat{\mathcal{R}}_{Z,x}$ are of dimension n, $\widehat{\mathcal{N}}_{M,x}$ is the quotient of $\widehat{\mathcal{R}}_{Z,x}$ by a minimal prime ideal. □

Proposition 8.4.2. *(With the same notations as in Proposition 8.4.1). If Z is normal (i.e. if, for every point y in Z, $\mathcal{R}_{Z,y}$ is integrally closed), then M is an open subset of Z and is contained in the set $\mathrm{Reg}(Z)$ of nonsingular points of Z. In this case, $\widehat{\mathcal{R}}_{Z,x} \simeq \widehat{\mathcal{N}}_{M,x}$, for every x in M.*

Proof. Zariski's "main theorem" ([350], Chap. 8, §13, Theorem 32) says that, if $\mathcal{R}_{Z,x}$ is integrally closed, then $\widehat{\mathcal{R}}_{Z,x}$ is also integrally closed. In particular, $\widehat{\mathcal{R}}_{Z,x}$ is an integral domain, and, hence, $\widehat{\mathcal{R}}_{Z,x}$ is isomorphic to $\widehat{\mathcal{N}}_{M,x}$ by Proposition 8.4.1. By Corollary 8.2.14, this implies that $\widehat{\mathcal{R}}_{Z,x}$ and $\mathcal{R}_{Z,x}$ are regular local rings of dimension n. Hence $M \subset \mathrm{Reg}(Z)$. Since the inclusion $M \hookrightarrow Z$ is a Nash diffeomorphism in a neighbourhood of x, M is open in Z. □

Proposition 8.4.3. *Let $M \subset R^p$ be a semi-algebraically connected Nash submanifold. Then the ring $\mathcal{N}(M)$ of Nash functions on M is integrally closed.*

Proof. By the identity theorem 8.1.9, $\mathcal{N}(M)$ is an integral domain. Let $f, g \in \mathcal{N}(M)$, with $g \neq 0$, such that f/g is integral over $\mathcal{N}(M)$. Then, for every point x in M, f/g is integral over $\mathcal{N}_{M,x}$. Since $\mathcal{N}_{M,x}$ is factorial (Corollary 8.2.8), it is integrally closed, and, hence, $f/g \in \mathcal{N}_{M,x}$. It is obvious that $\mathcal{N}(M)$, as a subset of its field of fractions $\mathrm{Fr}(\mathcal{N}(M))$, is the intersection of all $\mathcal{N}_{M,x} \cap \mathrm{Fr}(\mathcal{N}(M))$, for $x \in M$. Hence $f/g \in \mathcal{N}(M)$. □

Theorem 8.4.4 (Artin-Mazur Theorem). *Let $M \subset R^p$ be a semi-algebraically connected Nash submanifold of dimension n. Let $f : M \to R^k$ be a Nash mapping. There exist a nonsingular irreducible algebraic set $V \subset R^{p+q}$ of dimension n, an open semi-algebraic subset M' of V, a Nash diffeomorphism $\sigma : M \to M'$ and a polynomial mapping $g : V \to R^k$, such that the following diagram commutes*

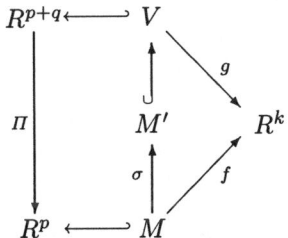

where Π is the projection onto the space of the first p coordinates. Moreover, M' is a semi-algebraically connected component of $\Pi^{-1}(M) \cap V$.

174   8. Nash Functions

Proof. Let G be the Zariski closure of the graph of f in R^{p+k}. The ideal $\mathcal{I}(G)$ is the kernel of the homomorphism

$$\rho : R[X_1, \ldots, X_p, Y_1, \ldots, Y_k] \to \mathcal{N}(M)$$

which sends Y_j to the jth component f_j of f. The algebraic set G is irreducible of dimension n, and ρ induces a homomorphism $\mu : \mathcal{P}(G) \to \mathcal{N}(M)$. Let B be the integral closure of $\mathcal{P}(G)$ in its field of fractions. We may assume $B = R[X_1, \ldots, X_p, Y_1, \ldots, Y_k, Z_{k+1}, \ldots, Z_q]/I$, where I is a real prime ideal. Set $V = \mathcal{Z}(I) \subset R^{p+q}$. The algebraic set V is irreducible of dimension n, and $\mathcal{P}(V) = B$. Since the homomorphism μ is injective and $\mathcal{N}(M)$ is integrally closed (Proposition 8.4.3), there is a commutative diagram of ring homomorphisms

$$\begin{array}{ccc} \mathcal{P}(G) & \xrightarrow{\mu} & \mathcal{N}(M) \\ \Pi_1^* \uparrow & \searrow \Pi_2^* & \uparrow \eta \\ R[X_1, \ldots, X_p] & \xrightarrow{\Pi_3^*} & \mathcal{P}(V) \end{array}$$

where $\Pi_1 : G \to R^p$, $\Pi_2 : V \to G$ and $\Pi_3 : V \to R^p$ are the projection mappings defined respectively by $\Pi_1(x, y) = x$, $\Pi_2(x, y, z) = (x, y)$ and $\Pi_3(x, y, z) = x$. Denote the classes of X_i, Y_j, Z_m in $\mathcal{P}(V)$ by the same symbols X_i, Y_j, Z_m, and define

$$\sigma = (\eta(X_1), \ldots, \eta(X_p), \eta(Y_1), \ldots, \eta(Y_k), \eta(Z_{k+1}), \ldots, \eta(Z_q)) \in \mathcal{N}(M)^{p+q}.$$

We see from the diagram that σ is a Nash mapping from M to V such that $\sigma(x) = (x, f(x), u(x))$, where $u = (\eta(Z_{k+1}), \ldots, \eta(Z_q))$. In particular, $\Pi \circ \sigma = \Pi_3 \circ \sigma = \text{Id}_M$. Let M' be the image of σ. Since M' is the graph of the Nash mapping $(f, u) : M \to R^q$, it is a Nash submanifold and $\sigma : M \to M'$ is a Nash diffeomorphism. By Proposition 8.4.2, M' is an open subset of V contained in $\text{Reg}(V)$. We may assume that V is nonsingular: if $\text{Sing}(V) \neq \emptyset$, then choose $h \in \mathcal{P}(V)$ such that $\text{Sing}(V) = \mathcal{Z}_V(h)$ and replace V with

$$V_1 = \{(x, y, z, t) \in R^{p+q+1} \mid (x, y, z) \in V \text{ and } t h(x, y, z) = 1\},$$

which is biregularly isomorphic to $\text{Reg}(V)$. Now, let $g : V \to R^k$ be defined by $g(x, y, z) = y$. Then $g \in \mathcal{P}(V)^k$, and $g \circ \sigma = f$. Finally, M' is closed in $\Pi^{-1}(M)$, and this proves the last statement of the theorem. $\square$

Remark 8.4.5. There is a clear similarity between the statement and the proof of the previous theorem and Theorem 8.1.4. The statement of Theorem 8.1.4 is, in fact, a local version of Theorem 8.4.4.

Proposition 8.4.6. *Let $M \subset R^p$ be a semi-algebraically connected Nash submanifold. Then M is Nash diffeomorphic to a semi-algebraically connected component of a nonsingular algebraic set $V \subset R^{p+q}$.*

8.5 The Substitution Theorem. The Positivstellensatz for Nash Functions 175

Proof. First we reduce the proof to the case where M is closed in R^p. By the definition of a Nash submanifold, M is locally closed. Let $K = \mathrm{clos}(M) \setminus M$. Then K is a closed semi-algebraic subset of R^p. By Proposition 2.7.5, there exists a function $f \in \mathcal{A}(R^p; R^p \setminus K)$ such that $f^{-1}(0) = K$. Define $h : R^p \setminus K \to R^{p+1}$ by $h(x) = (x, 1/f(x))$. Then h is a Nash diffeomorphism from $R^p \setminus K$ onto its image, and $h(M)$ is a closed Nash submanifold of R^{p+1}.

Hence, we may assume that M is closed in R^p. Then, by Theorem 8.4.4, M is Nash diffeomorphic to an open and closed semi-algebraic subset M' of a nonsingular algebraic set V. □

Remark 8.4.7. It is, in fact, true that every Nash submanifold M is Nash diffeomorphic to a nonsingular algebraic set. If $R = \mathbb{R}$ and M is compact, this is a result of Tognoli which will be proved in Chap. 14. For noncompact M, this is proved in [302]. Finally, the fact that the result holds for any real closed field R is proved in [103].

8.5 The Substitution Theorem. The Positivstellensatz for Nash Functions

In the proofs of the Positivstellensatz given in Chap.4, we used a property that is so obvious for polynomials that it is not even explicitly mentioned: if $F \supset R$ is a real closed extension and $\varphi : R[X_1, \ldots, X_n] \to F$ an R-algebra homomorphism, then, for every $f \in R[X_1, \ldots, X_n]$, $\varphi(f) = f_F(\varphi(X_1), \ldots, \varphi(X_n))$, where $f_F : F^n \to F$ is the extension of the polynomial function f to F. This substitution property is, however, not at all obvious for Nash functions. It will be proved in Theorem 8.5.2.

Let $f : S \to R$ be a semi-algebraic function defined on a semi-algebraic subset S of R^p and F, a real closed extension of R. The extension

$$f_F : S_F \to F$$

of f to F was defined in Definition 5.3.2.

Let $U \subset R^p$ be an open semi-algebraic set. The subring $\mathcal{A}(R^p; U)$ of the ring of continuous semi-algebraic functions on R^p was defined in Definition 2.7.4. Recall that, if $f \in \mathcal{A}(R^p; U)$, then $f|_U$ is Nash.

Lemma 8.5.1. *Let F be a real closed extension of R and $\varphi : \mathcal{N}(U) \to F$, an R-algebra homomorphism. If $f \in \mathcal{A}(R^p; U)$, then*

$$\varphi(f|_U) = f_F(\varphi(X_1), \ldots, \varphi(X_p)) \, .$$

Proof. The proof proceeds by induction on the construction of f, starting from the polynomials. The formula is obviously true when f is a polynomial. If it is true for g_1 and g_2, then it is true for $g_1 + g_2$ and $g_1 g_2$ as well. Assume that it is true for $g_1, \ldots, g_p$ and that $g = g_1^2 + \cdots + g_p^2$ is positive on U. If $f = \sqrt{g}$, then

$$\varphi(g|_U) = \varphi(f^2|_U) = \varphi(f|_U)^2 \,.$$

Since $f|_U$ is a square in $\mathcal{N}(U)$, $\varphi(f|_U) \geq 0$, and, hence,

$$\varphi(f|_U) = \sqrt{\varphi(g|_U)} = \sqrt{g_F(\varphi(X_1),\ldots,\varphi(X_p))} = f_F(\varphi(X_1),\ldots,\varphi(X_p))\,.$$

□

Theorem 8.5.2 (Substitution Theorem). *Let $V \subset R^p$ be an irreducible algebraic set and M, an open semi-algebraic subset of $\mathrm{Reg}(V)$. Let F be a real closed extension of R and $\varphi : \mathcal{N}(M) \to F$, an R-algebra homomorphism. Then:*
 (i) $\varphi(X) = (\varphi(X_1),\ldots,\varphi(X_p)) \in M_F \subset F^p$.
 (ii) *For every $f \in \mathcal{N}(M)$, $\varphi(f) = f_F(\varphi(X))$.*

Proof. (i) Let $U = M \cup (R^p \setminus V)$. By Proposition 2.7.5, there exists $h \in \mathcal{A}(R^p; U)$ such that $h > 0$ on U and $h = 0$ on $R^p \setminus U$. By the Tarski-Seidenberg principle (Corollary 5.2.4), since the formula

$$\forall x \in R^p \; x \in U \Leftrightarrow h(x) > 0$$

holds true, we have

$$\forall x \in F^p \; x \in U_F \Leftrightarrow h_F(x) > 0\,.$$

By Lemma 8.5.1 applied to the composition of φ with the restriction homomorphism $\mathcal{N}(U) \to \mathcal{N}(M)$,

$$\varphi(h|_M) = h_F(\varphi(X))\,.$$

On the other hand, since $h|_M$ is an invertible square in $\mathcal{N}(M)$, $\varphi(h|_M) > 0$. It follows that $\varphi(X) \in U_F$. Since $\varphi(X)$ satisfies the equations defining V, we have $\varphi(X) \in U_F \cap V_F = M_F$.

 (ii) If $M_1,\ldots,M_k$ are the semi-algebraically connected components of M, then $\mathcal{N}(M) \simeq \mathcal{N}(M_1) \times \cdots \times \mathcal{N}(M_k)$. Therefore, the general case can be reduced to the case where M is semi-algebraically connected. Let $f \in \mathcal{N}(M)$. By the Artin-Mazur description of Nash functions (Theorem 8.4.4), there exist a nonsingular irreducible algebraic set $W \subset R^{p+r}$, an open semi-algebraic subset M' of W and $q \in \mathcal{P}(W)$, such that:
 (a) $\Pi|_{M'}$ is a Nash diffeomorphism from M' onto M (where Π is the projection mapping from R^{p+r} onto the space of the first p coordinates), whose inverse is σ,
 (b) $f = q \circ \sigma$,
 (c) M' is a semi-algebraically connected component of $\Pi^{-1}(M) \cap W$.

Set $U = M \cup (R^p \setminus V)$ and $S = (\Pi^{-1}(M) \cap W) \setminus M'$. Then M' and S are two disjoint closed semi-algebraic subsets of $\Pi^{-1}(U)$ and, by the separation theorem 2.7.7, there exists $g \in \mathcal{A}(R^{p+r}; \Pi^{-1}(U))$ such that $g|_{M'} > 0$ and $g|_S < 0$. By the Tarski-Seidenberg principle 5.2.4, since the formula

8.5 The Substitution Theorem. The Positivstellensatz for Nash Functions

$$\forall y \in M \ \forall z \in R^{p+r} \ (z = \sigma(y) \Leftrightarrow z \in W \text{ and } \Pi(z) = y \text{ and } g(z) > 0)$$

holds true, we have

$$\forall y \in M_F \ \forall z \in F^{p+r} \ (z = \sigma_F(y) \Leftrightarrow z \in W_F \text{ and } \Pi_F(z) = y \text{ and } g_F(z) > 0).$$

Now we shall verify that the formula on the right side of the latter equivalence holds true for $y = \varphi(X)$ and $z = \varphi(\sigma) = (\varphi(\sigma_1), \ldots, \varphi(\sigma_{p+r}))$. Since $\sigma(M) \subset W$, $\varphi(\sigma)$ satisfies the equations of W, and $\varphi(\sigma) \in W_F$. Since $\Pi \circ \sigma = X$ on M and Π is a polynomial mapping, $\Pi_F(\varphi(\sigma)) = \varphi(\Pi \circ \sigma) = \varphi(X)$. Since $g \circ \sigma$ is an invertible square in $\mathcal{N}(M)$, $\varphi(g \circ \sigma) > 0$. By Lemma 8.5.1 applied to the homomorphism $\mathcal{N}(\Pi^{-1}(U)) \to F$ sending w to $\varphi(w \circ \sigma)$, we have $\varphi(g \circ \sigma) = g_F(\varphi(\sigma))$, and, hence, $g_F(\varphi(\sigma)) > 0$.

From the equivalence above, we get $\varphi(\sigma) = \sigma_F(\varphi(X))$. Therefore, since q is a polynomial,

$$\varphi(f) = \varphi(q \circ \sigma) = q_F(\varphi(\sigma)) = q_F \circ \sigma_F(\varphi(X)) = (q \circ \sigma)_F(\varphi(X)) = f_F(\varphi(X)).$$

$\square$

Theorem 8.5.3. *Let $M \subset R^p$ be a Nash submanifold. Let $(f_j)_{j=1,\ldots,s}$, $(g_k)_{k=1,\ldots,t}$, $(h_\ell)_{\ell=1,\ldots,u}$ be finite families of Nash functions in $\mathcal{N}(M)$. Let P be the cone generated by the $(f_j)_{j=1,\ldots,s}$, M, the multiplicative monoid generated by the $(g_k)_{k=1,\ldots,t}$ and I, the ideal generated by the $(h_\ell)_{\ell=1,\ldots,u}$. The following properties are equivalent:*

(i) *The set*

$$S = \{x \in M \mid f_j(x) \geq 0, \ j = 1, \ldots, s, \ g_k(x) \neq 0, \ k = 1, \ldots, t,$$
$$h_\ell(x) = 0, \ \ell = 1, \ldots, u\}$$

is empty.

(ii) *There exist $f \in P$, $g \in M$, $h \in I$ such that $f + g^2 + h = 0$.*

Proof. (ii) $\Rightarrow$ (i) is obvious.

(i) $\Rightarrow$ (ii) By Proposition 8.4.6, we may assume that M is an open semialgebraic subset of a nonsingular algebraic set. Suppose that there exists a homomorphism $\varphi : \mathcal{N}(M) \to F$ to a real closed field F, such that $\varphi(f_j) \geq 0$ for $j = 1, \ldots, s$, $\varphi(g_k) \neq 0$ for $k = 1, \ldots, t$, and $\varphi(h_\ell) = 0$ for $\ell = 1, \ldots, u$. Then we would have $\varphi(X) \in S_F$ by Theorem 8.5.2, and, hence, $S \neq \emptyset$ by the Tarski-Seidenberg principle 5.1.4. By assumption (i), there is, therefore, no such homomorphism, and we obtain (ii) by applying Proposition 4.4.1.

$\square$

Corollary 8.5.4. *Let $M \subset R^p$ be a Nash submanifold. Let $g, h_1, \ldots, h_u \in \mathcal{N}(M)$ such that, for every x in M,*

$$h_1(x) = \cdots = h_u(x) = 0 \ \Rightarrow \ g(x) = 0.$$

Then there exist an integer $n \in \mathbb{N}$ and $p_1, \ldots, p_k \in \mathcal{N}(M)$ such that

$$g^{2n} + p_1^2 + \cdots + p_k^2 \in (h_1, \ldots, h_u)\mathcal{N}(M).$$

Corollary 8.5.5 (Positivstellensatz). *Let $M \subset R^p$ be a Nash submanifold, $g_1, \ldots, g_t \in \mathcal{N}(M)$,*

$$W = \{x \in M \mid g_1(x) \geq 0, \ldots, g_t(x) \geq 0\}.$$

Let P be the cone of $\mathcal{N}(M)$ generated by $g_1, \ldots, g_t$. Let $f \in \mathcal{N}(M)$. Then:
  (i) $\forall x \in W \;\; f(x) \geq 0 \;\;\Leftrightarrow\;\; \exists m \in \mathbb{N} \;\exists g, h \in P \;\; fg = f^{2m} + h$.
  (ii) $\forall x \in W \;\; f(x) > 0 \;\;\Leftrightarrow\;\; \exists g, h \in P \;\; fg = 1 + h$.
  (iii) $\forall x \in W \;\; f(x) = 0 \;\;\Leftrightarrow\;\; \exists m \in \mathbb{N} \;\exists g \in P \;\; f^{2m} + g = 0$.

Proof. The proof is analogous to that of Corollary 4.4.3. □

Proposition 8.5.6 (Hilbert's 17th Problem for Nash Functions). *Let $M \subset R^p$ be a semi-algebraically connected Nash submanifold of dimension n. If $f \in \mathcal{N}(M)$ is nonnegative on M, then f is a sum of 2^n squares in the field of fractions of $\mathcal{N}(M)$.*

Proof. The fact that f is a sum of squares in the field of fractions of $\mathcal{N}(M)$ can be deduced from the Positivstellensatz 8.5.5 (i) (cf. Remark 6.1.2). We then apply Theorem 6.4.17 to prove that f is a sum of 2^n squares. We need to verify that the transcendence degree of the field of fractions of $\mathcal{N}(M)$ over R is equal to n. This field is contained in the field of fractions $\text{Fr}(\mathcal{N}_{M,x})$ of the local ring of germs of Nash functions at some point x of M. The field $\text{Fr}(\mathcal{N}_{M,x})$ is isomorphic to the field of fractions of the ring of algebraic power series in n variables, and, therefore, its transcendence degree over R is n. □

8.6 Nash Sets, Germs of Nash Sets

Throughout this section, M denotes a Nash submanifold of R^n.

Definition 8.6.1. *A Nash subset of M is a semi-algebraic subset of the form*

$$\mathcal{Z}_M(f_1, \ldots, f_p) = \{x \in M \mid f_1(x) = \cdots = f_p(x) = 0\},$$

where $f_1, \ldots, f_p$ are Nash functions from M to R.

More generally, if I is a subset of $\mathcal{N}(M)$, we set

$$\mathcal{Z}_M(I) = \{x \in M \mid \forall f \in I \;\; f(x) = 0\}$$

and, if A is a subset of M,

$$\mathcal{I}_{\mathcal{N}(M)}(A) = \{f \in \mathcal{N}(M) \mid \forall x \in A \;\; f(x) = 0\}.$$

The fact that $\mathcal{Z}_M(I)$ is a Nash set for every ideal I of $\mathcal{N}(M)$ does not follow immediately from the previous definition. This will be proved in the next proposition. This result can be considered as a weak form of the noetherian property. The noetherian property of the ring $\mathcal{N}(M)$ will be proved, in the case $R = \mathbb{R}$, in the next section.

8.6 Nash Sets, Germs of Nash Sets

Proposition 8.6.2. *Let I be an ideal of $\mathcal{N}(M)$. There exists a finite number of Nash functions $f_1, \ldots, f_p$ in I such that $\mathcal{Z}_M(I) = \mathcal{Z}_M(f_1, \ldots, f_p)$. The set $\mathcal{Z}_M(I)$ is thus a Nash subset of M.*

Proof. Let $J \subset I$ be an ideal of finite type of $\mathcal{N}(M)$. Assume that $\mathcal{Z}_M(J) \setminus \mathcal{Z}_M(I)$ is contained in a semi-algebraic subset N of M of dimension k. We claim that there exists an ideal of finite type J' of $\mathcal{N}(M)$, $J \subset J' \subset I$, such that $\mathcal{Z}_M(J') \setminus \mathcal{Z}_M(I)$ is contained in a semi-algebraic set of dimension $< k$. The proposition follows from this claim, by an easy descending induction on dimension. Let us prove the claim.

By Proposition 2.9.10, we have $N = \bigcup_{j=1}^{\ell} N_j$, where the N_j are semi-algebraically connected Nash submanifolds. We may assume that $N_j \not\subset \mathcal{Z}_M(I)$, for $j = 1, \ldots, \ell$. Hence, there exists $g_j \in I$ such that $N_j \not\subset g_j^{-1}(0)$ and, by Proposition 8.1.10, $\dim(\mathcal{Z}_M(g_j) \cap N_j) < \dim(N_j)$. We define J' as the ideal generated by J and the g_j, $j = 1, \ldots, \ell$. □

Corollary 8.6.3. *Every maximal ideal of $\mathcal{N}(M)$ is the ideal of Nash functions vanishing at some point of M.*

Proof. It is sufficient to prove that, for any proper ideal I of $\mathcal{N}(M)$, $\mathcal{Z}_M(I)$ is nonempty. By Proposition 8.6.2, there are $f_1, \ldots, f_p \in I$ such that $\mathcal{Z}_M(I) = \mathcal{Z}_M(f_1, \ldots, f_p)$. Suppose $\mathcal{Z}_M(I) = \emptyset$. Then the function $f_1^2 + \cdots + f_p^2 \in I$ would have an inverse in $\mathcal{N}(M)$, and, hence, the ideal I would not be proper. □

Theorem 8.6.4. *Let $\mathfrak{p}$ be a prime ideal of $\mathcal{N}(M)$. Then the set $\mathcal{Z}_M(\mathfrak{p})$ of zeros of $\mathfrak{p}$ is semi-algebraically connected.*

Proof. Suppose that $\mathcal{Z}_M(\mathfrak{p})$ is not semi-algebraically connected. Then there are two disjoint nonempty closed semi-algebraic subsets F_1 and F_2 of M such that $\mathcal{Z}_M(\mathfrak{p}) = F_1 \cup F_2$. By Proposition 8.4.6, we may assume that M is closed in R^n. The separation theorem 2.7.3 gives us a Nash function g on M (restriction of a Nash function on R^n), such that $g|_{F_1} > 0$ and $g|_{F_2} < 0$. By Proposition 8.6.2, there are $f_1, \ldots, f_p \in \mathfrak{p}$ such that $\mathcal{Z}_M(\mathfrak{p}) = \mathcal{Z}_M(f_1, \ldots, f_p)$. The functions

$$h_1 = \sqrt{f_1^2 + \cdots + f_p^2 + g^2} - g \quad \text{and} \quad h_2 = \sqrt{f_1^2 + \cdots + f_p^2 + g^2} + g$$

are Nash on M. Their product $h_1 h_2 = f_1^2 + \cdots + f_p^2$ belongs to $\mathfrak{p}$. On the other hand, $\mathcal{Z}_M(h_i) = F_i$ for $i = 1, 2$, and, hence, neither h_1 nor h_2 belongs to $\mathfrak{p}$. This contradicts the fact that $\mathfrak{p}$ is prime. □

Theorem 8.6.5 (Nullstellensatz for Nash Functions). *Let I be an ideal of $\mathcal{N}(M)$. Then $\mathcal{I}_{\mathcal{N}(M)}(\mathcal{Z}_M(I)) = \sqrt[R]{I}$. In other words, $f \in \mathcal{N}(M)$ vanishes on $\mathcal{Z}_M(I)$ if and only if there exist $m \in \mathbb{N}$ and $g_1, \ldots, g_q \in \mathcal{N}(M)$ such that $f^{2m} + g_1^2 + \cdots + g_q^2 \in I$.*

Proof. Follows from Proposition 8.6.2 and Corollary 8.5.4. □

Definition 8.6.6. *A Nash subset V in M is said to be* irreducible *if, whenever $V = V_1 \cup V_2$, with V_1 and V_2 Nash subsets of M, we have $V = V_1$ or $V = V_2$.*

Proposition 8.6.7. *Let V be a Nash subset of M. Then V is a finite union of irreducible Nash sets.*

Proof. Since V is semi-algebraic, $V = \bigcup_{i=1}^{q} N_i$, where the N_i are semi-algebraically connected Nash submanifolds (Proposition 2.9.10). By Proposition 8.6.2, the intersection of an arbitrary family of Nash sets is a Nash set. Hence, there is a smallest Nash subset V_i of M containing N_i. If $V_i = W_1 \cup W_2$ with W_1 and W_2 Nash sets, then $N_i = (N_i \cap W_1) \cup (N_i \cap W_2)$ and, hence, $W_1 \supset N_i$ or $W_2 \supset N_i$, by Proposition 8.1.9. This shows that every Nash set V_i is irreducible. Moreover, it is obvious that $V = \bigcup_{i=1}^{q} V_i$. □

Corollary 8.6.8. *Let V be a Nash subset of M. Then V admits a unique decomposition $V = \bigcup_{i=1}^{q} V_i$, where the V_i are irreducible Nash sets such that $V_i \not\subset V_j$ for $i \neq j$. The V_i are called the* Nash irreducible components *of V.*

Proof. Same as for algebraic sets ([295], Chap. 1, §3, Theorem 2). □

We now study germs of Nash sets. Let x be a point of a Nash submanifold M. We recall that two subsets V and W of M have the same germ at x if there is a neighbourhood U of x in M such that $V \cap U = W \cap U$. This equivalence relation is compatible with taking finite unions, finite intersections and complements. A *germ of a Nash set* at the point x is the equivalence class of a Nash subset V of an open semi-algebraic neighbourhood of x in M. We denote it by V_x. The germ V_x is said to be Nash irreducible if, whenever $V_x = V_{1,x} \cup V_{2,x}$, with $V_{1,x}$ and $V_{2,x}$ germs of Nash sets, we have $V_x = V_{1,x}$ or $V_x = V_{2,x}$.

Proposition 8.6.9. *($R = \mathbb{R}$) A germ of a Nash set is Nash irreducible if and only if it is analytically irreducible. The analytic irreducible components of a germ of a Nash set are germs of Nash sets.*

Proof. Let V_x be a Nash irreducible germ of a Nash set. Then the ideal $\mathcal{I}(V_x) \subset \mathcal{N}_{M,x}$ of germs of Nash functions vanishing on V_x is a real prime ideal. Let $\mathcal{O}_{M,x}$ be the local ring of germs of analytic functions on M at x. By Corollary 8.3.2, $\mathcal{I}(V_x)\mathcal{O}_{M,x}$ is a real prime ideal. Hence, by the Nullstellensatz for germs of real analytic functions [269], $\mathcal{I}(V_x)\mathcal{O}_{M,x}$ is the ideal of germs of analytic functions vanishing on V_x and V_x is analytically irreducible. The second part of the proposition is a straightforward consequence of the first part. □

Remark 8.6.10. The previous proposition raises the question whether the corresponding global result is true. A Nash set is a real global analytic set (or $\mathbb{C}$-analytic in the terminology of [75]), i.e. a set of the form $f^{-1}(0)$, with f real analytic. A real global analytic set can be decomposed into a locally finite union of irreducible global analytic sets, and such a decomposition, if it is not redundant, is unique (cf. [75], [245], p. 105, Proposition 17). For a Nash set V, this decomposition is finite. This can be proved as in Proposition 8.6.7, using the fact that an arbitrary intersection of global analytic sets is still a global analytic set ([245], p. 105, Proposition 16). Moreover, each irreducible global analytic component of V is semi-algebraic. This follows from the local result of Proposition 8.6.9 if V is compact. The case where V is not compact can be reduced to the compact case, by using an Alexandrov compactification.

If V is compact, then its irreducible global analytic components are Nash sets. This is proved in [101], as a consequence of a theorem of *global* approximation of analytic solutions of Nash equations. In the noncompact case, the problem is still open. For some partial results concerning this problem, see [102, 261].

The rest of this section is devoted to the question of algebraicity of germs of Nash sets. We shall consider germs of Nash sets at the origin 0 of R^m.

Definition 8.6.11. *Two germs of Nash sets V_0 and W_0 at the origin of R^m are said to be* Nash equivalent *if there exists a Nash diffeomorphism $\sigma : U \to U'$ between open semi-algebraic neighbourhoods of 0 in R^m, such that $\sigma(0) = 0$ and $\sigma(U \cap V) = U' \cap W$.*

Theorem 8.6.12. *Every germ of a Nash set at the origin of R^m is Nash equivalent to a germ of an algebraic set at the origin of R^m.*

The idea of the proof is to use the Artin-Mazur description of Nash functions. We then obtain an algebraic germ in some R^{m+p}, which we try to push down to R^m by means of a well-chosen projection mapping. First we deal with this last step. We work with the algebraically closed field $C = R[i]$.

Let $Z \subset \mathbb{P}_n(C)$ be a projective algebraic set and $L \subset \mathbb{P}_n(C)$ a projective subspace of dimension $n - m - 1$ such that $L \cap Z = \emptyset$. Denote by $\Pi_L : \mathbb{P}_n(C) \setminus L \to \mathbb{P}_m(C)$ the projection with centre L. We recall that $Z' = \Pi_L(Z)$ is a projective algebraic subset of $\mathbb{P}_m(C)$, and $\Pi = \Pi_L|_Z : Z \to Z'$, a finite algebraic morphism ([295], Chap. 1, §5, Theorem 7).

Given two distinct points x, y in $\mathbb{P}_n(C)$, we denote by $\overline{xy}$ the projective line connecting them.

Lemma 8.6.13. *Let $Z \subset \mathbb{P}_n(C)$ be a projective algebraic set, and x, a point of Z. Let L be a projective subspace of dimension $n - m - 1$ such that, for every y in $Z \setminus \{x\}$, $\overline{xy} \cap L = \emptyset$. Let $\Pi = \Pi_L|_Z : Z \to Z'$ and $x' = \Pi(x)$. Let $t_{Z,x}$ be the projective subspace that contains x and has the same Zariski tangent space as Z at x. If $L \cap t_{Z,x} = \emptyset$, then Π induces an isomorphism of local rings $\Pi^* : \mathcal{R}_{Z',x'} \to \mathcal{R}_{Z,x}$.*

Proof. Set $A = \mathcal{R}_{Z',x'}$ and $B = \mathcal{R}_{Z,x}$. Since Π is finite and $\Pi^{-1}(x') = \{x\}$, $\Pi^* : A \to B$ is injective, and B is an A-module of finite type. We can choose homogeneous coordinates $T_0, \ldots, T_n$ so that $x = (1 : 0 : \ldots : 0)$ and the equations of $t_{Z,x}$ (resp. L) are $T_{q+1} = \cdots = T_n = 0$ (resp. $T_0 = \cdots = T_m = 0$). Then $\mathfrak{m}_B = (t_1, \ldots, t_q)B$ and $\mathfrak{m}_A B = (t_1, \ldots, t_m)B$, where the t_i are the images of the affine coordinate functions T_i/T_0. If $L \cap t_{Z,x} = \emptyset$, then $m \geq q$, and, hence, $\mathfrak{m}_B = \mathfrak{m}_A B$. Since $A/\mathfrak{m}_A = B/\mathfrak{m}_B = C$, $\Pi^* : A \to B$ is an isomorphism by Nakayama's lemma. □

Note that, in the proof of the lemma, we needed to work with the algebraically closed field C: it is necessary that $\Pi^{-1}(x')$ contain no other point (real or complex) besides x.

Proposition 8.6.14. *Let $Z \subset \mathbb{P}_n(C)$ be a projective algebraic set, x a point of Z, and d the dimension of the Zariski tangent space to Z at x. Assume that $\max(d, \dim(Z)+1) \leq m < n$. Then there exists a nonempty Zariski open subset U of the grassmannian of projective subspaces of dimension $n - m - 1$ of $\mathbb{P}_n(C)$, such that, for every L in U,*

(i) $L \cap Z = \emptyset$ (hence $Z' = \Pi(Z)$ is a projective algebraic subset of $\mathbb{P}_m(C)$, with the same notations as above),

(ii) $\Pi^* : \mathcal{R}_{Z',x'} \to \mathcal{R}_{Z,x}$ (where $x' = \Pi(x)$) is an isomorphism of local rings.

Proof. Set $D_x = \{(z, y) \in Z \times \mathbb{P}_n(C) \mid y \in \overline{xz}, z \neq x\}$. Let $p_1 : Z \times \mathbb{P}_n(C) \to Z$ and $p_2 : Z \times \mathbb{P}_n(C) \to \mathbb{P}_n(C)$ be the two projection mappings. For every $z \in Z \setminus \{x\}$, $p_1^{-1}(z) \cap D_x$ is a projective line. Hence, the dimension of the Zariski closure of D_x is $1 + \dim(Z) \leq m$. Let A be the Zariski closure of $Z \cup t_{Z,x} \cup p_2(D_x)$, where $t_{Z,x}$ is as in Lemma 8.6.13. Then $\dim(A) \leq m$, and, hence, there exists a Zariski open subset U of the grassmannian of subspaces of dimension $n - m - 1$ of $\mathbb{P}_n(C)$, such that, for every $L \in U$, $L \cap A = \emptyset$. By the definition of A and Lemma 8.6.13, the properties (i) and (ii) are satisfied for such an L. □

The following corollary gives a minimal embedding of an algebraic singularity.

Corollary 8.6.15. *Let $Z \subsetneq R^n$ (resp. C^n) be an algebraic set, x a point of Z, d the dimension of the Zariski tangent space to Z at x, and $m = \max(d, \dim(Z) + 1)$ (if x is a singular point of Z, $m = d$). There exist an algebraic set $Z' \subset R^m$ (resp. C^m) and a biregular isomorphism f from a Zariski open neighbourhood of x in Z onto a Zariski open subset of Z'. In particular, the homomorphism of local rings $f^* : \mathcal{R}_{Z',f(x)} \to \mathcal{R}_{Z,x}$ induced by f is an isomorphism.*

Proof. Let $\overline{Z}$ be the Zariski closure of Z in $\mathbb{P}_n(C)$. We apply Proposition 8.6.14 to $\overline{Z}$. If Z is a real algebraic set, we may choose the centre of the projection L to be "real", i.e. the complexification of a subspace of $\mathbb{P}_n(R)$, since every nonempty Zariski open subset of the grassmannian contains such an L. □

Remark 8.6.16. If the point x is nonsingular, then $d = \dim(Z)$, and, in general, there is no biregular isomorphism from a Zariski open neighbourhood of x in Z onto a Zariski open subset of R^d.

Proof of Theorem 8.6.12. Let V_0 be the germ of a Nash set at the origin of R^m, $I \subset \mathcal{N}_{R^m,0}$ the ideal of germs of Nash functions vanishing on V_0. Let $f_1, \ldots, f_k$ be generators of I. By the Artin-Mazur theorem 8.4.4, there exist a nonsingular algebraic set $W \subset R^n$, a Nash diffeomorphism $\sigma : U \to W$ from an open semi-algebraic neighbourhood U of the origin of R^m onto an open semi-algebraic subset of W and a polynomial mapping $p : W \to R^k$, such that $p \circ \sigma = (f_1, \ldots, f_k)$. Set $x = \sigma(0)$. Then $\sigma^* : \mathcal{N}_{W,x} \to \mathcal{N}_{R^m,0}$ is an isomorphism, and $I = \sigma^*((p_1, \ldots, p_k)\mathcal{N}_{W,x})$. Let Z be the Zariski closure of $\mathcal{Z}_W(p_1, \ldots, p_k) \cap \sigma(U)$ in $\mathbb{P}_n(C)$. We can choose a projective subspace $L \subset \mathbb{P}_n(R)$ of dimension $n - m - 1$, disjoint from the projective subspace $t_{W,x}$ tangent to W at x and such that the complexification L_C satisfies conditions (i) and (ii) of Proposition 8.6.14. Set $Z' = \Pi_{L_C}(Z)$, $x' = \Pi_{L_C}(x)$ and $V' = Z' \cap \mathbb{P}_m(R)$. Then $\Pi = \Pi_{L_C}|_Z : Z \to Z'$ induces an isomorphism of local rings $\Pi^* : \mathcal{R}_{Z',x'} \to \mathcal{R}_{Z,x}$. There exists an open semi-algebraic neighbourhood $\Omega \subset \sigma(U)$ of x in W such that $\Pi_L|_\Omega$ is a Nash diffeomorphism onto an open semi-algebraic neighbourhood of x' in $\mathbb{P}_m(R)$. We may assume that $x' = 0 \in \Pi_L(\Omega) \subset R^m \subset \mathbb{P}_m(R)$. Then the Nash diffeomorphism $\Pi_L \circ \sigma|_{\sigma^{-1}(\Omega)}$ between neighbourhoods of the origin of R^m maps the germ of V onto the germ of V'. □

Remark 8.6.17. Artin's approximation theorem 8.3.1 and Theorem 8.6.12 are the main tools used to obtain results concerning algebraicity of germs of analytic sets (over $\mathbb{R}$ or $\mathbb{C}$). More precisely, Theorem 8.3.1 is used to show that, under some conditions, the germ of an analytic set in $\mathbb{R}^n$ (or $\mathbb{C}^n$) is analytically or C^ν-equivalent to the germ of a Nash set (for the equivalence defined as in 8.6.11). The algebraicity then follows by applying Theorem 8.6.12 (or the corresponding result over $\mathbb{C}$). The following results are obtained in this way.

a) Every germ of a complex analytic set with an isolated singularity is holomorphically equivalent to the germ of a complex algebraic set [17, 328].

b) For every $\nu \in \mathbb{N}$, every germ of a coherent real analytic set with an isolated singularity is C^ν-equivalent to the germ of a real algebraic set [328].

c) Every germ of a real or complex analytic set is homeomorphically equivalent to the germ of a real or complex algebraic set [239].

Remark 8.6.18. The proof of Theorem 8.6.12 gives a more precise result. If V_0 is the germ of a Nash set and $V_{1,0}, \ldots, V_{s,0}$ are its Nash irreducible components, then there exists a Nash diffeomorphism $\sigma : (R^m, 0) \to (R^m, 0)$ such that, for every $i = 1, \ldots, s$, the ideal of $\mathcal{N}_{R^m,0}$ of germs of Nash functions vanishing on $\sigma(V_{i,0})$ is generated by polynomials (and, in particular, each $\sigma(V_{i,0})$ is algebraic). This precision is useful for the study of algebraicity of analytic functions. It is used in [44], to prove a result (Theorem 4) implying that every germ of a Nash function $(\mathbb{R}^m, 0) \to \mathbb{R}$ can be transformed by a homeomorphism $(\mathbb{R}^m, 0) \to (\mathbb{R}^m, 0)$ into the germ of a polynomial function. The question, related to Theorem 8.6.12, whether this transformation can be performed by a Nash diffeomorphism remains open. This improvement of Theorem 8.6.12 is also necessary for the study of algebraicity of global analytic sets. In particular, it allows one to prove the fact that every semi-algebraic coherent compact analytic hypersurface of $\mathbb{R}^m$ with isolated singularities can be transformed, by an analytic diffeomorphism of $\mathbb{R}^m$, into an algebraic set [61]. We refer the reader to [41] for more information concerning the algebraicity of analytic sets or functions.

8.7 Henselian Properties. Noetherian Property of the Ring of Global Nash Functions

We begin by reviewing results concerning henselian local rings.

Proposition 8.7.1. *Let $\varphi : A \to B$ be a local homomorphism of local rings. The following properties are equivalent:*

(i) There is an A-algebra isomorphism from B to some algebra of the form $(A[X_1, \ldots, X_m]/(f_1, \ldots, f_m))_{\mathfrak{p}}$, where the prime ideal $\mathfrak{p}$ does not contain the class of the jacobian determinant $\det\left(\frac{\partial f_i}{\partial X_j}\right)$.

(ii) There is an A-algebra isomorphism from B to some algebra of the form $(A[X]/(f))_{\mathfrak{q}}$, where f is a monic polynomial and the prime ideal $\mathfrak{q}$ does not contain the class of the derivative f'.

Proof. (ii) ⇒ (i) is obvious. (i) ⇒ (ii) is an immediate consequence of the theorem of local structure of étale algebras and the jacobian criterion ([264], Chap. 5, Theorems 1 and 5). □

Definition 8.7.2. *If property (i) (or (ii)) of Proposition 8.7.1 is satisfied, B is said to be a* local-étale *A-algebra. If, moreover, the residue field of B is equal to the residue field of A, then B is said to be an* equiresidual local-étale *A-algebra.*

Note that if $\varphi : A \to B$ is such that B is a local-étale A-algebra, then φ is flat (this is an easy consequence of property (ii) of Proposition 8.7.1), and thus faithfully flat and injective, since it is local ([66], Chap. 1, §3, Proposition 9).

Proposition 8.7.3. *Let $A = \mathcal{R}_{R^n,0} = R[X_1, \ldots, X_n]_{(X_1, \ldots, X_n)}$. Then $\varphi : A \to B$ is an equiresidual local-étale A-algebra if and only if there exists an algebraic set $V \subset R^{n+m}$, a nonsingular point t in V and an isomorphism $\iota : B \to \mathcal{R}_{V,t}$ such that, if $\Pi : V \subset R^n \times R^m \to R^n$ is the projection mapping defined by $\Pi(x, y) = x$, then $\Pi(t) = 0$ (the origin of R^n), Π is étale at t (Definition 8.1.1), and $\iota \circ \varphi$ is the homomorphism induced by Π.*

8.7 Henselian Properties. Noetherian Property

Proof. If B is an equiresidual local-étale A-algebra, we may assume that $B = (A[Y_1, \ldots, Y_m]/(f_1, \ldots, f_m))_\mathfrak{p}$, where $f_i \in R[X_1, \ldots, X_n, Y_1, \ldots, Y_m]$. Let $V = \mathcal{Z}(f_1, \ldots, f_m) \subset R^{n+m}$. Since the residue field of $\mathfrak{p}$ is R, $\mathfrak{p}$ is the maximal ideal of a point $t \in V$. By Proposition 3.3.10 (more precisely, the proof of (iii)$\Rightarrow$(i)), t is a nonsingular point of V and $B = \mathcal{R}_{V,t}$. Since the canonical homomorphism $A \to B$ is local, $\Pi(t) = 0$. Since the jacobian determinant $\det\left(\frac{\partial f_i}{\partial Y_j}\right)$ does not vanish at t, Π is étale at t.

Now we prove the converse. By assumption, $\mathcal{R}_{V,t}$ is a regular local ring, and its dimension is n because Π is étale at t. By Proposition 3.3.7 (iii), there are $f_1, \ldots, f_m \in R[X_1, \ldots, X_n, Y_1, \ldots, Y_m]$ such that $\mathcal{R}_{V,t} = \mathcal{R}_{R^{n+m},t}/(f_1, \ldots, f_m)$ and the rank of the matrix $\left[\frac{\partial f_i}{\partial X_j}(t), \frac{\partial f_i}{\partial Y_l}(t)\right]$ is equal to m. Since Π is étale at t, the latter condition implies that the rank of the matrix $\left[\frac{\partial f_i}{\partial Y_l}(t)\right]$ is equal to m. Hence $\mathcal{R}_{V,t}$ is a local-étale A-algebra.
$\square$

Now we shall define a henselian local ring. If A is a local ring with residue field k_A, we say that a polynomial $f \in A[X]$ has *a simple root b in k_A* if $\overline{f}(b) = 0$ and $\overline{f}'(b) \neq 0$, where $\overline{f}$ is the image of f in $k_A[X]$. We denote by $\overline{a}$ the image of an element $a \in A$ in k_A.

Definition 8.7.4. *A local ring A with residue field k_A is said to be* henselian *if, for every monic polynomial $f \in A[X]$ having a simple root b in k_A, there exists a root c of f in A such that $\overline{c} = b$.*

Note that c is the unique root of f in A whose image in k_A is b.

Lemma 8.7.5. *Let A be a local ring, $f \in A[X]$, $b \in k_A$ a simple root of f. There is at most one root c of f in A such that $\overline{c} = b$.*

Proof. Assume that $c \in A$ is a root of f such that $\overline{c} = b$. There is $g \in A[X]$ such that $f = (X - c)g$. Let d be an element of A such that $\overline{d} = b$. Then $\overline{g(d)} = \overline{g}(b) = \overline{f}'(b) \neq 0$ and $g(d)$ is invertible in A. Since $f(d) = (d - c)g(d)$, d is a root of f if and only if $d = c$.
$\square$

Proposition 8.7.6. *Let $M \subset R^p$ be a Nash submanifold and $x \in M$. The local ring $\mathcal{N}_{M,x}$ of germs of Nash functions at x is henselian.*

Proof. The local ring $\mathcal{N}_{M,x}$ is isomorphic to $\mathcal{N}_{R^n,0}$ for some $n \in \mathbb{N}$, and the implicit function theorem 2.9.8 implies that $\mathcal{N}_{R^n,0}$ is henselian.
$\square$

Proposition 8.7.7. *A local ring A is henselian if and only if for every equiresidual local-étale A-algebra $\varphi : A \to B$, φ is an isomorphism.*

Proof. The fact that the condition is sufficient is obvious if one uses property (ii) of Proposition 8.7.1. Now we show that this condition is necessary. Let $(A[X]/(f))_\mathfrak{q} = B$ be an equiresidual local-étale A-algebra. We consider the following commutative diagram:

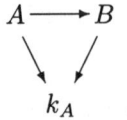

The image of X in k_A is a simple root of f in k_A. Since A is henselian, this simple root can be lifted to a root a of f in A. There is $g \in A[X]$ such that $f = (X-a)g$, and $g(a)$ is invertible in A, since $\overline{g(a)} = \overline{f}'(\overline{a}) \neq 0$. There is $h \in A[X]$ such that $g = (X-a)h + g(a)$. We have

$$1 = \frac{1}{g(a)} g - \frac{h}{g(a)} (X-a).$$

Hence, by the chinese remainder theorem, the canonical homomorphism $A[X]/(f) \to (A[X]/(X-a)) \times (A[X]/(g))$ is an isomorphism. The image of $\mathfrak{q}$ by this isomorphism is $\mathfrak{m} \times (A[X]/(g))$, where $\mathfrak{m}$ is the maximal ideal of $A[X]/(X-a) \simeq A$. Hence B is isomorphic to A. □

Definition 8.7.8. *Let A be a local ring. A henselization of A is a henselian local ring B with a local homomorphism $\varphi : A \to B$, such that, for every local homomorphism $\theta : A \to C$ to a henselian local ring C, there exists a unique local homomorphism $\psi : B \to C$ such that $\psi \circ \varphi = \theta$.*

We proceed to construct the henselization of a local ring A as the inductive limit of a directed system of equiresidual local-étale A-algebras. We need the following lemma.

Lemma 8.7.9. *Let A be a local ring, $\varphi : A \to B$ an equiresidual local-étale A-algebra and $\theta : A \to C$ a local homomorphism of local rings. There is at most one local homomorphism $\psi : B \to C$ such that $\psi \circ \varphi = \theta$.*

Proof. We may assume that $B = \big(A[X]/(f)\big)_{\mathfrak{q}}$, where f is a monic polynomial and the prime ideal $\mathfrak{q}$ does not contain the class of the derivative f'. Let $x \in B$ be the class of X. We have $\overline{x} \in k_B = k_A$. We still denote by f the image of f by the homomorphism $A[X] \to C[X]$ induced by θ. Note that if $\psi : B \to C$ is a local homomorphism such that $\psi \circ \varphi = \theta$, then $\psi(x)$ is a root of f in C and $\overline{\psi(x)} = \overline{\theta}(\overline{x})$ is a simple root of f in the residue field k_C, where $\overline{\theta} : k_A \to k_C$ denotes the homomorphism induced by θ. By Lemma 8.7.5, $\psi(x)$ is uniquely determined by $\overline{\theta}(\overline{x})$. Since $\psi(x)$ determines ψ, we deduce that there is at most one local homomorphism ψ such that $\psi \circ \varphi = \theta$. □

Let $\mathcal{D}$ be the set of isomorphism classes of equiresidual local-étale A-algebras. For each $\delta \in \mathcal{D}$, choose a representative A_δ. Set $\delta \leq \delta'$ if there is a local A-algebra homomorphism $\psi_{\delta,\delta'} : A_\delta \to A_{\delta'}$. By Lemma 8.7.9, the morphism $\psi_{\delta,\delta'}$ is unique.

Proposition 8.7.10.

(i) *The relation $\leq$ is an order relation on $\mathcal{D}$, and $(\mathcal{D}, \leq)$ is a directed set.*

(ii) *The inductive limit of the directed system $(A_\delta)_{\delta \in \mathcal{D}}$ with transition morphisms $\psi_{\delta, \delta'}$ is a henselization of A.*

Proof. (i) The reflexivity and transitivity of $\leq$ are obvious. If $\delta \leq \delta'$ and $\delta' \leq \delta$, then, by Lemma 8.7.9, $\psi_{\delta, \delta'}$ nd $\psi_{\delta', \delta}$ are inverse isomorphisms. Therefore $\delta = \delta'$.

Now we show that $(\mathcal{D}, \leq)$ is directed. Let $\delta, \delta' \in \mathcal{D}$. Consider the following commutative diagram:

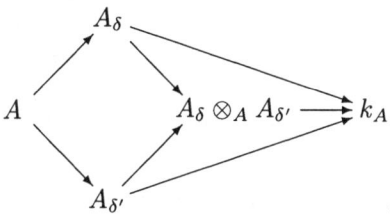

Let $\mathfrak{p}$ be the kernel of $A_\delta \otimes_A A_{\delta'} \to k_A$ and $B = (A_\delta \otimes_A A_{\delta'})_\mathfrak{p}$. We can verify, using, for example, property (i) of Proposition 8.7.1, that B is an equiresidual local-étale A-algebra. Let $\epsilon \in \mathcal{D}$ be the isomorphism class of B. We have $\delta \leq \epsilon$ and $\delta' \leq \epsilon$.

(ii) Let A' be the inductive limit of the directed system $(A_\delta)_{\delta \in \mathcal{D}}$. The A-algebra A' is local with residue field k_A, and the homomorphism $A \to A'$ is local.

Let $f \in A'[X]$ be a monic polynomial with a simple root b in k_A. Since $\mathcal{D}$ is directed, we can find $\delta \in \mathcal{D}$ and a monic polynomial $\varphi \in A_\delta[X]$ whose image in $A'[X]$ is f. Let $\mathfrak{p}$ be the kernel of the A_δ-algebra homomorphism $A_\delta[X]/(\varphi) \to k_A$ sending X to b, and let $B = (A_\delta[X]/(f))_\mathfrak{p}$. The A-algebra B is equiresidual local-étale. If ϵ is its isomorphism class, we may assume $B = A_\epsilon$. The image of the class of X in A_ϵ by the canonical homomorphism $A_\epsilon \to A'$ is a root of f in A', whose class in k_A is b. This shows that A' is henselian.

Let $\theta : A \to H$ be a local homomorphism to a henselian local ring H. Take $\delta \in \mathcal{D}$. We may assume that $A_\delta = (A[X]/(f))_\mathfrak{p}$, where f is monic and the prime ideal $\mathfrak{p}$ does not contain the class of the derivative f'. Let a be the image of X by the homomorphism $A_\delta \to k_A$. The image of f in $H[X]$ has a unique root in H whose class in the residue field k_H of H is the image of a. Hence, there is a unique local homomorphism $A_\delta \to H$ such that the diagram

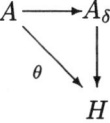

commutes. By the universal property of inductive limit, there is a unique local homomorphism $\psi : A' \to H$ such that the diagram

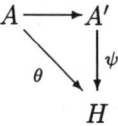

commutes. This proves that A' is a henselization of A. □

By Proposition 8.7.10, every local ring A has a henselization. By definition, this henselization is unique up to a unique A-algebra isomorphism. We denote by ^hA the henselization of A. We have already noted that, if $\varphi : A \to B$ is an equiresidual local-étale A-algebra, then φ is injective. Hence, the canonical homomorphism from a local ring A to its henselization ^hA is injective. This allows one to identify A with a subring of ^hA.

Corollary 8.7.11. *The ring $\mathcal{N}_{R^n,0} = R[[X_1, \ldots, X_n]]_{\mathrm{alg}}$ is the henselization of $\mathcal{R}_{R^n,0} = R[X_1, \ldots, X_n]_{(X_1,\ldots,X_n)}$.*

Proof. Theorem 8.1.4 and Proposition 8.7.3 imply (with the notations of this proposition) that $R[[X_1, \ldots, X_n]]_{\mathrm{alg}}$ is the inductive limit of those $\mathcal{R}_{V,t}$ which have the property that $\Pi^* : \mathcal{R}_{R^n,0} \to \mathcal{R}_{V,t}$ is an equiresidual local-étale $\mathcal{R}_{R^n,0}$-algebra. The corollary follows by applying Proposition 8.7.10. □

We shall use the following result.

Theorem 8.7.12. *Let A be a local ring with maximal ideal $\mathfrak{m}$, ^hA its henselization with maximal ideal $\mathfrak{n}$. Then:*

(i) *The henselization ^hA is faithfully flat over A and $\mathfrak{m}^q(^hA) = \mathfrak{n}^q$.*

(ii) *The henselization ^hA is reduced (resp. integrally closed) if and only if A is reduced (resp. integrally closed).*

(iii) *The henselization ^hA is noetherian if and only if A is noetherian and, in this case, for every prime ideal $\mathfrak{p}$ of A, $^hA \otimes_A k(\mathfrak{p})$ is a finite product of fields that are separable algebraic extensions of $k(\mathfrak{p})$.*

References for the proof: [264] Chap. 8, Theorem 3. □

Corollary 8.7.13. *Let A be a noetherian local ring and $\mathfrak{p}$ a prime ideal of A. There are finitely many prime ideals $\mathfrak{q}_1, \ldots, \mathfrak{q}_m$ of ^hA such that $\mathfrak{q}_i \cap A = \mathfrak{p}$. We have $\mathfrak{p}(^hA) = \mathfrak{q}_1 \cap \ldots \cap \mathfrak{q}_m$, and the $\mathfrak{q}_i$ are the minimal prime ideals belonging to $\mathfrak{p}(^hA)$. Moreover, the height of each $\mathfrak{q}_i$ is equal to the height of $\mathfrak{p}$.*

Proof. The first two statements are consequences of Theorem 8.7.12 (iii). Indeed, the canonical homomorphism from ^hA into $^hA \otimes_A k(\mathfrak{p})$ induces a bijection between the prime ideals $\mathfrak{q}$ of ^hA such that $\mathfrak{q} \cap A = \mathfrak{p}$ and the prime ideals of $^hA \otimes_A k(\mathfrak{p})$. The $\mathfrak{q}_1, \ldots, \mathfrak{q}_m$ are the inverse images in ^hA of the prime ideals of $^hA \otimes_A k(\mathfrak{p})$. This bijection also implies that if $\mathfrak{q}' \subset \mathfrak{q}$ are prime ideals of ^hA such that $\mathfrak{q} \cap A = \mathfrak{q}' \cap A$, then $\mathfrak{q} = \mathfrak{q}'$. Hence $\mathrm{ht}(\mathfrak{q}) \leq \mathrm{ht}(\mathfrak{q} \cap A)$. We prove the inverse inequality by induction on the height of $\mathfrak{q} \cap A$. Assume $\mathrm{ht}(\mathfrak{q} \cap A) > 0$, and let $\mathfrak{p}' \subset \mathfrak{q} \cap A$ be a prime ideal of A such that $\mathrm{ht}(\mathfrak{p}') = \mathrm{ht}(\mathfrak{q} \cap A) - 1$. Let $\mathfrak{q}'_1, \ldots, \mathfrak{q}'_n$ be the minimal prime ideals belonging to $\mathfrak{p}'(^hA)$. There exists some j such that $\mathfrak{q}'_j \subsetneq \mathfrak{q}$. By the inductive assumption, $\mathrm{ht}(\mathfrak{q}'_j) \geq \mathrm{ht}(\mathfrak{q} \cap A) - 1$, and, hence, $\mathrm{ht}(\mathfrak{q}) \geq \mathrm{ht}(\mathfrak{q} \cap A)$. $\square$

Corollary 8.7.14. *Let A be an integrally closed noetherian local ring and ^hA its henselization with maximal ideal $\mathfrak{n}$. Let B be an integrally closed local ring with maximal ideal $\mathfrak{m}$, such that $A \subset B \subset {}^hA$ and $\mathfrak{m} = \mathfrak{n} \cap B$. Then the homomorphism $^hB \to {}^hA$ is an isomorphism, and B is noetherian and faithfully flat over A.*

Proof. By the universal property of the henselization, we have the following commutative diagram

$$\begin{array}{ccc} A & \hookrightarrow B \hookrightarrow & {}^hA \\ \cap & \cap & {}^{\beta} \\ {}^hA & \xrightarrow{\alpha} {}^hB & \end{array}$$

where $\beta \circ \alpha = \mathrm{Id}_{{}^hA}$. Since A (resp. B) is integrally closed, ^hA (resp. ^hB) is an integral domain by Theorem 8.7.12 (ii), and, hence, the kernel of β is a prime ideal of ^hB whose intersection with B is the ideal (0). By Corollary 8.7.13, this kernel is the ideal (0) of ^hB. Hence, β is an isomorphism. By Theorem 8.7.12 (iii), ^hA is noetherian, and, hence, ^hB and B are noetherian. Since ^hA is faithfully flat over A and over B by Theorem 8.7.12 (i), B is faithfully flat over A. $\square$

Proposition 8.7.15. *Let $M \subset R^p$ be a Nash submanifold, V its Zariski closure in R^p and x a point of M. Let $\mathfrak{m}_x$ be the maximal ideal of $\mathcal{N}(M)$ of Nash functions vanishing at x. Then $^h(\mathcal{N}(M)_{\mathfrak{m}_x}) = \mathcal{N}_{M,x}$ and $\mathcal{N}(M)_{\mathfrak{m}_x}$ is a regular noetherian local ring of dimension $\dim(M)$. If $x \in \mathrm{Reg}(V)$ and $\mathfrak{p}$ is a prime ideal of $\mathcal{R}_{V,x}$, then $^h\mathcal{R}_{V,x} = \mathcal{N}_{M,x}$, and there are finitely many prime ideals $\mathfrak{q}_1, \ldots, \mathfrak{q}_q$ of $\mathcal{N}(M)_{\mathfrak{m}_x}$ such that $\mathfrak{q}_j \cap \mathcal{R}_{V,x} = \mathfrak{p}$. Each $\mathfrak{q}_j$ has the same height as $\mathfrak{p}$, and $\mathfrak{p}\mathcal{N}(M)_{\mathfrak{m}_x} = \mathfrak{q}_1 \cap \ldots \cap \mathfrak{q}_q$.*

Proof. We may assume that M is semi-algebraically connected. By Proposition 8.4.6, we may also assume that $x \in \mathrm{Reg}(V)$. Let $f_1, \ldots, f_n$ be a regular system of parameters of $\mathcal{R}_{V,x}$, where $n = \dim(M)$. By Corollary 8.7.11, $\mathcal{N}_{M,x}$ is the henselization of $R[f_1, \ldots, f_n]_{(f_1, \ldots, f_n)}$. Since

$$R[f_1, \ldots, f_n]_{(f_1, \ldots, f_n)} \subset \mathcal{R}_{V,x} \subset \mathcal{N}_{M,x}$$

and the assumptions of Corollary 8.7.14 are satisfied, $\mathcal{N}_{M,x} = {}^h\mathcal{R}_{V,x}$. Hence,

190   8. Nash Functions

$$\mathcal{R}_{V,x} \subset \mathcal{N}(M)_{\mathfrak{m}_x} \subset \mathcal{N}_{M,x} = {}^h\mathcal{R}_{V,x} \ .$$

Since $\mathcal{N}(M)$ is integrally closed (Proposition 8.4.3), its localization $\mathcal{N}(M)_{\mathfrak{m}_x}$ is integrally closed as well, and, hence, by Corollary 8.7.14,

$$\mathcal{N}_{M,x} = {}^h(\mathcal{N}(M)_{\mathfrak{m}_x}) = {}^h\mathcal{R}_{V,x} \ ,$$

and $\mathcal{N}(M)_{\mathfrak{m}_x}$ is noetherian and faithfully flat over $\mathcal{R}_{V,x}$. By Corollary 8.7.13,

$$\dim(\mathcal{N}(M)_{\mathfrak{m}_x}) = \dim(\mathcal{N}_{M,x}) = \dim(\mathcal{R}_{V,x}) = n \ .$$

Moreover, $f_1, \ldots, f_n$ generate the maximal ideal of $\mathcal{N}_{M,x}$, and, hence, by flatness, they also generate $\mathfrak{m}_x \mathcal{N}(M)_{\mathfrak{m}_x}$ (cf. Theorem 8.7.12 (i)). This implies that $\mathcal{N}(M)_{\mathfrak{m}_x}$ is regular. Corollary 8.7.13 and the faithful flatness of $\mathcal{N}_{M,x}$ over $\mathcal{N}(M)_{\mathfrak{m}_x}$ imply the last part of the proposition. □

Corollary 8.7.16. *Let $M \subset R^p$ be a Nash submanifold, V its Zariski closure in R^p, x a point in M such that $x \in \mathrm{Reg}(V)$. Then the local rings*

$$\mathcal{R}_{V,x} \hookrightarrow \mathcal{N}(M)_{\mathfrak{m}_x} \hookrightarrow \mathcal{N}_{M,x} \hookrightarrow \widehat{\mathcal{N}}_{M,x} = \widehat{\mathcal{R}}_{V,x}$$

are all regular of dimension $\dim(M)$, *and all inclusions are faithfully flat.*

We assume that $R = \mathbb{R}$ in the rest of this section. We shall prove the noetherian property of the ring of Nash functions (it is possible to prove the corresponding result for the ring of Nash functions over an arbitrary real closed field [104]).

Proposition 8.7.17. *Let $V \subset \mathbb{R}^p$ be a nonsingular irreducible algebraic set of dimension n and M a connected open semi-algebraic subset of V. Let $\mathfrak{p}$ be a prime ideal of $\mathcal{P}(V)$. There exist finitely many prime ideals $\mathfrak{q}_1, \ldots, \mathfrak{q}_k$ of $\mathcal{N}(M)$ such that $\mathfrak{q}_j \cap \mathcal{P}(V) = \mathfrak{p}$. Each $\mathfrak{q}_j$ has the same height as $\mathfrak{p}$ and $\mathfrak{p}\mathcal{N}(M) = \mathfrak{q}_1 \cap \ldots \cap \mathfrak{q}_k$.*

Proof. By Proposition 8.7.15, the conclusion of the proposition holds if we replace $\mathcal{N}(M)$ with any localization $\mathcal{N}(M)_{\mathfrak{m}_x}$, $x \in M$. On the other hand, by Corollary 8.6.3, every maximal ideal of $\mathcal{N}(M)$ is of the form $\mathfrak{m}_x$ for some $x \in M$. Hence, in order to prove Proposition 8.7.17, it suffices to prove that $\mathfrak{p}\mathcal{N}(M)$ is the intersection of a finite number of prime ideals of $\mathcal{N}(M)$.

Let $V_{\mathbb{C}}$ be the Zariski closure of V in $\mathbb{C}^p$ and $\mathcal{P}(V_{\mathbb{C}}) = \mathbb{C} \otimes_{\mathbb{R}} \mathcal{P}(V)$. Since V is nonsingular, there is a neighbourhood of V in $V_{\mathbb{C}}$ which is a complex analytic manifold. Hence, every $f \in \mathcal{N}(M)$ has a canonical holomorphic extension $f_{\mathbb{C}}$ to a neighbourhood U_f of M in $V_{\mathbb{C}}$.

Let $k(\mathfrak{p})$ be the residue field of $\mathcal{P}(V)$ at $\mathfrak{p}$. Then $\mathbb{C} \otimes_{\mathbb{R}} k(\mathfrak{p})$ is a field if -1 is not a square in $k(\mathfrak{p})$, and the product of two copies of $k(\mathfrak{p})$ if -1 is a square in $k(\mathfrak{p})$. Hence, there are two (not necessarily distinct) prime ideals $\mathfrak{p}_1$ and $\mathfrak{p}_2$ of $\mathcal{P}(V_{\mathbb{C}})$ above $\mathfrak{p}$, such that $\mathfrak{p}\mathcal{P}(V_{\mathbb{C}}) = \mathfrak{p}_1 \cap \mathfrak{p}_2$ and $\overline{\mathfrak{p}_1} = \mathfrak{p}_2$, where the bar denotes complex conjugation. Let $Z_i \subset V_{\mathbb{C}}$ be the set of zeros of $\mathfrak{p}_i$, $\Pi_i : W_i \to Z_i$ a

8.7 Henselian Properties. Noetherian Property   191

normalization of Z_i ([295], Chap. 2, §5, Theorem 4). Recall that $\mathcal{P}(W_i)$ is the integral closure of $\mathcal{P}(Z_i) = \mathcal{P}(V_\mathbb{C})/\mathfrak{p}_i$ in its field of fractions. We may assume that $W_i \subset \mathbb{C}^{p+r}$, $\overline{W_1} = W_2$ and Π_i is induced by the projection mapping $\Pi : \mathbb{C}^{p+r} \to \mathbb{C}^p$. The set $W_i \cap \Pi^{-1}(M)$ is semi-algebraic in $\mathbb{C}^{p+r} \simeq \mathbb{R}^{2(p+r)}$ and has, therefore, a finite number of connected components $W_{i,1}, \ldots, W_{i,s}$, where $\overline{W_{1,j}} = W_{2,j}$. For $i = 1, 2$ and $j = 1, \ldots, s$, define $\mathfrak{q}_{i,j}$ as the ideal consisting of those $f \in \mathcal{N}(M)$ which have the property that $f_\mathbb{C} \circ \Pi|_{\Pi^{-1}(U_f)}$ vanishes on a neighbourhood of $W_{i,j}$ in W_i. Note that $\mathfrak{q}_{1,j} = \mathfrak{q}_{2,j}$. Since W_i is normal, it is analytically normal, and every point of W_i has a fundamental system of neighbourhoods U_α in W_i such that the open subset of nonsingular points of U_α is connected ([242], Chap. 3, §9, p. 413). Hence, each $W_{i,j}$ has a fundamental system of neighbourhoods in W_i such that the open subset of nonsingular points of each of these neighbourhoods is connected. It follows that each $\mathfrak{q}_{i,j}$ is prime.

If $f \in \mathfrak{p}\mathcal{N}(M)$, then $f_\mathbb{C}$ vanishes on the intersection of Z_i with a neighbourhood of M in $V_\mathbb{C}$, and, hence, $f_\mathbb{C} \circ \Pi|_{\Pi^{-1}(U_f)}$ vanishes on a neighbourhood of each $W_{i,j}$ in W_i. Hence, $f \in \bigcap_{i,j} \mathfrak{q}_{i,j}$.

Conversely, let $f \in \bigcap_{i,j} \mathfrak{q}_{i,j}$. For every x in $\mathcal{Z}_M(\mathfrak{p})$, $f_\mathbb{C} \circ \Pi|_{\Pi^{-1}(U_f)}$ vanishes on a neighbourhood of $\Pi_i^{-1}(x)$ in W_i. Hence, $f_\mathbb{C}$ vanishes on a neighbourhood of x in Z_i. By the local Nullstellensatz ([142], Chap. 3, Section A, Theorem 7), $f \in \mathfrak{p}\widehat{\mathcal{N}}_{M,x}$. Using the property of faithful flatness (Corollary 8.7.16), one sees that $f \in \mathfrak{p}\mathcal{N}(M)_{\mathfrak{m}_x}$, for every x in $\mathcal{Z}_M(\mathfrak{p})$. Since the maximal ideals of $\mathcal{N}(M)$ containing $\mathfrak{p}$ are precisely the $\mathfrak{m}_x$ for x in $\mathcal{Z}_M(\mathfrak{p})$ (Corollary 8.6.3), we deduce that $f \in \mathfrak{p}\mathcal{N}(M)$. □

Theorem 8.7.18. *Let $M \subset \mathbb{R}^p$ be a Nash submanifold. Then the ring $\mathcal{N}(M)$ is noetherian.*

Proof. We use the following lemma.

Lemma 8.7.19. *Let A be a commutative ring. If every prime ideal of A of height $\geq h$ is finitely generated, then every ideal of A of height $\geq h$ is finitely generated.*

Proof. We suppose that the conclusion does not hold, and we obtain a contradiction by using Zorn's lemma for the family of ideals of A of height $\geq h$ that are not finitely generated (see [200], Chap. 2, Theorem II.2). □

We may assume, without loss of generality, that M is connected and that its Zariski closure V in $\mathbb{R}^p$ is nonsingular. We proceed by descending induction on the height of the prime ideals of $\mathcal{N}(M)$. Let m be the dimension of M.

The prime ideals of $\mathcal{N}(M)$ of height $\geq m$ are the maximal ideals of $\mathcal{N}(M)$, that is, the ideals of the form $\mathfrak{m}_x$ for $x \in M$ (Corollary 8.6.3). By Proposition 8.7.15, the local ring $\mathcal{N}(M)_{\mathfrak{m}_x}$ is noetherian, and, hence, $\mathfrak{m}_x$ is finitely generated.

For $h < m$, assume that every prime ideal of height $> h$ is finitely generated. Let $\mathfrak{q}$ be a prime ideal of $\mathcal{N}(M)$ of height h, $\mathfrak{p} = \mathfrak{q} \cap \mathcal{P}(V)$. Let

$\mathfrak{q}_1 = \mathfrak{q}, \mathfrak{q}_2, \ldots, \mathfrak{q}_k$ be the prime ideals of $\mathcal{N}(M)$ above $\mathfrak{p}$. They are all of height h, and $\mathfrak{p}\mathcal{N}(M) = \mathfrak{q} \cap \mathfrak{q}_2 \cap \cdots \cap \mathfrak{q}_k$ (Proposition 8.7.17). Set $\widetilde{\mathfrak{q}} = \mathfrak{q}_2 \cap \ldots \cap \mathfrak{q}_k$. There is an exact sequence

$$0 \to \mathfrak{p}\mathcal{N}(M) \to \mathfrak{q} \oplus \widetilde{\mathfrak{q}} \to \mathfrak{q} + \widetilde{\mathfrak{q}} \to 0 \ .$$

The ideal $\mathfrak{p}\mathcal{N}(M)$ is finitely generated. Since $\mathfrak{q}+\widetilde{\mathfrak{q}}$ strictly contains $\mathfrak{q}$, its height is $> h$, and hence $\mathfrak{q} + \widetilde{\mathfrak{q}}$ is finitely generated by the inductive assumption and Lemma 8.7.19. Therefore $\mathfrak{q} \oplus \widetilde{\mathfrak{q}}$ is a finitely generated $\mathcal{N}(M)$-module, and $\mathfrak{q}$ is finitely generated. $\square$

8.8 Nash Functions on the Real Spectrum. Efroymson's Approximation Theorem

Throughout this section, M denotes a Nash submanifold of R^p.

The first three results in this section concern the relationship between Nash functions and the real spectrum. They will not be used in the proof of the approximation theorem. First we give an interpretation of the substitution theorem 8.5.2 in terms of the real spectrum.

Proposition 8.8.1. *The canonical homomorphism $R[X_1, \ldots, X_p] \to \mathcal{N}(M)$ induces a homeomorphism from $\mathrm{Spec}_\mathrm{r}(\mathcal{N}(M))$ onto $\widetilde{M}$.*

Proof. Denote by $j : \mathrm{Spec}_\mathrm{r}(\mathcal{N}(M)) \to \widetilde{R^p}$ the mapping between the real spectra induced by $R[X_1, \ldots, X_p] \to \mathcal{N}(M)$. If $\alpha \in \widetilde{M}$, then by Proposition 7.3.1 there is an evaluation homomorphism $\mathrm{ev}_\alpha : \mathcal{N}(M) \to k(\alpha)$ defined by $\mathrm{ev}_\alpha(f) = f(\alpha)$. The prime cone $\beta = \{f \in \mathcal{N}(M) \mid f(\alpha) \geq 0\}$ satisfies $j(\beta) = \alpha$. Hence $\widetilde{M}$ is contained in the image of j. Part (i) of the substitution theorem 8.5.2 says precisely that the image of j is contained in $\widetilde{M}$. On the other hand, part (ii) (which says that a homomorphism $\varphi : \mathcal{N}(M) \to F$, with values in a real closed field, is completely determined by the composition $R[X_1, \ldots, X_p] \to \mathcal{N}(M) \to F$) implies that j is injective. Finally, j is open, since the image of an open subset $\widetilde{\mathcal{U}}(f)$ of $\mathrm{Spec}_\mathrm{r}(\mathcal{N}(M))$ by j is the open subset $\{x \in M \mid f(x) > 0\}\widetilde{}$. $\square$

The Nash functions form a sheaf on $\widetilde{M}$.

Proposition 8.8.2. *There exists a unique sheaf of rings $\widetilde{\mathcal{N}}_{\widetilde{M}}$ on $\widetilde{M}$ such that for every open semi-algebraic subset U of M, one has $\widetilde{\mathcal{N}}_{\widetilde{M}}(\widetilde{U}) = \mathcal{N}(U)$.*

Proof. Analogous to the proof of Proposition 7.3.2. $\square$

If $R = \mathbb{R}$, then the Nash functions also form a sheaf on M. But even in this case, the compactness of $\widetilde{M}$ is still useful in solving some problems (cf. Proposition 12.7.2 for instance).

Proposition 8.8.3. *Let* $\alpha \in \widetilde{M}$. *The stalk* $\widetilde{\mathcal{N}}_{\widetilde{M},\alpha} = \varinjlim_{\alpha \in \widetilde{U}} \mathcal{N}(U)$ *(for U an open semi-algebraic subset of M) is a henselian local ring whose residue field is $k(\alpha)$.*

Proof. The same proof as for Proposition 7.3.4 shows that $\widetilde{\mathcal{N}}_{\widetilde{M},\alpha}$ is a local ring whose residue field is $k(\alpha)$. It remains to prove that the stalk is henselian. Let $F(T) = T^n + f_1 T^{n-1} + \cdots + f_n \in \widetilde{\mathcal{N}}_{\widetilde{M},\alpha}[T]$. Let U be an open semi-algebraic subset of M such that $\alpha \in \widetilde{U}$ and $f_1, \ldots, f_n$ are defined on U. If $x \in U$, let
$$F_x(T) = T^n + f_1(x) T^{n-1} + \cdots + f_n(x) \in R[T] \, .$$

Define
$$\begin{aligned} U' &= \{x \in U \mid F_x(T) \text{ has a simple root}\} \\ V' &= \{(x,t) \in U' \times R \mid t \text{ is a simple root of } F_x(T)\} \, . \end{aligned}$$

By the Tarski-Seidenberg principle 2.2.4, U' and V' are semi-algebraic sets. By the implicit function theorem 2.9.8, U' is open in M, and the restriction of the projection mapping $\Pi : U' \times R \to U'$ to V' is a local diffeomorphism. By Proposition 9.3.9 (whose proof does not use any result of this section), there is a finite open semi-algebraic cover $V' = V_1 \cup \cdots \cup V_k$ such that, for $i = 1, \ldots, k$, $\Pi|_{V_i}$ is a Nash diffeomorphism onto its image. Hence, the last coordinate g_i of the inverse $s_i = (\Pi|_{V_i})^{-1}$ is a Nash function on $\Pi(V_i)$, such that, for every x in $\Pi(V_i)$, $F_x(g_i(x)) = 0$.

Now assume that $T^n + \widetilde{f_1(\alpha)} T^{n-1} + \cdots + f_n(\alpha) \in k(\alpha)[T]$ has a simple root a in $k(\alpha)$. The point β of $\widetilde{U' \times R}$ defined by $\widetilde{\Pi}(\beta) = \alpha$ and $T(\beta) = a$ belongs to $\widetilde{V}'$. By Theorem 7.2.3, we have $\widetilde{V'} = \widetilde{V_1} \cup \cdots \cup \widetilde{V_k}$. Therefore, β belongs to one of the $\widetilde{V}_i$, say $\widetilde{V}_1$, and $\widetilde{s}_1(\alpha) = \beta$. Hence, the germ of g_1 in $\widetilde{\mathcal{N}}_{\widetilde{M},\alpha}$ satisfies $g_1(\alpha) = a$ and $F(g_1) = 0$. $\square$

The rest of this section is devoted to the proof of the following result.

Theorem 8.8.4 (Efroymson's Approximation Theorem). *Let $M \subset R^p$ be a Nash submanifold. Let $g, \epsilon \in \mathcal{S}^0(M)$ be two continuous semi-algebraic functions from M into R, with $\epsilon > 0$. Then there exists $f \in \mathcal{N}(M)$ such that $|f - g| < \epsilon$ on M.*

This theorem can be compared with the Stone-Weierstrass approximation theorem. We recall the statement of this theorem.

Theorem 8.8.5 (Stone-Weierstrass Theorem). *Let K be a compact subset of $\mathbb{R}^n$. Let $g : K \to \mathbb{R}$ be a function of class C^k, $k \in \mathbb{N}$, on K. Then for every real number $\epsilon > 0$, there exists a polynomial $f \in \mathbb{R}[X_1, \ldots, X_n]$ such that, for every $a \in K$ and every multi-index $\alpha = (\alpha_1, \ldots, \alpha_n) \in \mathbb{N}^n$ of weight $|\alpha| = \alpha_1 + \cdots + \alpha_n \leq k$, one has*
$$\left| \frac{\partial^{|\alpha|} f}{\partial x^\alpha}(a) - \frac{\partial^{|\alpha|} g}{\partial x^\alpha}(a) \right| < \epsilon \, .$$

194   8. Nash Functions

Reference for the proof: [246]. □

The proof of the Stone-Weierstrass theorem uses the archimedean property of $\mathbb{R}$ in a crucial way, and cannot be extended to an arbitrary real closed field.

Example 8.8.6. Consider the field of Puiseux series $R = \mathbb{R}(X)^{\wedge}$ (Example 1.2.3). Recall that X is a positive element of R, smaller than every positive real number. There exists no polynomial $f \in R[T]$ such that $|f(t) - |t|| < X$ for all $t \in R$, $-1 \le t \le 1$. For, suppose that such a polynomial

$$f(T) = a_0(X) + a_1(X)T + \cdots + a_d(X)T^d$$

exists. The $a_i(X)$ are Puiseux series, and we denote by $a_{i,0}$ their constant terms. For every real number $t \in [0,1]$, the inequalities

$$|a_0(X) + (a_1(X) - 1)t + a_2(X)t^2 + \cdots + a_d(X)t^d| < X$$

imply

$$a_{0,0} + (a_{1,0} - 1)t + a_{2,0}t^2 + \cdots + a_{d,0}t^d = 0 \ .$$

Hence $a_{1,0} = 1$. But, considering the real numbers t of $[-1,0]$, we get $a_{1,0} = -1$.

However, Theorem 8.8.4 implies that there is a Nash function $\varphi : R \to R$ such that $|\varphi(t) - |t|| < X$ for all $t \in R$. Indeed, we can take φ defined by $\varphi(t) = \sqrt{t^2 + X^2/2}$.

We begin the proof of Theorem 8.8.4 with a few technical lemmas.

Lemma 8.8.7. *Let α be a closed point of $\widetilde{M}$, $a \in k(\alpha)$. Then there exists a continuous semi-algebraic function $h \in \mathcal{S}^0(M)$ such that $h(\alpha) = a$.*

Proof. There exists an open semi-algebraic subset U of M and $h_1 \in \mathcal{S}^0(U)$ such that $\alpha \in \widetilde{U}$ and $h_1(\alpha) = a$ (Corollary 7.3.5). By Proposition 7.1.25 (i), every closed point of $\widetilde{M}$ has a basis of closed constructible neighbourhoods. Hence, there is a closed semi-algebraic subset F of M such that $\alpha \in \widetilde{F} \subset \widetilde{U}$. By Proposition 2.6.9, there exists a function $h \in \mathcal{S}^0(M)$ such that $h|_F = h_1|_F$. We have $h(\alpha) = a$. □

If f and φ are semi-algebraic functions on M, and A is a subset of $\widetilde{M}$, the meaning of "$f < \varphi$ on A" is that, for every α in A, $f(\alpha) < \varphi(\alpha)$ in $k(\alpha)$.

Lemma 8.8.8. *Let A and B be two disjoint closed subsets of $\widetilde{M}$, $\varphi, \psi \in \mathcal{S}^0(M)$ two continuous semi-algebraic functions. Then there exists a Nash function $f \in \mathcal{N}(M)$ such that $f < \varphi$ on A and $f > \psi$ on B. If, in addition, φ and ψ are positive, then f can be chosen positive.*

8.8 Efroymson's Approximation Theorem   195

Proof. The closed subsets A and B are both intersections of constructible closed subsets of $\widetilde{M}$. By compactness of $\widetilde{M}$ for the constructible topology, we can find two disjoint closed semi-algebraic subsets F and G of M such that $A \subset \widetilde{F}$ and $B \subset \widetilde{G}$. We may assume, without loss of generality, that M is closed in R^p. By Mostowski's separation theorem 2.7.7, there is $g \in \mathcal{N}(M)$ negative on F and positive on G. By Proposition 2.6.2, there is a polynomial function h positive on M, with $h > |\varphi/g|$ on F and $h > |\psi/g|$ on G. We then set $f = gh$. The case of positive functions can be reduced to the preceding case using the Nash diffeomorphism $x \mapsto x + \sqrt{1 + x^2}$ that maps R onto $]0, +\infty[$. □

Lemma 8.8.9. *Let $M = U_1 \cup \cdots \cup U_k$ be a finite open semi-algebraic cover, and $\eta \in \mathcal{N}(M)$, $\eta > 0$. There exist functions $\varphi_i \in \mathcal{N}(M)$, $i = 1, \ldots, k$, such that $\varphi_i > 0$ on M, $\varphi_i < \eta$ on $M \setminus U_i$ and $\sum_{i=1}^{k} \varphi_i = 1$.*

Proof. Let $h_i(x) = \operatorname{dist}(x, M \setminus U_i)$ be the distance between $x \in M$ and $M \setminus U_i$, and $F_i = \{x \in M \mid h_i(x) = \max(h_1(x), \ldots, h_k(x))\}$. The F_i form a closed semi-algebraic cover of M, and $F_i \subset U_i$. By Lemma 8.8.8, we can find $\psi_i \in \mathcal{N}(M)$, $\psi_i > 0$, such that $\psi_i > 1$ on F_i and $\psi_i < \eta$ on $M \setminus U_i$. Set $\varphi_i = \psi_i(\sum_{j=1}^{k} \psi_j)^{-1}$. □

Proposition 8.8.10. *Let α be a closed point of $\widetilde{M}$. Then the image of the evaluation homomorphism ev_α from $\mathcal{N}(M)$ to $k(\alpha)$ is dense in $k(\alpha)$: for every $a \in k(\alpha)$ and every $\epsilon \in k(\alpha)$, $\epsilon > 0$, there exists a Nash function $f \in \mathcal{N}(M)$ such that $|f(\alpha) - a| < \epsilon$.*

Proof. The field $k(\alpha)$ is algebraic over the image of $R[X_1, \ldots, X_p]$ in $k(\alpha)$, and, a fortiori, over the image of $\mathcal{N}(M)$ in $k(\alpha)$. Let $p(T) = g_0 T^n + \cdots + g_n \in \mathcal{N}(M)[T]$ be a polynomial whose image in $k(\alpha)[T]$ is a minimal polynomial of a over $\operatorname{ev}_\alpha(\mathcal{N}(M))$ (we shall say that $p(T)$ is a minimal polynomial of a over $\mathcal{N}(M)$).

Lemma 8.8.11. *Assume that there exist two Nash functions $\rho, \theta \in \mathcal{N}(M)$ such that $\rho < \theta$ on M, $\rho(\alpha) < a < \theta(\alpha)$ and the derivative p' has no zero in the segment $[\rho(\alpha), \theta(\alpha)] \subset k(\alpha)$. Then there exists $f \in \mathcal{N}(M)$ such that $|f(\alpha) - a| < \epsilon$.*

Proof. We assume that p' is positive on $[\rho(\alpha), \theta(\alpha)]$ (the other case is similar). Then $p(\rho(\alpha)) < 0$ and $p(\theta(\alpha)) > 0$. Let F be the closed semi-algebraic subset of M defined as the complement of

$$\{x \in M \mid p(\rho(x)) < 0 \,, \; p(\theta(x)) > 0 \,, \text{ and } \forall y \in [\rho(x), \theta(x)] \; p'(y) > 0\} \,.$$

By Lemma 8.8.8, we can choose u and v in $\mathcal{N}(M)$, $u, v > 0$, such that

$$\max(u(\alpha), v(\alpha)) < \epsilon \gamma(\alpha) \,,$$

and, for all x in F,

$$u(x) > |p(\rho(x))|, \quad v(x) > |p(\theta(x))|, \quad v(x) > (\theta(x) - \rho(x))|\gamma(x)|,$$

where $\gamma(x) = \inf(\{p'(y) \mid y \in [\rho(x), \theta(x)]\})$. Set

$$q = p + (u+v)\frac{T-\rho}{\theta - \rho} - u \in \mathcal{N}(M)[T].$$

It follows that $q(\rho) = p(\rho) - u$, $q(\theta) = p(\theta) + v$ and $q'(y) = p'(y) + \frac{u+v}{\theta-\rho}$. Hence, for every x in M, we have $q(\rho(x)) < 0$, $q(\theta(x)) > 0$, and $q'(y) > 0$ for every $y \in [\rho(x), \theta(x)]$. Therefore, there is a unique function $f : M \to R$, $\rho < f < \theta$, such that $q(f(x)) = 0$. By the implicit function theorem 2.9.8, f is a Nash function. By the mean value theorem, $p(f(\alpha)) = (f(\alpha) - a)p'(c)$ for some $c \in [\rho(\alpha), \theta(\alpha)]$. Since

$$-v < p(f) = u - (u+v)\frac{f-\rho}{\theta - \rho} < u,$$

it follows that

$$|f(\alpha) - a| < (p'(c))^{-1}\max(u(\alpha), v(\alpha)) < \epsilon.$$

□

We return to the proof of Proposition 8.8.10. It suffices to prove the existence of the ρ and θ of the lemma, and we do this by induction on the degree of the minimal polynomial p of a. If the degree of p is 1, then let $h \in S^0(M)$ be such that $h(\alpha) = a$ (Lemma 8.8.7), and $\theta \in \mathcal{N}(M)$, $\theta > 0$, such that $\theta(\alpha) > |h(\alpha)|$ (Lemma 8.8.8). Set $\rho = -\theta$. Then the assumptions of the lemma are satisfied. Given $n > 1$, assume the proposition to be true for all $b \in k(\alpha)$ with minimal polynomial over $\mathcal{N}(M)$ of degree $< n$. Let $a \in k(\alpha)$ have minimal polynomial $p(T)$ of degree n. Let $b \in k(\alpha)$ be the smallest root of $p'(T)$ larger than a, if it exists. By Lemmas 8.8.7 and 8.8.8, there exists $\varphi \in \mathcal{N}(M)$, $\varphi > 0$, such that $\varphi(\alpha) < (b-a)/4$. By the inductive assumption we can find $\mu \in \mathcal{N}(M)$ such that $|\mu(\alpha) - b| < \varphi(\alpha)$. Set $\theta_1 = \mu - 2\varphi$. If b does not exist, take $\theta_1 \in \mathcal{N}(M)$ such that $\theta_1(\alpha) > a + 1$ (Lemmas 8.8.7 and 8.8.8). Similarly, we find $\rho \in \mathcal{N}(M)$ such that $\rho(\alpha) < a < \theta_1(\alpha)$ and p' has no zero in $[\rho(\alpha), \theta_1(\alpha)]$. Let H be the closed semi-algebraic subset of M consisting of all points where $\rho \geq \theta_1$. Choose $\eta \in \mathcal{N}(M)$, $\eta > 0$, such that $\eta > \rho - \theta_1$ on H and, if b exists, $\eta(\alpha) < b - \theta_1(\alpha)$. Set $\theta = \theta_1 + \eta \in \mathcal{N}(M)$. Then $\rho(\alpha) < a < \theta(\alpha)$, p' has no zero on $[\rho(\alpha), \theta(\alpha)]$, and $\theta > \rho$ on M. The assumptions of Lemma 8.8.11 are satisfied, and, hence, the conclusion of the proposition holds true for a. □

Proof of the approximation theorem 8.8.4. By Proposition 8.8.10, for every closed point α of $\widetilde{M}$, there exists $f_\alpha \in \mathcal{N}(M)$ such that $|f_\alpha(\alpha) - g(\alpha)| < \epsilon(\alpha)/2$. Let U_α be the open semi-algebraic subset of M consisting of all points where $|f_\alpha - g| < \epsilon/2$. The $\widetilde{U}_\alpha$ cover the closed points of $\widetilde{M}$, therefore they cover $\widetilde{M}$. By compactness of $\widetilde{M}$, we can extract a finite cover from $(\widetilde{U}_\alpha)_\alpha$.

Hence, there is a finite open semi-algebraic cover $M = U_1 \cup \cdots \cup U_k$ and functions $f_1, \ldots, f_k$ in $\mathcal{N}(M)$ such that $|f_i - g| < \epsilon/2$ on U_i. By Lemma 8.8.9, there are positive functions $\varphi_1, \ldots, \varphi_k \in \mathcal{N}(M)$ such that

$$\varphi_i < \frac{\epsilon}{2} \left(1 + k|g| + \sum_{i=1}^{k} |f_i| \right)^{-1} \quad \text{on } M \setminus U_i \,,$$

and $\sum_{i=1}^{k} \varphi_i = 1$. Set $f = \sum_{i=1}^{k} \varphi_i f_i \in \mathcal{N}(M)$. For every x in M we have

$$\sum_{i \in \{j | x \in U_j\}} \varphi_i(x) |f_i(x) - g(x)| < \frac{\epsilon(x)}{2} \,,$$

$$\sum_{i \in \{j | x \in M \setminus U_j\}} \varphi_i(x) |f_i(x) - g(x)| < \frac{\epsilon(x)}{2} \,.$$

Hence $|f - g| \leq \sum_{i=1}^{k} \varphi_i |f_i - g| < \epsilon$. □

Remark 8.8.12. Shiota's book [302] contains an approximation theorem for $\mathcal{S}^r$ functions together with their derivatives. In the case of an open semi-algebraic subset U of $\mathbb{R}^n$, this approximation theorem reads as follows: given an $\mathcal{S}^r$ function f on U and ε a positive continuous semi-algebraic function on U, there exists a Nash function φ on U such that, for every multi-index $\alpha \in \mathbb{N}^n$ of weight $|\alpha| \leq r$, one has $\left| \frac{\partial^\alpha}{\partial x^\alpha} (f - \varphi) \right| < \varepsilon$ on U. In the case of a Nash submanifold M, one has first to define an appropriate topology on the set of $\mathcal{S}^r$ functions on M. This theorem of approximation with derivatives is very important for the study of Nash manifolds. It allows one to make constructions in the $\mathcal{S}^r$ category, which is much more flexible, and then to approximate by Nash objects. See [302] for applications of this technique.

8.9 Tubular Neighbourhood. Approximation of $\mathcal{C}^\infty$ Functions. Extension Theorem

We begin by proving a few results concerning tubular neighbourhoods of Nash submanifolds. First we make precise the notion of a tangent space of a Nash submanifold. Let $M \subset R^n$ be a Nash submanifold of dimension m, and x a point of M. By the definition of Nash submanifolds, there exists a Nash diffeomorphism φ from an open semi-algebraic neighbourhood Ω of the origin in R^n onto an open semi-algebraic neighbourhood Ω' of x in M, such that $\varphi(0) = x$ and $\varphi((R^m \times \{0\}) \cap \Omega) = M \cap \Omega'$. One sees easily that the vector subspace $d\varphi_0(R^m \times \{0\}) \subset R^n$ does not depend on the choice of φ.

Definition 8.9.1. *The* tangent space *to M at x, denoted by $T_x(M)$, is the vector subspace $d\varphi_0(R^m \times \{0\}) \subset R^n$.*

198   8. Nash Functions

We denote by $V^\perp$ the orthogonal complement of the subspace $V \subset R^n$ with respect to the canonical scalar product on R^n.

Proposition 8.9.2. *Let $M \subset R^n$ be a Nash submanifold of dimension m. Define*
$$E = \{(x,y) \in M \times R^n \mid y \in T_x(M)^\perp\},$$
and let $\Pi : E \to M$ be the projection mapping defined by $\Pi(x,y) = x$. Then E is a Nash submanifold of dimension n and there exist a finite open semi-algebraic cover $U_1, \ldots, U_p$ of M and Nash diffeomorphisms
$$\theta_i : U_i \times R^{n-m} \longrightarrow \Pi^{-1}(U_i),$$
for $i = 1, \ldots, p$, such that $\Pi \circ \theta_i$ is the projection mapping $U_i \times R^{n-m} \to U_i$, and, for every x in U_i, $\theta_i|_{\{x\} \times R^{n-m}}$ is an R-linear isomorphism onto $\{x\} \times T_x(M)^\perp$

In the terminology that will be introduced in Chap. 12, (E, Π, M) is a Nash vector bundle (the normal bundle of M).

Proof. Corollary 9.3.10 (which will be proved in the next chapter) implies that there is a finite open semi-algebraic cover $M = U_1 \cup \cdots \cup U_p$ and, for $i = 1, \ldots, p$, Nash functions $f_{i,m+1}, \ldots, f_{i,n}$ on an open semi-algebraic neighbourhood V_i of U_i in R^n, such that $U_i = \{x \in V_i \mid f_{i,m+1}(x) = \cdots = f_{i,n}(x) = 0\}$, and the vectors $\operatorname{grad}(f_{i,m+1}(x)), \ldots, \operatorname{grad}(f_{i,n}(x))$ are linearly independent for every x in U_i. For $(x,z) \in U_i \times R^{n-m}$, set
$$\theta_i(x,z) = (x, z_1 \operatorname{grad}(f_{i,m+1}(x)) + \cdots + z_{n-m} \operatorname{grad}(f_{i,n}(x))).$$
Then $\theta_i(U_i \times R^{n-m}) = \Pi^{-1}(U_i)$. It is easy to verify that E is a Nash submanifold (E is semi-algebraic, since each $\Pi^{-1}(U_i)$ is semi-algebraic) and the θ_i are Nash diffeomorphisms. □

The next proposition uses the notations of Proposition 8.9.2.

Proposition 8.9.3. *Let $\varphi : E \to R^n$ be the Nash mapping defined by $\varphi(x,y) = x + y$. There exists an open semi-algebraic neighbourhood T of $M \times \{0\}$ in E, such that $\varphi|_T$ is a Nash diffeomorphism onto $\varphi(T)$, and, for every (x,y) in T and every $t \in [0,1]$, $(x, ty) \in T$.*

Proof. Let W be the open semi-algebraic subset consisting of the points $(x,y) \in E$ such that $d\varphi_{(x,y)}$ is injective. The set $M \times \{0\}$ is contained in W. For $x \in M$, define
$$\psi(x) = \inf \left(\{r \in R \mid \exists (z,u) \in E \; \exists y \in R^n \; ((x,y) \in E,\right.$$
$$\left. \|u\| \leq \|y\| = r, \; \varphi(x,y) = \varphi(z,u), \text{ and } x \neq z)\}\right).$$
The function ψ is semi-algebraic. On the other hand, the implicit function theorem implies that, for every x in M, φ is injective on a neighbourhood of $(x,0)$ in E, say $E \cap B_{2n}((x,0), c_x)$. Then ψ is bounded from below by $c_x/4$ on $M \cap B_n(x, c_x)$. We use the following lemma.

8.9 Tubular Neighbourhood. Extension Theorem 199

Lemma 8.9.4. *If M is a locally closed semi-algebraic set, and $\psi : M \to R$ is a semi-algebraic function locally bounded from below by positive constants, then ψ is bounded from below by a positive, continuous, semi-algebraic function on M.*

By this lemma, there exist a continuous positive semi-algebraic function $f : M \to R$ such that $f < \psi$. We may assume, moreover, that, for every $x \in M$, $f(x) < \text{dist}(x, E \setminus W)$. Take $T = \{(x,y) \in E \mid \|y\| < f(x)\}$. Then T is an open semi-algebraic neighbourhood of $M \times \{0\}$ in E, the mapping $\varphi|_T$ is injective and, by the inverse function theorem, it is a diffeomorphism onto its image. □

Proof of Lemma 8.9.4. We may assume, without loss of generality, that M is closed, $0 \in M$ and $\psi \leq 1$. We define the function $v : [0,+\infty[\to [0,+\infty[$ by

$$v(r) = \inf(\{\psi(x) \mid x \in M \text{ and } \|x\| \leq r\}) .$$

The function v is semi-algebraic and decreasing. We claim that v does not vanish. Suppose $v(r) = 0$, and let $\Pi : M \times R \to R$ be the projection mapping, and

$$A = \{(x, \epsilon) \in M \times R \mid \|x\| \leq r \text{ and } \psi(x) = \epsilon\} .$$

Since $0 \in \text{clos}(\Pi(A))$ and $\text{clos}(A)$ is closed and bounded, it follows from Theorem 2.5.8 that $0 \in \Pi(\text{clos}(A))$. Hence, there is $x \in M$ such that $(x, 0)$ belongs to the closure of the graph of ψ. This contradicts the fact that ψ is bounded from below by a positive constant on a neighbourhood of x, and thus the claim is proved. By Proposition 2.6.1 applied to $1/v$ and the fact that v is decreasing, there is a positive continuous semi-algebraic function g on $[0,+\infty[$ such that $g \leq v$. The positive continuous semi-algebraic function f defined on M by $f(x) = g(\|x\|)$ bounds ψ from below. □

Corollary 8.9.5 (Nash Tubular Neighbourhood). *Let $M \subset R^n$ be a Nash submanifold. There exists an open semi-algebraic neighbourhood U of M in R^n and a Nash retraction $\rho : U \to M$. Moreover, we may assume that $\text{dist}(x, M) = \|x - \rho(x)\|$ for every $x \in U$.*

Proof. Take $U = \varphi(T)$ and $\rho = \Pi \circ (\varphi|_T)^{-1}$. The last part of the corollary follows from the definition of ψ given in the proof of Proposition 8.9.3. □

Corollary 8.9.6. *Let $M \subset R^n, N \subset R^p$ be two Nash submanifolds. Let $g : M \to N$ be a continuous semi-algebraic mapping, $\epsilon : M \to R$ a positive continuous semi-algebraic function. There exists a Nash mapping $f : M \to N$ such that $\|f - g\| < \epsilon$ on M.*

Proof. Choose a Nash tubular neighbourhood $(U, \rho : U \to N)$ of N such that $\text{dist}(y, N) = \|y - \rho(y)\|$ for every $y \in U$. By the approximation theorem 8.8.4, there is a Nash mapping $\overline{f} : M \to R^p$ such that $\overline{f}(M) \subset U$ and $\|\overline{f} - g\| < \epsilon/2$. Then we define $f = \rho \circ \overline{f}$. □

Given two C^∞ manifolds M and N, we denote by $C^\infty(M,N)$ the set of C^∞ mappings from M into N, equipped with the C^∞ topology (cf. [160], Chap. 2).

Corollary 8.9.7. *($R = \mathbb{R}$) Let $M \subset \mathbb{R}^n$, $N \subset \mathbb{R}^p$ be two Nash submanifolds, with M compact. The set $\mathcal{N}(M,N)$ of Nash mappings from M into N is dense in the set of C^∞ mappings $C^\infty(M,N)$.*

Proof. Use a Nash tubular neighbourhood of N and the Stone-Weierstrass theorem 8.8.5. □

Remark 8.9.8. The last corollary is a density result, which allows the use of Nash mappings in the study of C^∞ mappings. Moreover, we shall see in Theorem 14.1.8 that every compact C^∞ manifold is C^∞-diffeomorphic to a Nash submanifold. This is used in [19] to study the periodic points of C^∞ diffeomorphisms of a compact C^∞ manifold to itself.

Corollary 8.9.9. *($R = \mathbb{R}$) Let $M \subset \mathbb{R}^n$, $N \subset \mathbb{R}^p$ be two compact Nash submanifolds. If M and N are C^∞-diffeomorphic, then they are Nash diffeomorphic.*

Proof. This is a consequence of Corollary 8.9.7 together with the fact that the set of C^∞ diffeomorphisms is open in $C^\infty(M,N)$. □

Remark 8.9.10. The previous result is false without the compactness assumption. There is an example in [300] of two nonsingular algebraic subsets of $\mathbb{R}^n$ that are C^∞-diffeomorphic but not Nash diffeomorphic.

The rest of this section is devoted to Efroymson's extension theorem.

Lemma 8.9.11. *Let $M \subset R^n$ be a Nash submanifold, $v \in \mathcal{N}(M)$, $V = v^{-1}(0)$. Let $f, \epsilon \in S^0(M)$ such that $f|_V = 0$ and $\epsilon > 0$. Then there exists $g \in v\mathcal{N}(M)$ such that $|f - g| < \epsilon$ on M.*

Proof. Replacing, if necessary, v with v^2, we may assume that $v \geq 0$ on M. By Theorem 8.8.4, there is $g_1 \in \mathcal{N}(M)$ such that $|f - g_1| < \epsilon/4$. The set $F = \{x \in M \mid |g_1(x)| \geq \epsilon/2\}$ is a closed semi-algebraic subset of M, disjoint from V. Hence $v/|g_1|$ vanishes nowhere on F. By Lemma 8.8.8, we can find $\eta \in \mathcal{N}(M)$, $\eta > 0$, such that $\eta < v\epsilon/|g_1|$ on F. An easy computation shows that $g = 2vg_1/(2v + \eta)$ satisfies $|f - g| < \epsilon$ on M. □

Theorem 8.9.12. *Let $M \subset R^n$ be a Nash submanifold, $v \in \mathcal{N}(M)$, $V = v^{-1}(0)$. Let $f : U \to R$ be a Nash function, where U is an open semi-algebraic neighbourhood of V in M. There exists a Nash function $g \in \mathcal{N}(M)$, such that $g - f \in v\mathcal{N}(U)$ (and, hence, $g|_V = f|_V$).*

Proof. Suppose that $V = V_1 \cup V_2$, where V_1 and V_2 are disjoint closed semi-algebraic subsets of M. Then, by the separation theorem 2.7.7, there exists $\varphi \in \mathcal{N}(M)$ such that $\varphi(V_1) > 0$ and $\varphi(V_2) < 0$. Set $v_1 = \sqrt{v^2 + \varphi^2} - \varphi$,

8.9 Tubular Neighbourhood. Extension Theorem

$v_2 = \sqrt{v^2 + \varphi^2} + \varphi$. We have $v_i^{-1}(0) = V_i$ for $i = 1, 2$. If $g_i \in \mathcal{N}(M)$ are such that $g_i - f \in v_i \mathcal{N}(U)$ for $i = 1, 2$, then $g = (g_1 v_2 + g_2 v_1)/(v_1 + v_2)$ satisfies $g - f \in v \mathcal{N}(U)$. Hence, we may assume that V is semi-algebraically connected. Therefore, U can also be assumed to be semi-algebraically connected.

By Theorem 8.4.4, there is a nonsingular irreducible algebraic set $Z \subset R^{n+p}$, an open semi-algebraic subset U' of Z, a Nash diffeomorphism $\sigma : U \to U'$ and a polynomial function $h : Z \to R$, such that the following diagram commutes

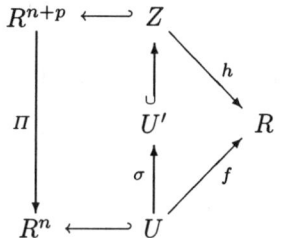

where Π is the projection onto the space of the first n coordinates. Let $p_1, \ldots, p_m$ be generators of the ideal $\mathcal{I}(Z) \subset R[X, Y]$. Let $q : M \times R^p \to R^{n+p+m}$ be the Nash mapping defined by

$$q(x, y) = (x, v(x)y, p_1(x, y), \ldots, p_m(x, y)).$$

Choose a closed semi-algebraic neighbourhood F of V in U, and extend the last p coordinates of $\sigma|_F$ to continuous semi-algebraic functions on M, using Proposition 2.6.9. We obtain, in this way, a continuous semi-algebraic section $\overline{\sigma} : M \to R^{n+p}$ of Π such that $\overline{\sigma}|_F = \sigma|_F$. Replacing U with $\mathrm{int}(F)$, we may assume $\overline{\sigma}|_U = \sigma$. Set $\Gamma = \overline{\sigma}(M) \subset R^{n+p}$. For every x in M, the rank of $dq_{\overline{\sigma}(x)}$ is $d + p$, where $d = \dim(M)$. Set

$$\epsilon_1(x) = \sup\left(\{r \in R \mid r \leq 1, \text{ and } q|_{M \times R^p \cap B_{n+p}(\overline{\sigma}(x), r)} \text{ is injective,}\right.$$
$$\left. \text{of rank } d + p \text{ and open as a mapping to its image}\}\right).$$

The function ϵ_1 is semi-algebraic on M and locally bounded from below by positive constants. By Lemma 8.9.4, there is a function $\epsilon \in \mathcal{S}^0(M)$ such that $0 < \epsilon < \epsilon_1$. Take

$$\Omega = \{(x, y) \in M \times R^p \mid \|(x, y) - \overline{\sigma}(x)\| < \epsilon(x)\}.$$

One verifies that Ω is an open semi-algebraic neighbourhood of Γ in $M \times R^p$, such that $q|_\Omega$ is a Nash embedding from Ω into R^{n+p+m}. Now we apply Corollary 8.9.5 to the Nash submanifold $q(\Omega)$ of R^{n+p+m}: there exists a semi-algebraic open neighbourhood W of $q(\Omega)$ in R^{n+p+m} and a Nash retraction $\rho : W \to q(\Omega)$. The last $p + m$ coordinates of the mapping $q \circ \overline{\sigma} : M \to q(\Omega) \subset R^{n+p+m}$ are continuous semi-algebraic functions vanishing on V. By Lemma 8.9.11, we obtain a Nash mapping $\tau : M \to W$ such that $\tau(x) = (x, \tau_1(x))$ and $\tau_1(x) \in \bigoplus_{p+m} v \mathcal{N}(M)$. Set

$$g = \overline{h} \circ (q|_\Omega)^{-1} \circ \rho \circ \tau : M \to R,$$

where $\overline{h} : R^{n+p} \to R$ is a polynomial function such that $\overline{h}|_Z = h$. Then $g \in \mathcal{N}(M)$, and, since $\tau - q \circ \sigma \in \bigoplus_{n+p+m} v\mathcal{N}(U)$, we have on U

$$g - f = \overline{h} \circ (q|_\Omega)^{-1} \circ \rho \circ \tau - \overline{h} \circ (q|_\Omega)^{-1} \circ \rho \circ q \circ \sigma \in v\mathcal{N}(U).$$

□

Corollary 8.9.13. *Let $M \subset R^n$ be a Nash submanifold, and V, a Nash subset of M that is also a Nash submanifold. Let $f : V \to R$ be a Nash function. There exists a Nash function $g \in \mathcal{N}(M)$ such that $g|_V = f$.*

Proof. Choose a Nash tubular neighbourhood (U, ρ) of V, and apply Theorem 8.9.12 to the composition $f \circ \rho|_{U \cap M}$. □

Remark 8.9.14. Let $M \subset R^n$ be a Nash submanifold. Then a Nash submanifold $N \subset M$ which is closed in M is a Nash subset of M ([302], Corollary II.5.4).

8.10 Families of Nash Functions

We continue the study begun in the fourth section of Chap. 7.

Proposition 8.10.1. *Let $S \subset R^p$ be a semi-algebraic set and $\alpha \in \widetilde{S}$. Let U be an open semi-algebraic subset of $R^m \times S$, and $f : U \to R \times S$, a semi-algebraic family of functions parametrized by S. If, for every $t \in S$, the fibre $f_t : U_t \to R$ is a Nash function, then the fibre $f_\alpha : U_\alpha \to k(\alpha)$ is also a Nash function.*

Proof. For every $k \in \mathbb{N}$, the fact that f_t is of class C^k on U_t can be expressed by a formula of the language of ordered fields with parameters in R (the proof of this claim is tedious but not difficult). By Proposition 7.2.2 (iii), f_α is of class C^k on U_α, for every k. Hence f_α is a Nash function. □

Lemma 8.10.2. *Let $\alpha \in \widetilde{R^p}$. The ultrafilter of semi-algebraic sets $S \subset R^p$ such that $\alpha \in \widetilde{S}$ has a basis of Nash submanifolds of R^p, i.e., for every semi-algebraic set $S \subset R^p$ such that $\alpha \in \widetilde{S}$, there exists a Nash submanifold $M \subset R^p$ such that $M \subset S$ and $\alpha \in \widetilde{M}$.*

Proof. Let $V = \mathcal{Z}(\text{supp}(\alpha))$. The algebraic set V is irreducible and of dimension $d = \dim(\alpha)$. Set $M = \text{int}(S \cap \text{Reg}(V))$, where the interior is taken in V. It is obvious that M is a Nash submanifold. Moreover, since $\dim((V \cap S) \setminus M) < d$, we have $\alpha \in \widetilde{M}$ by Proposition 7.5.8. □

8.10 Families of Nash Functions   203

Proposition 8.10.3. *Let $\alpha \in \widetilde{R^p}$, and let Ω be an open semi-algebraic subset of $k(\alpha)^m$, and $\varphi : \Omega \to k(\alpha)$, a Nash function. There exist a Nash submanifold $M \subset R^p$, with $\alpha \in \widetilde{M}$, an open semi-algebraic subset U of $R^m \times M$ and a semi-algebraic family of functions $f : U \to R \times M$ parametrized by M, such that $U_\alpha = \Omega$, $f_\alpha = \varphi$ and f is a Nash mapping.*

Proof. We may assume that Ω is connected. By the Artin-Mazur description of Nash functions (Theorem 8.4.4), there exist a nonsingular irreducible algebraic set $V \subset k(\alpha)^m \times k(\alpha)^q$ of dimension m, an open semi-algebraic subset Ω' of V, a Nash diffeomorphism $\sigma : \Omega \to \Omega'$ and a polynomial function $\gamma : V \to k(\alpha)$, such that the following diagram commutes

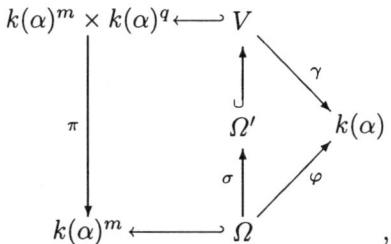

where Π is the projection onto the space of the first m coordinates. Let $\psi_1, \ldots, \psi_r \in k(\alpha)[X_1, \ldots, X_m, Y_1, \ldots, Y_q]$ be generators of the ideal $\mathcal{I}(V)$. Then $\psi_i(\sigma(x)) = 0$, and the matrix $\left[\frac{\partial \psi_i}{\partial Y_j}(\sigma(x))\right]_{i=1,\ldots,r; j=1,\ldots,q}$ has rank q for every $x \in \Omega$. We can write $\gamma = g(X, Y, a)$ and $\psi_i = h_i(X, Y, a)$, for $i = 1, \ldots, r$, where $a \in k(\alpha)^n$, $g, h_1, \ldots, h_r \in \mathbb{Z}[X, Y, T]$, $X = (X_1, \ldots, X_m)$, $Y = (Y_1, \ldots, Y_q)$ and $T = (T_1, \ldots, T_n)$. By Proposition 8.8.3, there is a Nash mapping $\theta : \Lambda \to R^n$ defined on an open semi-algebraic subset Λ of R^p, such that $\alpha \in \widetilde{\Lambda}$ and $a = \theta(\alpha)$. Applying Propositions 7.4.6 and 7.4.8 and Lemma 8.10.2, we get a Nash submanifold M of R^p such that $\alpha \in \widetilde{M}$ and $M \subset \Lambda$, an open semi-algebraic subset U of $R^m \times M$ and continuous semi-algebraic mappings $s : U \to R^m \times R^q \times R^p$ and $f : U \to R \times R^p$ that are semi-algebraic families of mappings parametrized by M, such that $U_\alpha = \Omega$, $s_\alpha = \sigma$ and $f_\alpha = \varphi$. By Proposition 7.2.2 (iii), we can take M small enough, so that $s(U)$ is contained in

$$W = \{(x, y, t) \in R^m \times R^q \times R^p \mid h_i(x, y, \theta(t)) = 0, \ i = 1, \ldots, r\},$$

the diagram

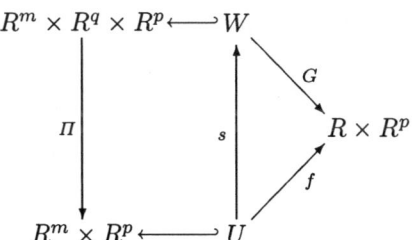

commutes (where Π is the projection onto the space of the first m and last p coordinates, and $G(x,y,t) = (g(x,y,\theta(t)),t))$) and the matrix

$$\left[\frac{\partial H_i}{\partial Y_j}(s(x,t))\right]_{i=1,\ldots,r;j=1,\ldots,q}$$

has rank q for every $(x,t) \in U$ (where $H_i(x,y,t) = h_i(x,y,\theta(t))$). Since the H_i are Nash functions, this last condition implies that s is a Nash mapping. Since G is the restriction of a Nash mapping, we conclude that f is Nash. $\square$

Remark 8.10.4. Let U be an open semi-algebraic subset of $R^m \times S$ and $f : U \to R \times S$ a semi-algebraic family of functions such that $f_t : U_t \to R$ is a Nash function for every $t \in S$. Then, by Propositions 8.10.1 and 8.10.3, S is a finite union $S = \bigcup_{i=1}^{k} M_i$ of Nash submanifolds, such that $f|_{U \cap (R^m \times M_i)}$ is Nash for $i = 1,\ldots,k$.

The rest of this section is devoted to the application of the study of semi-algebraic families of Nash functions to the following theorem. We say that a semi-algebraic function on a semi-algebraic subset S of R^n has *complexity* $\leq d$ if there is a nonzero polynomial P in $n+1$ variables with coefficients in R of total degree $\leq d$, such that $P(x, f(x)) = 0$ for all $x \in S$.

Theorem 8.10.5. *Given n, d in $\mathbb{N}$, there is a number $r = r(n,d) \in \mathbb{N}$ such that, for every open semi-algebraic subset U of R^n, every S^r function $f : U \to R$ of complexity $\leq d$ is Nash.*

Lemma 8.10.6. *Let $\Pi : R^{p+n} \to R^p$ be the projection onto the space of the first p coordinates. Let A be a semi-algebraic subset of R^{p+n}. There is a partition $A = \bigcup_{i=1}^{q} B_i$ into finitely many semi-algebraic subsets, such that, for every $i = 1,\ldots,q$ and every $t \in \Pi(B_i)$, $\Pi^{-1}(t) \cap B_i$ is semi-algebraically connected.*

Proof. By induction on n. If $n = 0$, there is no need to subdivide A. Given $n > 0$, assume that the lemma holds for $n-1$. We may slice the polynomials appearing in the definition of A (Theorem 2.3.1), and obtain a finite semi-algebraic partition $R^{p+n-1} = \bigcup_{j=1}^{m} C_j$ and semi-algebraic continuous functions

$$\xi_{j,1} < \ldots < \xi_{j,\ell_j} : C_j \to R,$$

such that A is the disjoint union of finitely many B_i which are graphs of $\xi_{j,k}$ or slices $]\xi_{j,k},\xi_{j,k+1}[$. Denote by $\Pi' : R^{p+n-1} \to R^p$ the projection onto the space of the first p coordinates. By the inductive assumption, we may assume that, for every $j = 1,\ldots,m$ and every $t \in \Pi'(C_j)$, $(\Pi')^{-1}(t) \cap C_j$ is semi-algebraically connected. If B_i projects onto C_j, it follows that $\Pi^{-1}(t) \cap B_i$ is semi-algebraically connected. $\square$

Lemma 8.10.7. *Let U be an open semi-algebraic subset of R^n. There is a semi-algebraic set Σ_d and a semi-algebraic family of functions $F_d : U \times \Sigma_d \to R \times \Sigma_d$ parametrized by Σ_d, such that a function $f : U \to R$ is a continuous semi-algebraic function of complexity $\leq d$ if and only if there is $s \in \Sigma_d$, such that f is equal to the fibre $(F_d)_s : U \to R$.*

Proof. The space of coefficients of polynomials in the $n+1$ variables $X = (X_1, \ldots, X_n)$ and Y of total degree $\leq d$ is the affine space R^p, where $p = \binom{n+1+d}{d}$. For $t \in R^p$, we denote by $P(t, X, Y) \in R[X, Y]$ the corresponding polynomial. We can consider P as a polynomial with integer coefficients in the $p + n + 1$ variables (T, X, Y). We slice the polynomial P with respect to Y (Theorem 2.3.1), and obtain a finite semi-algebraic partition $R^p \times U = \bigcup_{i=1}^{m} A_i$ and continuous semi-algebraic functions

$$\xi_{i,1} < \ldots < \xi_{i,\ell_i} : A_i \longrightarrow R,$$

such that either $A_i \times R$ is contained in $P^{-1}(0)$, or $P^{-1}(0) \cap (A_i \times R)$ is the union of the graphs of the $\xi_{i,j}$. Let $\Pi : R^{p+n} \to R^p$ be the projection onto the space of the first p coordinates. By Lemma 8.10.6, we may assume that, for every $i = 1, \ldots, m$ and every $t \in R^p$, $\Pi^{-1}(t) \cap A_i$ is connected. We may also assume that those A_i such that $A_i \times R \subset P^{-1}(0)$ are $A_{k+1}, \ldots, A_m$. Note that $\Pi^{-1}(t) \cap (\bigcup_{i=1}^{k} A_i)$ is dense in $\{t\} \times U$ for every $t \neq 0$. Denote by Φ the set of functions $\varphi : \{1, \ldots, k\} \to \mathbb{N}$ such that $0 < \varphi(i) \leq \ell_i$ for $i = 1, \ldots, k$. It is obvious that a continuous semi-algebraic function $f : U \to R$ is of complexity $\leq d$ if and only if there exists $t \in R^p \setminus \{0\}$ and $\varphi \in \Phi$ such that $f(x) = \xi_{i,\varphi(i)}(t, x)$, for every $i = 1, \ldots, k$ and every $x \in \Pi^{-1}(t) \cap A_i$. Let S be the product of $R^p \setminus \{0\}$ with the finite set Φ; this product may be considered as a semi-algebraic subset of R^{p+1}. Let $G \subset U \times R \times S$ be the semi-algebraic family parametrized by S whose fibres $G_{t,\varphi}$ are the closures in $U \times R$ of the unions of the graphs of the $\xi_{i,\varphi(i)}|_{\Pi^{-1}(t) \cap A_i}$ for $i = 1, \ldots, k$. Take for Σ_d the semi-algebraic subset of those $(t, \varphi) \in S$ for which $G_{(t,\varphi)}$ is the graph of a continuous semi-algebraic function from U to R. This function is the fibre $(F_d)_{(t,\varphi)}$ of a family F_d of continuous semi-algebraic functions from U to R parametrized by Σ_d, which satisfies the property of the lemma.
□

Proof of Theorem 8.10.5. Fix an open semi-algebraic subset U of R^n, and consider the semi-algebraic family of functions $F_d : U \times \Sigma_d \to R \times \Sigma_d$ parametrized by Σ_d of Lemma 8.10.7. Let T^r be the semi-algebraic subset of those $s \in \Sigma_d$ for which the function $(F_d)_s : U \to R$ is of class C^r (recall that the fact that a semi-algebraic function $U \to R$ is C^r can be expressed by a formula of the language of ordered fields with parameters in R). Set $C = \bigcap_{r \in \mathbb{N}} \widetilde{T^r}$, so that $\alpha \in C$ if and only if the fibre $(F_d)_\alpha : U_{k(\alpha)} \to k(\alpha)$ is Nash. By definition, C is closed for the constructible topology of $\widetilde{\Sigma_d}$. If $\alpha \in C$, then, by Proposition, 8.10.3 there is a constructible subset M of Σ_d containing α, such that $\widetilde{M} \subset C$. Therefore, C is also open for the constructible topology, and it is constructible. Hence, there is a semi-algebraic subset N of Σ_d such that $C = \tilde{N}$. By compactness of the constructible topology, there is an $r \in \mathbb{N}$ such that $\widetilde{T^r} = \tilde{N}$, or, equivalently, $T^r = N$. Therefore, every S^r function from U to R is Nash. It remains to verify that we may choose r independent of U. Since the fact that a semi-algebraic function is Nash is of a local nature, a positive integer r suitable for the open ball B_n is also suitable for any open semi-algebraic subset of R^n. □

Bibliographic Notes. Real analytic-algebraic functions are called Nash functions because they appear in Nash's paper [247]. Artin and Mazur [19] define the notion of a Nash manifold and give the description of Nash functions presented in the fourth section. For the local properties of Nash functions, we followed the study of algebraic power series in Lafon's paper [199]. We also refer the reader to [211] for these local properties. The results concerning the approximation of formal solutions of analytic equations can be found in Artin's paper [17] (cf. also [327, 328]). The paper of Mostowski [238] contains the Nullstellensatz (Corollary 8.5.4) and the positive answer to Hilbert's 17th problem for Nash functions (Proposition 8.5.6). Some of the proofs in [238] are not complete. Complete proofs are given in [126], which also contains the substitution theorem (see also [43]). The proof of Theorem 8.5.2 using the description of Artin-Mazur is taken from [94]. Theorem 8.6.4 can be found in Risler's paper [268]. The algebraicity of germs of Nash sets is taken from [44]. Lemma 8.6.13 can be found in [273]. The noetherian property of the ring of global Nash functions is proved in [268] and [125]. We followed the second proof. The construction of the sheaf of Nash functions on the real spectrum can be found in [280]. Efroymson's approximation and extension theorems were proved in [127]. We have followed the proofs of [251]. The counter-example of Remark 8.8.6 is in [77]. The proof of Theorem 8.10.5 is taken from [263].

Besides the topological or analytic applications of Nash functions mentioned in this chapter, we can cite the application to the study of composite $\mathcal{C}^\infty$ functions [38, 39, 329].

We only considered Nash submanifolds of R^n. General Nash manifolds, defined using charts, are rarely investigated. The results on general Nash manifolds concern mainly the existence of nonisomorphic Nash structures on a $\mathcal{C}^\infty$ manifold [84, 301].

The main reference for Nash manifolds is the book of Shiota [302] which contains, in particular, the theorem of approximation of $\mathcal{S}^r$ functions together with their derivatives by Nash functions, and results on compactification of Nash manifolds. For recent advances concerning the extension theorem and the analytic irreducibility of irreducible Nash sets, see [101, 102].

9. Stratifications

Abstract. In this chapter, we continue the study of semi-algebraic sets initiated in Chap. 2. In Section 1, we construct stratifications which have a cylindrical structure with respect to all successive projections $R^k \to R^{k-1}$, which is particularly useful in inductive arguments. In the second section, we prove that a closed and bounded semi-algebraic set can be semi-algebraically triangulated. In Section 3, we investigate the structure of continuous semi-algebraic mappings. As an application, we obtain the theorem of local conic structure of semi-algebraic sets. In Section 4 we prove that a continuous semi-algebraic function is triangulable. We obtain the finiteness of the number of topological types of polynomials in n variables of degree $\leq d$. Half-branches of algebraic curves are studied in Section 5. Semi-algebraic versions of Sard's and Bertini's theorems are contained in Section 6. The last section is devoted to Whitney's conditions a and b.

Throughout this chapter, R denotes a real closed field.

9.1 Stratifying Families of Polynomials

We recall that a family of polynomials $\mathcal{F}$ is said to be *stable under derivation with respect to the variable T* if, for every $f \in \mathcal{F}$, either $\frac{\partial f}{\partial T} = 0$, or $\frac{\partial f}{\partial T} \in \mathcal{F}$. We shall say that a polynomial $f \in R[X_1, \ldots, X_i]$ is *quasi-monic* with respect to X_i if

$$f = a_d X_i^d + g_{d-1}(X_1, \ldots, X_{i-1}) X_i^{d-1} + \cdots + g_0(X_1, \ldots, X_{i-1}),$$

where a_d is a nonzero element of R.

Definition 9.1.1. *A stratifying family of polynomials is a family of nonzero polynomials $(f_{i,j})_{i=1,\ldots,n; j=1,\ldots,l_i}$ with coefficients in R such that:*

(i) For i fixed, the family $(f_{i,j})_{j=1,\ldots,\ell_i}$ is a family of polynomials in $R[X_1, \ldots, X_i]$ which are quasi-monic with respect to the variable X_i. Moreover, this family is stable under derivation with respect to the variable X_i.

(ii) For $k > 1$ fixed, the family $(f_{i,j})_{i<k, j=1,\ldots,\ell_i}$ slices the family $(f_{k,j})_{j=1,\ldots,\ell_k}$ (cf. Definition 2.3.4).

Every family of polynomials can be completed, after a linear change of variables, to a stratifying family.

Proposition 9.1.2. *Let $g_1, \ldots, g_s$ be nonzero polynomials in $R[T_1, \ldots, T_n]$. There exist a linear automorphism $u : R^n \to R^n$ and a stratifying family $(f_{i,j})_{i=1,\ldots,n; j=1,\ldots,\ell_i}$, such that $f_{n,j}(X) = g_j(u(X))$ for $j = 1, \ldots, s$.*

Proof. The proof is by induction on n. If $n = 1$, we enlarge the family of polynomials $g_1, \ldots, g_s$ by adding all their nonzero derivatives of all orders and we take for $u : R \to R$ the identity mapping. Given $n > 1$, we assume the result holds true for $n - 1$. We can choose a linear change of variables $v : R^n \to R^n$ of the form

$$v(y_1, \ldots, y_n) = (y_1 + a_1 y_n, \ldots, y_{n-1} + a_{n-1} y_n, y_n)$$

such that all the polynomials $h_j(Y) = g_j(v(Y))$, for $j = 1, \ldots, s$, are quasi-monic with respect to Y_n. Then we add to the family of polynomials $h_1, \ldots, h_s$ all their nonzero derivatives of all orders with respect to the last variable Y_n. We obtain a family $h_1, \ldots, h_{\ell_n}$ of polynomials all of which are quasi-monic with respect to Y_n. By Theorem 2.3.1, there is a family of polynomials $p_1, \ldots, p_t \in R[Y_1, \ldots, Y_{n-1}]$ slicing the family $h_1, \ldots, h_{\ell_n}$. Applying the inductive assumption to this family $p_1, \ldots, p_t$, we find a linear automorphism $u' : R^{n-1} \to R^{n-1}$ and a stratifying family $(f_{i,j})_{i=1,\ldots,n-1; j=1,\ldots,\ell_i}$ such that $f_{n-1,j}(X_1, \ldots, X_{n-1}) = p_j(u'(X_1, \ldots, X_{n-1}))$, for $j = 1, \ldots, t$. Take $u = v \circ (u' \times \mathrm{Id}_R) : R^n \to R^n$, and $f_{n,j}(X) = h_j(u'(X_1, \ldots, X_{n-1}), X_n)$ for $j = 1, \ldots, \ell_n$. The family $(f_{i,j})_{i=1,\ldots,n; j=1,\ldots,\ell_i}$ is stratifying, and $f_{n,j}(X) = g_j(u(X))$ for $j = 1, \ldots, s$. □

We shall now analyze the partition of affine space determined by a stratifying family of polynomials.

Notation 9.1.3. *Let $(f_{i,j})_{i=1,\ldots,n; j=1,\ldots,\ell_i}$ be a stratifying family of polynomials. Let k be an integer, $1 \leq k \leq n$. We denote by $\mathcal{C}_k$ the family of nonempty semi-algebraic subsets of R^k of the form*

$$C = \bigcap_{i=1}^{k} \bigcap_{j=1}^{\ell_i} \{x \in R^k \mid \mathrm{sign}(f_{i,j}(x)) = \epsilon(i,j)\},$$

where ϵ is a mapping from the set $\{(i,j) \mid 1 \leq i \leq k, 1 \leq j \leq \ell_i\}$ to the set $\{-1, 0, +1\}$. Let $\mathcal{A}_k$ be the subfamily of those $C \in \mathcal{C}_k$ on which some $f_{k,j}$ vanishes. Let $\mathcal{B}_k = \mathcal{C}_k \setminus \mathcal{A}_k$. For $k > 1$, we denote by $\Pi_k : R^k \to R^{k-1}$ the projection onto the space of the first $k - 1$ coordinates.

The family $\mathcal{C}_k$ is a finite semi-algebraic partition of R^k.

Theorem 9.1.4. *Let $\mathcal{F} = (f_{i,j})_{i=1,\ldots,n; j=1,\ldots,\ell_i}$ be a stratifying family of polynomials.*

(i) For every k, $1 \leq k \leq n$, every element of $\mathcal{C}_k$ is a semi-algebraically connected Nash submanifold of R^k.

(ii) For every k, $1 < k \leq n$, and every $C \in \mathcal{C}_k$, $\Pi_k(C) \in \mathcal{C}_{k-1}$.

9.1 Stratifying Families of Polynomials   209

(iii) *For every k, $1 < k \leq n$, if $A \in \mathcal{A}_k$, A is the graph of a Nash function $\xi_A : \Pi_k(A) \to R$.*

(iv) *For every k, $1 < k \leq n$ and every $A \in \mathcal{A}_k$, ξ_A can be extended to a continuous semi-algebraic function $\overline{\xi}_A : \mathrm{clos}(\Pi_k(A)) \to R$. If $C \in \mathcal{C}_k$ and $C \subset \mathrm{clos}(A)$, then $C \in \mathcal{A}_k$ and $\xi_C = \overline{\xi}_A|_{\Pi_k(C)}$.*

(v) *For every k, $1 < k \leq n$ and every $C \in \mathcal{A}_k$, if $C' \in \mathcal{C}_{k-1}$ and $\mathrm{clos}(C') \supset \Pi_k(C)$, then there exists $A \in \mathcal{A}_k$ such that $\Pi_k(A) = C'$ and $\mathrm{clos}(A) \supset C$.*

(vi) *For every k, $1 < k \leq n$, if $B \in \mathcal{B}_k$, there exists a Nash diffeomorphism $\theta_B : \Pi_k(B) \times]0,1[\to B$ such that, for every $(x,t) \in \Pi_k(B) \times]0,1[$, $\Pi_k(\theta_B(x,t)) = x$. Moreover, if $C' \in \mathcal{C}_{k-1}$ is such that $\Pi_k(B) \subset \mathrm{clos}(C')$, then there exists a unique $C \in \mathcal{C}_k$ such that $\Pi_k(C) = C'$ and $B \subset \mathrm{clos}(C)$, and the mapping*
$$(C' \cup \Pi_k(B)) \times]0,1[\longrightarrow C \cup B$$
which coincides with θ_C on $C' \times]0,1[$ and θ_B on $\Pi_k(B) \times]0,1[$, is a semi-algebraic homeomorphism.

(vii) *For every k, $1 \leq k \leq n$, every $C \in \mathcal{C}_k$ is the image of a Nash embedding $\varphi_C :]0,1[^{\dim(C)} \to R^k$ (with $]0,1[^0 =$ a point). For $k > 1$, the embedding φ_C satisfies*
$$\varphi_C = (\mathrm{Id}_{R^{k-1}}, \xi_C) \circ \varphi_{\Pi_k(C)} \text{ and } \dim(C) = \dim(\Pi_k(C))$$
if $C \in \mathcal{A}_k$, and
$$\varphi_C = \theta_C \circ (\varphi_{\Pi_k(C)} \times \mathrm{Id}_{]0,1[}) \text{ and } \dim(C) = \dim(\Pi_k(C)) + 1$$
if $C \in \mathcal{B}_k$.

(viii) *For every k, $1 \leq k \leq n$ and every $C \in \mathcal{C}_k$, the closure of C in R^k is obtained by relaxing the inequalities defining C, i.e. if*
$$C = \bigcap_{i=1}^{k} \bigcap_{j=1}^{\ell_i} \{x \in R^k \mid \mathrm{sign}(f_{i,j}(x)) = \epsilon(i,j)\},$$
then
$$\mathrm{clos}(C) = \bigcap_{i=1}^{k} \bigcap_{j=1}^{\ell_i} \{x \in R^k \mid \mathrm{sign}(f_{i,j}(x)) \in \overline{\epsilon(i,j)}\},$$
where $\overline{0} = \{0\}$, $\overline{-1} = \{-1, 0\}$ and $\overline{+1} = \{0, +1\}$. In particular, the closure of C is the union of some elements of $\mathcal{C}_k$.

Proof. The proof is by induction on k. By Thom's lemma 2.5.4, the elements of $\mathcal{C}_1$ are either points or open intervals, and the theorem for $k = 1$ holds true.

Given $k > 1$, assume the theorem holds true for $k - 1$. The slicing (Theorem 2.3.1) gives (ii). By Thom's lemma (with respect to the variable X_k), every $A \in \mathcal{A}_k$ is the graph of a function $\xi_A : \Pi_k(A) \to R$ of the slicing and every $B \in \mathcal{B}_k$ is a slice cut out by the graphs of the functions ξ_A in the cylinder $\Pi_k(B) \times R$. Since ξ_A is a simple root of one of the polynomials $f_{k,j}$, the functions ξ_A are Nash. Properties (i) and (vii) are simply reformulations of 2.9.10. Property (iv) follows immediately from Lemma 2.5.6. Furthermore, if $D' \in \mathcal{C}_{k-1}$ and $D' \subset \text{clos}(\Pi_k(A))$, there exists $D \in \mathcal{A}_k$ such that $\xi_D = \overline{\xi_A}|_{D'}$ and $D = \text{clos}(A) \cap \Pi_k^{-1}(D')$. Property (v) is a consequence of the implicit function theorem and the fact that ξ_C is a simple root of some $f_{k,j}$.

The proof of Lemma 2.3.5 gives explicitly, for $B \in \mathcal{B}_k$, the Nash diffeomorphism θ_B. The second part of property (vi) is a consequence of the definition of θ_B and properties (iv) and (v) of the functions ξ_A.

Now we prove (viii). Define

$$E = \bigcap_{i=1}^{k} \bigcap_{j=1}^{\ell_i} \{x \in R^k \mid \text{sign}(f_{i,j}(x)) \in \overline{\epsilon(i,j)}\}$$

and

$$E' = \bigcap_{i=1}^{k-1} \bigcap_{j=1}^{\ell_i} \{x' \in R^{k-1} \mid \text{sign}(f_{i,j}(x')) \in \overline{\epsilon(i,j)}\}.$$

By the inductive assumption, $E' = \text{clos}(\Pi_k(C))$. It is obvious that $\text{clos}(C) \subset E$. Let x' be a point of E'. If $C \in \mathcal{A}_k$, by Thom's lemma, $\Pi_k^{-1}(x') \cap E$ is either empty or a point, and, hence, it coincides with the point $(x', \overline{\xi}_C(x'))$, which is the intersection of $\Pi_k^{-1}(x')$ with $\text{clos}(C)$. This implies (viii) for $C \in \mathcal{A}_k$. If $C \in \mathcal{B}_k$, then either C is limited by $A \in \mathcal{A}_k$ such that $\Pi_k(A) = \Pi_k(C)$, or $C = \Pi_k(C) \times R$. In both cases $\Pi_k^{-1}(x') \cap \text{clos}(C)$ is nonempty. By Thom's lemma again, $\Pi_k^{-1}(x') \cap E$ is either empty (which is impossible), or a point (coinciding with $\Pi_k^{-1}(x') \cap \text{clos}(C)$), or a closed interval not reduced to a point. In this last case, if (x', x_n) is in the interior of the interval, then $\text{sign}(f_{k,j}(x', x_n)) = \epsilon(k,j) \neq 0$ for $j = 1, \ldots, \ell_n$. Hence $(x'', x_n) \in C$ for every $x'' \in \Pi_k(C)$ close enough to x', so that $(x', x_n) \in \text{clos}(C)$. Therefore $E = \text{clos}(C)$. □

Definition 9.1.5. *Let E be a semi-algebraic subset of R^n given as the union of a subfamily of $\mathcal{C}_n$. The family of the $C \in \mathcal{C}_n$ contained in E is called the stratification of E induced by the stratifying family of polynomials $\mathcal{F}$. If $C \in \mathcal{C}_n$, $C \subset E$ and $\dim(C) = d$, then C is said to be a d-stratum of this stratification.*

Theorem 9.1.6. *Let E be a semi-algebraic subset of R^n of dimension d. There exist a linear automorphism $v : R^n \to R^n$ and a stratifying family of polynomials $\mathcal{F} = (f_{i,j})_{i=1,\ldots,n; j=1,\ldots,\ell_i}$, such that*

  (i) *$v(E)$ admits a stratification induced by the family $\mathcal{F}$.*

  (ii) *If $\Pi_{n,d} : R^n \to R^d$ is the projection onto the space of the first d coordinates, then, for every stratum $C \in \mathcal{C}_n$ contained in $v(E)$, $\Pi_{n,d}|_{\text{clos}(C)}$ is a homeomorphism onto $\text{clos}(\Pi_{n,d}(C))$.*

9.1 Stratifying Families of Polynomials

Moreover, given a finite family $(F_\lambda)_{\lambda \in \Lambda}$ of semi-algebraic subsets of E, we can choose v and the family $\mathcal{F}$ such that every $v(F_\lambda)$ is the union of some strata of the stratification induced by the family $\mathcal{F}$.

Proof. (i) and the claim concerning the F_λ are easy consequences of Proposition 9.1.2. In order to prove (ii), we follow the proof of Proposition 9.1.2. We remark that for $d < k \leq n$, $\Pi_{n,k}(v(E)) \subset R^k$ is contained in the union of the sets of zeros of the polynomials $f_{k,1}, \ldots, f_{k,\ell_k}$. Hence, for every $C \in \mathcal{C}_k$ contained in $\Pi_{n,k}(v(E))$, we have $C \in \mathcal{A}_k$. The result follows by applying Theorem 9.1.4, (iii) and (iv). $\square$

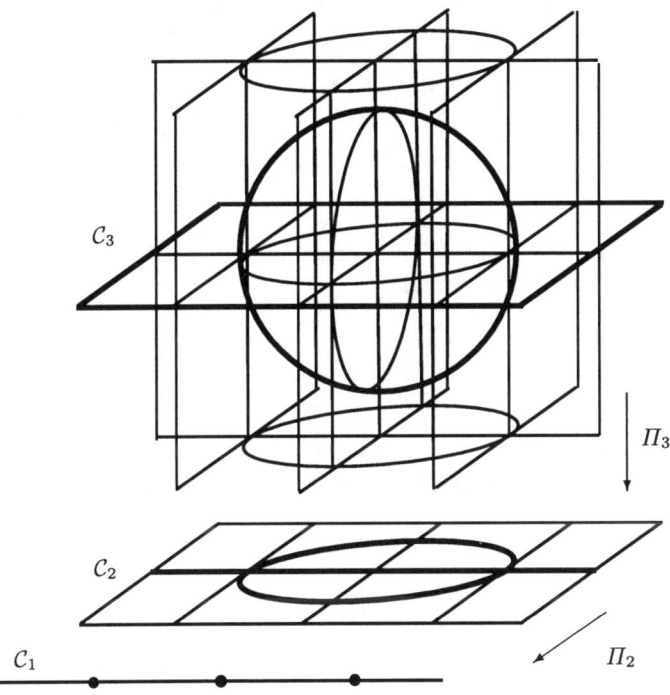

Figure 9.1. Stratification of the sphere induced by a stratifying family of polynomials

Consider the example of a sphere. The family consisting of the unique element $g = X^2 + Y^2 + Z^2 - 1$ can be completed to a stratifying family in the following way:

$$g = f_{3,1} = X^2 + Y^2 + Z^2 - 1 \qquad f_{3,2} = 2Z$$
$$f_{2,1} = X^2 + Y^2 - 1 \qquad f_{2,2} = 2Y$$
$$f_{1,1} = X^2 - 1 \qquad f_{1,2} = 2X$$

The stratifications $\mathcal{C}_3, \mathcal{C}_2, \mathcal{C}_1$ are represented in Fig. 9.1

The stratifications induced by a stratifying family of polynomials are stratifications in the sense of [340], with strata which are Nash manifolds.

Definition 9.1.7. *Let E be a semi-algebraic subset of R^n. A Nash stratification of E is a finite partition $(E_\alpha)_{\alpha \in A}$ of E, where each E_α is a semi-algebraically connected Nash submanifold of R^n, such that, if $E_\alpha \cap \mathrm{clos}(E_\beta) \neq \emptyset$ and $\alpha \neq \beta$, then $E_\alpha \subset \mathrm{clos}(E_\beta)$ and $\dim(E_\alpha) < \dim(E_\beta)$. The E_α are called the strata of the stratification, and, if $d = \dim(E_\alpha)$, then E_α is said to be a d-stratum.*

Note that the condition $\dim(E_\alpha) < \dim(E_\beta)$ is a consequence of $E_\alpha \subset \mathrm{clos}(E_\beta)$ and $E_\alpha \cap E_\beta = \emptyset$, by Proposition 2.8.13.

Proposition 9.1.8. *Let E be a semi-algebraic subset of R^n, and $(F_\lambda)_{\lambda \in \Lambda}$, a finite family of semi-algebraic subsets of E. There exists a Nash stratification $(E_\alpha)_{\alpha \in A}$ of E, such that each F_λ is the union of some strata of this stratification.*

Proof. Follows directly from Theorem 9.1.6. □

We shall need some special properties of the stratifications induced by stratifying families of polynomials.

Proposition 9.1.9. *Let $(f_{i,j})_{i=1,\ldots,n;j=1,\ldots,\ell_i}$ be a stratifying family of polynomials. For every k, $1 \leq k \leq n$, and every d, $0 \leq d \leq k$, the union $W_{k,d}$ of the strata C of $\mathcal{C}_k$, such that $\dim(C) \leq d$, is an algebraic subset of R^k.*

Proof. The proof is by induction on k. The result is obvious for $k = 1$. Given $k > 1$, assume the proposition holds true for $k - 1$. It is obvious that $W_{k,0}$ and $W_{k,k}$ are algebraic. Now consider d such that $0 < d < k$. Let V be the union of all strata $C \in \mathcal{A}_k$. By definition of $\mathcal{A}_k$, V is an algebraic subset of R^k. Since
$$W_{k,d} = \Pi_k^{-1}(W_{k-1,d-1}) \cup \left(V \cap \Pi_k^{-1}(W_{k-1,d})\right),$$
it follows from the inductive assumption that $W_{k,d}$ is algebraic. □

The following proposition will not be used later, but its geometric meaning is quite interesting: the stratification induced by a stratifying family of polynomials is locally trivial. This means that each stratum C has an open neighbourhood semi-algebraically homeomorphic to a product $C \times F$.

Proposition 9.1.10. *Let $(f_{i,j})_{i=1,\ldots,n;j=1,\ldots,\ell_i}$ be a stratifying family of polynomials. Given k, $1 \leq k \leq n$, and $C \in \mathcal{C}_k$, define*
$$U_C = \bigcup \{D \in \mathcal{C}_k \mid C \subset \mathrm{clos}(D)\}.$$

The semi-algebraic set U_C is an open neighbourhood of C in R^k. There exists a semi-algebraic set F, with:
  (i) *a distinguished point $x_C \in F$,*

(ii) *for each $D \in \mathcal{C}_k$ such that $D \subset U_C$ and $D \neq C$, a semi-algebraic subset $F_D \subset F$,*

(iii) *a semi-algebraic homeomorphism $\rho : C \times F \to U_C$ such that $\rho(C \times \{x_C\}) = C$ and $\rho(C \times F_D) = D$.*

Proof. The proof is by induction on k. If $k = 1$, then C is either a point or an open interval. In the first case, U_C is the union of the point and two open intervals containing C in their closures, and $F = U_C$. In the second case, $U_C = C$ and $F = \{x_C\}$.

Given $k > 1$, assume Proposition 9.1.10 holds true for $k - 1$. Set $C' = \Pi_k(C)$. By the inductive assumption, we have F', $x'_{C'} \in F'$, $F'_{D'} \subset F'$ for $D' \in \mathcal{C}_{k-1}$ such that $D' \neq C'$ and $\text{clos}(D') \supset C'$, and $\rho' : C' \times F' \to U_{C'}$. Note that $\Pi_k(U_C) = U_{C'}$. We consider two cases.

a) $C \in \mathcal{B}_k$. We first prove that, if $D \subset U_C$, then $D \in \mathcal{B}_k$. Suppose, for instance, that C is bounded in $\Pi_k^{-1}(C')$ by two strata A_1 and A_2 in $\mathcal{A}_k$, $\xi_{A_1} < \xi_{A_2}$. Let $D' \in \mathcal{C}_{k-1}$ such that $D' \subset U_{C'}$ and $D' \neq C'$. Let E_1 (resp. E_2) be the highest (resp. the lowest) stratum in $\mathcal{A}_k$ contained in $\Pi_k^{-1}(D')$, such that $A_1 \subset \text{clos}(E_1)$ (resp. $A_2 \subset \text{clos}(E_2)$). Then $U_C \cap \Pi_k^{-1}(D')$ is the stratum of $\mathcal{B}_k$ lying between E_1 and E_2. Hence, there is a semi-algebraic homeomorphism $\theta : U_{C'} \times]0, 1[\to U_C$ such that, for every $D \subset U_C$, $\theta|_{\Pi_k(D) \times]0,1[} = \theta_D$ (with the notation of Theorem 9.1.4 (vi)). Take $F = F'$, $x_C = x'_{C'}$, $F_D = F'_{\Pi_k(D)}$ and

$$\rho : C \times F \xrightarrow{\alpha} (C' \times]0, 1[) \times F \simeq (C' \times F') \times]0, 1[\xrightarrow{\beta} U_{C'} \times]0, 1[\xrightarrow{\theta} U_C ,$$

where $\alpha = \theta_C^{-1} \times \text{Id}_F$ and $\beta = \rho' \times \text{Id}_{]0,1[}$.

b) $C \in \mathcal{A}_k$. We may assume that $x'_{C'} \in C'$, $F' \subset U_{C'}$ and, for every y' in F', $\rho'(x'_{C'}, y') = y'$. Define $F = \Pi_k^{-1}(F') \cap U_C$, $x_C = (x'_{C'}, \xi_C(x'_{C'}))$, $F_D = F \cap D$, for $D \subset U_C$ such that $D \neq C$. We now construct $\rho : C \times F \to U_C$. Let $(x, y) \in C \times F$ and $x' = \Pi_k(x) \in C'$, $y' = \Pi_k(y) \in F'$. If $y = x_C$, then $\rho(x, x_C) = x$. If $y \in D \in \mathcal{A}_k$, $D \neq C$, then $\rho(x, y) = (\rho'(x', y'), \xi_D(\rho'(x', y')))$. If $y \in D \in \mathcal{B}_k$, then $\rho(x, y) = \theta_D(\rho'(x', y'), t)$, where $(y', t) = \theta_D^{-1}(y)$. The mapping ρ satisfies the properties of the proposition. □

Now we show that the partition of the space induced by a stratifying family of polynomials defines a semi-algebraic CW complex. The concept of a topological CW complex is defined in [236], Definition 6.1.

Definition 9.1.11. *Let S be a closed and bounded semi-algebraic set. A semi-algebraic cellular decomposition of S is a finite partition of S into semi-algebraic sets $C_1, \ldots, C_p$, together with, for each $i = 1, \ldots, p$, a continuous semi-algebraic mapping $f_i : \overline{B}_{\dim(C_i)} \to S$ from a closed ball to S, such that*

(i) $f_i|_{B_{\dim(C_i)}}$ *is a homeomorphism onto C_i,*

(ii) *if $x \in \text{clos}(C_i) \setminus C_i$, then $x \in C_j$ with $\dim(C_j) < \dim(C_i)$.*

The C_i are called the cells *of the cellular decomposition, and f_i, the* characteristic mapping *of the cell C_i.*

Proposition 9.1.12. *Let S be a closed and bounded semi-algebraic set. Then S admits a semi-algebraic cellular decomposition such that the closure of each cell is the union of cells. Moreover, given a finite family $(S_j)_{j=1,\ldots,k}$ of semi-algebraic subsets of S, one can choose the cellular decomposition such that each S_j is the union of cells.*

Proof. By Proposition 9.1.2, we may assume that S and the S_j are given by boolean combinations of sign conditions on polynomials (in $R[X_1,\ldots,X_n]$) of a stratifying family. The cells of the semi-algebraic cellular decomposition of S are the strata C of $\mathcal{C}_n$ (with Notation 9.1.3) contained in S. By Theorem 9.1.4 (viii), condition (ii) of Definition 9.1.11 is satisfied, which also shows that the closure of a cell is a union of cells. By Theorem 9.1.4 (vii), for each cell C, there exists a semi-algebraic homeomorphism $\varphi_C :]0,1[^{\dim(C)} \to C$. We use the following lemma to define a characteristic mapping of C.

Lemma 9.1.13. *For every k, $1 \leq k \leq n$, if $C \in \mathcal{C}_k$ is bounded, the homeomorphism $\varphi_C :]0,1[^{\dim(C)} \to C$ can be extended to a continuous semi-algebraic mapping $\overline{\varphi}_C : [0,1]^{\dim(C)} \to \mathrm{clos}(C)$.*

Proof. The proof is by induction on k. The result is obvious for $k = 1$, since C is either a point or a bounded open interval. Given $k > 1$, assume that the lemma holds true for $k - 1$. If $C \in \mathcal{A}_k$, take

$$\overline{\varphi}_C = (\mathrm{Id}_{R^{k-1}}, \overline{\xi}_C) \circ \overline{\varphi}_{\Pi_k(C)}.$$

If $C \in \mathcal{B}_k$, then, since C is bounded, C is a slice lying between two strata A_1 and A_2 of $\mathcal{A}_k$, with $\Pi_k(C) = \Pi_k(A_1) = \Pi_k(A_2)$. We take

$$\overline{\varphi}_C : [0,1]^{\dim(C)-1} \times [0,1] \longrightarrow \mathrm{clos}(C)$$
$$(t, u) \longmapsto (x, (1-u)\overline{\xi}_{A_1}(x) + u\overline{\xi}_{A_2}(x)),$$

where $x = \overline{\varphi}_{\Pi_k(C)}(t)$. It is easy to verify that, in both cases, $\overline{\varphi}_C|_{]0,1[^{\dim(C)}} = \varphi_C$. □

Now we can complete the proof of Proposition 9.1.12. Let $h_d : \overline{B}_d \to [0,1]^d$ be a semi-algebraic homeomorphism. If $d = \dim(C)$, we take $\overline{\varphi}_C \circ h_d$ as the characteristic mapping of the cell C. □

The next proposition describes the conic structure of stratifications induced by stratifying families of polynomials,. It will be used in Chap. 11 to compute the local Euler-Poincaré characteristic.

Proposition 9.1.14. *Let $\{a\}$ be a 0-stratum of $\mathcal{C}_k$, and let*

$$U_a = \bigcup \{C \in \mathcal{C}_k \mid a \in \mathrm{clos}(C)\}.$$

Then:

9.1 Stratifying Families of Polynomials   215

(i) *There exists a closed and bounded semi-algebraic set $G \subset U_a$ and a continuous semi-algebraic surjection $\eta : G \times [0,1[\to U_a$, such that $\eta(z,0) = a$ and $\eta(z, 1/2) = z$ for every z in G, $\eta|_{G \times]0,1[}$ is a semi-algebraic homeomorphism onto $U_a \setminus \{a\}$ and $\eta((G \cap C) \times]0, 1[) = C$ for each $C \in \mathcal{C}_k$ such that $C \subset U_a$ and $C \neq \{a\}$.*

(ii) *The sets $G \cap C$, as above, are the cells of a semi-algebraic cellular decomposition of G, and $\dim(G \cap C) = \dim(C) - 1$.*

Proof. The proof is by induction on k. If $k = 1$, U_a is the union of the point a and two open intervals on each side of a. The set G consists of two points, one in each open interval, and η is easily constructed.

Given $k > 1$, assume the proposition holds true for $k - 1$. Let $a' = \Pi_k(a) \in R^{k-1}$. By the inductive assumption, we have $U'_{a'} \subset R^{k-1}$, $G' \subset U'_{a'}$ and $\eta' : G' \times [0,1[\to U'_{a'}$ satisfying the properties of the proposition. We shall construct, for each stratum $C \neq \{a\}$ of $\mathcal{C}_k$ contained in U_a, a semi-algebraic set $G_C \subset C$ and a semi-algebraic homeomorphism $\eta_C : G_C \times]0, 1[\to C$. The set G and the mapping η will be obtained by gluing all these G_C and η_C together. We use the notations of Theorem 9.1.4. We consider several cases:

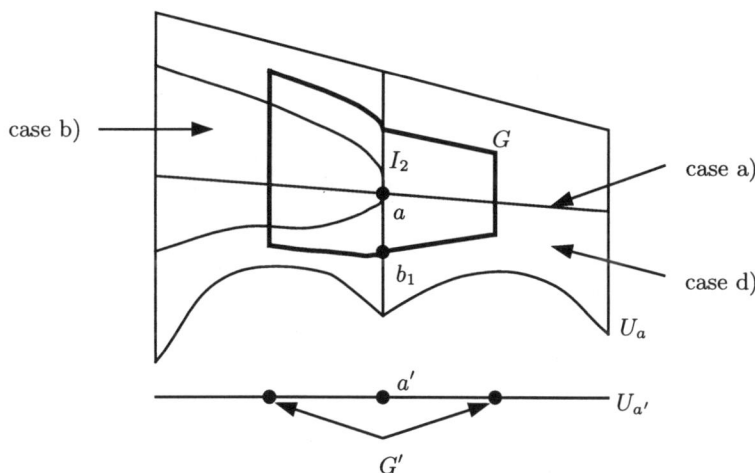

Figure 9.2. Construction of G

a) The stratum C belongs to $\mathcal{A}_k$. One takes

$$G_C = \{(x', \xi_C(x')) \mid x' \in G' \cap \Pi_k(C)\}$$

and η_C defined by

$$\eta_C((x', \xi_C(x')), t) = (\eta'(x', t), \xi_C(\eta'(x', t))) \ .$$

b) The stratum C belongs to $\mathcal{B}_k$, $\Pi_k(C) \neq \{a'\}$ and $\text{clos}(C) \cap (\{a'\} \times R) = \{a\}$. Then C is bounded by two strata A_1 and A_2 of $\mathcal{A}_k$ contained in U_a, with $\Pi_k(A_1) = \Pi_k(A_2) = \Pi_k(C)$. One takes

$$G_C = \{\theta_C(x', u) \mid x' \in G' \cap \Pi_k(C) \text{ and } u \in]0,1[\}$$

and η_C defined by

$$\eta_C(\theta_C(x', u), t) = \theta_C(\eta'(x', t), u) .$$

c) The stratum C is the open interval I_1 (resp. I_2) of $\{a'\} \times R$ lying exactly below (resp. above) a. One takes

$$G_{I_1} = \{b_1\} = \{\theta_{I_1}(a', 1/2)\} \qquad (\text{resp. } G_{I_2} = \{b_2\} = \{\theta_{I_2}(a', 1/2)\})$$

and η_{I_1} (resp. η_{I_2}) defined by

$$\eta_{I_1}(b_1, t) = \theta_{I_1}(a', 1-t) \qquad (\text{resp. } \eta_{I_2}(b_2, t) = \theta_{I_2}(a', t)) .$$

d) The stratum C belongs to $\mathcal{B}_k$, $\Pi_k(C) \neq \{a'\}$ and $\text{clos}(C) \cap (\{a'\} \times R)$ contains I_1 (resp. I_2). If $I_1 \subset \text{clos}(C)$, take

$$G_C = \{\theta_C(\eta'(x', u), v) \mid x' \in G' \cap \Pi_k(C) ,$$
$$(0 < u \leq 1/2 \text{ and } v = 1/2) \text{ or } (u = 1/2 \text{ and } 1/2 \leq v < 1)\}$$

and η_C defined by

$$\eta_C(\theta_C(\eta'(x', u), v), t) = \theta_C(\eta'(x', 2tu), 1 - 2t(1-v)) .$$

The case where $I_2 \subset \text{clos}(C)$ is left to the reader.

This construction is better understood by looking at Figure 9.2. The set G is the union of the G_C, and we define $\eta|_{G \times]0,1[}$ by gluing the η_C together. It is easy to verify that properties (i) and (ii) are satisfied, although the details are tedious. $\square$

9.2 Triangulation of Semi-algebraic Sets

This section is devoted to the proof of the fact that a closed and bounded semi-algebraic set can be triangulated. We first fix some notation. Let $a_0, \ldots, a_k$ be $k+1$ points affinely independent in R^n, which means that the affine subspace they span is of dimension k. The k-*simplex* $[a_0, \ldots, a_k]$ is the set of $x \in R^n$ such that there exist nonnegative $\lambda_0, \ldots, \lambda_k \in R$ with $\sum_{i=0}^{k} \lambda_i = 1$ and $x = \sum_{i=0}^{k} \lambda_i a_i$. The numbers $\lambda_0, \ldots, \lambda_k$ are the barycentric coordinates of x. If $\{a_{i_0}, \ldots, a_{i_\ell}\}$ is a nonempty subset of $\{a_0, \ldots, a_k\}$, then the ℓ-simplex $[a_{i_0}, \ldots, a_{i_\ell}]$ is called a *face* (an ℓ-*face* if one needs to specify the dimension) of $[a_0, \ldots, a_k]$. If σ is a simplex, then σ^0 is the subset of points of

σ whose barycentric coordinates are all positive (i.e. the points not belonging to any proper face of σ). We call σ^0 an *open simplex*. A *finite simplicial complex* of R^n is a finite collection of simplices $K = (\sigma_i)_{i=1,\ldots,p}$ such that the faces of every σ_i belong to K and such that, for every i and j between 1 and p, $\sigma_i \cap \sigma_j$ is either empty, or a common face of σ_i and σ_j. The realization of the complex is $|K| = \bigcup_{i=1}^{p} \sigma_i$. The open simplices σ_i^0 form a partition of $|K|$.

Theorem 9.2.1 (Triangulation). *Every closed and bounded semi-algebraic set $S \subset R^n$ is semi-algebraically triangulable, i.e. there exists a finite simplicial complex $K = (\sigma_i)_{i=1,\ldots,p}$ and a semi-algebraic homeomorphism $\Phi : |K| \to S$. Moreover, given a finite family $(S_j)_{j=1,\ldots,q}$ of semi-algebraic subsets of S, we can choose a finite simplicial complex $K = (\sigma_i)_{i=1,\ldots,p}$ in R^n and a semi-algebraic triangulation $\Phi : |K| \to S$, such that every S_j is the union of some $\Phi(\sigma_i^0)$.*

Proof. The proof is by induction on n. The triangulation is obvious for $n = 1$.

Given $n > 1$, assume that every closed and bounded semi-algebraic subset of R^{n-1} is semi-algebrically triangulable. By Proposition 9.1.2, we may assume that S and the S_j are given by boolean combinations of sign conditions on polynomials of a stratifying family. Let $\Pi_n : R^n \to R^{n-1}$ be the projection onto the space of the first $n-1$ coordinates and $T = \Pi_n(S)$. We can apply the inductive assumption to the closed and bounded semi-algebraic set $T \subset R^{n-1}$ and the family of strata of $\mathcal{C}_{n-1}$ contained in T. We obtain a finite simplicial complex $L = (\tau_k)_{k=1,\ldots,r}$ in R^{n-1} and a semi-algebraic homeomorphism $\Psi : |L| \to T$ such that every stratum of $\mathcal{C}_{n-1}$ contained in T is the union of some $\Psi(\tau_k^0)$.

The inductive step is a consequence of the following technical lemma. This lemma contains more information than is needed for the proof of Theorem 9.2.1. However, it will also be used for the proof of the triangulability of semi-algebraic functions.

Lemma 9.2.2. *Let $S \subset R^n$ (with $n > 1$) be a closed and bounded semi-algebraic set, $(S_j)_{j=1,\ldots,q}$ a finite family of semi-algebraic subsets of S. Let $\Pi_n : R^n \to R^{n-1}$ be the projection onto the space of the first $n-1$ coordinates and $T = \Pi_n(S)$. Assume that there exist a finite simplicial complex $L = (\tau_k)_{k=1,\ldots,r}$ in R^{n-1}, a semi-algebraic triangulation $\Psi : |L| \to T$ and, for every $k = 1, \ldots, r$, a finite collection of continuous semi-algebraic functions*

$$\xi_{k,1} < \ldots < \xi_{k,\ell_k} : \Psi(\tau_k^0) \to R,$$

such that:

i) *S and every S_j are the unions of graphs of some $\xi_{i,j}$ and some slices of cylinders $\Psi(\tau_k^0) \times R$ bounded by the graphs of consecutive functions $\xi_{k,j}$ and $\xi_{k,j+1}$.*

ii) *Every function $\xi_{k,j}$ can be continuously extended to a function $\overline{\xi}_{k,j} : \Psi(\tau_k) \to R$ and, if $\tau_{k'}$ is a face of τ_k, then $\overline{\xi}_{k,j}|_{\Psi(\tau_{k'}^0)}$ is equal to some function $\xi_{k',j'}$.*

Then there exist a finite simplicial complex $K = (\sigma_i)_{i=1,\ldots,p}$ in R^n such that $\Pi_n(\sigma_i)$ is a simplex of the barycentric subdivision L' of L for $i = 1,\ldots,p$, and a semi-algebraic homeomorphism $\Phi : |K| \to S$ such that every S_j is the union of some $\Phi(\sigma_i^0)$ and $\Pi_n \circ \Phi = \Psi \circ \Pi_n$.

Proof. Denote by $b(\tau)$ the barycentre of a simplex τ of L. Recall that L' is the collection of all simplices $[b(\tau_{k_1}),\ldots,b(\tau_{k_m})]$, where $\tau_{k_{i+1}}$ is a proper face of τ_{k_i}.

If C is the graph of a function $\xi_{k,j} : \tau_k \to R$ (we call C an $\mathcal{A}$-cell), we set $b(C) = (b(\tau_k), j) \in R^n$. If C is the slice of a cylinder $\Psi(\tau_k^0) \times R$ bounded by the graphs of consecutive functions $\xi_{k,j}$ and $\xi_{k,j+1}$ (we call C a $\mathcal{B}$-cell), we set $b(C) = (b(\tau_k), j + \frac{1}{2}) \in R^n$. Define the simplicial complex K as the collection of the simplices $[b(C_0), b(C_1),\ldots,b(C_p)]$ for every sequence $(C_0,\ldots,C_p)$ of cells contained in S such that $C_i \subset \mathrm{clos}(C_{i+1})$, for $i = 0,\ldots,p-1$. By definition, the image by Π_n of every simplex of K is a simplex of L'.

If C is a cell contained in S, we denote by $P(\mathrm{clos}(C))$ the polytope which is the union of all simplices $[b(C_0), b(C_1),\ldots,b(C_p)]$ of K such that $C_p \subset \mathrm{clos}(C)$. Assume $\Pi_n(C) = \Psi(\tau_k^0)$. If C is an $\mathcal{A}$-cell, then $P(\mathrm{clos}(C))$ is the graph of a continuous function $\omega_C : \tau_k \to R$, which is affine on each simplex of the subdivision of τ_k. If C is a $\mathcal{B}$-cell bounded in the cylinder $\Psi(\tau_k^0) \times R$ by the $\mathcal{A}$-cells C_- and C_+, define $h_C : \tau_k \to R$ by $h_C(x) = \omega_{C_+}(x) - \omega_{C_-}(x)$.

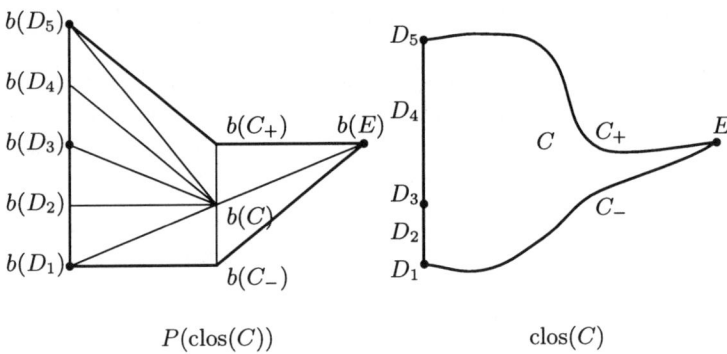

Figure 9.3. Construction of K

Now we define a semi-algebraic homeomorphism $\Phi : |K| \to S$ such that $\Pi_n \circ \Phi = \Psi \circ \Pi_n$ and $\Phi(P(\mathrm{clos}(C))) = \mathrm{clos}(C)$, for every cell $C \subset S$. We proceed by induction on the dimension of the simplex τ_k such that $\Pi_n(C) = \Psi(\tau_k^0)$. If C is an $\mathcal{A}$-cell which is the graph of $\xi_{k,j} : \Psi(\tau_k^0) \to R$, then, for every $(x, \omega_C(x)) \in P(\mathrm{clos}(C))$, we set $\Phi(x, \omega_C(x)) = (\Psi(x), \bar{\xi}_{k,j}(\Psi(x)))$. Now we consider the case where C is a $\mathcal{B}$-cell bounded in $\Psi(\tau_k^0) \times R$ by the $\mathcal{A}$-cells C_- and C_+. The cells C_- and C_+ are the graphs of $\xi_- : \tau_k^0 \to R$ and $\xi_+ : \tau_k^0 \to R$, with $\xi_- < \xi_+$. By the inductive assumption, Φ is already

9.2 Triangulation of Semi-algebraic Sets

defined on $W(\text{clos}(C)) = P(\text{clos}(C)) \cap \Pi_n^{-1}(\tau_k \setminus \tau_k^0)$. We have defined Φ on $P(\text{clos}(C_-))$ and $P(\text{clos}(C_+))$. The polytope $P(\text{clos}(C))$ is the cone with vertex $b(C)$ and base $W(\text{clos}(C)) \cup P(\text{clos}(C_-)) \cup P(\text{clos}(C_+))$. Hence every point $x = (\Pi_n(x), x_n) \in P(\text{clos}(C))$ can be represented as $x = (1-r) b(C) + ry$, where $0 \leq r \leq 1$ and $y \in W(\text{clos}(C)) \cup P(\text{clos}(C_-)) \cup P(\text{clos}(C_+))$. This representation is unique if $x \neq b(C)$.

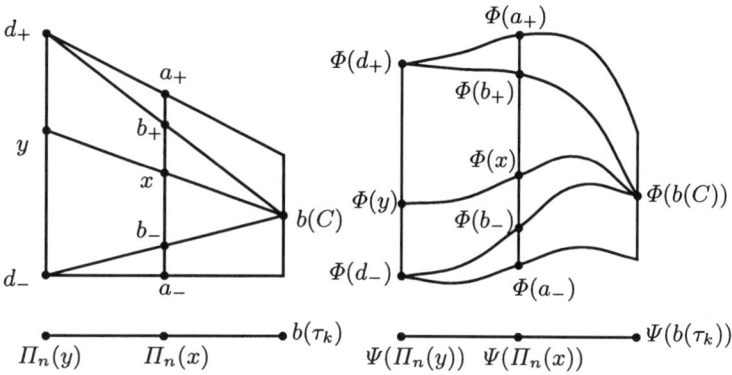

Figure 9.4. Construction of Φ

First we consider the case where $y \in P(\text{clos}(C_-)) \cup P(\text{clos}(C_+))$. This corresponds to the case where x belongs to the segment $[a_-, b_-]$ or $[b_+, a_+]$ (see Figure 9.4). Then we define $\Phi(x)$ as the image of x by the affine mapping of the segment $[a_-, a_+]$ which sends a_- to $\Phi(a_-)$ and a_+ to $\Phi(a_+)$. Explicitly, $\Phi(x) = (\Psi(\Pi_n(x)), t)$, where $t = \overline{\xi}_+(\Psi(\Pi_n(x))) = \overline{\xi}_-(\Psi(\Pi_n(x)))$ if $h_C(\Pi_n(x)) = 0$, and

$$t = \frac{x_n - \omega_{C_-}(\Pi_n(x))}{h_C(\Pi_n(x))} \overline{\xi}_+(\Psi(\Pi_n(x))) + \frac{\omega_{C_+}(\Pi_n(x)) - x_n}{h_C(\Pi_n(x))} \overline{\xi}_-(\Psi(\Pi_n(x)))$$

if $h_C(\Pi_n(x)) \neq 0$. Finally, we consider the case where $y \in W(\text{clos}(C))$. This corresponds to the case where x belongs to the segment $[b_-, b_+]$ (see Figure 9.4). The images $\Phi(b_-)$ and $\Phi(b_+)$ have already been defined above. We take as $\Phi(x)$ the point which divides the segment $[\Phi(b_-), \Phi(b_+)]$ in the same way as $\Phi(y)$ divides the segment $[\Phi(d_-), \Phi(d_+)]$ (see Figure 9.4). If $h_C(\Pi_n(y)) = 0$ (note that in this case $x_n = \frac{1}{2}(\omega_{C_+}(\Pi_n(x)) + \omega_{C_-}(\Pi_n(x)))$), we set

$$\Phi(x) = \left(\Psi(\Pi_n(x)), \frac{1}{2}\big(\xi_+(\Psi(\Pi_n(x))) + \xi_-(\Psi(\Pi_n(x))) \big) \right).$$

If $h_C(\Pi_n(y)) \neq 0$, then $\overline{\xi}_+(\Psi(\Pi_n(y))) \neq \overline{\xi}_-(\Psi(\Pi_n(y)))$, and we define $s \in [0,1]$ by $u_n = s\overline{\xi}_+(\Psi(\Pi_n(y))) + (1-s)\overline{\xi}_-(\Psi(\Pi_n(y)))$, where $\Phi(y) = (\Psi(\Pi_n(y)), u_n)$. We set $\Phi(x) = (\Psi(\Pi_n(x)), t)$, where

$$t = \frac{r\,s\,h_C(\Pi_n(y)) + \frac{1-r}{2} h_C(\Pi_n(b(C)))}{h_C(\Pi_n(y))} \bar{\xi}_+(\Psi(\Pi_n(x)))$$
$$+ \frac{r(1-s)h_C(\Pi_n(y)) + \frac{1-r}{2} h_C(\Pi_n(b(C)))}{h_C(\Pi_n(y))} \bar{\xi}_-(\Psi(\Pi_n(x))).$$

From the geometric description of Φ, it should be clear that $\Phi : |K| \to S$ is a semi-algebraic homeomorphism. □ □

Remark 9.2.3. a) The triangulation $\Phi : |K| \to S$ constructed in the previous proof satisfies, in addition, the following property: the restriction of Φ to each σ_i^0, for σ_i a simplex of K, is a Nash embedding.

b) We can choose the complex K such that its vertices have coordinates in $\mathbb{Q}$. Indeed, this is obvious in the case $n = 1$, and the case $n > 1$ follows from the method of construction of K in Lemma 9.2.2.

The following version of Lemma 2.5.6 will be needed in Sections 9.3 and 9.4.

Lemma 9.2.4. *Let τ be a simplex in R^d, $\Psi : \tau \to R^n$ a semi-algebraic mapping which induces a homeomorphism of τ onto $\Psi(\tau)$. Let $f(X,Y)$ be a polynomial with coefficients in R in the variables $X = (X_1, \ldots, X_n)$ and Y. Let $\xi : \Psi(\tau^0) \to R$ be a bounded continuous semi-algebraic function such that $f(\Psi(x), \xi(\Psi(x))) = 0$ for every $x \in \tau^0$. Assume that, for every $x \in \tau$, $f(\Psi(x), Y)$ is a nonzero polynomial in $R[Y]$. Then ξ extends continuously to $\Psi(\tau)$.*

Proof. Let a be a point in $\tau \setminus \tau^0$. The polynomial $f(\Psi(a), Y)$ has finitely many roots $z_1, \ldots, z_r$ in R. Choose $\epsilon > 0$ in R so small that the intervals $]z_i - \epsilon, z_i + \epsilon[$ are disjoint. For $i = 1, \ldots, r$, define
$$A_i = \{x \in \tau^0 \mid |\xi(\Psi(x)) - z_i| < \epsilon\}.$$

We claim that there is $\delta > 0$ in R such that the intersection of the ball $B(a, \delta)$ with τ^0 is contained in $\bigcup_{i=1}^r A_i$. Otherwise, by the curve selection lemma 2.5.5, there would be a continuous mapping $\gamma : [0,1] \to \tau$ such that $\gamma(0) = a$, $\gamma(]0,1]) \subset \tau^0$ and $|\xi(\Psi(\gamma(t))) - z_i| \geq \epsilon$, for every $t \in]0,1]$ and every $i = 1, \ldots, r$. Since ξ is bounded, $\xi(\Psi(\gamma(t)))$ has a limit $\ell \in R$, as t tends to 0. This limit necessarily satisfies $f(\Psi(a), \ell) = 0$ and $|\ell - z_i| \geq \epsilon$ for every $i = 1, \ldots, r$, which is impossible.

Since $B(a, \delta) \cap \tau^0$ is semi-algebraically connected and contained in the disjoint union of the open subsets A_i of τ^0, there is an i such that $B(a, \delta) \cap \tau^0 \subset A_i$. Since we can find such a $\delta > 0$ for every ϵ small enough, we conclude that ξ extends continuously to a. □

9.3 Semi-algebraic Triviality of Semi-algebraic Mappings

Definition 9.3.1. *Let S, T and T' be semi-algebraic sets, $T' \subset T$, and let $f : S \to T$ be a continuous semi-algebraic mapping. A semi-algebraic trivialization of f over T', with fibre F, is a semi-algebraic homeomorphism $\theta : T' \times F \to f^{-1}(T')$, such that $f \circ \theta$ is the projection mapping $T' \times F \to T'$. We say that the semi-algebraic trivialization θ is compatible with a subset S' of S if there is a subset F' of F such that $\theta(T' \times F') = S' \cap f^{-1}(T')$.*

Theorem 9.3.2 (Semi-algebraic Triviality). *Let S and T be two semi-algebraic sets, $f : S \to T$ a continuous semi-algebraic mapping, $(S_j)_{j=1,\ldots,q}$ a finite family of semi-algebraic subsets of S. There exist a finite partition of T into semi-algebraic sets $T = \bigcup_{\ell=1}^{r} T_\ell$ and, for each ℓ, a semi-algebraic trivialization $\theta_\ell : T_\ell \times F_\ell \to f^{-1}(T_\ell)$ of f over T_ℓ, compatible with S_j, for $j = 1, \ldots, q$.*

Proof. We may assume that S is a bounded semi-algebraic subset of R^{m+n} and f is the restriction of the projection $\Pi : R^{m+n} \to R^m$ onto the space of the first m coordinates. Indeed, we can make the sets S and T bounded using semi-algebraic homeomorphisms of the form $x \mapsto x/(1 + \|x\|)$, and then replace S with the graph of f. The proof proceeds by induction on the lexicographic ordering of the pairs (m, n) : we assume that the semi-algebraic triviality holds for all projections $R^{m'+n'} \to R^{m'}$ with (m', n') smaller than (m, n) with respect to the lexicographic order.

The sets S and S_j are given by boolean combinations of sign conditions on a finite number of polynomials $f_i(X, Y)$, where $X = (X_1, \ldots, X_m)$ and $Y = (Y_1, \ldots, Y_n)$. Making, if necessary, a linear change of the variables Y, we may assume that each f_i is of the form

$$f_i(X, Y) = Y_n^{d_i} g_{0,i}(X) + Y_n^{d_i-1} g_{1,i}(X, Y') + \cdots + g_{d_i,i}(X, Y'),$$

where $Y' = (Y_1, \ldots, Y_{n-1})$ and $g_{0,i} \neq 0$. The dimension of the semi-algebraic set $A = \{x \in T \mid \prod_i g_{0,i}(x) = 0\}$ is smaller than m. By Theorem 2.3.6, this set can be represented as a finite disjoint union of subsets of the form $\varphi(]0, 1[^k)$, where φ is a semi-algebraic homeomorphism and $k < m$. By considering the pullback

$$\{(u, y) \in]0, 1[^k \times R^n \mid (\varphi(u), y) \in S\},$$

we reduce the study of f over A to the case of the projection $R^{k+n} \to R^k$, to which we can apply the inductive assumption.

Hence, it suffices to consider the mapping $f|_{S \setminus (A \times R^n)}$. The change of variables $(X, Y) \mapsto (X, Y', Z_n)$, where $Z_n = Y_n \prod_i g_{0,i}(X)$, induces a homeomorphism from $S \setminus (A \times R^n)$ onto a bounded semi-algebraic subset S' of R^{m+n}. We denote by S'_j the image of $S_j \setminus (A \times R^n)$ by this homeomorphism. The sets S' and S'_j are given by boolean combinations of sign conditions on polynomials $h_1, \ldots, h_s$ which are monic with respect to the last variable Z_n and other polynomials $h_{s+1}, \ldots, h_t$ which do not depend on Z_n.

Let $\Pi_{m+n} : R^{m+n} \to R^{m+n-1}$ be the projection onto the space of the coordinates X, Y'. Let $(A_i, (\xi_{i,j})_{j=1,\ldots,\ell_i})_{i=1,\ldots,m}$ be a slicing of the polynomials $h_1, \ldots, h_s$ with respect to the variable Z_n. We may assume that the polynomials $h_{s+1}, \ldots, h_t$ have constant signs -1, 0 or $+1$ on every A_i and that $\Pi_{m+n}(\mathrm{clos}(S'))$ is the union of some A_i. We can choose a finite simplicial complex L in R^{m+n-1} and a semi-algebraic homeomorphism $\Psi : |L| \to \Pi_{m+n}(\mathrm{clos}(S'))$, such that every A_i contained in $\Pi_{m+n}(\mathrm{clos}(S'))$ is the union of images by Ψ of some open simplices of L. Let τ be a simplex of L such that $\tau \subset A_i$, and $\xi_{i,j} : A_i \to R$, a function of the slicing whose graph is contained in S' or is the limit of a slice contained in S'. Then, by Lemma 9.2.4, $\xi_{i,j}|_{\Psi(\tau^0)}$ extends continuously to $\Psi(\tau)$. Note that $\mathrm{clos}(S')$ is the union of the graphs of these extensions and some slices of cylinders $\Psi(\tau) \times R$ bounded by these graphs. Applying Lemma 9.2.2 to $\mathrm{clos}(S')$ with subsets S' and the S'_j, we obtain a finite simplicial complex $K = (\sigma_i)_{i=1,\ldots,p}$ in R^{m+n}, such that every $\Pi_{m+n}(\sigma_i)$ is a simplex of the barycentric subdivision L', and a semi-algebraic homeomorphism $\Phi : |K| \to \mathrm{clos}(S')$, such that $\Pi_{m+n} \circ \Phi = \Psi \circ \Pi_{m+n}|_{|K|}$ and S' and the S'_j are the unions of some $\Phi(\sigma_i^0)$. It suffices to prove the theorem for the projection of $\mathrm{clos}(S')$ onto R^m, with S' and the S'_j as semi-algebraic subsets. Hence, we have reduced the problem to the following situation: S is closed and bounded, and there are semi-algebraic triangulations:

$$\Phi : |K| = \bigcup_{i=1}^{p} \sigma_i \longrightarrow S, \quad |K| \subset R^{m+n}$$

$$\Psi : |L'| = \bigcup_{k=1}^{s} \tau_k \longrightarrow \Pi_{m+n}(S), \quad |L'| \subset R^{m+n-1},$$

such that the image of every simplex σ_i of K by Π_{m+n} is a simplex τ_k of L', $\Pi_{m+n} \circ \Phi = \Psi \circ \Pi_{m+n}|_{|K|}$ and each S_j is the union of some $\Phi(\sigma_i^0)$.

We apply the inductive assumption to $\Pi_{m+n}(S)$, with the subsets $\Psi(\tau_k^0)$, and to the projection mapping $\Pi' : R^{m+n-1} \to R^m$. We obtain a finite partition of R^m into semi-algebraic sets $(T_\ell)_{\ell=1,\ldots,r}$ and, for $\ell = 1, \ldots, r$, semi-algebraic trivializations $\rho_\ell : T_\ell \times G_\ell \to \Pi'^{-1}(T_\ell) \cap \Pi_{m+n}(S)$ of $\Pi'|_{\Pi_{m+n}(S)}$ over T_ℓ, compatible with $\Psi(\tau_k^0)$, for $k = 1, \ldots, s$. Fix ℓ, and let x^1 be a point of T_ℓ. We may assume that $G_\ell = \Pi'^{-1}(x^1) \cap \Pi_{m+n}(S)$ and $\rho_\ell(x^1, (x^1, y')) = (x^1, y')$ for every $(x^1, y') \in G_\ell$. Set $F_\ell = \Pi^{-1}(x^1) \cap S$ and $F_{\ell,j} = \Pi^{-1}(x^1) \cap S_j$. It remains to construct $\theta_\ell : T_\ell \times F_\ell \to \Pi^{-1}(T_\ell) \cap S$. The construction is illustrated by Fig. 9.5.

Let $x \in T_\ell$ and $(x^1, y') \in G_\ell$. Then (x^1, y') belongs to one of the $\Psi(\tau_k^0)$, say $\Psi(\tau_1^0)$, and $\rho_\ell(x, (x^1, y')) \in \Psi(\tau_1^0)$. The triangulations Φ and Ψ are such that the $\Phi(\sigma_i)$ induce homeomorphic subdivisions of $\Pi_{m+n}^{-1}(x^1, y') \cap S$ and $\Pi_{m+n}^{-1}(\rho_\ell(x, (x^1, y'))) \cap S$. The function θ_ℓ maps affinely each nonempty segment $\{x\} \times (\Pi_{m+n}^{-1}(x^1, y') \cap \Phi(\sigma_i)) \subset T_\ell \times F_\ell$ onto the corresponding segment $\Pi_{m+n}^{-1}(\rho_\ell(x, (x^1, y'))) \cap \Phi(\sigma_i)$. We leave to the reader the verification that

9.3 Semi-algebraic Triviality of Semi-algebraic Mappings   223

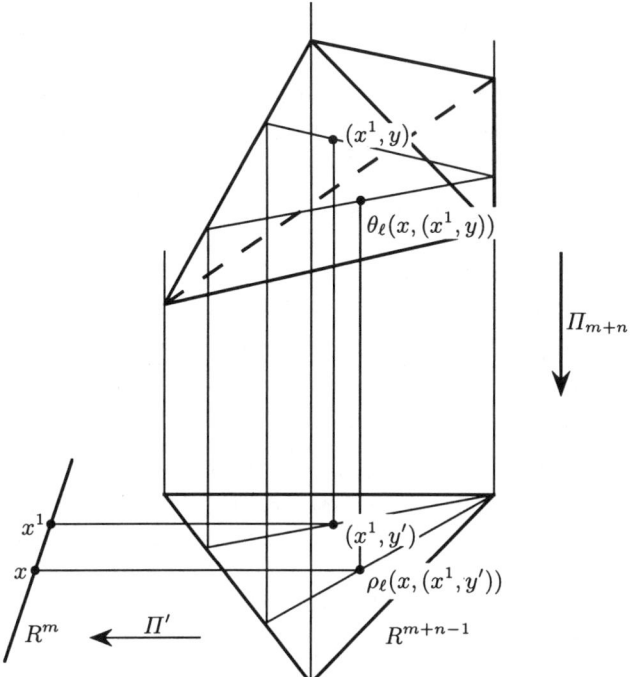

Figure 9.5. Construction of θ_ℓ

θ_ℓ is a semi-algebraic homeomorphism and $\theta_\ell(T_\ell \times F_{\ell,j}) = \Pi^{-1}(T_\ell) \cap S_j$. □

We now give another proof, which relies on the study of semi-algebraic families in Chap. 7, Section 4.

Second proof of Theorem 9.3.2. As in the previous proof, we may assume that S is a bounded semi-algebraic subset of $R^m \times R^n$ and that f is the restriction of the projection mapping $R^m \times R^n \to R^m$. Replacing S and its subsets $S_1, \ldots, S_q$ with $\operatorname{clos}(S)$ and its subsets $S, S_1, \ldots, S_q$, we may assume that S is closed and bounded.

Let $\alpha \in \widetilde{R^m}$. The fibre $S_\alpha \subset k(\alpha)^n$ is closed (Proposition 7.4.6) and bounded, and, by Theorem 9.2.1, there is a finite simplicial complex $K = (\sigma_i)_{i=1,\ldots,p}$ in $k(\alpha)^n$ and a semi-algebraic homeomorphism $\Phi : |K| \to S_\alpha$, such that, for each fibre $(S_j)_\alpha$, the inverse image $G_j = \Phi^{-1}((S_j)_\alpha)$ is the union of some open simplices σ_i^0. By Remark 9.2.3 b), we can choose the vertices of the complex K with coordinates in $\mathbb{Q}$. Hence $|K|$ and the G_j are the extensions to $k(\alpha)$ of semi-algebraic subsets F and F_j of R^n. By Example 7.4.2, $|K|$ and the G_j are the fibres at α of the constant families $R^m \times F$ and $R^m \times F_j$. By Corollary 7.4.10, there is a semi-algebraic set $T^\alpha \subset R^m$ such that $\alpha \in \widetilde{T^\alpha}$, and a semi-algebraic homeomorphism $\theta : T^\alpha \times F \to S \cap (T^\alpha \times R^n)$

commuting with the projection mapping onto T^α and such that Φ is the fibre θ_α. Shrinking T^α, if necessary, we have $\theta(T^\alpha \times F_j) = S_j \cap (T^\alpha \times R^n)$ for $j = 1, \ldots, q$, since the equality is satisfied for the fibres at α. The $\widetilde{T^\alpha}$ cover $\widetilde{R^m}$. Hence, by the compactness of $\widetilde{R^m}$ for the constructible topology (7.1.12), R^m is covered by a finite number of T^α. $\square$

Corollary 9.3.3. *Let S and T be two semi-algebraic sets and $f : S \to T$ a continuous semi-algebraic mapping. There exists a closed semi-algebraic subset V of T, $\dim(V) < \dim(T)$, such that f has a semi-algebraic trivialization over each semi-algebraically connected component of $T \setminus V$.*

Proof. By Proposition 9.1.8, the semi-algebraic partition of T in Theorem 9.3.2 can be refined to a Nash stratification. We take for V the union of the strata of T of dimension smaller than $\dim(T)$. $\square$

Remark 9.3.4. a) We keep the notations of Theorem 9.3.2. This theorem implies that if y and z belong to the same subset T_ℓ of T, there exists a semi-algebraic homeomorphism $\varphi : f^{-1}(y) \to f^{-1}(z)$ such that $\varphi(f^{-1}(y) \cap S_j) = f^{-1}(z) \cap S_j$.

b) Theorem 9.3.2 and its corollary 9.3.3 no longer hold true if the mapping f is not continuous. For example, let $f : R^2 \to R$ be the semi-algebraic function defined by $f(x,y) = 1/y$ if $y \neq 0$ and $f(x,y) = x$ if $y = 0$. Corollary 9.3.3 would give us an $M > 0$ such that $f^{-1}([M, +\infty[)$ is semi-algebraically homeomorphic to $[M, +\infty[\times f^{-1}(M)$. But $f^{-1}([M, +\infty[)$ is semi-algebraically connected while $f^{-1}(M)$ is the disjoint sum of a line and a point.

The only place in the proof where we have used the continuity of f is the replacement of S with the graph of f. This graph is not necessarily homeomorphic to S if f is not continuous. However, for every $x \in T$, the fibre $f^{-1}(x)$ is homeomorphic to the fibre $\Pi^{-1}(x) \cap \text{graph}(f)$ of the projection. Hence, if f is not continuous, the following weaker result holds: there exists a finite partition of T into semi-algebraic subsets T_ℓ such that the fibres $f^{-1}(x)$ and $f^{-1}(x')$ are semi-algebraically homeomorphic whenever x and x' belong to the same T_ℓ.

An easy consequence of the theorem of semi-algebraic triviality 9.3.2 is the fact that algebraic subsets of R^n defined by equations of degree $\leq d$ have a finite number of topological types.

Theorem 9.3.5. *Let n and d be two positive integers. Let $\mathcal{M}(n,d)$ be the family of algebraic subsets of R^n defined by any number of polynomials in $R[X_1, \ldots, X_n]$ of degree $\leq d$. There exists a finite number of algebraic subsets $V_1, \ldots, V_s$ of R^n in $\mathcal{M}(n,d)$, such that, for every V in $\mathcal{M}(n,d)$, there exist i, $1 \leq i \leq s$, and a semi-algebraic homeomorphism $\varphi : R^n \to R^n$ with $\varphi(V_i) = V$.*

Proof. The family $\mathcal{M}(n,d)$ is contained in the family $\mathcal{F}$ of algebraic subsets of R^n defined by a single equation of degree $\leq 2d$, because $\mathcal{Z}(f_1, \ldots, f_k) =$

$Z(f_1^2 + \cdots + f_k^2)$. The family $\mathcal{F}$ is parametrized by the space R^N of coefficients of the equation, where $N = \binom{n+2d}{d}$. By abuse of notation, we denote by f the point of R^N whose coordinates are the coefficients of f. Define

$$S = \{(f, x) \in R^N \times R^n \mid f(x) = 0\}.$$

The set S is algebraic. Let $\Pi : R^N \times R^n \to R^N$ be the canonical projection. Then $\Pi^{-1}(f) \cap S = \{f\} \times Z(f)$. The proof is completed by applying Theorem 9.3.2 and Remark 9.3.4 a) to $S \subset R^N \times R^n$ and Π. □

Another consequence of Theorem 9.3.2 is the result concerning the local conic structure of semi-algebraic sets, illustrated by Figure 9.6.

Theorem 9.3.6 (Local Conic Structure). *Let E be a semi-algebraic subset of R^n and x a nonisolated point of E. There exist $\epsilon \in R$, $\epsilon > 0$, and a semi-algebraic homeomorphism $\varphi : \overline{B}_n(x, \epsilon) \to \overline{B}_n(x, \epsilon)$ such that:*
  (i) $\|\varphi(y) - x\| = \|y - x\|$ *for every* $y \in \overline{B}_n(x, \epsilon)$,
  (ii) $\varphi|_{S^{n-1}(x,\epsilon)}$ *is the identity mapping,*
  (iii) $\varphi^{-1}(E \cap \overline{B}_n(x, \epsilon))$ *is the cone with vertex x and basis $E \cap S^{n-1}(x, \epsilon)$.*

Proof. Applying Theorem 9.3.2 with $S = R^n$, $S_1 = E$ and $f : S \to R$ defined by $f(y) = \|y - x\|$, we find $\epsilon > 0$ and a semi-algebraic homeomorphism $\theta :]0, \epsilon] \times S^{n-1}(x, \epsilon) \to \overline{B}_n(x, \epsilon) \setminus \{x\}$ such that $\theta(]0, \epsilon] \times (E \cap S^{n-1}(x, \epsilon))) = E \cap \overline{B}_n(x, \epsilon) \setminus \{x\}$ and, for every $y \in S^{n-1}(x, \epsilon)$, $\theta(\epsilon, y) = y$ and $\|\theta(t, y) - x\| = t$ for $t \in]0, \epsilon]$. We define the mapping φ by $\varphi(x) = x$ and $\varphi(x + \lambda(y - x)) = \theta(\lambda \epsilon, y)$, for $y \in S^{n-1}(x, \epsilon)$ and $\lambda \in]0, 1]$. □

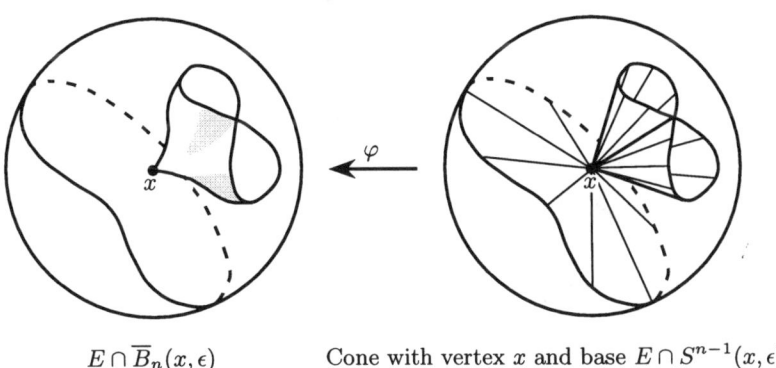

$E \cap \overline{B}_n(x, \epsilon)$ Cone with vertex x and base $E \cap S^{n-1}(x, \epsilon)$

Figure 9.6. Local conic structure

Corollary 9.3.7. *Let S be a semi-algebraic subset of R^n. There exists $r \in R$, $r > 0$, such that $S \cap \overline{B}_n(0, r)$ is a semi-algebraic deformation retract of S, i.e. there exists a continuous semi-algebraic mapping $h : [0, 1] \times S \to S$ such* that, for every $x \in S$, $h(0,x) = x$ and $h(1,x) \in S \cap \overline{B_n}(0,r)$, and, for every $t \in [0,1]$ and every x in $S \cap \overline{B_n}(0,r)$, $h(t,x) = x$.

Proof. We may assume that S is not bounded. We construct h using the inversion mapping $\varphi : R^n \setminus \{0\} \to R^n \setminus \{0\}$ defined by $\varphi(x) = x/\|x\|^2$ and the local conic structure of $\varphi(S) \cup \{0\}$ at 0. $\square$

Remark 9.3.8. The previous corollary combined with Proposition 2.2.9 and Theorem 2.5.8 implies that every locally closed semi-algebraic subset S of R^n has a closed and bounded semi-algebraic subset K which is a semi-algebraic deformation retract of S. This result can also be proved without assuming S locally closed, using the triangulation of semi-algebraic subsets of R^n (cf. [109]).

Proposition 9.3.9. *Let $f : S \to T$ be be a semi-algebraic local homeomorphism. There exists a finite open semi-algebraic cover $S = \bigcup_{i=1}^n U_i$ such that, for every i, $f|_{U_i}$ is a homeomorphism onto $f(U_i)$.*

Proof. We may assume that T is bounded. Applying the semi-algebraic trivialization theorem 9.3.2 and the triangulation theorem 9.2.1, we may assume that $\mathrm{clos}(T) = |K|$, where $K = (\sigma_i)_{i=1,\ldots,p}$ is a finite simplicial complex such that T is the union of some open simplices σ_ℓ^0 and f has a semi-algebraic trivialization $\theta_\ell : \sigma_\ell^0 \times F_\ell \to f^{-1}(\sigma_\ell^0)$ over each σ_ℓ^0. Since f is a local semi-algebraic homeomorphism, each F_ℓ consists of a finite number of points $x_{\ell,1}, \ldots, x_{\ell,p_\ell}$. Fix $\sigma_\ell^0 \subset T$, and let V_ℓ be the union of all $\sigma_m^0 \subset T$ such that σ_ℓ is a face of σ_m. For every $x_{\ell,\lambda} \in F_\ell$ and every $\sigma_m^0 \subset V_\ell$, there exists a unique point $x_{m,\mu} = \beta_{\ell,m}(x_{\ell,\lambda}) \in F_m$ such that

$$\theta_\ell(\sigma_\ell^0 \times \{x_{\ell,\lambda}\}) = \mathrm{clos}(\theta_m(\sigma_m^0 \times \{x_{m,\mu}\})) \cap f^{-1}(\sigma_\ell^0) .$$

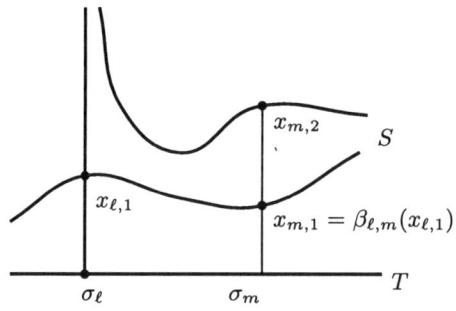

Figure 9.7.

Set

$$U_{\ell,\lambda} = \bigcup \{\theta_m(\sigma_m^0 \times \{\beta_{\ell,m}(x_{\ell,\lambda})\}) \mid \sigma_m^0 \subset V_\ell\} .$$

Then $f|_{U_{\ell,\lambda}}$ is a homeomorphism onto V_ℓ, and the $U_{\ell,\lambda}$ form a finite open semi-algebraic cover of S. $\square$

Corollary 9.3.10. *Let M be a Nash submanifold of R^n of dimension d. There exists a finite open semi-algebraic cover of $M = \bigcup_{i=1}^{p} M_i$ such that, for each M_i, there are $1 \le j_1 < \ldots < j_d \le n$ such that the restriction of the projection mapping $(x_1, \ldots, x_n) \mapsto (x_{j_1}, \ldots, x_{j_d})$ to M_i is a Nash diffeomorphism onto its image.*

Proof. Let $\Pi : R^n \to R^d$ be the projection onto the space of the first d coordinates, and $M' \subset M$ the set of points x such that Π induces an isomorphism from the tangent space $T_x(M)$ onto R^d. The mapping $\Pi|_{M'}$ is a local Nash diffeomorphism. Hence, by Proposition 9.3.9, we can cover M' by finitely many open semi-algebraic subsets M_i such that $\Pi|_{M_i}$ is a Nash diffeomorphism onto its image. We can repeat the same construction for the other projections $R^n \to R^d$ onto spaces of coordinates, finally getting a finite open semi-algebraic cover of M with the required property. □

9.4 Triangulation of Semi-algebraic Functions

Continuous semi-algebraic functions on a closed and bounded semi-algebraic set are semi-algebraically triangulable. The following theorem makes this statement precise.

Theorem 9.4.1 (Triangulability of Semi-algebraic Functions). *Let S be a closed and bounded semi-algebraic subset of R^n and $f : S \to R$ a continuous semi-algebraic function. There exist a finite simplicial complex K in R^{n+1} and a semi-algebraic homeomorphism $\Phi : |K| \to S$ such that $f \circ \Phi : |K| \to R$ is affine on every simplex of K. Moreover, given a finite collection $S_1, \ldots, S_p$ of semi-algebraic subsets of S, we may choose K and Φ such that every S_i is the union of images by Φ of some open simplices of K.*

Replacing S with the graph of the function f, we may assume that f is the restriction to S of the projection mapping $R^n \times R \to R$. Hence Theorem 9.4.1 is a consequence of the following proposition.

Proposition 9.4.2. *Let S be a closed and bounded semi-algebraic subset of $R^n \times R$, and $S_1, \ldots, S_p$ semi-algebraic subsets of S. Let $\Pi : R^n \times R \to R$ be the projection onto the last factor. Then there exist a finite simplicial complex K in $R^n \times R$, whose vertices have coordinates in $\mathbb{Q}$, and semi-algebraic homeomorphisms $\Phi : |K| \to S$ and $\Theta : R \to R$, such that $\Theta \circ \Pi \circ \Phi = \Pi|_{|K|}$ and every S_i is the union of images by Φ of some open simplices of K.*

If we drop the requirement that the vertices of K have coordinates in $\mathbb{Q}$, then we can choose $\Theta = \mathrm{Id}_R$.

Proof. We proceed by induction on n. The case of $n = 0$ is easy. We can subdivide R using finitely many points $x_1 < \ldots < x_p$ such that S and the S_i are the unions of some points x_j and some open intervals $]x_j, x_{j+1}[$. If we want a complex K with vertices in $\mathbb{Q}$, we can choose a semi-algebraic homeomorphism $\Phi : R \to R$ such that $\Phi(i) = x_i$ for $i = 1, \ldots, p$, and take $\Theta = \Phi^{-1}$.

Given $n > 0$, assume the proposition holds true for $n - 1$. The semi-algebraic sets S and S_i are defined by boolean combinations of sign conditions on polynomials $f_1, \ldots, f_q$ in $R[X, T]$, where $X = (X_1, \ldots, X_n)$. Replacing the polynomials f_i with their irreducible factors, we may assume without loss of generality that $f_i = T - c_i$, for $i = 1, \ldots, k$, and, for $i = k + 1, \ldots, q$ and for every $t \in R$, $f_i(X, t) \neq 0$ in $R[X]$. Moreover, we claim that we may assume that, for $i = k+1, \ldots, q$, for every $x' = (x_2, \ldots, x_n) \in R^{n-1}$ and every $t \in R$, $f_i(X_1, x', t) \neq 0$ in $R[X_1]$. This claim is a consequence of the following lemma, applied to $g = \prod_{i=k+1}^{q} f_i$.

Lemma 9.4.3. *Let g be a polynomial in $R[X, T]$, where $X = (X_1, \ldots, X_n)$, such that, for every $t \in R$, $f(X, t) \neq 0$ in $R[X]$. Then there exists a polynomial automorphism $v : R^n \to R^n$ such that, for every $x' \in R^{n-1}$ and $t \in R$, $g(v(X_1, x'), t) \neq 0$ in $R[X_1]$.*

Proof. Let $v' : R \to R^{n-1}$ be a polynomial mapping. If ℓ is a nonnegative integer, we set

$$W_\ell = \bigcup_{\lambda=0}^{\ell} \{(x', t) \in R^{n-1} \times R \mid g(\lambda, x' + v'(\lambda), t) = 0\}.$$

Since the W_ℓ are algebraic subsets of $R^{n-1} \times R$ and $W_{\ell+1} \subset W_\ell$, the sequence $(W_\ell)_{\ell \in \mathbb{N}}$ is stationary by the noetherian property of $R[X]$. Note that W_ℓ depends only on the values $v'(0), \ldots, v'(\ell)$. Assume $W_\ell \neq \emptyset$, and choose $(x', t) \in W_\ell$. Since $g(X, t) \neq 0$, we can find an integer $m > \ell$ and $y' \in R^{n-1}$ such that $g(m, y', t) \neq 0$. If we have $v'(m) = y' - x'$, we obtain $W_m \subsetneq W_\ell$. Hence, by assigning the values $v'(\lambda)$ for finitely many integers λ, we can choose v' so that $W_\ell = \emptyset$ for some integer ℓ. It follows that for every $(x', t) \in R^{n-1} \times R$, $g(X_1, x' + v'(X_1), t) \neq 0$ in $R[X_1]$. We define the polynomial automorphism $v : R^n \to R^n$ by $v(x_1, x') = (x_1, x' + v'(x_1))$. □

We return to the proof of Proposition 9.4.2. Let $\Pi_{n+1} : R^n \times R \to R^{n-1} \times R$ (resp $\Pi' : R^{n-1} \times R \to R$) be the projection mapping defined by $\Pi_{n+1}(x_1, x', t) = (x', t)$ (resp $\Pi'(x', t) = t$). Choose a slicing $(A_\lambda, (\xi_{\lambda, \mu}))$ of the polynomials $f_{k+1}, \ldots, f_q$, with respect to the variable X_1. We may assume that the polynomials $T - c_i$, for $i = 1, \ldots, k$, have constant signs on each A_λ. Hence S and the S_i are the unions of some "graphs"

$$\{(\xi_{\lambda, \mu}(x', t), x', t) \mid (x', t) \in A_\lambda\}$$

and some slices of cylinders $R \times A_\lambda$ bounded by these graphs. By the inductive assumption, we can choose a simplicial complex L in R^{n-1}, whose vertices have coordinates in $\mathbb{Q}$, and semi-algebraic homeomorphisms $\Psi : |L| \to$

9.4 Triangulation of Semi-algebraic Functions 229

$\Pi_{n+1}(S)$ and $\Theta : R \to R$, such that $\Theta \circ \Pi' \circ \Psi = \Pi'|_{|L|}$ and such that the A_λ contained in $\Pi_{n+1}(S)$ are the unions of images by Ψ of some open simplices of L. By Lemma 9.2.4, if τ is a simplex of L such that $\Psi(\tau^0) \subset A_\lambda$, and $\xi_{\lambda,\mu} : A_\lambda \to R$, a function of the slicing whose graph is contained in S, then $\xi_{\lambda,\mu}|_{\Psi(\tau^0)}$ continuously extends to $\Psi(\tau)$. Hence, we can apply Lemma 9.2.2, and we obtain a simplicial complex K in $R^n \times R$, whose vertices have coordinates in $\mathbb{Q}$, and a semi-algebraic homeomorphism $\Phi : |K| \to S$, such that $\Theta \circ \Pi \circ \Phi = \Pi|_{|K|}$ and every S_i is the union of images by Φ of some open simplices of K.

The fact that we can choose $\Theta = \mathrm{Id}_R$, if we drop the requirement that the vertices of K have coordinates in $\mathbb{Q}$, follows in an obvious way from the construction. $\square$

A semi-algebraic mapping is not always triangulable. Consider for instance, the mapping $f : [0,1]^2 \to \mathbb{R}^2$ defined by $f(x,y) = (x, xy)$. There is no way to choose triangulations $\Phi : |K| \to [0,1]^2$ and $\Psi : |L| \to f([0,1]^2)$ such that $\Psi^{-1} \circ f \circ \Phi$ is affine on every simplex of K.

The following proposition gives information concerning neighbourhoods of closed and bounded semi-algebraic sets.

Proposition 9.4.4. *Let $Z \subset S$ be two closed and bounded semi-algebraic sets. Let f be a nonnegative continuous semi-algebraic function on S such that $f^{-1}(0) = Z$. Then there are $\delta > 0$ and a continuous semi-algebraic mapping $h : f^{-1}(\delta) \times [0, \delta] \to f^{-1}([0,\delta])$, such that $f(h(x,t)) = t$ for every $(x,t) \in f^{-1}(\delta) \times [0,\delta]$, $h(x,\delta) = x$ for every $x \in f^{-1}(\delta)$, and $h|_{f^{-1}(\delta) \times]0,\delta]}$ is a homeomorphism onto $f^{-1}(]0,\delta])$.*

Proof. By triangulating f we obtain a finite simplicial complex K and a semi-algebraic homeomorphism $\Phi : |K| \to S$, such that $f \circ \Phi$ is affine on every simplex of K. Note that Z is the union of images by Φ of some simplices of K. Choose $\delta > 0$ so small that, for every vertex a of K such that $\Phi(a) \notin Z$, $\delta < f(\Phi(a))$. Let $x \in f^{-1}(\delta)$, $y = \Phi^{-1}(x)$. The point y belongs to a simplex $[a_0, \ldots, a_d]$ of K. We may assume that $\Phi(a_i) \in Z$, for $i = 0, \ldots, k$, and $\Phi(a_i) \notin Z$, for $i = k+1, \ldots, d$. Let $(\lambda_0, \ldots, \lambda_d)$ be the barycentric coordinates of y in the simplex $[a_0, \ldots, a_d]$. Note that $\delta = f(x) = \sum_{i=k+1}^d \lambda_i f(\Phi(a_i))$. Hence, if we set $\alpha = \sum_{i=1}^k \lambda_i$, we have necessarily $0 < \alpha < 1$. For $t \in [0,\delta]$, we define $h(x,t)$ as the image by Φ of the point of $[a_0, \ldots, a_d]$ with barycentric coordinates $(\mu_0, \ldots, \mu_d)$, where

$$\mu_i = \begin{cases} \frac{t\alpha + \delta - t}{\delta \alpha} \lambda_i & \text{for } i = 0, \ldots, k, \\ \frac{t}{\delta} \lambda_i & \text{for } i = k+1, \ldots, d. \end{cases}$$

It is easily verified that h is well defined and satisfies the required properties. $\square$

We shall use the semi-algebraic triangulation of continuous semi-algebraic functions to obtain a result concerning the semi-algebraic triviality of semi-algebraic families of continuous functions. First we fix some terminology.

Definition 9.4.5.
i) Let $f : R^n \to R^p$ and $f' : R^n \to R^p$ be two continuous semi-algebraic mappings. We say that f and f' have the same semi-algebraic topological type if there are semi-algebraic homeomorphisms $\lambda : R^n \to R^n$ and $\theta : R^p \to R^p$ such that $f = \theta \circ f' \circ \lambda$.

ii) Let
$$F : R^n \times S \longrightarrow R^p \times S$$
$$(x,t) \longmapsto (F_t(x),t)$$
be a semi-algebraic family of continuous mappings parametrized by S, and T a semi-algebraic subset of S. The family F is said to be semi-algebraically trivial over T if there are $t_0 \in T$ and semi-algebraic homeomorphisms

$$\Lambda : R^n \times T \longrightarrow R^n \times T \qquad \Theta : R^p \times T \longrightarrow R^n \times T$$
$$(x,t) \longmapsto (\Lambda_t(x),t) \qquad (y,t) \longmapsto (\Theta_t(y),t),$$

such that $F_{t_0} \circ \Lambda_t = \Theta_t \circ F_t$ for every $t \in T$

It follows from the definition that, if the family F is semi-algebraically trivial over T, then, for every t and t' in T, the mappings $F_t : R^n \to R^p$ and $F_{t'} : R^n \to R^p$ have the same semi-algebraic topological type.

Theorem 9.4.6. Let
$$F : R^n \times S \longrightarrow R \times S$$
$$(x,t) \longmapsto (F_t(x),t)$$
be a semi-algebraic family of continuous functions parametrized by S. Then there is a finite semi-algebraic partition $S = \bigcup_{i=1}^k T_i$ of S such that the family F is semi-algebraically trivial over each T_i.

Proof. Let α be a point of $\widetilde{S}$ and consider the fibre $F_\alpha : \kappa(\alpha)^n \to \kappa(\alpha)$. The function F_α is continuous and semi-algebraic (cf. Remark 7.4.9). Let $h :\ ]-1,1[_{\kappa(\alpha)} \to \kappa(\alpha)$ be the semi-algebraic homeomorphism defined by $h(x) = x/(1-x^2)$. Set
$$\Gamma = \left\{ (x_1,\ldots,x_n, h^{-1}(F_\alpha(h(x_1),\ldots,h(x_n)))) \mid (x_1,\ldots,x_n) \in\ ]-1,1[_{\kappa(\alpha)}^n \right\}.$$

Let X be the closure of Γ in $\kappa(\alpha)^{n+1}$ and $\Pi_{\kappa(\alpha)} : \kappa(\alpha)^{n+1} \to \kappa(\alpha)$ the projection mapping defined by $\Pi_{\kappa(\alpha)}(x_1,\ldots,x_n,t) = t$. We can apply Proposition 9.4.2 to X and Γ. We obtain a finite simplicial complex K, whose vertices have coordinates in $\mathbb{Q}^{n+1}$, and semi-algebraic homeomorphisms $\Phi : |K|_{\kappa(\alpha)} \to X$ and $\Theta : \kappa(\alpha) \to \kappa(\alpha)$, such that $\Theta \circ \Pi_{\kappa(\alpha)} \circ \Phi = \Pi_{\kappa(\alpha)}|_{|K|_{\kappa(\alpha)}}$ and

9.4 Triangulation of Semi-algebraic Functions   231

$\Gamma = \Phi(V_{\kappa(\alpha)})$, where V is the union of some open simplices of K. Note that the fact that the vertices of K have coordinates in $\mathbb{Q}$ allows us to consider $|K|$ and V as subsets of R^{n+1}. Moreover, we may assume that $\Pi_{\kappa(\alpha)}(V_{\kappa(\alpha)}) \subset \;]-1,1[_{\kappa(\alpha)}$, $\Theta(-1) = -1$ and $\Theta(1) = 1$.

Let $\lambda : V_{\kappa(\alpha)} \to \kappa(\alpha)^n$ be the semi-algebraic homeomorphism defined by $\lambda(v) = (h(x_1), \ldots, h(x_n))$, where $\Phi(v) = (x_1, \ldots, x_n, t)$. Let $\rho : \;]-1,1[_{\kappa(\alpha)} \to \kappa(\alpha)$ be the semi-algebraic homeomorphism defined by $\rho(t) = h(\Theta^{-1}(t))$. We have a commutative diagram of semi-algebraic mappings

$$\begin{array}{ccc} V_{\kappa(\alpha)} & \xrightarrow{\Pi_{\kappa(\alpha)}} &]-1,1[_{\kappa(\alpha)} \\ \lambda \downarrow & & \downarrow \rho \\ \kappa(\alpha)^n & \xrightarrow{F_\alpha} & \kappa(\alpha) \end{array}$$

By the results of Chap. 7, Section 4, we deduce that there are a semi-algebraic subset T^α of S such that $\alpha \in \widetilde{T^\alpha}$ and a commutative diagram of semi-algebraic families of mappings parametrized by T^α

$$\begin{array}{ccc} V \times T^\alpha & \xrightarrow{\Pi_R \times S^\alpha} &]-1,1[_R \times T^\alpha \\ \downarrow & & \downarrow \\ R^n \times T^\alpha & \xrightarrow{F|_{R^n \times T^\alpha}} & R \times T^\alpha \end{array},$$

where the vertical arrows are homeomorphisms. This shows that the family F is semi-algebraically trivial over T^α. The $\widetilde{T^\alpha}$ cover $\widetilde{S}$. Since $\widetilde{S}$ is compact for the constructible topology, we can extract a finite semi-algebraic cover $\widetilde{T_1}, \ldots, \widetilde{T_k}$, and we may, moreover, assume that the T_i form a partition of S. The family F is trivial over each T_i.   □

Corollary 9.4.7. *Let n and d be positive integers. The family $\mathcal{P}(n,d)$ of polynomial functions $R^n \to R$ of degree $\leq d$ has only a finite number of semi-algebraic topological types.*

Proof. Let R^N, where $N = \binom{n+d}{n}$, be the space of coefficients of polynomials in n variables of degree $\leq d$. If $a \in R^N$, let $\mathcal{P}_a$ be the polynomial with coefficients a. The corollary follows from Theorem 9.4.6 applied to the family

$$\mathcal{P}(n,d) : R^n \times R^N \longrightarrow R \times R^N$$
$$(x,a) \longmapsto (\mathcal{P}_a(x), a)$$

parametrized by the space R^N.   □

Remark 9.4.8. The finiteness of the number of topological types does not hold, in general, for polynomial mappings $\mathbb{R}^n \to \mathbb{R}^m$ of bounded degree. The first counter-examples were given by Thom in [322], and there is a counter-example for polynomial mappings from $\mathbb{R}^3$ to $\mathbb{R}^2$ in [244]. In the case of two variables, the finiteness is proved for polynomial germs $(K^2, 0) \to (K^2, 0)$ (where $K = \mathbb{R}$ or $\mathbb{C}$) in [14] and for polynomial mappings $\mathbb{C}^2 \to \mathbb{C}^m$ in [283]. An example of a semi-algebraic family of continuous mappings $\mathbb{R}^2 \to \mathbb{R}^2$ having infinitely many topological types is given in [98].

9.5 Half-branches of Algebraic Curves

It is easy to describe the topological structure of the germ of an algebraic curve (i.e. an algebraic set of dimension 1) at one of its points.

Proposition 9.5.1. *Let $\Gamma \subset R^n$ be an algebraic curve and $a \in \Gamma$. For every sufficiently small open ball U with centre a, $\Gamma \cap (U \setminus \{a\})$ has a finite number of semi-algebraically connected components $B_1, \ldots, B_k$ and there are semi-algebraic homeomorphisms $f_i : [0, 1[\to B_i \cup \{a\}$ with $f_i(0) = a$, for $i = 1, \ldots, k$.*

Proof. The statement is an immediate consequence of the theorem of local conic structure 9.3.6. Note that the germs $(B_i)_a$ do not depend on the radius of the ball, if it is small enough. □

Definition 9.5.2. *With the notation of the previous proposition, the germs $(B_i)_a$ are called the half-branches of the curve Γ centred at a.*

Now we shall prove that the number of half-branches of Γ centred at a is even. We shall reduce the problem to the case of a plane algebraic curve. This reduction requires a result concerning projections that is interesting not only for curves. First we introduce some notation. If x and y are two distinct points of R^n, we denote by $R(y - x) \in \mathbb{P}_{n-1}(R)$ the line generated by the vector $y - x$. If $v \in \mathbb{P}_{n-1}(R)$, we denote by Π_v the orthogonal projection of R^n onto the vector hyperplane of R^n orthogonal to v.

Lemma 9.5.3. *Let $A \subset R^n$ be a semi-algebraic set of dimension $d > 0$. Let S be the set of $v \in \mathbb{P}_{n-1}(R)$ such that there exist infinitely many couples $(x, y) \in A \times A$ with $x \neq y$ and $\Pi_v(x) = \Pi_v(y)$. Then S is a semi-algebraic subset of $\mathbb{P}_{n-1}(R)$ of dimension $< 2d$.*

Proof. Let $\Delta \subset A \times A$ be the diagonal and $\varphi : A \times A \setminus \Delta \to \mathbb{P}_{n-1}(R)$ the continuous semi-algebraic mapping defined by $\varphi(x, y) = R(y - x)$. Note that
$$S = \{v \in \mathbb{P}_{n-1}(R) \mid \dim(\varphi^{-1}(v)) \geq 1\}.$$
Theorem 9.3.2, applied to the mapping φ, shows that S is a semi-algebraic set and $\dim(\varphi^{-1}(S)) \geq \dim(S) + 1$, or $S = \emptyset$. Since
$$\dim(\varphi^{-1}(S)) \leq \dim(A \times A \setminus \Delta) = 2d,$$
it follows that $\dim(S) < 2d$. □

9.5 Half-branches of Algebraic Curves   233

Lemma 9.5.4. *Let $A \subset R^n$ be a semi-algebraic set of dimension $d > 0$ and $a \in A$. Define*

$$T = \{v \in \mathbb{P}_{n-1}(R) \mid \exists \epsilon > 0 \; \forall \eta > 0 \; \exists x \in A$$
$$(\|a - x\| \geq \epsilon \text{ and } \|\Pi_v(a) - \Pi_v(x)\| < \eta)\}.$$

Then T is a semi-algebraic subset of $\mathbb{P}_{n-1}(R)$ of dimension $\leq d$.

Proof. Define $B = \text{clos}(\{R(x - a) \in \mathbb{P}_{n-1}(R) \mid x \in A \setminus \{a\}\})$. The dimension of this set, which is the closure of the image of $\{a\} \times (A \setminus \{a\})$ by the mapping φ of the previous lemma, is $\leq d$. Let $v \in T$. Then

$$\exists \epsilon > 0 \; \forall \eta > 0 \; \exists x \in A \; (\|a - x\| \geq \epsilon \text{ and } \|\Pi_v(a) - \Pi_v(x)\| < \eta).$$

When $\Pi_v(x)$ tends to $\Pi_v(a)$ with $\|a - x\| \geq \epsilon$, then $R(x - a)$ tends to v. This proves that v belongs to B. Hence $T \subset B$ and $\dim(T) \leq \dim(B) \leq d$. □

Proposition 9.5.5. *Let $A \subset R^n$ be a semi-algebraic set of dimension d with $0 < 2d < n$ and $a \in A$. There exist a linear mapping $\Pi : R^n \to R^{2d}$, a semi-algebraic neighbourhood U of a in A, a semi-algebraic neighbourhood V of $\Pi(a)$ in $\Pi(A)$ and a finite set of points $B \subset A$, such that $\Pi|_U$ is a homeomorphism onto V and $\Pi|_{A \setminus B}$ is injective.*

Proof. By induction on n, it is enough to find a linear mapping $\Pi_1 : R^n \to R^{n-1}$, a semi-algebraic neighbourhood U_1 of a in A, a semi-algebraic neighbourhood V_1 of $\Pi_1(a)$ in $\Pi_1(A)$ and a finite set of points $B_1 \subset A$, such that $\Pi_1|_{U_1}$ is a homeomorphism onto V_1 and $\Pi_1|_{A \setminus B_1}$ is injective.

Replacing A with its closure, we may assume that A is closed in R^n. By Lemmas 9.5.3 and 9.5.4 and the assumption $0 < 2d < n$, there is $v \in \mathbb{P}_{n-1}(R) \setminus (S \cup T)$, where S and T are defined in 9.5.3 and 9.5.4, respectively. Since $v \notin S$, there exists a finite set of points $B_1 \subset A$ such that $\Pi_v|_{A \setminus B_1}$ is injective. We may assume that $a \notin B_1$. Since A is closed, there is a closed and bounded semi-algebraic neighbourhood U_0 of a in A such that $U_0 \subset A \setminus B_1$. Since $v \notin T$, there exists a semi-algebraic neighbourhood V_0 of $\Pi_v(a)$ in the hyperplane orthogonal to v such that $\Pi_v^{-1}(V_0) \cap A \subset U_0$. We may take V_0 to be closed. Let $U_1 = \Pi_v^{-1}(V_0) \cap A$, $V_1 = \Pi_v(U_1) = V_0 \cap \Pi_v(A)$. Then U_1 is a closed and bounded semi-algebraic neighbourhood of a in A. Since $\Pi_v|_{U_1}$ is a continuous semi-algebraic bijection onto V_1, it is a homeomorphism onto V_1. □

Proposition 9.5.6. *Let $\Gamma \subset R^n$ be an algebraic curve and $a \in \Gamma$. There exist a linear mapping $\Pi : R^n \to R^2$, an algebraic curve $C \subset R^2$ such that $\Pi(a) \in C$, a neighbourhood U of a in Γ and a neighbourhood V of $\Pi(a)$ in C such that $\Pi|_U$ is a homeomorphism onto V.*

Proof. By Proposition 9.5.5, there is a linear mapping $\Pi : R^n \to R^2$, a neighbourhood U of a in Γ, a neighbourhood V of $\Pi(a)$ in $\Pi(\Gamma)$ and a finite set of points $B \subset \Gamma$, such that $\Pi|_U$ is a homeomorphism onto V and such that $\Pi|_{\Gamma \setminus B}$ is injective. By Lemma 11.4.3, this last property implies that $\mathrm{clos}_{\mathrm{Zar}}(\Pi(\Gamma)) \setminus \Pi(\Gamma)$ has only finitely many points. Hence V is a neighbourhood of $\Pi(a)$ in the plane algebraic curve $C = \mathrm{clos}_{\mathrm{Zar}}(\Pi(\Gamma))$. □

Theorem 9.5.7. *Let $\Gamma \subset R^n$ be an algebraic curve and $a \in \Gamma$. The number of half-branches of Γ centred at a is even.*

Proof. Using Proposition 9.5.6, we may assume that Γ is a plane algebraic curve, i.e. $n = 2$. If a is a nonsingular point of Γ, then a neighbourhood of a in Γ is Nash diffeomorphic to an open interval of R, and hence the number of half-branches is equal to two. Now suppose that $a = (b, c)$ is a singular point of Γ. Let $f(X, Y)$ be an equation of Γ without a multiple factor. Changing, if necessary, the Y axis, we may assume that f is monic with respect to Y and a is the only singular point of Γ on the line $X = b$. Choose $\epsilon > 0$ such that the discriminant of f with respect to Y has no zero in the intervals $]b - \epsilon, b[$ and $]b, b + \epsilon[$.

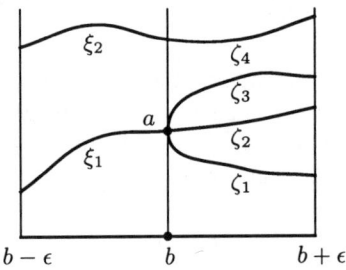

Figure 9.8.

The real roots of $f(x, Y)$ are given by Nash functions $\xi_1 < \cdots < \xi_r$ for $x \in]b - \epsilon, b[$ and $\varsigma_1 < \cdots < \varsigma_s$ for $x \in]b, b + \epsilon[$. Both r and s have the same parity as the degree of f with respect to Y. Hence $r + s$ is even.

Since f is monic with respect to Y, every function ξ_i and ς_j has a finite limit as x tends to b. If this limit is $d \ne c$, then the point (b, d) is a nonsingular point of Γ. Since there are two half-branches centred at each nonsingular point, the number of half-branches centred at a is even. □

Remark 9.5.8. a) The *half-branches of a germ of a Nash curve* can be defined as for algebraic curves, and their number is also even. Indeed, Theorem 8.6.12 permits a reduction to the case of an algebraic curve.

b) The *half-branches at infinity* of an algebraic curve Γ can be defined as follows. Choose an algebraic Alexandrov compactification $\dot{\Gamma} \simeq \Gamma \cup P$ of Γ (cf. Proposition 3.5.3), and consider the half-branches of $\dot{\Gamma}$ centred at P. The number of half-branches at infinity is also even.

9.6 The Theorems of Sard and Bertini

First, we state the constant rank theorem for Nash functions which follows from the implicit function theorem, in the same way as for C^∞ functions (cf. [115]).

Theorem 9.6.1. *Let f be a Nash mapping from an open semi-algebraic subset A of R^n to R^m, such that the rank of $d_x f$ is constant and equal to p for every $x \in A$. Let a be a point in A. There exist Nash diffeomorphisms $u :]-1,1[^n \to U$ and $v : V \to]-1,1[^m$, where $U \subset A$ is an open semi-algebraic neighbourhood of a and $V \subset R^m$ an open semi-algebraic neighbourhood of $f(a)$ containing $f(U)$, such that $v \circ f|_U \circ u$ is the projection mapping $(x_1, \ldots, x_n) \mapsto (x_1, \ldots, x_p, 0, \ldots, 0)$.*

Next we state the semi-algebraic version of Sard's theorem. If $f : N \to M$ is a Nash mapping between two Nash manifolds N and M, a *critical point* of f is a point x of N where the rank of $d_x f : T_x(N) \to T_{f(x)}(M)$ is smaller than the dimension of M. A *critical value* of f is the image by f of a critical point.

Theorem 9.6.2 (Sard's Theorem). *Let $f : N \to M$ be a Nash mapping between two Nash manifolds. The set of critical values of f is a semi-algebraic subset of M, of dimension smaller than the dimension of M.*

Proof. By Corollary 9.3.10, we may assume that M (resp. N) is an open semi-algebraic subset of R^m (resp. R^n). Let $S \subset N$ be the set of critical points of f. The set S is semi-algebraic since the partial derivatives of f are Nash functions. By Proposition 2.9.10, S is the finite union of semi-algebraic sets S_i which are the images of Nash embeddings $\varphi_i :]0,1[^{d_i} \to N$. The rank of the composite function $f \circ \varphi_i$ is $< m$. By the following lemma, the dimension of the image of $f \circ \varphi_i$ is $< m$.

Lemma 9.6.3. *Let $g :]0,1[^d \to R^m$ be a Nash mapping such that the rank of $d_x g$ is $< m$, for every $x \in]0,1[^d$. Then the dimension of the image of g is $< m$.*

Proof. Suppose that $\dim(g(]0,1[^d)) = m$. Applying Corollary 9.3.3 to g, we can find an open semi-algebraic subset U of R^m contained in $g(]0,1[^d)$, and a semi-algebraic trivialization of g over U. Hence, g is open on $g^{-1}(U)$. Choose $a \in g^{-1}(U)$ such that the rank of $d_a g$ is maximal. Then the rank of $d_x g$ is constant and smaller than m on a neighbourhood of a. By the constant rank theorem 9.6.1, there is a semi-algebraic neighbourhood of a whose image by g has dimension $< m$, which contradicts the fact that g is open. □ □

Finally, we give a "real" version of Bertini's theorem.

Theorem 9.6.4 (Bertini's Theorem). *Let $V \subset R^n$ be a nonsingular algebraic set (resp. $M \subset R^n$ a Nash submanifold) and L an affine subspace of R^n of dimension $n - 2$, disjoint from V (resp. M). Then, except for a finite number of hyperplanes, the intersection of V (resp. M) with a hyperplane H of R^n containing L is nonsingular (resp. is a Nash submanifold).*

Proof. By Proposition 3.3.8, we may assume that V is defined by equations $f_1(x) = \cdots = f_k(x) = 0$, $f_i \in R[X_1, \ldots, X_n]$, such that the rank of $(d_x f_1, \ldots, d_x f_k)$ is equal to k for every $x \in V$. Indeed, R^n can be covered by a finite number of Zariski open subsets where this is true. Let $\Phi_0^{-1}(0)$ and $\Phi_\infty^{-1}(0)$ be two distinct hyperplanes containing L. The family of hyperplanes containing L, except $\Phi_\infty^{-1}(0)$, can be parametrized by $(\Phi_0 + t\Phi_\infty)^{-1}(0)$, where $t \in R$. The point y is a singular point of the intersection of V with the hyperplane $(\Phi_0 + t\Phi_\infty)^{-1}(0)$ if and only if $d_y\Phi_0 + t d_y\Phi_\infty$ vanishes on the kernel of $(d_y f_1, \ldots, d_y f_k)$, i.e. if and only if y is a critical point of the function $x \mapsto -\Phi_0(x)/\Phi_\infty(x)$ from $\{x \in V \mid \Phi_\infty(x) \neq 0\}$ to R. Sard's theorem 9.6.2 implies that the number of critical values of this function is finite. The proof in the case of a Nash submanifold M is similar, except that Corollary 9.3.10 is used for the initial reduction and $f_1, \ldots, f_k$ are Nash functions. □

9.7 Whitney's Conditions a and b

In this section, we prove the existence of Nash stratifications satisfying Whitney's conditions a and b. These conditions will not be needed later in the book. However, they are important for problems which are beyond the scope of this book, for example in the proof of the theorem stating that the topologically stable C^∞ functions are dense in the space of $\mathcal{C}^\infty$ functions [135, 226, 227].

Whitney's conditions are formulated in terms of limits of tangent spaces. These limits are taken with respect to the euclidean topology of the grassmannians $\mathbb{G}_{n,k}(R)$, which is induced by their structure of real algebraic manifold (Remark 3.2.15). First, we give the classical statements of conditions a and b over $\mathbb{R}$.

Definition 9.7.1. *Let X and Y be two disjoint connected Nash submanifolds of $\mathbb{R}^n$ such that $Y \subset \mathrm{clos}(X)$. Let $y \in Y$ and $k = \dim(X)$.*

a) The pair (X, Y) is said to satisfy condition a at the point y if, for every sequence $(x_\nu)_{\nu \in \mathbb{N}}$ of points of X such that $\lim_{\nu \to \infty} x_\nu = y$ and $\lim_{\nu \to \infty} T_{x_\nu}(X) = \tau \in \mathbb{G}_{n,k}(\mathbb{R})$, τ contains $T_y(Y)$.

b) The pair (X, Y) is said to satisfy condition b at the point y if, for every sequences $(x_\nu)_{\nu \in \mathbb{N}}$ of points of X and $(y_\nu)_{\nu \in \mathbb{N}}$ of points of Y such that $\lim_{\nu \to \infty} x_\nu = \lim_{\nu \to \infty} y_\nu = y$, $\lim_{\nu \to \infty} T_{x_\nu}(X) = \tau \in \mathbb{G}_{n,k}(\mathbb{R})$ and $\lim_{\nu \to \infty} \mathbb{R}(x_\nu - y_\nu) = \delta \in \mathbb{P}_{n-1}(\mathbb{R})$, one has $\delta \subset \tau$.

9.7 Whitney's Conditions a and b

Conditions a and b are better understood by looking at examples where they are not satisfied.

Example 9.7.2. a) Let $X = \{(x,y,z) \in \mathbb{R}^3 \mid x^3 - z(x^2+y^2) = 0 \text{ and } y > 0\}$. The set X is an open subset of the cloth of the umbrella of Figure 3.2. Let $Y \subset \mathbb{R}^3$ be the line $x = z$, $y = 0$. The pair (X,Y) does not satisfy condition a at the origin. Indeed, when approaching the origin along the half axis y contained in X, the tangent plane to X is always the plane $z = 0$ which does not contain the line Y.

b) Let $X = \{(x,y,z) \in \mathbb{C}^3 \mid x^3 + y^2 - z^2 x^2 = 0 \text{ and } x \neq 0\}$ and $Y \subset \mathbb{C}^3$ be the z axis. The sets X and Y are *connected* Nash submanifolds of $\mathbb{C}^3 \simeq \mathbb{R}^6$. The pair (X,Y) satisfies condition a at the origin, but not condition b, as can be seen in Figure 9.9 which shows the real parts of X and Y.

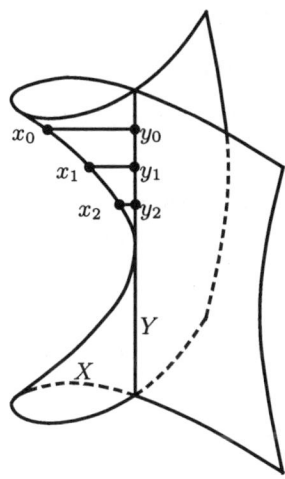

Figure 9.9.

We give now a formulation of conditions a and b which does not use sequences and which allows one to study these conditions over an arbitrary real closed field R.

Definition 9.7.3. *Let X and Y be two disjoint semi-algebraically connected Nash submanifolds of R^n such that $Y \subset \text{clos}(X)$. Let $k = \dim(X)$. Define*

$$F_a(X) = \{(x,T) \in R^n \times \mathbb{G}_{n,k}(R) \mid x \in X \text{ and } T = T_x(X)\},$$
$$F_b(X,Y) = \{(x,T,y,d) \in F_a(X) \times R^n \times \mathbb{P}_{n-1}(R) \mid y \in Y \text{ and } d = R(x-y)\}.$$

a) The pair (X,Y) is said to satisfy condition a *at the point $y \in Y$ if, for every $\tau \in \mathbb{G}_{n,k}(R)$,*

$$(y, \tau) \in \text{clos}(F_a(X)) \Rightarrow T_y(Y) \subset \tau.$$

b) The pair (X,Y) is said to satisfy condition b at the point $y \in Y$ if, for every $\tau \in \mathbb{G}_{n,k}(R)$ and $\delta \in \mathbb{P}_{n-1}(R)$,

$$(y, \tau, y, \delta) \in \mathrm{clos}(F_b(X,Y)) \Rightarrow \delta \subset \tau .$$

It is easy to verify that Definitions 9.7.1 and 9.7.3 coincide in the case $R = \mathbb{R}$.

Proposition 9.7.4. *We use the notation of Definition 9.7.3.*
  (i) *The sets $F_a(X)$ and $F_b(X,Y)$ are semi-algebraic.*
  (ii) *The set of points $y \in Y$ where the pair (X,Y) does not satisfy condition a (resp. b) is semi-algebraic.*

Proof. (i) follows immediately from the fact that the mappings $x \mapsto T_x(X)$ from X into $\mathbb{G}_{n,k}(R)$ and $(x,y) \mapsto R(x-y)$ from $X \times Y$ into $\mathbb{P}_{n-1}(R)$ are Nash. The assertion concerning the first mapping is proved as in Corollary 3.4.8.
  (ii) is an immediate consequence of (i) and Definition 9.7.3. □

The main point of this section is to prove that the set of points where conditions a and b are not satisfied is "small".

Theorem 9.7.5. *Let X and Y be two disjoint semi-algebraically connected Nash submanifolds of R^n such that $Y \subset \mathrm{clos}(X)$. Then the set of points of Y where (X,Y) does not satisfy condition a (resp. condition b) is of dimension smaller than $\dim(Y)$.*

The proof requires some preparation. First, let us examine how conditions a and b are related.

Definition 9.7.6. *Let X and Y be two disjoint semi-algebraically connected Nash submanifolds of R^n such that $Y \subset \mathrm{clos}(X)$. Let V be a Nash tubular neighbourhood of Y with the orthogonal retraction $\rho : V \to Y$ (cf. Corollary 8.9.5). Define*

$$F_{b'}(X,Y) = \{(x, T, d) \in F_a(X) \times \mathbb{P}_{n-1}(R) \mid x \in V \text{ and } d = R(x - \rho(x))\} .$$

The pair (X,Y) is said to satisfy condition b' at the point $y \in Y$ if, for every $\tau \in \mathbb{G}_{n,k}(R)$ and $\delta \in \mathbb{P}_{n-1}(R)$,

$$(y, \tau, \delta) \in \mathrm{clos}(F_{b'}(X,Y)) \Rightarrow \delta \subset \tau .$$

Proposition 9.7.7. *The pair (X,Y) satisfies condition b at the point $y \in Y$ if and only if it satisfies conditions a and b' at this point.*

Proof. It is obvious that condition b is stronger than condition b'. We shall prove that condition b implies condition a. Suppose that condition a is not satisfied at the point $y \in Y$. Then there exists a continuous semi-algebraic mapping $\gamma : [0,1] \to R^n \times \mathbb{G}_{n,k}(R)$ such that, for every $t \in {]0,1]}$, $\gamma(t) =$

$(\gamma_1(t), T_{\gamma_1(t)}(X))$ with $\gamma_1(t) \in X$, and $\gamma(0) = (y, \tau)$ with $\tau \not\supset T_y(Y)$. Let $\varphi :]-\epsilon, \epsilon[\to Y$ be a Nash mapping such that $\varphi(0) = y$ and $\varphi'(0)$ is a vector of $T_y(Y)$ which does not belong to τ. We denote by δ the line generated by $\varphi'(0)$. Changing, if necessary, the parametrization of γ, we may assume that the exponent of the first term of the Puiseux series expansion of $\gamma_1(t) - y$ is > 2. Hence $\|\gamma_1(t) - y\| \leq t^2$ for t small enough, and the limit of the lines $R(\gamma_1(t) - \varphi(t))$, as t tends to 0, is δ. Then $(y, \tau, y, \delta) \in \mathrm{clos}(F_b(X,Y))$ and $\delta \not\subset \tau$, and, hence, condition b is not satisfied at the point y.

Now we prove the converse. Suppose that conditions a and b' are satisfied at the point $y \in Y$, and let $(y, \tau, y, \delta) \in \mathrm{clos}(F_b(X,Y))$. Then there is a continuous semi-algebraic mapping

$$\begin{aligned}\gamma : [0,1] &\longrightarrow R^n \times \mathbb{G}_{n,k}(R) \times R^n \times \mathbb{P}_{n-1}(R) \\ t &\longmapsto (\gamma_1(t), T_{\gamma_1(t)}(X), \gamma_2(t), R(\gamma_1(t) - \gamma_2(t))),\end{aligned}$$

where $\gamma_1(t) \in X$ and $\gamma_2(t) \in Y$ for $t > 0$ and $\gamma(0) = (y, \tau, y, \delta)$. By condition a, $\tau \supset T_y(Y)$, and, by condition b', τ also contains the limit of the lines $R(\gamma_1(t) - \rho(\gamma_1(t)))$, as t tends to 0 (this limit exists by Proposition 2.5.3). Moreover, if $\rho(\gamma_1(t))$ and $\gamma_2(t)$ do not coincide for small values of t, then the limit of the lines $R(\rho(\gamma_1(t)) - \gamma_2(t))$, as t tends to 0, exists by Proposition 2.5.3 and is contained in $T_y(Y)$. This shows that δ, which is the limit of the lines $R(\gamma_1(t) - \gamma_2(t))$, is contained in τ. Hence, condition b is satisfied at the point y. $\square$

Note that the previous proposition shows, in particular, that condition b implies condition a for pairs of Nash submanifolds.

The proof of Theorem 9.7.5 will use "Nash wings".

Definition 9.7.8. *Let Y be a Nash submanifold of R^n and T, a Nash tubular neighbourhood of Y, with the orthogonal retraction $\rho : T \to Y$. A Nash wing with axis Y is a Nash mapping*

$$w :]-1,1[\times Y \to R^n$$

which is a homeomorphism onto its image and satisfies the following properties for every $y \in Y$:

(i) $w(0, y) = y$,
(ii) *for every* $t \in]-1, 1[$, $w(t, y) \in T$ *and* $\rho(w(t, y)) = y$,
(iii) *for every* $t \in]-1, 1[$, *if* $t \neq 0$ *then* $\frac{\partial w}{\partial t}(t, y) \neq 0$,
(iv) $\frac{\partial^i w}{\partial t^i}(0, y) = 0$, *for* $0 < i < q$, *and* $\frac{\partial^q w}{\partial t^q}(0, y) \neq 0$, *where q does not depend on y.*

Proposition 9.7.9. *Let w be a Nash wing with axis Y and $X = w(]0, 1[\times Y)$. Then the pair (X, Y) satisfies conditions a and b at every point $y \in Y$.*

Proof. Taking local coordinates, we may assume that Y is an open subset of R^p and ρ is the restriction of the projection of R^n onto R^p, where R^p is identified with the space $\{0\} \times R^p \subset R^{n-p} \times R^p$. Note that X is a Nash submanifold of dimension $p+1$. The tangent space $T_x(X)$ at the point $x = w(t,y)$ is generated by the vectors $\frac{\partial w}{\partial t}(t,y), \frac{\partial w}{\partial y_1}(t,y), \ldots, \frac{\partial w}{\partial y_p}(t,y)$. When x tends to $y^0 \in Y$, i.e. when (t,y) tends to $(0, y^0)$, the vector $\frac{(q-1)!}{t^{q-1}} \frac{\partial w}{\partial t}(t,y)$ tends to $\frac{\partial^q w}{\partial t^q}(0, y^0)$, while the vectors $\frac{\partial w}{\partial y_i}(t,y)$ tend to the vectors of the canonical basis of R^p. Hence the unique vector subspace $\tau \in \mathbb{G}_{n,p+1}(R)$ such that $(y^0, \tau) \in \text{clos}(F_a(X))$ is generated by $\frac{\partial^q w}{\partial t^q}(0, y^0)$ and R^p. This already shows that condition a is satisfied at the point y^0. Finally, since the vector $\frac{q!}{t^q}(\rho(x) - x) = \frac{q!}{t^q}(w(0,y) - w(t,y))$ tends to $-\frac{\partial^q w}{\partial t^q}(0, y^0)$ as x tends to y^0, condition b' is also verified at the point y^0. Application of Proposition 9.7.7 completes the proof. □

Now we need a theorem about the existence of Nash wings. This theorem is a parametrized version of the Nash curve selection lemma. The proof relies on the study of families of Nash mappings made in Chap. 8, Section 10.

Theorem 9.7.10 (Wing Lemma). *Let $Y \subset R^n$ be a Nash submanifold and $S \subset R^n$ a semi-algebraic set such that $Y \cap S = \emptyset$ and $Y \subset \text{clos}(S)$. There exist a finite number of Nash wings*

$$w_i :]-1,1[\times U_i \to R^n, \quad i = 1, \ldots, k,$$

where $U_1, \ldots, U_k$ are open semi-algebraic subsets of Y, such that $w_i(]0,1[\times U_i)$ is contained in S, for $i = 1, \ldots, k$, and $\dim\left(Y \setminus \bigcup_{i=1}^k U_i\right) < \dim(Y)$.

Proof. By taking a finite number of Nash charts (cf. Corollary 9.3.10), we may assume that Y is a semi-algebraic open subset of R^p which can be identified with $\{0\} \times R^p \subset R^m \times R^p = R^n$. Let α be a point of dimension p of $\widetilde{Y}$. We consider the set $S \subset R^m \times R^p$ as a semi-algebraic family of sets parametrized by R^p, and we take its fibre $S_\alpha \subset k(\alpha)^m$. Then $0 \in k(\alpha)^m$ belongs to the closure of S_α. Otherwise, we could find an open semi-algebraic subset $W \subset k(\alpha)^m$ such that $0 \in W$ and $W \cap S_\alpha = \emptyset$. By Propositions 7.4.6 and 7.4.1, there would be an open semi-algebraic subset $T \subset Y$, with $\alpha \in \widetilde{T}$, and an open semi-algebraic subset $V \subset R^m \times T$, such that $\{0\} \times T \subset V$ and $V \cap S = \emptyset$, which contradicts the assumption $Y \subset \text{clos}(S)$.

Hence, we can apply the Nash curve selection lemma 8.1.13. We obtain a Nash mapping $\omega :]-1,1[_{k(\alpha)} \to k(\alpha)^m$ such that $\omega(0) = 0$ and $\omega(t) \in S_\alpha$ for every $t \in]0,1[_{k(\alpha)}$. Since ω is not constant, we may assume that ω is a homeomorphism onto its image, $\frac{d\omega}{dt}(t) \neq 0$ for every $t \neq 0$, $\frac{d^i\omega}{dt^i}(0) = 0$ for $0 < i < q$ and $\frac{d^q\omega}{dt^q}(0) \neq 0$. By Proposition 8.10.3, there is an open semi-algebraic subset U^α of R^p such that $\alpha \in \widetilde{U^\alpha}$, and a Nash mapping

$$w :]-1,1[\times U^\alpha \to R^m \times R^p = R^n$$

whose fibre w_α is equal to w. If we shrink U^α sufficiently, w becomes a homeomorphism onto its image, and, for every $y \in U^\alpha$, $w(0,y) = y$, $\frac{\partial w}{\partial t}(t,y) \neq 0$ for $t \neq 0$, $\frac{\partial^i w}{\partial t^i}(0,y) = 0$ for $0 < i < q$ and $\frac{\partial^q w}{\partial t^q}(0,y) \neq 0$. Hence, w is a Nash wing with axis U^α. Shrinking U^α further, we have also $w(t,y) \in S$, for every $t \in\,]0,1[$ and every $y \in U^\alpha$. The $\widetilde{U^\alpha}$ cover the set of points of dimension p of $\widetilde{Y}$. This set is a closed and open subset of the space of orderings of the field of rational functions on R^p, and, hence, it is compact. Therefore we can choose finitely many open subsets $U_1, \ldots, U_k$ among the U^α such that $\dim\left(Y \setminus \bigcup_{i=1}^{k} U_i\right) < p$. □

Proof of theorem 9.7.5. Denote by $S_a(X,Y)$ (resp. $S_{b'}(X,Y)$) the set of points of Y where condition a (resp. b') is not satisfied. It is obvious that $S_{b'}(X,Y)$ is also semi-algebraic. By Proposition 9.7.7, it suffices to prove that the dimensions of $S_a(X,Y)$ and $S_{b'}(X,Y)$ are smaller than the dimension of Y.

We may assume, as in the proof of 9.7.10, that Y is a semi-algebraic open subset of R^p identified with $\{0\} \times R^p \subset R^n$. The orthogonal retraction ρ of the Nash tubular neighbourhood of Y is then the restriction of the projection mapping $R^n \to R^p$.

The identification of $\mathbb{G}_{n,\ell}(R)$ with an algebraic subset of R^{n^2} (Theorem 3.4.4) gives a metric $d : \mathbb{G}_{n,\ell}(R) \times \mathbb{G}_{n,\ell}(R) \to R$. If $\ell \leq k$, we define a function, also denoted by $d : \mathbb{G}_{n,\ell}(R) \times \mathbb{G}_{n,k}(R) \to R$, by

$$d(\delta, \tau) = \inf(\{d(\delta, \gamma) \mid \gamma \in \mathbb{G}_{n,\ell}(R), \gamma \subset \tau\}).$$

Note that $\{\gamma \in \mathbb{G}_{n,\ell}(R) \mid \gamma \subset \tau\}$ is closed and bounded. The function d is semi-algebraic and continuous, and $d(\delta, \tau) = 0$ if and only if $\delta \subset \tau$.

For $y \in Y$ and $t \in R$, $t > 0$, set

$$\varphi(y,t) = \sup(\{d(R^p, T_x(X)) \mid x \in X \text{ and } \|x - y\| \leq t\}).$$

For y fixed, the function $t \mapsto \varphi(y,t)$ is semi-algebraic and bounded and, by Proposition 2.5.3, it has a limit $f(y) \in R$, as t tends to 0. The function $f : Y \to R$ is semi-algebraic, and $y \in S_a(X,Y)$ if and only if $f(y) \neq 0$. Suppose that $\dim(S_a(X,Y)) = p$. Since f is piecewise continuous on Y, there are $c \in R$, $c > 0$, and a semi-algebraic open subset $U \subset Y$ such that $f > c$ on U. Define

$$S = \{x \in X \mid d(R^p, T_x(X)) \geq c\}.$$

Since $U \subset \text{clos}(S)$, the wing lemma 9.7.10 can be applied. This gives a Nash wing $w :\,]-1,1[\times U' \to R^n$ whose axis U' is an open subset of U, and such that $w(]0,1[\times U') \subset S$. By Proposition 9.7.9, the pair $(U', w(]0,1[\times U'))$ satisfies condition a at every point of U', which contradicts the definition of S. Therefore $\dim(S_a(X,Y)) < p$.

To prove that $\dim(S_{b'}(X,Y)) < p$, we proceed in the same way, replacing the function φ with the function ψ defined by:

$$\psi(y,t) = \sup(\{d(R(x - \rho(x)), T_x(X)) \mid x \in X \text{ and } \|x - y\| \le t\}) \,.$$

□

Now, let $(E_i)_{i\in I}$ be a Nash stratification of a semi-algebraic subset $E = \bigcup_{i\in I} E_i$ of R^n. The stratification $(E_i)_{i\in I}$ is said to satisfy condition a (resp. b) if every pair of distinct strata (E_i, E_j), with $E_j \subset \mathrm{clos}(E_i)$, satisfies condition a (resp. b) at every point of E_j.

Theorem 9.7.11. *Let $(E_i)_{i\in I}$ be a Nash stratification of the semi-algebraic set $E = \bigcup_{i\in I} E_i$. There exists a Nash stratification $(F_\lambda)_{\lambda \in \Lambda}$ of E satisfying conditions a and b and finer than the stratification $(E_i)_{i\in I}$ (i.e. each stratum E_i is the union of some strata F_λ).*

Proof. The proof is by induction. Assume that for $k \in \mathbb{N}$ the following property (∗) holds true:

Every pair of distinct strata (E_i, E_j), such that $E_j \subset \mathrm{clos}(E_i)$ and $\dim(E_j) > k$, satisfies conditions a and b at every point of E_j.

If E_j is a k-stratum, define

$$T_j = \mathrm{clos}(\bigcup \{S_b(E_i, E_j) \mid E_j \subset \mathrm{clos}(E_i), \, E_j \ne E_i\}) \cap E_j \,,$$

where $S_b(E_i, E_j)$ is the set of points of E_j where the pair (E_i, E_j) does not satisfy condition b. By Theorem 9.7.5, we have $\dim(T_j) < k$. With an appropriate Nash stratification of the union of the T_j and the strata E_ℓ of dimension $< k$ (Proposition 9.1.8), we obtain a Nash stratification of E finer than $(E_i)_{i\in I}$, with the same strata of dimension $> k$ as $(E_i)_{i\in I}$ and whose strata of dimension k are the connected components of the $E_j \setminus T_j$. This new stratification satisfies the property (∗) for $k - 1$. □

Bibliographic Notes. The notion of a stratifying family of polynomials and the properties of the stratification given by such a family have two origins: one is the separation lemma (or the generalized Thom's lemma) in Efroymson's paper [126], reviewed by Houdebine [95], and the other is the stratification technique used by Hardt [146] to study semi-algebraic sets.

The first result about the triangulation of algebraic sets can be found in van der Waerden's paper [333]. Lojasiewicz [216] and Giesecke [136] deal with the triangulation of semi-analytic sets. Hironaka [159] gives a simpler proof in the semi-algebraic framework. The triangulation of semi-algebraic sets over an arbitrary real closed field appears in the unpublished thesis of Brakhage [67]. The construction in Lemma 9.2.2 is taken from [147]. The semi-algebraic triviality of semi-algebraic mappings, due to Hardt [148], extends some previous results by Varčenko [330] and Wallace [336]. Delfs and Knebusch [109] give proofs of triangulation and semi-algebraic triviality valid for an arbitrary real closed field. The local conic structure of algebraic sets can be found in

Milnor's book [233], where it is used for the study of isolated singularities. The paper of Shiota and Yokoi [304] contains an important theorem on the uniqueness of semi-algebraic triangulation : If K and L are finite simplicial complexes in $\mathbb{R}^n$ and there is a semi-algebraic homeomorphism from $|K|$ to $|L|$, then there is a piecewise linear homeomorphism from $|K|$ to $|L|$. As a consequence one obtains, in particular, an example of two compact semi-algebraic subsets of $\mathbb{R}^n$ which are homeomorphic but not semi-algebraically homeomorphic.

The triangulation of semi-algebraic functions is proved in [303]. The uniqueness of the semi-algebraic neighbourhoods considered in Proposition 9.4.4 is proved in [124] (it can also be deduced from the uniqueness of semi-algebraic triangulation). The finiteness of the number of topological types of polynomials of bounded degree was conjectured by Thom and first proved by Fukuda [132]. See also [31] for the effective aspect of this question.

A quantitative version of the semi-algebraic Sard theorem is due to Yomdin, who applies it in the study of singularities of differentiable mappings [347, 348]. Conditions a and b for stratifications were introduced by Whitney in [340], who also proved a wing lemma. The version of this lemma given in Theorem 9.7.10 is closer to one published in [277].

10. Real Places

Abstract. In this chapter, we study the relation between the orderings of a field and its real places (or, equivalently, its valuation rings with real residue field). The main result, known as the Baer-Krull theorem, is the following: if B is a valuation ring of a field K, and γ an ordering of the residue field of B, then γ can be lifted to an ordering of K for which B is convex, and the number of such liftings depends only on the value group of B. From this classical theorem, we deduce results concerning specializations in the real spectrum, more particularly specializations of codimension 1. These results will be used later to obtain some information about the topology of real algebraic sets. In Section 3, we explain how the half-branches of real algebraic curves can be considered as points of the real spectrum. We finish the chapter with a glimpse of the theory of fans, presenting a few applications to the study of basic semi-algebraic sets.

Throughout this chapter, R is a fixed real closed field.

10.1 Real Places and Orderings

We start by recalling elementary notions of valuation theory. We refer the reader to [210], Chap. 13 §4, or [66], Chap. 6, for more details.

If k is a field, we denote by k^* the multiplicative group of its nonzero elements. We partially extend the addition and the multiplication to $k \cup \{\infty\}$ by setting $x + \infty = \infty + x = \infty$ if $x \in k$, and $x \times \infty = \infty \times x = \infty$ if $x \in k^* \cup \{\infty\}$. The sum $\infty + \infty$ and the products $0 \times \infty$ and $\infty \times 0$ are not defined.

Definition 10.1.1. *Let K be a field.*
  (i) A valuation ring of K is a subring B of K such that
$$\forall x \in K^* \quad x \in B \text{ or } x^{-1} \in B.$$

  (ii) A place of K is a mapping $\lambda : K \to k \cup \{\infty\}$, where k is a field, such that $\lambda(1) = 1$, $\lambda(x+y) = \lambda(x) + \lambda(y)$ and $\lambda(xy) = \lambda(x)\lambda(y)$, whenever the expressions on the right-hand side of these formulas are defined.

  (iii) A valuation of K is a group homomorphism v from K^* to an ordered commutative group Γ such that

$$\forall x \in K^* \ \forall y \in K^* \quad x+y \neq 0 \Rightarrow v(x+y) \geq \inf(v(x),v(y)) \ .$$

The three concepts, a valuation ring, a place and a valuation of a field, are essentially equivalent. We now recall this equivalence and fix the notations.

A valuation ring B of a field K is a local ring. We denote by $\mathfrak{m}_B$ its maximal ideal, $B^* = B \setminus \mathfrak{m}_B$ the multiplicative group of its invertible elements and $k_B = B/\mathfrak{m}_B$ its residue field. The mapping $\lambda_B : K \to k_B \cup \{\infty\}$ such that $\lambda_B|_B$ is the canonical surjection from B onto k_B, and $\lambda_B(x) = \infty$ if $x \notin B$, is a place of K. The canonical mapping $v_B : K^* \to \Gamma_B = K^*/B^*$, where Γ_B is ordered by $v_B(x) \leq v_B(y) \Leftrightarrow yx^{-1} \in B$, is a valuation of K. The ordered group Γ_B is called the *value group of B*.

If $\lambda : K \to k \cup \{\infty\}$ is a place of K, then $B = \{x \in K \mid \lambda(x) \neq \infty\}$ is a valuation ring of K, called the *valuation ring of the place* λ. There exists a unique field homomorphism $i : k_B \to k$ such that $\lambda|_B = i \circ \lambda_B|_B$.

If $v : K^* \to \Gamma$ is a valuation of K, then

$$B = \{x \in K \mid x = 0 \text{ or } v(x) \geq 0\}$$

is a valuation ring of K, called the *valuation ring of* v. There exists a unique injective homomorphism of ordered groups $j : \Gamma_B \to \Gamma$ such that $v = j \circ v_B$.

Example 10.1.2.

a) A *discrete rank 1 valuation ring* B is a valuation ring whose value group is $\mathbb{Z}$. A *uniformizing parameter of B* is an element with valuation 1. An example of a discrete rank 1 valuation ring is $\mathbb{R}[X]_{(X)} \subset \mathbb{R}(X)$. The element X is a uniformizing parameter and the valuation of $P(X)/Q(X) \in \mathbb{R}(X)$ is $m \in \mathbb{Z}$ if and only if $P(X)/Q(X) = X^m p(X)/q(X)$ with $p(0) \neq 0$, $q(0) \neq 0$. More generally, a regular local ring (see Section 3.3) of dimension 1 is a discrete rank 1 valuation ring, with uniformizing parameter a generator of the maximal ideal.

b) A *discrete rank n valuation ring* is a valuation ring with value group $\mathbb{Z}^n$, with the lexicographic order. Let V be an irreducible real algebraic set of dimension n, x a nonsingular point of V and $f_1,\ldots,f_n$ polynomial functions on V inducing a regular system of parameters of the regular local ring $\mathcal{R}_{V,x} \subset \mathcal{K}(V)$ (see Section 3.3). We associate to this regular system of parameters a discrete rank n valuation ring B of $\mathcal{K}(V)$ *dominating* $\mathcal{R}_{V,x}$ (i.e., B contains $\mathcal{R}_{V,x}$ and $\mathfrak{m}_B \cap \mathcal{R}_{V,x}$ is the maximal ideal of $\mathcal{R}_{V,x}$) and having residue field R. We proceed by induction on n. From Example 10.1.2 a), it follows that, for $n = 1$, we can take $B = \mathcal{R}_{V,x}$. Now assume $n > 1$. Define V_1 as the codimension 1 irreducible algebraic subset of V coinciding with $f_n^{-1}(0)$ in a Zariski neighbourhood of x. The point x is a nonsingular point of V_1 and $f_1,\ldots,f_{n-1}$ is a regular system of parameters of $\mathcal{R}_{V_1,x}$. Assume that we have defined a discrete rank $n-1$ valuation ring B_1 of $\mathcal{K}(V_1)$ whose residue field is R and which dominates $\mathcal{R}_{V_1,x}$. Now, the ring $\mathcal{R}_{V,V_1}$ of rational functions with denominator not vanishing on V_1 is a regular local ring of dimension 1, and therefore a discrete rank 1 valuation ring of $\mathcal{K}(V)$ with residue field

$\mathcal{K}(V_1)$. The inverse image of B_1 by the canonical surjection $\mathcal{R}_{V,V_1} \to \mathcal{K}(V_1)$ is a discrete rank n valuation ring B of $\mathcal{K}(V)$, dominating $\mathcal{R}_{V,x}$ and having residue field R. Note that $(v_B(f_n), \ldots, v_B(f_1))$ is a basis of Γ_B as a $\mathbb{Z}$-module.

Definition 10.1.3.
(i) A valuation ring B of a field K is said to be real *if its residue field k_B is real. A place (resp. a valuation) of K is said to be* real *if its valuation ring is real.*

(ii) *Let β be the positive cone of an ordering of K, and let $\leq_\beta$ denote this ordering. A subring A of K is said to be β-convex if, for all $x, y \in \beta$ such that $x + y \in A$, we have $x \in A$ (this is equivalent to requiring that, for all $z \in A$ and $x \in K$, $0 \leq_\beta x \leq_\beta z \Rightarrow x \in A$). The ordering $\leq_\beta$ is said to be* compatible with a place λ of K *if the valuation ring of λ is β-convex.*

Proposition 10.1.4. *Let $(K, \leq_\beta)$ be an ordered field and B a valuation ring of K. The following properties are equivalent:*
  (i) B *is β-convex,*
  (ii) $\mathfrak{m}_B$ *is $(\beta \cap B)$-convex (cf. Definition 4.2.3),*
  (iii) $\forall x \in \mathfrak{m}_B, \quad 1 + x >_\beta 0$.

Proof. (i) $\Rightarrow$ (iii) If $1 + x \leq_\beta 0$, then $0 \leq_\beta -x^{-1} \leq_\beta 1$ and $x^{-1} \in B$, since B is β-convex. Hence $x \notin \mathfrak{m}_B$.

(iii) $\Rightarrow$ (ii) Let $x, y \in \beta \cap B$ with $x + y \in \mathfrak{m}_B$. If $x \notin \mathfrak{m}_B$, then $x^{-1} \in B$ and $x^{-1}(x + y) = 1 + x^{-1}y \in \mathfrak{m}_B$. By (iii), $1 - (1 + x^{-1}y) = -x^{-1}y >_\beta 0$, which is impossible. Hence $x \in \mathfrak{m}_B$.

(ii) $\Rightarrow$ (i) If B is not β-convex, then there are $x \in B$, $y \notin B$ such that $0 \leq_\beta y \leq_\beta x$. It follows that $y^{-1} \in \mathfrak{m}_B$ and $1 \leq_\beta y^{-1}x \in \mathfrak{m}_B$. Hence, by (ii), $1 \in \mathfrak{m}_B$, which is impossible. $\square$

Example 10.1.5. Let β be the positive cone of the ordering 0_+ of $\mathbb{R}(X)$ (Example 1.1.2). Let B be the localization $\mathbb{R}[X]_{(X)} = \mathcal{R}_{\mathbb{R},0}$. Using Proposition 10.1.4 (iii), it is easy to prove that B is a β-convex valuation ring.

Proposition 10.1.6. *Let $(K, \leq_\beta)$ be an ordered field and B a β-convex valuation ring of K. There exists a unique ordering $\leq_{\overline{\beta}}$ of the residue field k_B such that, for every x in B^*, $\lambda_B(x) >_{\overline{\beta}} 0$ if and only if $x >_\beta 0$. In particular, B is a real valuation ring.*

Proof. We first prove that, if $x, x' \in B^*$ and $\lambda_B(x) = \lambda_B(x')$, then x and x' have the same sign with respect to $\leq_\beta$. Indeed, $y = x' - x \in \mathfrak{m}_B$ and $x' = x(1 + x^{-1}y)$, where, by Proposition 10.1.4 (iii), $1 + x^{-1}y >_\beta 0$. Thus it is clear that we define an ordering $\leq_{\overline{\beta}}$ of k_B by setting $\lambda_B(x) >_{\overline{\beta}} 0 \Leftrightarrow x >_\beta 0$. $\square$

Definition 10.1.7. *The ordering $\leq_{\overline{\beta}}$ of Proposition 10.1.6 is called* the ordering of k_B induced by $\leq_\beta$. *Its positive cone is*

$$\overline{\beta} = \{y \in k_B \mid y = 0 \text{ or } (\exists x \in B^* \ y = \lambda_B(x) \text{ and } x >_\beta 0)\}.$$

Proposition 10.1.8. *Let K be a field and B a real valuation ring of K. Let γ be the positive cone of an ordering of k_B. There exists at least one ordering of K, with positive cone β, compatible with the place λ_B and such that $\gamma = \overline{\beta}$ (i.e. the ordering $\leq_\gamma$ is the ordering induced by $\leq_\beta$).*

Proof. Let
$$P = \{x \in K \mid \exists y \in K \; \exists z \in B^* \; \lambda_B(z) >_\gamma 0 \text{ and } x = y^2 z\}.$$

First we show that P is a proper cone of K. The properties $x \in P, y \in P \Rightarrow xy \in P$ and $x \in K \Rightarrow x^2 \in P$ are obvious. If $-1 \in P$, then there exist $y \in K$, $z \in B^*$ such that $\lambda_B(z) >_\gamma 0$ and $z = -y^{-2}$. Since $\lambda_B(-y^{-2}) \leq_\gamma 0$ or $\lambda_B(-y^{-2}) = \infty$, we obtain a contradiction. It remains to prove that $P + P \subset P$. Let $x_i = z_i y_i^2$ with $z_i \in B^*$, $\lambda_B(z_i) > 0$ for $i = 1, 2$. Assume that $y_2 y_1^{-1} \in B$ (otherwise $y_1 y_2^{-1} \in B$). Then, $x_1 + x_2 = z_1 y_1^2 (1 + z_1^{-1} z_2 y_1^{-2} y_2^2)$. Set $z = 1 + z_1^{-1} z_2 y_1^{-2} y_2^2$. Since
$$\lambda_B(z_1^{-1} z_2 y_1^{-2} y_2^2) = (\lambda_B(z_1))^{-1} \lambda_B(z_2)(\lambda_B(y_2 y_1^{-1}))^2,$$
we have $\lambda_B(z) > 0$ and therefore $x_1 + x_2 = (z_1 z) y_1^2 \in P$.

Thus there exists an ordering of K whose positive cone β contains P (Lemma 1.1.7). It is obvious that B is β-convex (Proposition 10.1.4 (iii)) and $\overline{\beta} = \gamma$. □

Corollary 10.1.9. *A field K having a real valuation ring is real.*

Theorem 10.1.10 (Baer-Krull Theorem). *Let B be a real valuation ring of the field K and γ the positive cone of an ordering of k_B. The set $F(\gamma)$ of orderings $\leq_\beta$ of K compatible with the place λ_B and inducing the ordering $\leq_\gamma$ of k_B has a canonical structure of affine space over $\mathbb{Z}/2$, with underlying vector space $\mathrm{Hom}(\Gamma_B, \mathbb{Z}/2)$.*

Proof. By Proposition 10.1.8, there is an ordering $\leq_{\beta_0}$ of K such that B is β_0-convex and $\gamma = \overline{\beta_0}$. If β is another element of $F(\gamma)$, we define $\langle \beta_0, \beta \rangle \in \mathrm{Hom}(\Gamma_B, \mathbb{Z}/2)$ by $\langle \beta_0, \beta \rangle(v_B(x)) = 0$ if x has the same sign with respect to $\leq_\beta$ and $\leq_{\beta_0}$, and $\langle \beta_0, \beta \rangle(v_B(x)) = 1$ otherwise. It is obvious that $x \mapsto \langle \beta_0, \beta \rangle(v_B(x))$ is a homomorphism from the multiplicative group K^* into $\mathbb{Z}/2$. The kernel of this homomorphism contains B^*. Indeed, if $x \in B^*$, then either $\lambda_B(x) >_\gamma 0$ or $\lambda_B(x) <_\gamma 0$, so that, for every $\beta \in F(\gamma)$, we have $x >_\beta 0$ in the first case and $x <_\beta 0$ in the second case. Hence, $\langle \beta_0, \beta \rangle : \Gamma_B \to \mathbb{Z}/2$ is a well-defined group homomorphism. The mapping $\beta \mapsto \langle \beta_0, \beta \rangle$ is injective since $\langle \beta_0, \beta \rangle$ and β_0 entirely determine β. Next we prove that this mapping is surjective. Let $\varphi \in \mathrm{Hom}(\Gamma_B, \mathbb{Z}/2)$ and define
$$\beta = \{x \in K \mid x = 0 \text{ or } (\varphi(v_B(x)) = 0 \text{ and } x \in \beta_0)$$
$$\text{or } (\varphi(v_B(x)) = 1 \text{ and } x \in -\beta_0)\}.$$

It is sufficient to prove that β is an element of $F(\gamma)$. First we prove that β is the positive cone of an ordering. The only property which is not obvious is $\beta + \beta \subset \beta$. Let $x, y \in \beta \setminus \{0\}$ and assume that $x^{-1}y \in B$ (otherwise, $y^{-1}x \in B$). If $x^{-1}y \in \mathfrak{m}_B$, then $v_B(1 + x^{-1}y) = 0$ and $1 + x^{-1}y \in \beta_0$. Hence, $x + y = x(1 + x^{-1}y) \in \beta$. If $x^{-1}y \in B^*$, then $x^{-1}y \in \beta$ and $v_B(x^{-1}y) = 0$. Therefore $x^{-1}y \in \beta_0$. It follows that $1 + x^{-1}y \in B^* \cap \beta_0$. Since $B^* \cap \beta_0 \subset \beta$, we conclude that $x + y = x(1 + x^{-1}y) \in \beta$. The fact that B is β-convex is easily proved using property (iii) of Proposition 10.1.4, and $\overline{\beta} = \gamma$ is obvious.

Finally, if β_0, β and β' are three elements of $F(\gamma)$, the equality $\langle \beta_0, \beta \rangle + \langle \beta, \beta' \rangle = \langle \beta_0, \beta' \rangle$ is immediate. □

Example 10.1.11. Let V be an irreducible real algebraic set of dimension n, x a nonsingular point of V and $f_1, \ldots, f_n$ polynomial functions on V inducing a regular system of parameters of the regular local ring $\mathcal{R}_{V,x} \subset \mathcal{K}(V)$. In Example 10.1.2 b), we have associated to the regular system of parameters $(f_1, \ldots, f_n)$ a discrete rank n valuation ring B of $\mathcal{K}(V)$ which dominates $\mathcal{R}_{V,x}$. The residue field R has a unique ordering. Theorem 10.1.10 shows that there are 2^n orderings of $\mathcal{K}(V)$ compatible with the place λ_B. Moreover, since the classes of $v_B(f_1), \ldots, v_B(f_n)$ form a basis of the $\mathbb{Z}/2$-vector space $\Gamma/2\Gamma$, these 2^n orderings are characterized by the signs that they give to $f_1, \ldots, f_n$.

Definition 10.1.12. *Let $(K, \leq_\beta)$ be an ordered field and A a subring of K. The β-convex envelope of A in K is*

$$B = \{x \in K \mid \exists a \in A, \ -a \leq_\beta x \leq_\beta a\}.$$

Proposition 10.1.13. *The β-convex envelope B of A in K is the smallest β-convex valuation ring of K containing A.*

Proof. It is easy to check that B is a subring of K. Since $-1 \leq_\beta x \leq_\beta 1$ or $-1 \leq_\beta x^{-1} \leq_\beta 1$ for every x in K^*, B is a valuation ring of K. The ring B is β-convex by its very definition, and it is obvious that every β-convex valuation ring of K containing A also contains B. □

10.2 Real Places and Specialization in the Real Spectrum

First we recall the notion of centre of a place.

Definition 10.2.1. *Let $\lambda : K \to k \cup \{\infty\}$ be a place of the field K and A a subring of K. The place λ is said to be finite on A if A is contained in the valuation ring of the place λ. If λ is finite on A, the centre of λ (in A) is the prime ideal $\lambda^{-1}(0) \cap A$ of A.*

Remark 10.2.2. a) If λ is a real place which is finite on A, its centre is a real prime ideal of A.

b) A prime ideal $\mathfrak{p}$ of A is the centre of a place λ if and only if the valuation ring B of λ dominates the local ring $A_\mathfrak{p}$ (i.e. $\mathfrak{m}_B \cap A_\mathfrak{p} = \mathfrak{p}A_\mathfrak{p}$).

Proposition 10.2.3. *Let A be an integral domain, K its field of fractions, β the positive cone of an ordering of K and $\beta' = \beta \cap A \in \mathrm{Spec}_r(A)$.*

(i) *Let $\alpha \in \mathrm{Spec}_r(A)$ be a specialization of β', and B, the β-convex envelope of the local ring $A_{\mathrm{supp}(\alpha)}$ in K. The centre of the place λ_B in A is $\mathrm{supp}(\alpha)$, and the ordering $\leq_{\overline{\beta}}$ of k_B induced by $\leq_\beta$ extends the ordering $\leq_\alpha$ of $k(\mathrm{supp}(\alpha))$. Moreover,*

$$\forall x \in k_B \ \exists y \in k(\mathrm{supp}(\alpha)) , \ -y \leq_{\overline{\beta}} x \leq_{\overline{\beta}} y .$$

(ii) *Let B be a β-convex valuation ring containing A and $\mathfrak{p}$ the centre of the place λ_B in A. The inverse image of the positive cone $\overline{\beta}$ by the homomorphism $\lambda_B|_A : A \to k_B$ is a specialization of β' in $\mathrm{Spec}_r(A)$ whose support is $\mathfrak{p}$.*

Proof. (i) We prove that $\mathfrak{m}_B \cap A = \mathrm{supp}(\alpha)$. Since $B \supset A_{\mathrm{supp}(\alpha)}$, we have $\mathfrak{m}_B \cap A \subset \mathrm{supp}(\alpha)$. Let $x \in \mathrm{supp}(\alpha)$. We may assume that $x >_\beta 0$. Every element of $A_{\mathrm{supp}(\alpha)}$ is of the form yz^{-1} with $y, z \in A$, $z \notin \mathrm{supp}(\alpha)$, and we may assume $z(\alpha) > 0$. Since $xy \in \mathrm{supp}(\alpha)$ we have $xy(\alpha) = 0 < z(\alpha)$. Since α is a specialization of β', Proposition 7.1.18 implies that $xy(\beta') < z(\beta')$ and $0 < z(\beta')$, i.e. $xy <_\beta z$ and $0 <_\beta z$. Hence $yz^{-1} <_\beta x^{-1}$. This proves that $x^{-1} \notin B$ and therefore $x \in \mathfrak{m}_B$.

Since $z(\alpha) > 0$ implies $z \in B^*$ and $z >_\beta 0$, and thus $\lambda_B(z) >_{\overline{\beta}} 0$, the ordering $\leq_{\overline{\beta}}$ extends the ordering $\leq_\alpha$. The property

$$\forall x \in k_B \ \exists y \in k(\mathrm{supp}(\alpha)) , \ -y \leq_{\overline{\beta}} x \leq_{\overline{\beta}} y$$

is obvious from the definition of the β-convex envelope.

(ii) Set $\alpha = \lambda_B^{-1}(\overline{\beta}) \cap A \in \mathrm{Spec}_r(A)$. If $x \in A$ is such that $x(\alpha) > 0$, then $\lambda_B(x) >_{\overline{\beta}} 0$ and therefore $x >_\beta 0$. This shows that α is a specialization of β' in $\mathrm{Spec}_r(A)$. □

Note that the notion of central point (Definition 7.6.3) is related to the notion of centre of a place.

Proposition 10.2.4. *Let $V \subset R^n$ be an irreducible algebraic set. Let $\mathfrak{p}$ be a prime ideal of $\mathcal{P}(V)$. Then the following properties are equivalent:*

(i) *$\mathfrak{p}$ is the centre in $\mathcal{P}(V)$ of a real place of $\mathcal{K}(V)$ which is finite on $\mathcal{P}(V)$,*

(ii) *$\mathfrak{p} = \mathcal{I}_{\mathcal{P}(V)}(\mathcal{Z}_V(\mathfrak{p}) \cap \mathrm{Cent}(V))$.*

In particular, $x \in \mathrm{Cent}(V)$ if and only if the maximal ideal of x is the centre of a real place of $\mathcal{K}(V)$ which is finite on $\mathcal{P}(V)$.

Proof. By Corollary 7.6.7, property (ii) is equivalent to the fact that $\mathfrak{p}$ is $(\Sigma \mathcal{K}(V)^2 \cap \mathcal{P}(V))$-convex. By Proposition 4.2.9, this is equivalent to the existence of an ordering of $\mathcal{K}(V)$, with positive cone β, such that $\mathfrak{p}$ is $(\beta \cap \mathcal{P}(V))$-convex. By Proposition 4.3.8, this means that there exist $\alpha \in \mathrm{Spec}_r(\mathcal{P}(V))$ and an ordering of $\mathcal{K}(V)$, with positive cone β, such that $\mathrm{supp}(\alpha) = \mathfrak{p}$ and α is a specialization of $\beta \cap \mathcal{P}(V)$. Finally, by Propositions 10.1.8 and 10.2.3, this last property is equivalent to (i). □

10.2 Real Places and Specialization in the Real Spectrum

Proposition 10.2.5. Cent(V) *is closed and bounded if and only if every real place of* $\mathcal{K}(V)$ *finite on* R *is finite on* $\mathcal{P}(V)$.

Proof. Assume that Cent(V) is closed and bounded, and let $f \in \mathcal{P}(V)$. There exists $c \in R$ such that $-c \leq f \leq c$ on Cent(V). If λ is a real place of $\mathcal{K}(V)$ and β is the positive cone of an ordering compatible with λ, then $-c \leq_\beta f \leq_\beta c$. Hence, if λ is finite on R, $\lambda(f) \neq \infty$.

Conversely, assume that every real place of $\mathcal{K}(V)$ which is finite on R is finite on $\mathcal{P}(V)$. Let β be the positive cone of an ordering of $\mathcal{K}(V)$. The β-convex envelope of R is also the β-convex envelope of $\mathcal{P}(V)$. Therefore, if $f \in \mathcal{P}(V)$, there exists $c_\beta \in R$ such that $-c_\beta \leq_\beta f \leq_\beta c_\beta$. The set of $\gamma \in \mathrm{Spec}_r(\mathcal{K}(V))$ such that $-c_\beta \leq_\gamma f \leq_\gamma c_\beta$ is a closed and open subset of $\mathrm{Spec}_r(\mathcal{K}(V))$. By compactness of $\mathrm{Spec}_r(\mathcal{K}(V))$, there exists $c \in R$ such that, for every $\beta \in \mathrm{Spec}_r(\mathcal{K}(V))$, we have $-c \leq_\beta f \leq_\beta c$. By Proposition 7.6.4, $-c \leq f \leq c$ on Cent(V). This shows that Cent(V) is closed and bounded. □

We now analyze the case of codimension 1 in detail, first for affine space and then for an arbitrary irreducible algebraic set.

Let R be a real closed field, $\mathfrak{p}$ a prime ideal of height 1 (of dimension $n-1$) of $R[X_1, \ldots, X_n]$. The ideal $\mathfrak{p}$ is principal, generated by an irreducible polynomial f. If $g \in R(X_1, \ldots, X_n)^*$, then there is a unique $v(g) \in \mathbb{Z}$ such that $g = f^{v(g)} p/q$, where $p, q \in R[X_1, \ldots, X_n]$ and f divides neither p nor q. The mapping $v : R(X_1, \ldots, X_n)^* \to \mathbb{Z}$ is a valuation (called the $\mathfrak{p}$-*adic valuation*), and its valuation ring is the localization $R[X_1, \ldots, X_n]_\mathfrak{p}$. This valuation is real if $\mathfrak{p}$ is real, or, equivalently (by Theorem 4.5.1), if the sign of f changes on R^n.

Proposition 10.2.6. *Let* $\alpha \in \widetilde{R^n} = \mathrm{Spec}_r(R[X_1, \ldots, X_n])$, $\dim(\alpha) = n - 1$. *There exist exactly two* β *in* $\widetilde{R^n}$ *such that* α *is a specialization of* β *and* $\dim(\beta) = n$.

Proof. First recall that the equality $\dim(\beta) = n$ holds if and only if there exists an ordering of $R(X)$ (where $X = (X_1, \ldots, X_n)$) with positive cone γ such that $\beta = \gamma \cap R[X]$.

Let $F(\alpha)$ be the set of orderings of $R(X)$ compatible with the supp(α)-adic valuation and inducing the ordering $\leq_\alpha$ of the residue field $k(\mathrm{supp}(\alpha))$. If γ is the positive cone of an element of $F(\alpha)$, Proposition 10.2.3 (ii) implies that α is a specialization of $\gamma \cap R[X]$. Conversely, let γ be the positive cone of an ordering of $R(X)$ such that α is a specialization of $\gamma \cap R[X]$. Let B be the γ-convex envelope of $R[X]_{\mathrm{supp}(\alpha)}$. By Proposition 10.2.3 (i), B dominates $R[X]_{\mathrm{supp}(\alpha)}$. Since $R[X]_{\mathrm{supp}(\alpha)}$ is a valuation ring, we have $B = R[X]_{\mathrm{supp}(\alpha)}$ and therefore $R[X]_{\mathrm{supp}(\alpha)}$ is γ-convex. It follows that γ is the positive cone of an element of $F(\alpha)$.

Since the value group of $R[X]_{\mathrm{supp}(\alpha)}$ is $\mathbb{Z}$, the set $F(\alpha)$ has, by Theorem 10.1.10, exactly two elements. □

Example 10.2.7. Consider R^{n-1} embedded in R^n as the hyperplane with equation $X_n = 0$. Let $\alpha \in \widetilde{R^{n-1}}$ be a point of dimension $n-1$. Let $\mathcal{F}$ be the ultrafilter of semi-algebraic subsets S of R^{n-1} such that $\alpha \in \widetilde{S}$. Let $\mathcal{F}_\uparrow$ (resp. $\mathcal{F}_\downarrow$) be the ultrafilter of semi-algebraic subsets of R^n containing a semi-algebraic subset of the form

$$\{(y, u) \in S \times R \mid 0 < u < f(y) \text{ (resp. } f(y) < u < 0)\},$$

where $S \in \mathcal{F}$ and $f : S \to R$ is semi-algebraic and positive (resp. negative) (see Fig. 7.2). The points $\alpha_\uparrow$ and $\alpha_\downarrow$ of $\widetilde{R^n}$ determined by the ultrafilters $\mathcal{F}_\uparrow$ and $\mathcal{F}_\downarrow$ are the two points of dimension n specializing to α.

Corollary 10.2.8. *Let $V \subset R^p$ be an irreducible algebraic set of dimension n and $\alpha \in (\mathrm{Reg}(V))\widetilde{}$ of dimension $n-1$. Then there exist exactly two β in $\widetilde{V}$ such that $\dim(\beta) = n$ and α is a specialization of β.*

Proof. By Corollary 9.3.10, there is a finite open semi-algebraic cover $\mathrm{Reg}(V) = U_1 \cup \cdots \cup U_m$ and, for $i = 1, \ldots, m$, there are semi-algebraic homeomorphisms $\varphi_i : U_i \to W_i$, where W_i is an open semi-algebraic subset of R^n. Since $(\mathrm{Reg}(V))\widetilde{} = \widetilde{U}_1 \cup \ldots \cup \widetilde{U}_m$, there exists $i \in \{1, \ldots, m\}$ such that $\alpha \in \widetilde{U}_i$. The homeomorphism $\widetilde{\varphi}_i : \widetilde{U}_i \to \widetilde{W}_i$ preserves the relation of specialization and the dimension (by Proposition 7.5.8 (ii)). Finally we apply Proposition 10.2.6 to $\widetilde{\varphi}_i(\alpha)$. □

We now turn to the case where $\alpha \notin (\mathrm{Reg}(V))\widetilde{}$. First we recall a small part of the principal result concerning the extensions of a valuation of a field K to a finite algebraic extension L ([66], Chap. 6, §8, Theorem 1).

Let B' be a valuation ring of K. Let $I(B')$ be the set of valuation rings B of L such that $B \cap K = B'$. For $B \in I(B')$, denote by $e(B/B')$ the *ramification index*, i.e. the index of the group $\Gamma_{B'}$ in the group Γ_B, and denote by $f(B/B')$ the *residual degree*, i.e. the degree $[k_B : k_{B'}]$ of the extension of the residue fields.

Proposition 10.2.9.
(i) *The set $I(B')$ is finite and, for every $B \in I(B')$, the numbers $e(B/B')$ and $f(B/B')$ are finite.*
(ii) *If B_1 and B_2 are two distinct elements of $I(B')$, then B_1 is not contained in B_2 and B_2 is not contained in B_1.*

Theorem 10.2.10. *Let $V \subset R^p$ be an irreducible algebraic set of dimension n and $\alpha \in \widetilde{V}$ of dimension $n-1$. Denote by $g(\alpha)$ the number of $\beta \in \widetilde{V}$ such that $\dim(\beta) = n$ and α is a specialization of β. Denote by $\mathcal{B}(\mathrm{supp}(\alpha))$ the set of valuation rings B of $\mathcal{K}(V)$ containing $\mathcal{P}(V)$ and such that the centre of the place λ_B in $\mathcal{P}(V)$ is $\mathrm{supp}(\alpha)$. For $B \in \mathcal{B}(\mathrm{supp}(\alpha))$, denote by $m(B, \alpha)$ the number of orderings of k_B which extend the ordering $\leq_\alpha$ of $k(\mathrm{supp}(\alpha))$. Then $\mathcal{B}(\mathrm{supp}(\alpha))$ is finite, the number $m(B, \alpha)$ is finite for each $B \in \mathcal{B}(\mathrm{supp}(\alpha))$ and*

$$g(\alpha) = \sum_{B \in \mathcal{B}(\mathrm{supp}(\alpha))} 2\, m(B, \alpha).$$

In particular, $g(\alpha)$ is even.

Proof. By Emmy Noether's normalization lemma ([349], Chap. 5 §4), there is an injective R-algebra homomorphism $i : R[Y] \hookrightarrow \mathcal{P}(V)$ such that $\mathcal{P}(V)$ is finite over $R[Y]$ (where $Y = (Y_1, \ldots, Y_n)$). Let $\alpha' = i^{-1}(\alpha) \in \mathrm{Spec}_r(R[Y])$. Since $k(\mathrm{supp}(\alpha))$ is an algebraic extension of $k(\mathrm{supp}(\alpha'))$, α' is also of dimension $n-1$. We consider $\mathcal{K}(V)$ as a finite extension of $R(Y)$. Let $B \in \mathcal{B}(\mathrm{supp}(\alpha))$ and $B' = B \cap R(Y)$. The ring B' is a valuation ring which dominates $R[Y]_{\mathrm{supp}(\alpha')}$, and, therefore, $B' = R[Y]_{\mathrm{supp}(\alpha')}$. Proposition 10.2.9 (i) has several consequences. First, $\mathcal{B}(\mathrm{supp}(\alpha))$ is finite. Second, $\Gamma_{B'} = \mathbb{Z}$ has finite index in Γ_B, and, therefore, Γ_B is also isomorphic to $\mathbb{Z}$. Third, $[k_B : k(\mathrm{supp}(\alpha'))]$ is finite, and, therefore, $[k_B : k(\mathrm{supp}(\alpha))]$ is finite. By Proposition 1.3.7, $m(B, \alpha)$ is finite. Moreover, by Proposition 10.2.9 (ii), if β is the positive cone of an ordering of $\mathcal{K}(V)$ such that α is a specialization of $\beta \cap \mathcal{P}(V)$, then there exists only one β-convex valuation ring in $\mathcal{B}(\mathrm{supp}(\alpha))$. By Theorem 10.1.10, given $B \in \mathcal{B}(\mathrm{supp}(\alpha))$ and an ordering $\leq_\gamma$ of k_B, there exist exactly two orderings of $\mathcal{K}(V)$ compatible with λ_B and inducing $\leq_\gamma$. From these facts and Proposition 10.2.3, we deduce the formula $g(\alpha) = \sum_{B \in \mathcal{B}(\mathrm{supp}(\alpha))} 2\, m(B, \alpha)$. □

Corollary 10.2.11. *Let α and α' be two points of dimension $n-1$ of $\widetilde{V}$ such that $\mathrm{supp}(\alpha) = \mathrm{supp}(\alpha')$. Then $g(\alpha) \equiv g(\alpha') \pmod{4}$.*

Proof. The equality $\mathcal{B}(\mathrm{supp}(\alpha)) = \mathcal{B}(\mathrm{supp}(\alpha'))$ holds true, and if $B \in \mathcal{B}(\mathrm{supp}(\alpha))$, then $m(B, \alpha) \equiv [k_B : k(\mathrm{supp}(\alpha))] \equiv m(B, \alpha') \pmod 2$ by Proposition 1.3.7. □

Example 10.2.12. The geometric meaning of these results will appear more clearly in Chap. 11. We now give an example, illustrated by Fig. 10.1. Let $V \subset R^3$ be the umbrella given by the equation $z^2 x = y^2$. Choose α (resp. α') in $\widetilde{V}$ such that $\mathrm{supp}(\alpha) = \mathrm{supp}(\alpha')$ is the ideal of the x-axis and $x(\alpha) > 0$ (resp. $x(\alpha') < 0$). In this case, $\mathcal{B}(\mathrm{supp}(\alpha))$ has only one element. This can be seen by considering the parabolic cylinder given by the equation $x = t^2$ in the space of coordinates (x, t, z), which is birationally equivalent to V by setting $t = y/z$. This cylinder is obtained from V by blowing up, with the stick as centre (cf. Example 3.5.10). The valuation ring belonging to $\mathcal{B}(\mathrm{supp}(\alpha))$ is the local ring of the parabola P which is the intersection of the cylinder with the plane $z = 0$. Let B be this ring. The ordering $\leq_\alpha$ can be extended to two orderings of k_B corresponding to two points α_1 and α_2 of $\widetilde{P}$ above α. Each α_i is the specialization of two points of dimension 2. Therefore $g(\alpha) = 4$. On the other hand, $\leq_{\alpha'}$ cannot be extended to k_B: there exists no point of $\widetilde{P}$ above α'. Therefore $g(\alpha') = 0$.

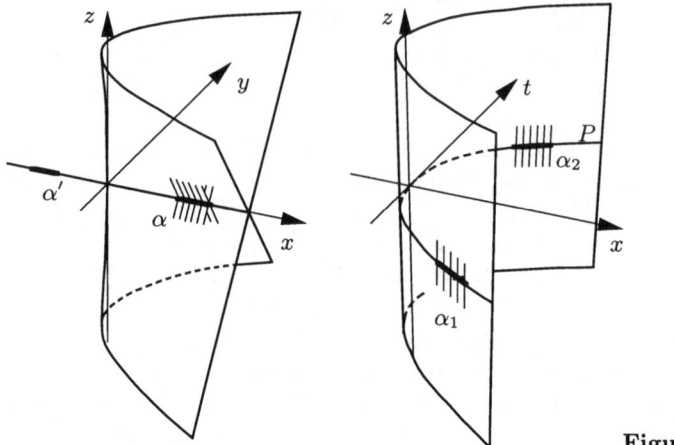

Figure 10.1.

10.3 Half-branches of Algebraic Curves Again

We prove, in this section, that the half-branches of algebraic curves (defined in Section 4 of Chap. 9) can be identified with points of the real spectrum.

Proposition 10.3.1. *Let $\Gamma \subset R^n$ be an algebraic curve and $x \in \Gamma$. There exists a canonical bijection between the set of half-branches of Γ centred at x and the set of points $\alpha \in \widetilde{\Gamma}$ different from x and specializing to x.*

Proof. Recall that there exists an open semi-algebraic neighbourhood U of x in Γ such that $U \setminus \{x\}$ has a finite number of semi-algebraically connected components $B_1, \ldots, B_k$, with semi-algebraic homeomorphisms

$$\varphi_i : [0,1[\to \{x\} \cup B_i , \quad \varphi_i(0) = x .$$

The half-branches of Γ centred at x are the germs $(B_i)_x$ (cf. Proposition 9.5.1). If x is a specialization of $\alpha \in \widetilde{\Gamma} \setminus \{x\}$, then $\alpha \in \widetilde{U} \setminus \{x\}$. Hence, α belongs to a unique $\widetilde{B_i}$. By Example 10.2.7, $0_+ \in]0,1[^{\sim}$ is the only point of $]0,1[^{\sim}$ specializing to 0. The semi-algebraic homeomorphism φ_i induces a homeomorphism $\widetilde{\varphi_i} : [0,1[^{\sim} \to \{x\} \cup \widetilde{B_i}$ (Proposition 7.2.8). Hence $\widetilde{B_i}$ contains exactly one point specializing to x. □

Using Proposition 10.3.1, we can give another proof of the fact that the number of half-branches of Γ centred at x is even (Theorem 9.5.7). Without loss of generality, we may assume that the curve Γ is irreducible. Theorem 10.2.10 implies that the number of points of dimension 1 of $\widetilde{\Gamma}$ specializing to x is even. We then apply Proposition 10.3.1.

In the case where $R = \mathbb{R}$, a simple description of all points of $\widetilde{\Gamma}$ can be given.

Proposition 10.3.2. *Let $\Gamma \subset \mathbb{R}^n$ be an algebraic curve. There is a canonical bijection between the set of points of dimension 1 of $\widetilde{\Gamma}$ and the set of all half-branches (centred at a point of Γ or at infinity) of Γ.*

Proof. If Γ is not compact, replacing it with its algebraic Alexandrov compactification (Definition 3.5.4) changes neither the set of points of dimension 1 of $\widetilde{\Gamma}$ nor the set of half-branches of Γ. So we may assume that Γ is compact. By Proposition 10.3.1, it suffices to prove that every point of dimension 1 of $\widetilde{\Gamma}$ specializes to a point of Γ (this point is unique by Proposition 7.1.23). Without loss of generality, we may assume that Γ is irreducible. Then, a point $\alpha \in \widetilde{\Gamma}$ of dimension 1 can be identified with an ordering of the field of rational functions $\mathcal{K}(\Gamma)$. By Proposition 10.2.5, the α-convex envelope B of $\mathbb{R}$ in $\mathcal{K}(\Gamma)$ contains $\mathcal{P}(\Gamma)$, which, by Proposition 10.2.3 (ii), gives a specialization of α in $\widetilde{\Gamma}$. Since the residue field of B is an archimedean ordered extension of $\mathbb{R}$, it is equal to $\mathbb{R}$, which implies that this specialization is a point of Γ. □

Remark 10.3.3. In the case where $R \neq \mathbb{R}$, the last argument of the previous proof is no longer valid. It could happen that $B = \mathcal{K}(\Gamma)$ and that $\mathcal{K}(\Gamma)$, with the ordering α, is an ordered extension of R where every element is bounded by an element of R. For example, this is the case if $R = \mathbb{R}_{\mathrm{alg}}$ and α is an ordering of $\mathbb{R}_{\mathrm{alg}}(X)$ given by the evaluation at a transcendental number (cf. Example 7.5.2). The points of dimension 1 of $\widetilde{\mathbb{R}_{\mathrm{alg}}}$ correspond to points of $\mathbb{R} \setminus \mathbb{R}_{\mathrm{alg}}$ and to half-branches of $\mathbb{R}_{\mathrm{alg}}$ centred at points of $\mathbb{R}_{\mathrm{alg}}$ or at infinity.

Remark 10.3.4. We return to the case $R = \mathbb{R}$. In $\widetilde{\mathbb{R}^n}$ the points of dimension 0 are the points of $\mathbb{R}^n$. By Proposition 10.3.2, a point $\alpha \in \widetilde{\mathbb{R}^n}$ of dimension 1 can be identified with a half-branch of the algebraic curve $\mathcal{Z}(\mathrm{supp}(\alpha)) \subset \mathbb{R}^n$. An example of these points is shown in Fig. 10.2(a), where the four half-branches of the cubic curve $x^3 + x^2 = y^2$, centred at the node, are represented.

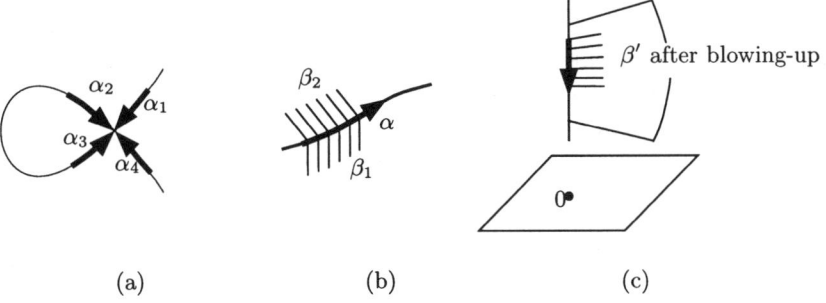

Figure 10.2.

The situation is far more complicated when the dimension is 2 or greater. We give a few examples of points of dimension 2 in $\widetilde{\mathbb{R}^2}$.

First consider the points which have a specialization of dimension 1. There are two of them specializing to the same half-branch of an algebraic curve, and they can be viewed as the two "sides" of this half-branch, as shown in Fig. 10.2(b). But there are many points of dimension 2 besides these. For example
$$\beta = \{f \in \mathbb{R}[X,Y] \mid \exists \epsilon > 0 \ \forall t \in]0, \epsilon[\ f(t, e^t) \geq 0\},$$
which can be viewed as a half-branch of the graph of the exponential function, or
$$\beta' = \{f \in \mathbb{R}[X,Y] \mid \exists \epsilon > 0 \ \forall u \in]0, \epsilon[\ \exists \eta > 0 \ \forall x \in]0, \eta[\ f(x, ux) \geq 0\}$$
which, after blowing-up, with the origin as centre, specializes to a half-branch of the exceptional divisor above the origin (Fig. 10.2(c)).

Neither β nor β' has a specialization of dimension 1 in $\widetilde{\mathbb{R}^2}$. This can be seen by considering the nontrivial β-convex (resp. β'-convex) valuation rings of $\mathbb{R}(X,Y)$. In the first case, the only such valuation ring, which is also discrete rank 1, is the one for which the valuation of f is the order of the series $f(t, e^t)$. The only proper specialization of β is the point $(0,1) \in \mathbb{R}^2$. In the second case, there are valuation rings of the discrete rank 1 place
$$\lambda_1 : \mathbb{R}(X,Y) = \mathbb{R}(X, Y/X) \to \mathbb{R}(Y/X) \cup \{\infty\}$$
and of the composite discrete rank 2 place
$$\lambda_2 : \mathbb{R}(X,Y) \to \mathbb{R}(Y/X) \cup \{\infty\} \to \mathbb{R} \cup \{\infty\}.$$
The places λ_1 and λ_2 have the same centre in $\mathbb{R}[X,Y]$, namely the maximal ideal (X,Y). The only proper specialization of β' is the point $(0,0)$ of $\mathbb{R}^2$.

10.4 Fans and Basic Semi-algebraic Sets

Let K be a real field, B a real valuation ring of the field K, and $\overline{F}$ a set of orderings of the residue field k_B. Let F be the set of orderings of K, compatible with the place λ_B and inducing an ordering of k_B which belongs to $\overline{F}$. We call F the *lifting of $\overline{F}$ to K along λ_B*.

Proposition 10.4.1. *Let K be a real field, and let $\overline{F}$ be a subset of the set of orderings of K with one or two elements. Let F be the lifting of $\overline{F}$ to K along λ_B. Then F is endowed with the structure of an affine space over $\mathbb{Z}/2$ satisfying the following property: for every element $f \in K$, the set of $\beta \in F$ such that $f >_\beta 0$ is either equal to F, or empty, or an affine hyperplane of F.*

10.4 Fans and Basic Semi-algebraic Sets

Proof. First, consider the case where $\overline{F}$ has one element α. We already know by Theorem 10.1.10 that $F = F(\alpha)$ is endowed with the structure of an affine space over $\mathbb{Z}/2$. We use the notations of the proof of this theorem. Choose $\beta_0 \in F$. Recall that we defined a bijection $\beta \mapsto \langle \beta_0, \beta \rangle$ from F to $\mathrm{Hom}(\Gamma_B, \mathbb{Z}/2)$. Let f be a nonzero element of K. Let $\delta_0 \in \mathbb{Z}/2$ be such that $(-1)^{\delta_0}$ is the sign of f for β_0. Then $\beta \in F$ is such that $f >_\beta 0$ if and only if

$$\langle \beta_0, \beta \rangle (v_B(f)) + \delta_0 = 0 \,.$$

The set of $\beta \in F$ satisfying this equation is either F or empty if $v_B(f) \in 2\Gamma_B$, and an affine hyperplane of F if $v_B(f) \notin 2\Gamma_B$.

Second, consider the case where $\overline{F}$ has two elements α_0 and α_1. Then F is the disjoint union $F(\alpha_0) \cup F(\alpha_1)$. Choose $\beta_i \in F(\alpha_i)$ for $i = 0, 1$. Define

$$\nu : F \longrightarrow \mathrm{Hom}(\Gamma_B, \mathbb{Z}/2) \times \mathbb{Z}/2$$
$$\beta \longmapsto \begin{cases} (\langle \beta_0, \beta \rangle, 0) & \text{if } \beta \in F(\alpha_0) \\ (\langle \beta_1, \beta \rangle, 1) & \text{if } \beta \in F(\alpha_1) \end{cases}.$$

The mapping ν is a bijection which endows F with the structure of an affine space over $\mathbb{Z}/2$. Let f be a nonzero element of K. Let $\delta_i \in \mathbb{Z}/2$ be such that $(-1)^{\delta_i}$ is the sign of f for β_i, for $i = 0, 1$. Then $\beta \in F$ is such that $f >_\beta 0$ if and only if

$$\varphi(v_B(f)) + (\delta_0 + \delta_1)\epsilon + \delta_0 = 0 \,,$$

where $(\varphi, \epsilon) = \nu(\beta)$. The set of $\beta \in F$ satisfying this equation is either F or empty if $v_B(f) \in 2\Gamma_B$ and f has the same sign for β_0 and β_1, and an affine hyperplane of F in the other case. □

Definition 10.4.2. *Let K be a real field.*

A trivial fan of K is a subset of the set of orderings of K with one or two elements.

A fan F of K is a subset of the set of orderings of K such that there exist a real valuation ring B of K and a trivial fan G of the residue field k_B, such that F is an affine subspace of the lifting of G to K along the place λ_B.

Example 10.4.3. a) The set $\{0_+, 1_+\}$ is a trivial fan of $R(X)$. Recall that if α is an ordering of $R(X)$, considered as a point of the real spectrum $\widetilde{R^2}$ of $R[X, Y]$, we denote by $\alpha_\uparrow$ and $\alpha_\downarrow$ the two points of dimension 2 of $\widetilde{R^2}$ specializing to α (see 10.2.7). We can identify $\alpha_\uparrow$ with the ordering of $R(X, Y)$ extending the ordering α of $R(X)$ and such that Y is positive and smaller than every positive element of $R(X)$. Then $\{0_{+,\uparrow}, 0_{+,\downarrow}, 1_{+,\uparrow}, 1_{+,\downarrow}\}$ is a fan of $R(X, Y)$ compatible with the valuation ring $R[X, Y]_{(Y)}$ and inducing the trivial fan $\{0_+, 1_+\}$ of $R(X)$. The same construction can be carried out for any two half-branches of an irreducible algebraic curve in R^2 (corresponding to a trivial fan with two elements of the field of rational functions of the curve), and also gives a fan with four elements.

b) Let x be a nonsingular point of an n-dimensional irreducible algebraic set $V \subset R^p$. The 2^n orderings of $\mathcal{K}(V)$ compatible with the discrete rank n valuation associated to a regular system of parameters of $\mathcal{R}_{V,x}$ (see Example 10.1.11) form a fan.

Recall that a semi-algebraic subset S of an irreducible real algebraic set V of dimension n is called generically basic if there exist polynomial functions $f_1, \ldots, f_k$ on V such that the symmetric difference of S and

$$\mathcal{U}(f_1, \ldots, f_k) = \{x \in V \mid f_1(x) > 0, \ldots, f_k(x) > 0\}$$

is of dimension smaller than n (cf. 6.2.1).

We consider a fan of $\mathcal{K}(V)$ as a subset of $\widetilde{V} = \mathrm{Spec}_r(\mathcal{P}(V))$. The main result relating basic semi-algebraic sets and fans is the following.

Corollary 10.4.4. *Let V be an irreducible real algebraic set of dimension n. Let S be a semi-algebraic subset of V, generically equal to $\mathcal{U}(f_1, \ldots, f_k)$. Let F be a fan of $\mathcal{K}(V)$ with 2^d elements. Then $F \cap \widetilde{S}$ is either empty or has exactly 2^ℓ elements, where ℓ is a nonnegative integer such that $\ell \geq d - k$.*

Proof. By Proposition 7.6.1, we have $\widetilde{S} \cap F = \widetilde{\mathcal{U}}(f_1, \ldots, f_k) \cap F$. By Proposition 10.4.1, $\widetilde{\mathcal{U}}(f_1, \ldots, f_k) \cap F$ is either empty or an affine subspace of F of codimension at most k. □

Example 10.4.5. a) The union of a rectangle and two half-discs shown in Fig. 10.3 is not generically basic. The fan with four elements F considered here is one of those constructed in 10.4.3 a). The intersection $\widetilde{S} \cap F$ has three elements.

Figure 10.3.

b) Consider the curve with equation

$$f(x,y) = (y - \frac{x}{5})(y + \frac{x}{5})(\frac{y}{5} - x)(\frac{y}{5} + x) + ((x-2)^2 + y^2 - 1) .$$

This curve with its four asymptotes and the circle which appear in its equation are represented in Fig. 10.4(a). The semi-algebraic subset where $f(x,y)$ is positive has two connected components S (the one which contains the negative x-axis) and T.

Let us prove that S is not generically basic. We embed the affine plane in the projective plane. Let E be the line at infinity. In a neighbourhood of E we obtain the picture shown in Fig. 10.4(b). We see a fan F of the field of rational functions on the plane such that $F \cap \widetilde{S}$ has three elements.

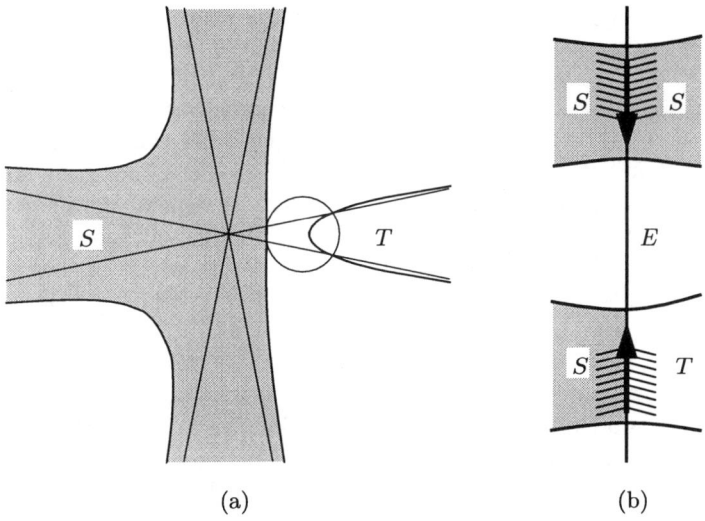

(a)   (b)

Figure 10.4.

The following result generalizes Proposition 6.5.11.

Proposition 10.4.6. *Let V be an irreducible real algebraic set of dimension n, x a nonsingular point of V and $f_1, \ldots, f_n$ polynomial functions on V which form a regular system of parameters of $\mathcal{R}_{V,x}$. There is no family $g_1, \ldots, g_k$ of polynomial functions on V, with $k < n$, such that the basic open semi-algebraic subset $\mathcal{U}(f_1, \ldots, f_n)$ is generically equal to $\mathcal{U}(g_1, \ldots, g_k)$.*

Proof. The regular system of parameters $(f_1, \ldots, f_n)$ determines a fan F of $\mathcal{K}(V)$ with 2^n elements (Example 10.1.11). Each ordering $\alpha \in F$ is characterized in F by the signs of $f_1(\alpha), \ldots, f_n(\alpha)$. In particular, $\widetilde{\mathcal{U}}(f_1, \ldots, f_n) \cap F$ has exactly one element. The proposition is thus a consequence of Corollary 10.4.4. □

In the case of basic closed sets, we have the following result.

Proposition 10.4.7. *Each irreducible real algebraic set V of dimension $n \geq 1$ contains a basic closed semi-algebraic subset which cannot be described by fewer than $n(n+1)/2$ polynomial inequalities on V.*

Proof. We shall prove the following, more precise statement. Let a be a nonsingular point of V. There exist polynomial functions $h_1, \ldots, h_{n(n+1)/2}$ on V such that $h_1(a) > 0, \ldots, h_{n(n+1)/2}(a) > 0$ and there exists no family of polynomial functions $g_1, \ldots, g_k$ on V, with $k < n(n+1)/2$, such that the basic closed semi-algebraic subset

$$T = \{x \in V \mid h_1(x) \geq 0, \ldots, h_{n(n+1)/2}(x) \geq 0\}$$

is equal to
$$\{x \in V \mid g_1(x) \geq 0, \ldots, g_k(x) \geq 0\}.$$

We proceed by induction on n. For $n = 1$, just choose a polynomial function h_1 such that $h_1(a) > 0$ and $h_1(b) < 0$, for some other point $b \in V$. Now assume $n \geq 2$ and the proposition holds true for $n - 1$. Choose polynomial functions $\varphi_1, \ldots, \varphi_n$ on V which form a regular system of parameters of $\mathcal{R}_{V,a}$. We have a Nash chart $(\varphi_1, \ldots, \varphi_n)$ from a neighbourhood of a in V to a neighbourhood of the origin in R^n, $\varphi_i(a) = 0$, $i = 1, \ldots, n$. Let a_1 be the point with coordinates $(-\epsilon, \ldots, -\epsilon)$ in this chart, with $\epsilon > 0$ sufficiently small. Set $f_i = \varphi_i + \epsilon$ for $i = 1, \ldots, n$. The point a_1 is nonsingular in V and $(f_1, \ldots, f_n)$ is a regular system of parameters of $\mathcal{R}_{V,a_1}$. Let V_1 be the codimension 1 irreducible algebraic subset of V which coincides with $f_n^{-1}(0)$ in a neighbourhood of a_1. Then a_1 is a nonsingular point of V_1, and, by the inductive assumption, there are polynomial functions $h_1, \ldots, h_{n(n-1)/2}$ on V_1 such that $h_i(a_1) > 0$, for $i = 1, \ldots, n(n-1)/2$, and the basic closed semi-algebraic subset

$$T_1 = \{x \in V_1 \mid h_1(x) \geq 0, \ldots, h_{n(n-1)/2}(x) \geq 0\}$$

cannot be described by fewer than $n(n-1)/2$ polynomial inequalities in V_1. We can extend the h_i to polynomial functions on V, which we also denote by h_i, and we may moreover assume $h_i(a) > 0$, by adding rf_n^2 with $r > 0$ sufficiently large. Define $h_{(n(n-1)/2)+i} = f_i f_n^2$ for $i = 1, \ldots, n-1$ and $h_{n(n+1)/2} = f_n$. We have $h_i(a) > 0$ for $i = 1, \ldots, n(n+1)/2$.

Now we prove that the basic closed semi-algebraic subset

$$T = \{x \in V \mid h_1(x) \geq 0, \ldots, h_{n(n+1)/2}(x) \geq 0\}$$

cannot be defined by fewer than $n(n+1)/2$ polynomial inequalities in V. Note that $T \cap V_1 = T_1$. We consider the fan F of $\mathcal{K}(V)$ (resp. F_1 of $\mathcal{K}(V_1)$) associated to the regular system of parameters $(f_1, \ldots, f_n)$ (resp. $(f_1, \ldots, f_{n-1})$) as in Example 10.1.11. Note that F is the lifting of F_1 along the canonical discrete rank 1 place of $\mathcal{K}(V)$ with residue field $\mathcal{K}(V_1)$. We consider F and F_1 as subsets of the real spectrum $\widetilde{V}$. Every α in F or F_1 specializes to a_1 and therefore $h_i(\alpha) > 0$ for $i = 1, \ldots, n(n-1)/2$. There is a unique $\alpha \in F$ such that $h_i(\alpha) > 0$ for $i = (n(n-1)/2)+1, \ldots, n(n+1)/2$. Hence $F_1 \subset \widetilde{T}$ and $F \cap \widetilde{T}$ has exactly one element.

Suppose that

$$T = \{x \in V \mid g_1(x) \geq 0, \ldots, g_k(x) \geq 0\},$$

where $g_1, \ldots, g_k$ are polynomial functions on V. From Proposition 10.4.1 and the fact that $F \cap \widetilde{T}$ has exactly one element, we deduce that there are at least n of these functions, say $g_1, \ldots, g_n$, such that $\{\alpha \in F \mid g_i(\alpha) > 0\}$ is a hyperplane of F. This implies that $g_i|_{V_1} = 0$ for $i = 1, \ldots, n$. Otherwise, we

would have $g_i(\alpha) > 0$ for every $\alpha \in F_1$ and therefore also for every $\alpha \in F$. Hence
$$T_1 = T \cap V_1 = \{x \in V_1 \mid g_{n+1}(x) \geq 0, \ldots, g_k(x) \geq 0\}.$$
Since T_1 cannot be described by fewer than $n(n-1)/2$ polynomial inequalities in V_1, we deduce that $k \geq n(n+1)/2$. □

Figure 10.5 shows the construction of a T as in Proposition 10.4.7 for $V = R^3$.

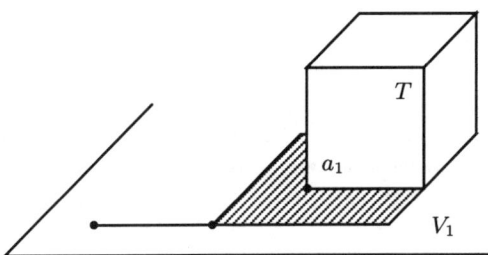

Figure 10.5.

The lower bound of Proposition 10.4.7 is the best possible, as the following result shows.

Theorem 10.4.8. *Let $V \subset R^p$ be an algebraic set of dimension $n \geq 1$, and T, a basic closed semi-algebraic subset of V. There exist $n(n+1)/2$ polynomial functions $f_1, \ldots, f_{n(n+1)/2}$ on V such that*
$$T = \{x \in V \mid f_1(x) \geq 0, \ldots, f_{n(n+1)/2}(x) \geq 0\}.$$

Proof. Let $W_1, \ldots, W_\ell$ be the irreducible components of dimension n of V. By Corollary 6.5.6, for every $i = 1, \ldots, \ell$, there are $g_{i,1}, \ldots, g_{i,n}$ in $\mathcal{P}(V)$, with $g_{i,j}|_{W_i}$ not identically zero, such that $T \cap W_i$ is generically equal in W_i to
$$\{x \in W_i \mid g_{i,1}(x) \geq 0, \ldots, g_{i,n}(x) \geq 0\}.$$
Let h_i be a polynomial function on V such that $h_i \geq 0$ and $h_i^{-1}(0)$ is the union of the irreducible components of V different from W_i. Set $g_j = \sum_{i=1}^{\ell} h_i g_{i,j}$. Then the symmetric difference of T and the basic closed semi-algebraic subset
$$C = \{x \in V \mid g_1(x) \geq 0, \ldots, g_n(x) \geq 0\}$$
is contained in an algebraic subset Z of dimension $< n$ of V. By construction, we have $\dim(g_j^{-1}(0)) < n$ for $j = 1, \ldots, n$. Moreover, multiplying the polynomial functions g_j by a nonnegative polynomial on V whose zero set is Z, we may assume that C contains T.

First we consider the case $n = 1$. Then we have $C = \{x \in V \mid g_1(x) \geq 0\}$ and $C \setminus T$ consists of a finite number of points. We may assume that $g_1^{-1}(0)$ contains these points. The set $T \cap g_1^{-1}(0)$ also consists of a finite number of points. Choose a nonnegative polynomial p on V with $p^{-1}(0) = T \cap g_1^{-1}(0)$. By Lojasiewicz's inequality (2.6.2 and 2.6.6), we have a positive integer ν and a polynomial function k, which is positive on V, such that $p^\nu \le k g_1$ on T. Define $f_1 = k g_1 - p^\nu$. Then $T = \{x \in V \mid f_1(x) \ge 0\}$.

Now assume $n > 1$ and the proposition holds true for $n - 1$. By the inductive assumption, there are polynomial functions $h_1, \ldots, h_{n(n-1)/2}$ on V such that

$$T \cap Z = \{x \in Z \mid h_1(x) \ge 0, \ldots, h_{n(n-1)/2} \ge 0\}.$$

Choose a nonnegative polynomial p on V such that $p^{-1}(0) = Z$. For $i = 1, \ldots, n(n-1)/2$, let F_i be the closed semi-algebraic subset of $x \in T$ such that $h_i(x) \le 0$. Since $p^{-1}(0) \cap F_i \subset h_i^{-1}(0) \cap F_i$, we have, by Lojasiewicz's inequality, a positive integer ν_i and a polynomial function k_i, which is positive on V, such that $|h_i|^{\nu_i} \le k_i p$. We may assume that ν_i is odd. The polynomial function $f_i = h_i^\nu + k_i p$ has the same sign as h_i on Z and is nonnegative on T. Hence, setting $f_{(n(n-1)/2)+i} = g_i$ for $i = 1, \ldots, n$, we obtain

$$T = \{x \in V \mid f_1(x) \ge 0, \ldots, f_{n(n+1)/2}(x) \ge 0\}.$$

$\square$

Bibliographic Notes. The relation between orderings and valuations of a field were studied in the papers of Baer [21] and Krull [195] when they initiated the general theory of valuation. The concept of a real place can be found in [209]. Theorem 10.1.10 appears in the works of Baer and Krull, but the formulation given in this chapter is taken from [73]. We refer the reader to the survey of Becker [25] for more information concerning real places and their role in real algebraic geometry. A geometrical description of orderings of $R(X,Y)$, more complete than the one sketched in Remark 10.3.4, can be found in [76], §8.12, and [12]. The notion of a fan was introduced in [27] for the study of quadratic forms. The definition given here is based on the "fan trivialization theorem" of Bröcker [68]. The geometric applications of fans (including Example 10.4.5) are described in [72]. Example 10.4.5 b) was already used in [238] for the problem of separation of semi-algebraic sets. Proposition 10.4.7 and Theorem 10.4.8 are in [287]. The results concerning basic semi-algebraic sets have "abstract" counterparts in the theory of spaces of signs developed in [13] and [224], where the notion of a fan plays a crucial role. Many other interesting geometric results (in the analytic as well as algebraic case) are obtained as consequences of this theory.

11. Topology of Real Algebraic Varieties

Abstract. In the first section, we prove some combinatorial topological properties of real algebraic sets; the simplest and most important of these properties is the fact that, for every semi-algebraic triangulation of a bounded algebraic set of dimension d and every $(d-1)$-simplex σ of such a triangulation, the number of d-simplices of the triangulation having σ as a face is even. In the second section, we use this property and an appropriate stratification to prove that, for every point a of an algebraic set V, the local Euler-Poincaré characteristic $\chi(V, V \setminus a)$ is odd; this result gives a necessary combinatorial condition for a polyhedron to be homeomorphic to a real algebraic set. In Section 3 we define the fundamental $\mathbb{Z}/2$-homology class of a real algebraic variety. This leads to the concept of algebraic homology groups of a real algebraic variety, consisting of the homology classes represented by algebraic subsets. These groups, which are basic invariants, will be used in Chap. 12 and 13. We construct examples of nonsingular algebraic sets whose homology is not totally algebraic. In Section 4, we use the Borel-Moore fundamental classes to prove that an injective regular mapping from a nonsingular irreducible algebraic set to itself is surjective. The analogous result in complex algebraic geometry (without the assumption of nonsingularity) is well known, but the methods of proof are completely different. Section 5 contains an upper bound for the sum of the Betti numbers of an algebraic set. Section 6 is devoted to algebraic curves in the real projective plane. We prove Harnack's theorem concerning the maximum number of connected components of a nonsingular curve of given degree and some results concerning the first part of Hilbert's 16th problem (without proving the crucial Rokhlin congruence, Theorem 11.6.4).

For algebraic subsets of $\mathbb{R}^n$, the homology H_* (resp. H_*^{BM}) used in this chapter is the usual singular homology (resp. the Borel-Moore homology for locally compact spaces). In order to extend the results to the case of an arbitrary real closed field, we need a homology theory for semi-algebraic sets, with the properties of the usual singular homology. We explain how to construct such a theory in the appendix to this chapter. This construction is only sketched. Nevertheless, we hope to convince the reader that this homology theory behaves like singular homology theory in those situations we are interested in. For this reason, in Sections 2 to 5 we use a homology theory

11.1 Combinatorial Properties of Algebraic Sets

Theorem 11.1.1. *Let $V \subset R^n$ be an algebraic set of dimension d.*

(i) *Assume that V is bounded, and let $\Phi : |K| \to V$ be a semi-algebraic triangulation of V. If σ is a $(d-1)$-simplex of K, denote by $g(\sigma)$ the number of d-simplices τ of K, such that σ is a face of τ. Then $g(\sigma)$ is even.*

(ii) *Assume that V has a stratification induced by a stratifying family of polynomials (cf. Definition 9.1.5 and Theorem 9.1.6 (i)). If C is a stratum of dimension $d-1$ of V, let $g(C)$ denote the number of strata D of dimension d of V, such that $C \subset \mathrm{clos}(D)$. Then $g(C)$ is even.*

Proof. We first show that statements (i) and (ii) are equivalent. We may assume, without loss of generality, that V is bounded, since we can replace V with its algebraic Alexandrov compactification (Definition 3.5.4). If $\Phi : |K| \to V$ is a semi-algebraic triangulation of V, there is a stratification of V induced by a stratifying family of polynomials that refines the triangulation (Theorem 9.1.6), and conversely (Theorem 9.2.1). We shall show that, if σ is a $(d-1)$-simplex of K, and C is a $(d-1)$-stratum of the stratification with $C \subset \sigma^0$ (resp. $\sigma^0 \subset C$), then $g(C) = g(\sigma)$. The case $C \subset \sigma^0$ is easy. For the case $\sigma^0 \subset C$, note that, if C and D are strata, of dimension $d-1$ and d, respectively, of a stratification of R^n induced by a stratifying family of polynomials, with $C \subset \mathrm{clos}(D)$, then there exists a semi-algebraic homeomorphism $\psi : [0,1[\times]0,1[^{d-1} \to C \cup D$, such that $C = \psi(\{0\} \times]0,1[^{d-1})$. The homeomorphism ψ is constructed by induction on n, by gluing together the homeomorphisms φ_C and φ_D of Theorem 9.1.4, after possibly permuting the coordinates.

In order to prove the theorem, it now suffices to prove either (i) or (ii). Two different proofs will be given: a proof of (i), using the real spectrum and the results of the previous chapter, and a proof of (ii) by reduction to the case of half-branches of an algebraic curve.

(i) Using the tilde operation, we get a homeomorphism $\widetilde{\Phi} : \widetilde{|K|} \to \widetilde{V}$ (Proposition 7.2.8). Since $\dim(\sigma) = d-1$, there exists $\alpha \in \widetilde{\sigma}$ of dimension $d-1$ (Proposition 7.5.8 (i)). By Proposition 10.2.6 and Example 10.2.7, for every d-simplex τ such that σ is a face of τ, there is exactly one β of dimension d such that $\beta \in \widetilde{\tau}$ and α is a specialization of β. Since, by Proposition 7.5.8 (ii), $\dim(\widetilde{\Phi}(\gamma)) = \dim(\gamma)$ for every $\gamma \in \widetilde{|K|}$, $g(\sigma)$ is equal to the number of points of dimension d of $\widetilde{V}$ having $\widetilde{\Phi}(\alpha)$ as a specialization. Let $V_1, \ldots, V_k$

11.1 Combinatorial Properties of Algebraic Sets

be the irreducible components of dimension d of V, and denote by $g_i(\widetilde{\Phi}(\alpha))$ ($i = 1, \ldots, k$) the number of points of dimension d of $\widetilde{V}_i$ having $\widetilde{\Phi}(\alpha)$ as a specialization. By Theorem 10.2.10, $g_i(\widetilde{\Phi}(\alpha))$ is even, for $i = 1, \ldots, k$. Moreover, it is obvious that a point of dimension d of $\widetilde{V}$ belongs to a unique $\widetilde{V}_i$. Hence, $g(\sigma) = \sum_{i=1}^{k} g_i(\widetilde{\Phi}(\alpha))$ is even.

(ii) Denote by $\Pi : R^n \to R^d$ the projection onto the space of the first d coordinates. Without loss of generality, we may assume that, for every stratum A of V, the restriction $\Pi|_{\text{clos}(A)}$ is a homeomorphism onto $\text{clos}(\Pi(A))$ (cf. Theorem 9.1.6). Choose $x \in C$ and an affine line $L \subset R^d$ containing $\Pi(x)$ and transversal to $\Pi(C)$ (which is a Nash submanifold of R^d of dimension $d-1$, cf. Theorem 9.1.4 (i), (ii)). Then $\Gamma = \Pi^{-1}(L) \cap V$ is an algebraic set of dimension ≤ 1. If $\dim(\Gamma) = 0$, then $g(C) = 0$. If Γ is a curve, there is a canonical bijection between the set of half-branches of Γ centred at x and the set of d-strata D of the stratification of V, such that $C \subset \text{clos}(D)$. By Theorem 9.5.7, $g(C)$ is even. $\square$

Theorem 11.1.2. *Let $V \subset R^n$ be an algebraic set of dimension d and $W \subset V$, an irreducible algebraic subset of dimension $d-1$. Then, with the assumptions and notation of Theorem 11.1.1, one has:*

(i) If σ and σ' are two $(d-1)$-simplices of K, such that $\Phi(\sigma \cup \sigma') \subset W$, then

$$g(\sigma) \equiv g(\sigma') \pmod{4}.$$

(ii) If C and C' are two $(d-1)$-strata of V contained in W, then

$$g(C) \equiv g(C') \pmod{4}.$$

Proof. (i) Choose α in $\widetilde{\sigma}$ and α' in $\widetilde{\sigma}'$, both of dimension $d-1$. If $\mathfrak{p}$ is the ideal of W in $\mathcal{P}(V)$, then $\text{supp}(\widetilde{\Phi}(\alpha)) = \text{supp}(\widetilde{\Phi}(\alpha')) = \mathfrak{p}$. We use the notation of the proof of Theorem 11.1.1 (i). By Corollary 10.2.11, $g_i(\widetilde{\Phi}(\alpha)) \equiv g_i(\widetilde{\Phi}(\alpha'))$ (mod 4), for $i = 1, \ldots, k$. Hence, $g(\sigma) \equiv g(\sigma') \pmod{4}$.

(ii) is proved in the same way. $\square$

Example 11.1.3. Let a, b, c, d, e, f be points in R^3 such that

$$|K| = [b, c, d] \cup [c, d, a] \cup [d, a, b] \cup [a, b, c] \cup [b, e] \cup [e, a] \cup [b, f] \cup [f, a]$$

is a simplicial complex and let $|L| = [a, b] \cup [b, e] \cup [e, a]$. Then $|L|$ is semi-algebraically homeomorphic to a circle, and $|K|$ is semi-algebraically homeomorphic to the union of a sphere and a circle which intersect in two points (cf. Figure 11.1).

Nevertheless, there is no semi-algebraic homeomorphism Φ from $|K|$ onto an algebraic set V such that $\Phi(|L|)$ is algebraic. For, suppose such a Φ exists; then $\Phi(|L|)$ is irreducible (by Theorem 9.5.7, an algebraic curve which is semi-algebraically homeomorphic to a circle is irreducible), which is impossible, by Theorem 11.1.2, since $g([e, a]) = 0$ and $g([a, b]) = 2$.

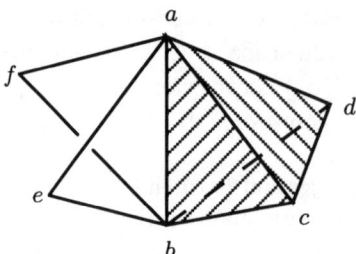

Figure 11.1.

Theorem 11.1.2 has another consequence concerning this example. For every semi-algebraic homeomorphism $\Phi : |K| \to V$ onto an algebraic set V, the set $A = \Phi([b,e] \cup [e,a] \cup [b,f] \cup [f,a])$ is almost algebraic, in the sense that $\mathrm{clos}_{\mathrm{Zar}}(A) \setminus A$ consists of a finite number of points. If this were not the case, we could choose a semi-algebraic triangulation $\Psi : |M| \to V$, such that A and $\mathrm{clos}_{\mathrm{Zar}}(A)$ are the unions of some simplices of M, and $\Psi(\sigma^0) \subset \mathrm{clos}_{\mathrm{Zar}}(A) \setminus A$ for some 1-simplex σ of M. Then there is necessarily another 1-simplex σ' of M, whose image is contained in the same irreducible component of $\mathrm{clos}_{\mathrm{Zar}}(A)$ as the image of σ, and such that $\Psi(\sigma') \subset A$. It follows that $g(\sigma) = 2$ and $g(\sigma') = 0$, which contradicts Theorem 11.1.2.

In the statements of the previous example, we can replace "semi-algebraic homeomorphism" with "homeomorphism" in the case $R = \mathbb{R}$. We mentioned in the bibliographic notes of Chap. 9 that there exist semi-algebraic sets that are homeomorphic but not semi-algebraically homeomorphic. Hence, a claim concerning semi-algebraic homeomorphisms may not be valid in the case of arbitrary homeomorphisms. However, the statements of Example 11.1.3 are also valid for arbitrary homeomorphisms: by counting the connected components of pointed neighbourhoods, we see that, if $h : |K| \to V$ is any homeomorphism and $x \in |K|$, then $\dim(|K|_x) = \dim(V_{h(x)})$.

Remark 11.1.4. Theorems 11.1.1 and 11.1.2 can also be proved by taking the normalization of V. One then uses the fact, that a normal algebraic set is nonsingular in codimension 1, to obtain results which hold generically along codimension 1 algebraic subsets of V. In the proofs of Theorems 11.1.1 (i) and 11.1.2 given above, the use of places plays the role of normalization and the points of maximal dimension of the real spectrum are, in a sense, generic points.

11.2 Local Euler-Poincaré Characteristic of Algebraic Sets

Let A be a semi-algebraic subset of R^n. The r-th Betti number $b_r(A) = \dim_{\mathbb{Q}}(H_r(A, \mathbb{Q}))$ of A is finite and $b_r(A) = 0$ if $r > \dim(A)$. Recall that the

11.2 Local Euler-Poincaré Characteristic of Algebraic Sets

Euler-Poincaré characteristic of A is defined to be $\chi(A) = \sum_r (-1)^r b_r(A)$. If $v \in A$, the *local Euler-Poincaré characteristic* $\chi(A, A \setminus v)$ is, by definition, the alternating sum

$$\chi(A, A \setminus v) = \sum_r (-1)^r \dim_{\mathbb{Q}}(H_r(A, A \setminus v; \mathbb{Q}))$$

(it is also $\chi(A) - \chi(A \setminus v)$; see 11.7.16). We shall now explain how this number can be computed from a triangulation of A or from a stratification induced by a stratifying family of polynomials.

Proposition 11.2.1.
(i) Let K be a finite simplicial complex in R^n and v a vertex of K. Let m_r be the number of r-simplices of K having v as a vertex. Then

$$\chi(|K|, |K| \setminus v) = \sum_r (-1)^r m_r .$$

(ii) Let A be a locally closed semi-algebraic subset of R^n, and $v \in A$. Assume that A has a stratification induced by a stratifying family of polynomials, such that v is a 0-stratum. Let m_r be the number of r-strata C of A such that $v \in \mathrm{clos}(C)$. Then

$$\chi(A, A \setminus v) = \sum_r (-1)^r m_r .$$

Proof. (i) It suffices to note that, if L is the subcomplex of all simplices that do not contain v, then $|L|$ is a semi-algebraic deformation retract of $|K| \setminus v$ (see Figure 11.2).

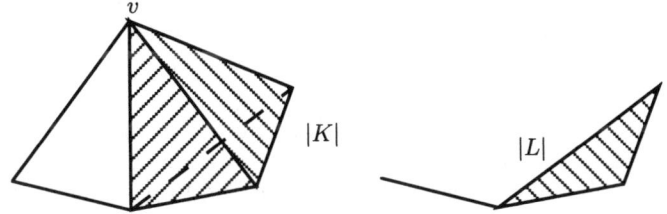

Figure 11.2.

Then $\chi(|K|, |K| \setminus v) = \chi(|K|) - \chi(|L|)$, and the Euler-Poincaré characteristic of $|K|$ (or $|L|$) can be computed as the alternating sum of the numbers of r-simplices.

(ii) We use the conic structure of the stratification (Proposition 9.1.14). If U_v is the union of all strata C of A such that $v \in \mathrm{clos}(C)$, there exist a closed and bounded semi-algebraic set G contained in U_v and a continuous semi-algebraic surjective mapping $\eta : G \times [0, 1[\to U_v$, such that $\eta|_{G \times]0,1[}$

is a homeomorphism onto $U_v \setminus v$, and, for every $z \in G$, $\eta(z,0) = v$ and $\eta(z,1/2) = z$. Then $H_0(A, A \setminus v; \mathbb{Q}) = \mathbb{Q}$, and $H_{r+1}(A, A \setminus v; \mathbb{Q}) = H_r(G, \mathbb{Q})$ (cf. 11.7.16). By Proposition 9.1.14, the collection of $G \cap C$, for all strata C of A contained in U_v, $C \neq \{v\}$, is a semi-algebraic cellular decomposition of G, with $\dim(G \cap C) = \dim(C) - 1$. The combinatorial computation of $1 - \chi(G) = \chi(A, A \setminus v)$ gives the result $\sum_r (-1)^r m_r$. $\square$

Theorem 11.2.2. *Let $V \subset R^n$ be an algebraic set, and $a \in V$. Then $\chi(V, V \setminus a)$ is odd.*

Proof. We shall proceed by induction on the dimension $d = \dim(V)$. We shall first explain the meaning of the theorem in the case of a stratification of V induced by a stratifying family of polynomials (resp. a triangulation of V) having a as a 0-stratum (resp. as a vertex). With the notation of Proposition 11.2.1, we have $m_0 = 1$; this corresponds to the 0-stratum (resp. vertex) a. Hence, the assertion that $\chi(V, V \setminus a)$ is odd is equivalent to the assertion that $\sum_{r>0}(-1)^r m_r$ is even, or, equivalently, $\sum_{r>0} m_r$ is even: the total number of strata C of the stratification (resp. simplices of the triangulation), such that $a \in \text{clos}(C) \setminus C$, is even. If the dimension d is 1, Theorem 11.2.2 is equivalent to Theorem 9.5.7.

Now assume that $d > 1$ and the theorem holds true for algebraic sets of dimension $< d$. By Theorem 9.1.6, we may assume that V has a stratification induced by a stratifying family of polynomials, such that a is a 0-stratum. Moreover, if $\Pi_{n,d} : R^n \to R^d$ is the projection onto the space of the first d coordinates, we may assume that, for every stratum C of V, $\Pi_{n,d}|_{\text{clos}(C)}$ is a homeomorphism onto $\text{clos}(\Pi_{n,d}(C))$. Recall that the stratifying family of polynomials also induces a stratification $\mathcal{C}_d$ of R^d, and $\Pi_{n,d}(C)$ is a stratum of $\mathcal{C}_d$ (cf. Theorem 9.1.4). For $k \leq d$, let V_k be the union of all strata of V of dimension at most k. By Proposition 9.1.9, each V_k is algebraic. By the inductive assumption, $\chi(V_k, V_k \setminus a)$ is odd, for $k < d$, which implies that $\sum_{r=1}^{k} m_r$ is even. Therefore, m_r is even, for $0 < r < d$. We still have to prove that m_d is even.

Let $b = \Pi_{n,d}(a)$, and let U_b be the union of strata C' of the stratification $\mathcal{C}_d$ of R^d, such that $b \in \text{clos}(C')$. Let $\mathcal{E}_r$ be the set of r-strata of $\mathcal{C}_d$ contained in U_b, and e_r, the number of elements of $\mathcal{E}_r$, for $0 \leq r \leq d$. Proceeding as above, we can prove that e_r is even, for $0 < r < d$. Since the sphere S^{d-1} of dimension $d-1$ is a semi-algebraic deformation retract of $R^d \setminus b$, and $\chi(S^{d-1}) = 0$ (resp. 2) if d is even (resp. odd), one obtains $\chi(R^d, R^d \setminus b) = (-1)^d$. This implies that $\sum_{0 < r \leq d} e_r$ is even, and therefore e_d is even. For $C' \in \mathcal{E}_d$, denote by $n(C')$ the number of d-strata C of V, such that $\Pi_{n,d}(C) = C'$ and $a \in \text{clos}(C)$. Since the image of a d-stratum of V by $\Pi_{n,d}$ is a d-stratum, one has $m_d = \sum_{C' \in \mathcal{E}_d} n(C')$. Since e_d is even, it suffices to prove that all $n(C')$ have the same parity.

First, we consider the case of two strata C'_1 and C'_2 of $\mathcal{E}_d$, such that there exists a $(d-1)$-stratum D', with $D' \subset \text{clos}(C'_1) \cap \text{clos}(C'_2)$ (we say that C'_1 and

C_2' are *contiguous*). Let D be a $(d-1)$-stratum of V such that $\Pi_{n,d}(D) = D'$. If C is a d-stratum of V such that $D \subset \mathrm{clos}(C)$, then $\Pi_{n,d}(C) = C_1'$ or C_2'. By Theorem 11.1.1, the number of such d-strata C is even. By considering all strata D projecting onto D' and such that $a \in \mathrm{clos}(D)$, it follows that $n(C_1') + n(C_2')$ is even. Therefore, $n(C_1')$ and $n(C_2')$ have the same parity (see Figure 11.3).

Next, we consider the case of any two strata C' and C'' of $\mathcal{E}_d$. There is a sequence $C' = C_0', C_1', \ldots, C_k' = C''$ in $\mathcal{E}_d$, such that C_{i-1}' and C_i' are contiguous for $i = 1, \ldots, k$. Otherwise, $\mathcal{E}_d$ would be the disjoint union of two nonempty subsets $\mathcal{X}$ and $\mathcal{Y}$, such that no element of $\mathcal{X}$ is contiguous to an element of $\mathcal{Y}$. The closures of $\bigcup_{D \in \mathcal{X}} D$ and $\bigcup_{D \in \mathcal{Y}} D$ in U_b would then be two semi-algebraic sets, whose union is U_b and whose intersection is a semi-algebraic subset of dimension $< d - 1$, which is impossible by Lemma 4.5.2. Hence, $n(C')$ and $n(C'')$ have the same parity. This completes the proof. □

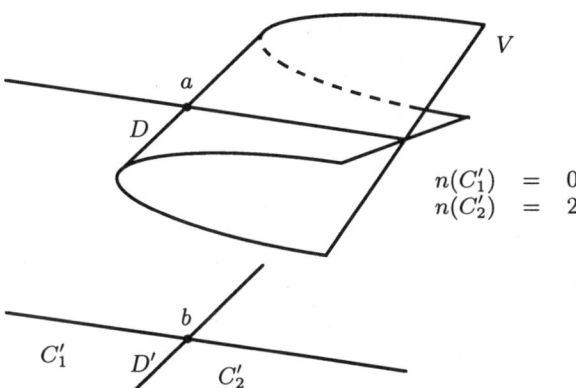

Figure 11.3.

Example 11.2.3. If K is the simplicial complex in Figure 11.1, then $\chi(|K|, |K| \setminus a) = -1$. We know that $|K|$ is homeomorphic to an algebraic set. On the other hand, if K is the simplicial complex in Figure 11.2, then $\chi(|K|, |K| \setminus a) = 0$. Hence $|K|$ is not homeomorphic to an algebraic set.

Corollary 11.2.4. *Let $V \subset R^n$ be an algebraic set, and $a \in V$. There exists an $\epsilon > 0$, such that $V \cap \overline{B}_n(a, \epsilon)$ is semi-algebraically homeomorphic to a cone, whose base $V \cap S^{n-1}(a, \epsilon)$ has even Euler-Poincaré characteristic.*

Proof. Follows immediately from Theorems 9.3.6 and 11.2.2. □

Remark 11.2.5. The condition concerning the local Euler-Poincaré characteristic given in Theorem 11.2.2 is necessary for a simplicial complex to be homeomorphic to an algebraic set. It can be proved ([8, 29]) that this condition is sufficient for simplicial complexes of dimension at most 2, but is not sufficient in dimension 3. The following counter-example was given in [187]. Consider the set $X \subset \mathbb{R}^3$ of dimension 2 shown in Figure 11.4. The set X is homeomorphic to an algebraic Alexandrov compactification (Definition 3.5.4) of the union Z of a circle and the umbrella given by the equation $y^2 x = z^2$ (see Figure 11.4).

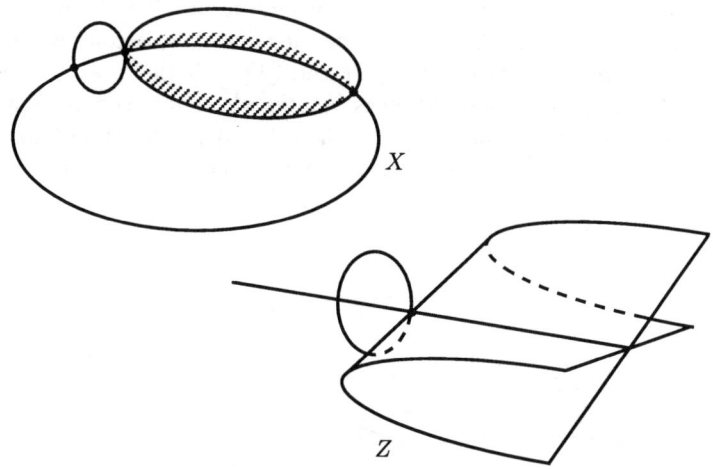

Figure 11.4. X is an algebraic Alexandrov compactification of Z

However, the suspension of X is not homeomorphic to an algebraic set, although it satisfies the condition concerning the local Euler-Poincaré characteristic. This can be seen by using the following theorem in [97], that we state without proof.

Theorem 11.2.6. *Let $V \subset R^n$ be an algebraic set, $W \subset V$ an irreducible algebraic subset. There is a proper algebraic subset Z of W, such that, for all a and b in $W \setminus Z$,*

$$\chi(V, V \setminus a) \equiv \chi(V, V \setminus b) \pmod 4 .$$

Note that, if $\dim(W) = \dim(V) - 1$, this theorem is equivalent to Theorem 11.1.2.

We now return to the counter-example of [187]. Recall that the suspension of $X \subset \mathbb{R}^3$ may be constructed in the following way. Let α and β be the points $(0, 0, 0, 1)$ and $(0, 0, 0, -1)$, respectively, and identify $\mathbb{R}^3$ with $\mathbb{R}^3 \times \{0\} \subset \mathbb{R}^4$. The suspension ΣX of X is the union of all segments $[\alpha, x]$ and $[\beta, x]$, for $x \in X$. Suppose that there exists a homeomorphism $\Phi : \Sigma X \to V$, where V is a real algebraic set. The set of points $x \in \Sigma X$ such that $\chi(\Sigma X, \Sigma X \setminus x) \equiv 1 \pmod 4$ is the suspension $\Sigma(Y \setminus \{b, c\})$, where Y is the 1-dimensional subset

11.3 Fundamental Class of a Real Algebraic Variety. Algebraic Homology

of X indicated in Figure 11.5 and b and c are the points indicated in the same figure. Applying the semi-algebraic trivialization theorem 9.3.2 to the

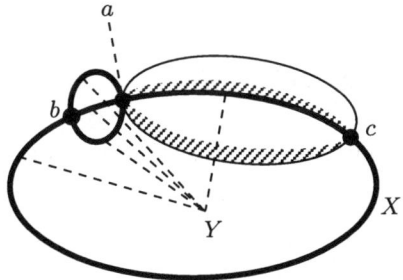

Figure 11.5.

projection $V \times V \to V$ on the first factor and the subset $\{(x,x) \mid x \in V\} \subset V \times V$, we see that the image $S = \Phi(\Sigma(Y \setminus \{b,c\}))$ would be semi-algebraic. The Zariski closure $T = \text{clos}_{\text{Zar}}(S)$ would be an algebraic set of dimension 2 and, by Theorem 11.2.6, $\dim(T \setminus S) \leq 1$. Let U be the set of points $t \in T$ such that $\chi(V, V \setminus t) \equiv 1 \pmod 4$ and $\chi(T, T \setminus t) \equiv 3 \pmod 4$. Then $U = \Phi([\alpha, a] \cup [a, \beta])$, except for finitely many points. The Zariski closure $\text{clos}_{\text{Zar}}(U)$ would be an algebraic curve and, by Theorem 11.2.6, $\text{clos}_{\text{Zar}}(U) \setminus U$ would consist of finitely many points. This is impossible, since the curve $\text{clos}_{\text{Zar}}(U)$ would have only one half-branch at $\Phi(\alpha)$.

11.3 Fundamental Class of a Real Algebraic Variety. Algebraic Homology

Proposition 11.3.1. *Let X be an affine complete real algebraic variety of dimension d, and $\Phi : |K| \to X$, a semi-algebraic triangulation of X. The sum of all d-simplices of K is a cycle with coefficients in $\mathbb{Z}/2$, representing a nonzero element of $H_d(X, \mathbb{Z}/2)$. This element, which is independent of the choice of the triangulation, is called the* **fundamental class** *of X and is denoted by $[X]$.*

Proof. The fact, that the sum of all d-simplices of K is a cycle with coefficients in $\mathbb{Z}/2$, is a straightforward consequence of Theorem 11.1.1 (i), and it is obvious that this cycle is not a boundary. Note that, for every point x in X which is nonsingular in dimension d, the image of $[X]$ in $H_d(X, X \setminus x; \mathbb{Z}/2) \simeq \mathbb{Z}/2$ is the nonzero element. This implies that $[X]$ does not depend on the choice of the triangulation. □

Definition 11.3.2. *Let X be an affine complete real algebraic variety. Given a k-dimensional Zariski closed subset Y of X, we call the element $i_*([Y])$ of $H_k(X, \mathbb{Z}/2)$, where $i_* : H_k(Y, \mathbb{Z}/2) \to H_k(X, \mathbb{Z}/2)$ is the homomorphism* induced by the inclusion map $i : Y \hookrightarrow X$, the homology class of X represented by Y. By abuse of notation, we denote $i_*([Y])$ by $[Y]$.

Denote by $H_k^{\mathrm{alg}}(X, \mathbb{Z}/2)$ the subgroup of $H_k(X, \mathbb{Z}/2)$ consisting of all homology classes represented by Zariski closed k-dimensional subsets of X, and set
$$H_*^{\mathrm{alg}}(X, \mathbb{Z}/2) = \bigoplus_{k \geq 0} H_k^{\mathrm{alg}}(X, \mathbb{Z}/2) .$$

If $H_*^{\mathrm{alg}}(X, \mathbb{Z}/2) = H_*(X, \mathbb{Z}/2)$, the homology of X is said to be totally algebraic.

Proposition 11.3.3. *The homology of the grassmannians is totally algebraic:* $H_*^{\mathrm{alg}}(\mathbb{G}_{n,k}(R), \mathbb{Z}/2) = H_*(\mathbb{G}_{n,k}(R), \mathbb{Z}/2)$.

Proof. First we recall the cellular decomposition of $\mathbb{G}_{n,k}(R)$ given by the Schubert cells (cf. [236], §6). Consider the chain of inclusions $R \subset R^2 \subset \cdots \subset R^n$, where each R^j is embedded in R^n as $R^j \simeq R^j \times \{0\} \hookrightarrow R^n$. Let $\sigma = (\sigma_1, \ldots, \sigma_k)$ be a sequence of integers with $1 \leq \sigma_1 < \sigma_2 < \cdots < \sigma_k \leq n$. Let $e(\sigma) \subset \mathbb{G}_{n,k}(R)$ be the set of vector subspaces X of dimension k of R^n such that
$$\dim(X \cap R^{\sigma_i - 1}) = i - 1, \quad \dim(X \cap R^{\sigma_i}) = i \quad \text{for } i = 1, \ldots, k .$$

The $e(\sigma)$ form a finite partition of $\mathbb{G}_{n,k}(R)$ into semi-algebraic subsets, and each $e(\sigma)$ is semi-algebraically homeomorphic to an open ball $B_{d(\sigma)}$ of dimension $d(\sigma) = (\sigma_1 - 1) + \cdots + (\sigma_k - k)$ (the semi-algebraicity is a straightforward consequence of the explicit computations in [236]). The homeomorphism $q : B_{d(\sigma)} \to e(\sigma)$ can be extended to a continuous surjective mapping $\overline{q} : \overline{B}_{d(\sigma)} \to \overline{e}(\sigma) = \mathrm{clos}(e(\sigma))$. It can be checked that
$$\overline{e}(\sigma) = \bigcup \{e(\tau) \mid \forall i = 1, \ldots, k \;\; \tau_i \leq \sigma_i\} .$$

We shall prove that the set $\overline{e}(\sigma)$ is algebraic, using the identification (Theorem 3.4.4)
$$\mathbb{G}_{n,k}(R) = \{A \in \mathbb{M}_{n,n}(R) \mid A = A^2 = {}^tA, \; \mathrm{trace}(A) = k\} .$$

A vector subspace X of dimension k of R^n is identified with the matrix A of the orthogonal projection onto X. Denote by $A'_j \in \mathbb{M}_{n,j}(R)$ the matrix consisting of the first j columns of the matrix $A' = \mathrm{Id}_n - A$. Then
$$\overline{e}(\sigma) = \{A \in \mathbb{G}_{n,k}(R) \mid \mathrm{rank}(A'_{\sigma_i}) \leq \sigma_i - i \text{ for } i = 1, \ldots, k\} ,$$

which is clearly algebraic. The homology of the grassmannian can be computed from the cellular decomposition given by the Schubert cells. Since the closure $\overline{e}(\sigma)$ of each Schubert cell is algebraic, every homology class in $H_*(\mathbb{G}_{n,k}(R), \mathbb{Z}/2)$ can be represented by an algebraic set.

11.3 Fundamental Class of a Real Algebraic Variety. Algebraic Homology 273

The case of projective space $\mathbb{P}_n(R) = \mathbb{G}_{n+1,1}(R)$ is particularly simple. There is exactly one cell in dimension k, $0 \leq k \leq n$, and the closures of these cells form a chain of inclusions $\mathbb{P}_0(R) \subset \mathbb{P}_1(R) \subset \cdots \subset \mathbb{P}_n(R)$. It follows that $H_k(\mathbb{P}_n(R), \mathbb{Z}/2) = H_k^{\text{alg}}(\mathbb{P}_n(R), \mathbb{Z}/2) = \mathbb{Z}/2$ for $k = 0, \ldots, n$, the nonzero homology class being represented by a projective subspace. $\square$

The group $H_*^{\text{alg}}(X, \mathbb{Z}/2)$ is a basic invariant of X. As one might expect, the homology of X is not always totally algebraic.

In the remainder of this section, we consider only the case $R = \mathbb{R}$.

If X is a compact affine nonsingular real algebraic variety of dimension d, we set

$$H_{\text{alg}}^k(X, \mathbb{Z}/2) = D^{-1}(H_{d-k}^{\text{alg}}(X, \mathbb{Z})) ,$$
$$H_{\text{alg}}^*(X, \mathbb{Z}/2) = \bigoplus_{k \geq 0} H_{\text{alg}}^k(X, \mathbb{Z}/2) ,$$

where $D : H^*(X, \mathbb{Z}/2) \longrightarrow H_*(X, \mathbb{Z}/2)$ is the Poincaré duality isomorphism. We denote by f_* (resp. f^*) the homomorphism of homology (resp. cohomology) groups induced by a map f.

Basic properties of H_*^{alg} and H_{alg}^* are stated in the following two theorems, which are consequences of the results in [64], Section 5. Other proofs are given in [59].

Theorem 11.3.4. *For every regular map $f : X \to Y$ between compact affine real algebraic varieties,*
  (i) $f_*(H_*^{\text{alg}}(X, \mathbb{Z}/2)) \subset H_*^{\text{alg}}(Y, \mathbb{Z}/2)$,
  (ii) $f^*(H_{\text{alg}}^*(Y, \mathbb{Z}/2)) \subset H_{\text{alg}}^*(X, \mathbb{Z}/2)$, *provided that X and Y are nonsingular.*

Theorem 11.3.5. *For every compact nonsingular affine real algebraic variety X, the group $H_{\text{alg}}^*(X, \mathbb{Z}/2)$ is a subring of the cohomology ring $H^*(X, \mathbb{Z}/2)$.*

We shall use Theorem 11.3.4 (i) in the construction of varieties with $H_k^{\text{alg}} \neq H_k$. A proof of Theorem 11.3.4 (ii) for H_{alg}^1 will be given in Chap. 12, which also contains an important application of this theorem. We shall not use Theorem 11.3.5 in this book, but its content is clearly enlightening and important.

Definition 11.3.6. *A nonsingular affine real algebraic variety X diffeomorphic to a C^∞ manifold M is called an* algebraic model *of M.*

A remarkable theorem of Tognoli (cf. Theorem 14.1.10) asserts that every compact C^∞ manifold has an algebraic model.

We are now concerned with algebraic models of C^∞ manifolds with some subgroups H_{alg}^k given in advance. The next two results clarify the situation for $k = 1$.

Given a compact C^∞ manifold M, we denote by $w_1(M)$ its first Stiefel-Whitney class. For the definition and basic properties of Stiefel-Whitney classes, we refer the reader to the excellent book by Milnor and Stasheff [236] (see also Section 12.1). Here, we merely recall that $w_1(M)$ is an element of $H^1(M, \mathbb{Z}/2)$, which is zero if and only if M is orientable.

Theorem 11.3.7. *Let M be a compact connected C^∞ manifold of dimension at least 3, and let G be a subgroup of $H^1(M, \mathbb{Z}/2)$. Then the following conditions are equivalent:*

(i) *There exist an algebraic model X of M and a diffeomorphism $\varphi : X \longrightarrow M$ such that $\varphi^*(G) = H^1_{\mathrm{alg}}(X, \mathbb{Z}/2)$;*

(ii) *$w_1(M) \in G$.*

Reference for the proof. The implication (i)$\Rightarrow$(ii) follows from the fact that, for every compact nonsingular affine real algebraic variety X, the class $w_1(X)$ is in $H^1_{\mathrm{alg}}(X, \mathbb{Z}/2)$ (cf. Theorem 12.1.10). The proof of (ii)$\Rightarrow$(i), which is rather involved, is given in [48]. □

For surfaces, only a somewhat weaker result is known. Given a compact nonsingular real algebraic surface X, we set

$$\beta(X) = \dim_{\mathbb{Z}/2} H^1_{\mathrm{alg}}(X, \mathbb{Z}/2) ,$$
$$\delta(X) = \dim_{\mathbb{Z}/2} \{v \in H^1_{\mathrm{alg}}(X, \mathbb{Z}/2) \mid v \cup v = 0\} ,$$

where $v \cup v \in H^2(X, \mathbb{Z}/2)$ is the cup product. If X is connected and orientable (resp. nonorientable of odd topological genus), then $\beta(X) = \delta(X)$ (resp. $\beta(X) = \delta(X) + 1$). For X connected, nonorientable and of even topological genus, one has either $\beta(X) = \delta(X)$ or $\beta(X) = \delta(X) + 1$. These statements are consequences of elementary topological arguments and of the fact that $w_1(X) \in H^1_{\mathrm{alg}}(X, \mathbb{Z}/2)$. The next result shows that all topologically possible cases can be realized algebraically.

Theorem 11.3.8. (i) *Let M be a compact connected C^∞ surface of genus g, and let β be an integer satisfying*

$$0 \leq \beta \leq 2g \quad \text{for } M \text{ orientable,}$$
$$1 \leq \beta \leq g \quad \text{for } M \text{ nonorientable.}$$

Then there exists an algebraic model X_β of M with $\beta(X_\beta) = \beta$.

(ii) *Let M be a compact connected nonorientable C^∞ surface of even genus g, and let β and δ be integers satisfying either*

$$\beta = \delta + 1, \quad 2 \leq \beta \leq g$$

or

$$\beta = \delta, \quad 1 \leq \beta \leq g - 1.$$

Then there exists an algebraic model $X_{\beta,\delta}$ of M such that $\beta(X_{\beta,\delta}) = \beta$ and $\delta(X_{\beta,\delta}) = \delta$.

11.3 Fundamental Class of a Real Algebraic Variety. Algebraic Homology 275

Reference for the proof: [48]. □

Since the proofs of Theorems 11.3.7 and 11.3.8 are omitted, we shall now explain the construction of varieties X with $H_k^{\text{alg}}(X, \mathbb{Z}/2) \neq H_k(X, \mathbb{Z}/2)$, in order to illustrate, in some detail, the kind of phenomena that can occur.

Example 11.3.9. Denote by S^p the unit sphere in $\mathbb{R}^{p+1}$. Let M be a $\mathcal{C}^\infty$ manifold which is the connected sum of s copies of $S^n \times S^k$, $s \geq 1$, $n \geq 1$, $k \geq 1$ (note that, in particular, every compact connected orientable $\mathcal{C}^\infty$ surface of positive genus g can be obtained in this fashion, with $s = g$ and $n = k = 1$). Below we construct an algebraic model X of M satisfying $H_{\text{alg}}^k(X, \mathbb{Z}/2) \neq H^k(X, \mathbb{Z}/2)$.

Let $C = \{(x_1, \ldots, x_{n+1}) \in \mathbb{R}^{n+1} \mid x_1^4 - 4x_1^2 + 1 + x_2^2 + \cdots + x_{n+1}^2 = 0\}$. Clearly, C is a nonsingular irreducible algebraic subset of $\mathbb{R}^{n+1}$ with two connected components C_1 and C_2, each of which is diffeomorphic to S^n. Set $W = C \times S^{k+1}$. Choose a compact $\mathcal{C}^\infty$ submanifold B of S^{k+1} whose boundary ∂B has s connected components, each of which is diffeomorphic to S^k. Choose a point y in ∂B. By joining the connected components of $C_1 \times \partial B$ with "tubes" in W, we can construct a compact $\mathcal{C}^\infty$ submanifold V of W, with boundary $\partial V = N$ diffeomorphic to M and containing $C_1 \times \{y\}$. Since N bounds in W, there exists a $\mathcal{C}^\infty$ function $f : W \longrightarrow \mathbb{R}$, such that $0 \in \mathbb{R}$ is a regular value of f and $N = f^{-1}(0)$. The Stone-Weierstrass approximation theorem allows us to find a polynomial function $p : W \longrightarrow \mathbb{R}$ close to f in the $\mathcal{C}^\infty$ topology. If p is sufficiently close to f, then $X = p^{-1}(0)$ is a nonsingular algebraic subset of W, and there exists a $\mathcal{C}^\infty$ diffeomorphism $\varphi : W \longrightarrow W$, close in the $\mathcal{C}^\infty$ topology to the identity map of W, with $\varphi(N) = X$ (in particular, X is an algebraic model of M). Furthermore, the $\mathcal{C}^\infty$ submanifold $D = \varphi(C_1 \times \{y\})$ of X is mapped diffeomorphically onto C_1 by the canonical projection $\pi : W \longrightarrow C$. Let u be the homology class in $H_n(X, \mathbb{Z}/2)$ represented by D. Then $\pi_*(u)$ is the homology class in $H_n(C, \mathbb{Z}/2)$ represented by C_1, and, hence, $\pi_*(u)$ is not in $H_n^{\text{alg}}(C, \mathbb{Z}/2)$ (C is irreducible of dimension n). Thus, in view of Theorem 11.3.4(i), u is not in $H_n^{\text{alg}}(X, \mathbb{Z}/2)$, which implies $H_{\text{alg}}^k(X, \mathbb{Z}/2) \neq H^k(X, \mathbb{Z}/2)$.

Another proof of the fact that $[D]$ is not in $H_n^{\text{alg}}(X, \mathbb{Z}/2)$, for $k = 1$, based on a different argument, will be given in Chap. 12 (cf. Example 12.4.12).

The next two theorems deal with H_{alg}^2. Given a $\mathcal{C}^\infty$ manifold M, denote by $W^2(M)$ the subset of $H^2(M, \mathbb{Z}/2)$ consisting of all elements of the form $w_2(\xi)$, for some topological $\mathbb{R}$-vector bundle ξ on M, where $w_2(\xi)$ is the second Stiefel-Whitney class of ξ. The second Stiefel-Whitney class of the tangent bundle of M is denoted by $w_2(M)$ (cf. Section 12.1 and [236]). One easily checks that $W^2(M)$ is a subgroup of $H^2(M, \mathbb{Z}/2)$.

Theorem 11.3.10. *Let M be a compact connected orientable $\mathcal{C}^\infty$ manifold of dimension at least 5 and let G be a subgroup of $H^2(M, \mathbb{Z}/2)$. Then the following conditions are equivalent:*

(i) *There exist an algebraic model X of M and a diffeomorphism* $\varphi : X \longrightarrow M$ *such that* $\varphi^*(G) = H^2_{\text{alg}}(X, \mathbb{Z}/2)$;
(ii) $w_2(M) \in G$ *and* $G \subseteq W^2(M)$.

Reference for the proof: [54]. □

Theorem 11.3.10 is nicely complemented by the following result.

Theorem 11.3.11. *Given an integer $n \geq 6$, there exists a compact connected orientable $\mathcal{C}^\infty$ manifold M of dimension n such that $W^2(M) \neq H^2(M, \mathbb{Z}/2)$. Hence, there is a cohomology class $\eta \in H^2(M, \mathbb{Z}/2)$ such that, for every algebraic model X of M and every diffeomorphism $\varphi : X \to M$, one has*

$$\varphi^*(\eta) \notin H^2_{\text{alg}}(X, \mathbb{Z}/2) .$$

Reference for the proof: [319]. □

A manifold M as in Theorem 11.3.11 has no algebraic model with total algebraic homology. In dimension less than 6 the situation is quite different.

Theorem 11.3.12. *For every compact $\mathcal{C}^\infty$ manifold M of dimension less than 6, there exists an algebraic model X of M satisfying $H^*_{\text{alg}}(X, \mathbb{Z}/2) = H^*(X, \mathbb{Z}/2)$.*

Sketch of proof. By Thom's theorem (cf. [320] Theorem II.26), there exist compact $\mathcal{C}^\infty$ submanifolds $M_1, \ldots, M_k$ of M, whose homology classes generate $H_*(M, \mathbb{Z}/2)$. We may assume that the M_i are in general position. Then it follows from [32] Theorem 4 that there exist an algebraic model X of M and a diffeomorphism $h : M \longrightarrow X$ such that $h(M_i)$ is a Zariski closed nonsingular subvariety of X for $i = 1, \ldots, k$. This implies the conclusion. □

It should be remarked that no general result analogous to Theorems 11.3.7 and 11.3.10 is known for H^k_{alg}, $k \geq 3$.

Several interesting explicit examples of varieties with $H^{\text{alg}}_* \neq H_*$ are obtained by investigating the underlying real algebraic structure of complex projective varieties. Every complex projective variety V of dimension n can be regarded as an affine real algebraic variety of dimension $2n$, and, as such, will be denoted by $V_\mathbb{R}$ (cf. Proposition 3.4.6). Assume that V is nonsingular and irreducible. A method for computing the group $H^1_{\text{alg}}(V_\mathbb{R}, \mathbb{Z}/2)$ is well understood [56, 145, 171, 172]. We shall now present a few results in this direction, concentrating on the comparison between the real algebraic geometry invariant

$$d(V) = \dim_{\mathbb{Z}/2} H^1_{\text{alg}}(V_\mathbb{R}, \mathbb{Z}/2)$$

and the topological invariants

$$h(V) = \dim_{\mathbb{Z}/2} H^1(V, \mathbb{Z}/2)$$
$$b_1(V) = \text{the first Betti number of } V = \dim_\mathbb{Q} H_1(V, \mathbb{Q}) .$$

11.3 Fundamental Class of a Real Algebraic Variety. Algebraic Homology

Note that we have $b_1(V) = \dim_{\mathbb{Z}/2}(H^1(V,\mathbb{Z}) \otimes_{\mathbb{Z}} \mathbb{Z}/2)$ since $H^1(V,\mathbb{Z})$ is torsion-free.

Recall first that to V we can associate functorially a complex Abelian variety, the Albanese variety $\mathrm{Alb}(V)$ of V, and a regular map $\alpha : V \longrightarrow \mathrm{Alb}(V)$, the Albanese map (cf. [207]). For curves ($\dim V = 1$), $\mathrm{Alb}(V)$ is simply the Jacobian variety of V. The dimension of $\mathrm{Alb}(V)$ is equal to $\frac{1}{2}b_1(V)$ (it is well known that $b_1(V)$ is even).

The following result, proved in [145], reduces the computation of $H^1_{\mathrm{alg}}(V_{\mathbb{R}}, \mathbb{Z}/2)$ to the case of Abelian varieties, which was earlier investigated in [171].

Theorem 11.3.13. *The Albanese map $\alpha : V \longrightarrow \mathrm{Alb}(V)$ induces an isomorphism*

$$\alpha^* : H^1_{\mathrm{alg}}((\mathrm{Alb}(V))_{\mathbb{R}}, \mathbb{Z}/2) \longrightarrow H^1_{\mathrm{alg}}(V_{\mathbb{R}}, \mathbb{Z}/2).$$

For a complex Abelian variety A, the invariant $d(A)$ can be computed in terms of a period matrix of A. We shall describe this method in the case of 1-dimensional complex Abelian varieties, that is, elliptic curves.

Example 11.3.14. Let $E_\tau = \mathbb{C}/(\mathbb{Z}+\mathbb{Z}\tau)$ be a complex elliptic curve, $\tau \in \mathbb{C}\setminus\mathbb{R}$. We shall express $d(E_\tau)$ as an explicitly computable function of τ. Let S_τ and T_τ be the subgroups of $\mathbb{Z}^2$ defined as follows:

$$S_\tau = \{(m,n) \in \mathbb{Z}^2 \mid m + n|\tau|^2 \in 2(\mathrm{Re}\,\tau)\mathbb{Z}\}$$
$$T_\tau = S_\tau \cap 2\mathbb{Z}^2.$$

Clearly, S_τ/T_τ is a $\mathbb{Z}/2$-vector space. It is shown in [171] that

$$d(E_\tau) = \dim_{\mathbb{Z}/2} S_\tau/T_\tau.$$

For example, if $\tau = \alpha\sqrt{-1}$, where $\alpha \in \mathbb{R} \setminus \{0\}$, one has

$$d(E_\tau) = \begin{cases} 1 & \text{if } \alpha^2 \in \mathbb{Q}, \\ 0 & \text{if } \alpha^2 \notin \mathbb{Q}. \end{cases}$$

$\square$

Observe that the homomorphism

$$\alpha^* : H^1(\mathrm{Alb}(V), \mathbb{Z}/2) \longrightarrow H^1(V, \mathbb{Z}/2),$$

induced by the Albanese map $\alpha : V \longrightarrow \mathrm{Alb}(V)$, need not be surjective. In fact, it is surjective if and only if $H^2(V,\mathbb{Z})$ has no 2-torsion. Therefore the topology of V imposes strong restrictions on the existence of homology classes in codimension 1 represented by real algebraic hypersurfaces of $V_{\mathbb{R}}$. Theorem 11.3.13 implies the next two corollaries, in which V is a nonsingular irreducible complex projective variety.

Corollary 11.3.15. *The group $H^1_{\text{alg}}(V_{\mathbb{R}}, \mathbb{Z}/2)$ is contained in the image of the canonical injective homomorphism*

$$H^1(V, \mathbb{Z}) \otimes_{\mathbb{Z}} \mathbb{Z}/2 \longrightarrow H^1(V, \mathbb{Z}/2).$$

In particular, $d(V) \leq b_1(V)$.

Corollary 11.3.16. *If the group $H^2(V, \mathbb{Z})$ has 2-torsion, then*

$$H^1_{\text{alg}}(V_{\mathbb{R}}, \mathbb{Z}/2) \neq H^1(V_{\mathbb{R}}, \mathbb{Z}/2).$$

As a special case one obtains the following.

Example 11.3.17. Each complex Enriques surface F has $b_1(F) = 0$ and, hence, $H^1_{\text{alg}}(F_{\mathbb{R}}, \mathbb{Z}/2) = 0$. On the other hand, $H^1(F, \mathbb{Z}/2) = \mathbb{Z}/2$.

More generally, a construction of complex algebraic varieties V with prescribed $\dim V \geq 2$, $d(V)$, $b_1(V)$, and $h(V)$ is given in [145].

Theorem 11.3.18. *Given nonnegative integers $n, a, g, h, n \geq 2$, the following conditions are equivalent:*

(i) *There exists a nonsingular irreducible complex projective variety V such that $\dim V = n$, $d(V) = a$, $b_1(V) = 2g$, and $h(V) = h$;*

(ii) $0 \leq a \leq 2g \leq h$.

For further results concerning the group $H^1_{\text{alg}}(V_{\mathbb{R}}, \mathbb{Z}/2)$, we refer the reader to [56, 145, 171, 172].

11.4 Injective Regular Self-Mappings of an Algebraic Set

In this section we use Borel-Moore homology, denoted by H^{BM}_*. For semi-algebraic sets, this homology can be defined and computed in a combinatorial way, without referring to the general theory presented in [65]. This can be done using the semi-algebraic Alexandrov compactification of locally closed semi-algebraic sets Y. Definition 11.7.13 shows immediately that, for such a set Y, $\dim_{\mathbb{Z}/2} H^{\text{BM}}_*(Y, \mathbb{Z}/2) < \infty$.

Borel-Moore homology allows one to define the fundamental class of an unbounded algebraic set.

Definition 11.4.1. *Let $V \subset \mathbb{R}^n$ be an unbounded algebraic set of dimension d, and $(\dot{V}, i : V \to \dot{V})$, its algebraic Alexandrov compactification (Definition 3.5.4). The fundamental class of V in $H^{\text{BM}}_d(V, \mathbb{Z}/2)$, denoted by $[V]$, is the image of $[\dot{V}]$ by the canonical homomorphism*

$$H_d(\dot{V}, \mathbb{Z}/2) \to H_d(\dot{V}, \dot{V} \setminus i(V); \mathbb{Z}/2) = H^{\text{BM}}_d(V, \mathbb{Z}/2).$$

11.4 Injective Regular Self-Mappings of an Algebraic Set

This section is devoted to the following result.

Theorem 11.4.2. *Let $V \subset R^n$ be a nonsingular irreducible algebraic set, and $f : V \to V$, a regular mapping. If f is injective, then f is surjective.*

In the proof we shall use the following lemma.

Lemma 11.4.3. *Let V be an irreducible algebraic set, and $f : V \to R^p$, a regular mapping. Assume that there is a proper algebraic subset $Z \subset V$, such that $f|_{V \setminus Z}$ is injective. Let W be the Zariski closure of $f(V)$ in R^p. Then $\dim(W \setminus f(V)) < \dim(V) = \dim(W)$.*

Proof. The set W is an irreducible algebraic set of the same dimension as V, since, by Theorem 2.8.8, $\dim(V) = \dim(V \setminus Z) = \dim(f(V \setminus Z))$. The mapping f induces an injection of fields $\mathcal{K}(W) \hookrightarrow \mathcal{K}(V)$ and $[K(V) : K(W)]$ is finite. Replacing V with the graph of f, we may assume that V is an algebraic subset of $R^n \times R^p$, and f is the restriction of the canonical projection $R^n \times R^p \to R^p$. Denote by V_C (resp. W_C) the Zariski closure of V (resp. W) in $C^n \times C^p$ (resp. C^p), where $C = R[i]$, and let $f_C : V_C \to W_C$ be the restriction of the projection $C^n \times C^p \to C^p$. The set of points $y \in W_C$, such that the number of points of $f_C^{-1}(y)$ is equal to the degree of the extension $[\mathcal{K}(V_C) : \mathcal{K}(W_C)] = [\mathcal{K}(V) : \mathcal{K}(W)]$, contains a nonempty Zariski open subset of W_C ([295], Theorem 7, p. 117). If $y \in W$, the points of $f_C^{-1}(y) \cap (V_C \setminus V)$ are pairwise conjugate. Hence, there exists a nonempty Zariski open subset U of W, such that, for every $y \in W$, the number of points of $f^{-1}(y)$ is congruent to $[\mathcal{K}(V) : \mathcal{K}(W)]$ modulo 2. By assumption, the set of $y \in W$ such that $f^{-1}(y)$ contains exactly one point is Zariski-dense in W. Thus, for $y \in U$, the number of elements of $f^{-1}(y)$ is odd. Hence $U \subset f(V)$, which proves the lemma. □

Another proof of Lemma 11.4.3, avoiding complexification, can also be given. First, note that the assumption implies that the mapping from $\mathrm{Spec}_r(\mathcal{K}(V))$ to $\mathrm{Spec}_r(\mathcal{K}(W))$, induced by f, is injective. Therefore, by Proposition 1.3.7, every $\beta \in \mathrm{Spec}_r(\mathcal{K}(W))$ admits an extension in $\mathrm{Spec}_r(\mathcal{K}(V))$. By Proposition 7.5.8, since $(W \setminus f(V))^\sim$ contains no ordering of $\mathcal{K}(W)$, one has $\dim(W \setminus f(V)) < \dim(W)$.

Proof of Theorem 11.4.2. Set $X = f(V)$ and $Y = V \setminus X$. Suppose $Y \neq \emptyset$. The set Y has the following properties:
(i) Y is semi-algebraic.
(ii) Y is closed in V. This is a consequence of the theorem of invariance of domain, which says, for $R = \mathbb{R}$, that a continuous injective mapping from $\mathbb{R}^n$ to $\mathbb{R}^n$ is open [243]. The proof for an arbitrary R and a continuous injective semi-algebraic mapping is given in [109], Theorem 5.13.
(iii) $\dim(Y) < \dim(V)$. Indeed, by Theorem 2.8.8, $\dim(f(V)) = \dim(V)$, since f is injective, and V is the Zariski closure of $f(V)$, since V is irreducible. The claim follows by applying Lemma 11.4.3.

(iv) If Z is the Zariski closure of Y, then $\dim(Z \setminus Y) < \dim(Y)$. Choose an irreducible component Z' of Z of dimension equal to $\dim(Y)$. Suppose $\dim(f^{-1}(Z')) = \dim(Y)$ and let T be an irreducible component of $f^{-1}(Z')$ of dimension $\dim(Y)$. Hence Z' would be the Zariski closure of $f(T)$, and we would have $\dim(Z' \setminus f(T)) < \dim(Y) = \dim(Z')$. Then $Y \cap Z'$, which is contained in $Z' \setminus f(T)$, would not be Zariski-dense in Z'. This contradiction implies that $\dim(f^{-1}(Z')) < \dim(Y)$, which proves the claim.

Let $f^k : V \to V$ be the composition of f with itself k times. Set $X_k = f^k(V) \subset V$ and $Y_k = V \setminus X_k$ (we have $X_1 = X$ and $Y_1 = Y$). Since f^k is regular and injective, Y_k also has properties (i)-(iv) listed above for Y. Furthermore, since $Y_{k+1} = f^k(Y) \cup Y_k$, where the union is disjoint, $\dim(Y_{k+1} \setminus Y_k) = \dim(f^k(Y)) = \dim(Y)$. Denote by d the dimension of Y. Let Z_k be the Zariski closure of Y_k. If Z_k is bounded, consider a semi-algebraic triangulation of Z_k, such that Y_k is the union of images of some simplices. Since $\dim(Z_k \setminus Y_k) < d$, the image of every d-simplex of the triangulation is contained in Y_k, and the sum of these d-simplices represents a nonzero homology class in $H_d(Y_k, \mathbb{Z}/2)$, denoted by $[Y_k]$, whose image in $H_d(Z_k, \mathbb{Z}/2)$ is the class $[Z_k]$. If Z_k is unbounded, we take its algebraic Alexandrov compactification and obtain a nonzero class $[Y_k]$ in $H_d^{\mathrm{BM}}(Y_k, \mathbb{Z}/2)$. The inclusions $H_d^{\mathrm{BM}}(Y_j, \mathbb{Z}/2) \hookrightarrow H_d^{\mathrm{BM}}(Y_{j+1}, \mathbb{Z}/2)$ allow one to identify $[Y_\ell]$ with a nonzero element of $H_d^{\mathrm{BM}}(Y_k, \mathbb{Z}/2)$ for $k \geq l$.

Lemma 11.4.4. *The classes $[Y_1], \ldots, [Y_k]$ are linearly independent (over $\mathbb{Z}/2$) in $H_d^{\mathrm{BM}}(Y_k, \mathbb{Z}/2)$. Hence $\dim_{\mathbb{Z}/2}(H_d^{\mathrm{BM}}(Y_k, \mathbb{Z}/2)) \geq k$.*

Proof. Consider a semi-algebraic triangulation of Z_k (or of its Alexandrov compactification $\dot{Z}_k$), such that the Y_ℓ (or their closures in $\dot{Z}_k$) are the unions of images of some simplices, for $\ell \leq k$. We know that the $Y_{\ell+1} \setminus Y_\ell$ are all disjoint and of dimension d, and, therefore, $[Y_1], [Y_2] - [Y_1], \ldots, [Y_k] - [Y_{k-1}]$ are linearly independent. □

Now we can complete the proof of Theorem 11.4.2. The long exact sequence for Borel-Moore homology (Proposition 11.7.15) gives

$$\cdots \to H_{d+1}^{\mathrm{BM}}(X_k, \mathbb{Z}/2) \xrightarrow{u} H_d^{\mathrm{BM}}(Y_k, \mathbb{Z}/2) \xrightarrow{v} H_d^{\mathrm{BM}}(V, \mathbb{Z}/2) \to \cdots,$$

where $\mathrm{Im}(u) = \mathrm{Ker}(v)$. Since f^k is a continuous and open bijection from V onto X_k, X_k is semi-algebraically homeomorphic to V, and $H_{d+1}^{\mathrm{BM}}(X_k, \mathbb{Z}/2) \simeq H_{d+1}^{\mathrm{BM}}(V, \mathbb{Z}/2)$. Therefore

$$\begin{aligned}
\dim_{\mathbb{Z}/2}(H_d^{\mathrm{BM}}(Y_k, \mathbb{Z}/2)) &= \dim_{\mathbb{Z}/2}(\mathrm{Im}(v)) + \dim_{\mathbb{Z}/2}(\mathrm{Ker}(v)) \\
&= \dim_{\mathbb{Z}/2}(\mathrm{Im}(v)) + \dim_{\mathbb{Z}/2}(\mathrm{Im}(u)) \\
&\leq \dim_{\mathbb{Z}/2}(H_d^{\mathrm{BM}}(V, \mathbb{Z}/2)) + \dim_{\mathbb{Z}/2}(H_{d+1}^{\mathrm{BM}}(V, \mathbb{Z}/2)).
\end{aligned}$$

Since $\dim_{\mathbb{Z}/2}(H_j^{\mathrm{BM}}(V, \mathbb{Z}/2))$ is finite for every j, this contradicts Lemma 11.4.4. Hence the set Y is empty and f is surjective. □

Remark 11.4.5. Theorem 11.4.2 is no longer true over an arbitrary real field, as is shown by the example of the function $f : \mathbb{Q} \to \mathbb{Q}$ defined by $f(x) = x^3$.

11.5 Upper Bound for the Sum of the Betti Numbers of an Algebraic Set

The object of this section is to give an upper bound for the sum of the Betti numbers of an algebraic subset of R^n defined by equations of degree at most d. This bound is given as an explicit function of n and d. First we shall find this bound for $R = \mathbb{R}$, using Morse theory. Next we prove that the same bound is still valid for an arbitrary real closed field.

As mentioned in Section 11.2, if X is a semi-algebraic subset of R^n, the sum of the Betti numbers of X, i.e., the dimension of the $\mathbb{Q}$-vector space $H_*(X, \mathbb{Q})$, is finite. All results of this section remain valid if the field of coefficients $\mathbb{Q}$ is replaced by an arbitrary commutative field. We shall abbreviate $H_i(X, \mathbb{Q})$ to $H_i(X)$.

We shall need a version of Bezout's theorem. Let

$$\Pi : \mathbb{C}^{n+1} \setminus \{0\} \longrightarrow \mathbb{P}_n(\mathbb{C})$$
$$x = (x_0, \ldots, x_n) \longmapsto \Pi(x) = (x_0 : \ldots : x_n)$$

be the canonical surjection. Let $P_1, \ldots, P_n$ be homogeneous polynomials in $\mathbb{C}[X_0, \ldots, X_n]$. We say that $\Pi(x)$ is a *nonsingular projective solution* of $P_1 = \ldots = P_n = 0$ if $P_i(x) = 0$ for $i = 1, \ldots, n$ and

$$\mathrm{rank}\left(\left[\frac{\partial P_i}{\partial X_j}(x)\right]_{1 \leq i \leq n,\ 0 \leq j \leq n}\right) = n.$$

Note that $(x_1, \ldots, x_n)$ is a nonsingular solution of

$$P_1(1, X_1, \ldots, X_n) = \ldots = P_n(1, X_1, \ldots, X_n) = 0$$

if and only if $(1 : x_1 : \ldots : x_n)$ is a nonsingular projective solution of

$$P_1 = \ldots = P_n = 0.$$

Lemma 11.5.1. *Let $P_1, \ldots, P_n$ be homogeneous polynomials in $\mathbb{C}[X_0, \ldots, X_n]$ of degrees $d_1, \ldots, d_n$, respectively. Then the number of nonsingular projective solutions of $P_1 = \ldots = P_n = 0$ is at most $\prod_{i=1}^{n} d_i$.*

Proof. Set

$$H_{i,\lambda,\mu}(X_0, \ldots, X_n) = \lambda P_i + \mu(X_i - X_0)(X_i - 2X_0) \cdots (X_i - d_i X_0),$$

for $i = 1, \ldots, n$ and $(\lambda, \mu) \in \mathbb{C}^2 \setminus \{0\}$. We denote by $\mathcal{S}_{(\lambda : \mu)}$ the system of equations $H_{1,\lambda,\mu} = \ldots = H_{n,\lambda,\mu} = 0$ (we identify equations which differ only by a nonzero constant factor). Note that the system $\mathcal{S}_{(0:1)}$ has $d_1 \cdots d_n$ nonsingular projective solutions and $\mathcal{S}_{(1:0)}$ is $P_1 = \ldots = P_n = 0$. The subset of $(\tau, x) \in \mathbb{P}_1(\mathbb{C}) \times \mathbb{P}_n(\mathbb{C})$, such that x is a singular projective solution of the system $\mathcal{S}_\tau$, is clearly Zariski closed. Therefore, its projection Δ on $\mathbb{P}_1(\mathbb{C})$ is a Zariski closed subset of $\mathbb{P}_1(\mathbb{C})$. Since $(0:1) \notin \Delta$, the set Δ consists of finitely many points. It follows that there is a continuous path $\gamma : [0,1] \to \mathbb{P}_1(\mathbb{C})$ such that $\gamma(0) = (1:0)$, $\gamma(1) = (0:1)$ and $\gamma(]0,1]) \subset \mathbb{P}_1(\mathbb{C}) \setminus \Delta$. Note that $\tau \in \mathbb{P}_1(\mathbb{C}) \setminus \Delta$ if and only if all projective solutions of $\mathcal{S}_\tau$ are nonsingular. By the implicit function theorem, for every nonsingular projective solution x of $\mathcal{S}_{(0:1)}$, there exists a continuous path $\sigma_x : [0,1] \to \mathbb{P}_n(\mathbb{C})$ such that $\sigma_x(0) = x$ and, for every $t \in {]0,1]}$, $\sigma_x(t)$ is a (nonsingular) projective solution of $\mathcal{S}_{\gamma(t)}$. Moreover, if y is another nonsingular projective solution of $\mathcal{S}_{(0:1)}$, then $\sigma_x(t) \neq \sigma_y(t)$ for every $t \in [0,1]$. From this we conclude that the number of nonsingular projective solutions of $P_1 = \ldots = P_n = 0$ is less than or equal to the number of projective solutions of $\mathcal{S}_{(1:0)}$, which is $d_1 \cdots d_n$. □

Proposition 11.5.2. *Let $W \subset \mathbb{R}^n$ be a compact nonsingular algebraic hypersurface defined by the equation $f = 0$, where f is a polynomial of degree $2d$. Then the sum of the Betti numbers of W is less than or equal to $2d(2d-1)^{n-1}$.*

Proof. Let $\eta : W \to S^{n-1}$ be the mapping defined by

$$\eta(x) = \mathrm{grad}(f(x))/\|\mathrm{grad}(f(x))\| \ .$$

According to Sard's theorem 9.6.2, the dimension of the set of critical values of η is at most $n-2$. Hence, there are two antipodal points of S^{n-1} such that neither is a critical value of η. After rotating the coordinate system, we may assume that $(0,\ldots,0,1)$ and $(0,\ldots,0,-1)$ are not critical values of η.

The points $y \in W$ such that $\eta(y) = (0,\ldots,0,\pm 1)$ are the critical points of the "height" function $h : W \to \mathbb{R}$ defined by $h(x) = x_n$. We claim that h is a Morse function, i.e., all its critical points are nondegenerate. Let y be a critical point of h. We can choose local coordinates $(u_1,\ldots,u_{n-1})$ defined in a neighbourhood U of y in W, so that $u_1 = x_1,\ldots,u_{n-1} = x_{n-1}$. Differentiating the identity

$$f(u_1,\ldots,u_{n-1},h(u_1,\ldots,u_{n-1})) = 0 \ ,$$

we obtain, for $i = 1,\ldots,n-1$, the equations

$$\frac{\partial f}{\partial x_i} + \frac{\partial f}{\partial x_n}\frac{\partial h}{\partial u_i} = 0 \quad \text{on } U \ . \tag{$*$}$$

Hence,

$$\eta(u_1,\ldots,u_{n-1}) = \pm\left(\frac{\partial h}{\partial u_1},\ldots,\frac{\partial h}{\partial u_{n-1}},-1\right)\bigg/\sqrt{1+\sum_{i=1}^{n-1}\left(\frac{\partial h}{\partial u_i}\right)^2} \ .$$

11.5 Upper Bound for the Sum of the Betti Numbers of an Algebraic Set

Taking the partial derivative with respect to u_i of the j-th coordinate η_j of η, $1 \leq i, j \leq n-1$, and evaluating at y, we obtain

$$\frac{\partial \eta_j}{\partial u_i}(y) = \pm \frac{\partial^2 h}{\partial u_j \partial u_i}(y).$$

Since $(0, \ldots, 0, \pm 1)$ are not critical values of η, the matrix

$$\left[\frac{\partial \eta_i}{\partial u_j}(y)\right]_{1 \leq i,j \leq n-1}$$

is nonsingular, and, hence, y is a nondegenerate critical point of h. The function h is, therefore, a Morse function.

Applying Morse theory to the function $h : W \to \mathbb{R}$, it follows that the sum of the Betti numbers of W is less than or equal to the number of critical points of h ([230], Theorem 5.2, p.29). The critical points of h can be characterized as the real solutions of the system of n polynomial equations in n variables

$$\frac{\partial f}{\partial x_1} = 0, \ldots, \frac{\partial f}{\partial x_{n-1}} = 0, \ f = 0.$$

We claim that every real solution y of this system is nonsingular, i.e., the jacobian matrix

$$\begin{bmatrix} \frac{\partial^2 f}{\partial x_1 \partial x_1}(y) & \cdots & \frac{\partial^2 f}{\partial x_{n-1} \partial x_1}(y) & \frac{\partial f}{\partial x_1}(y) \\ \vdots & \cdots & \vdots & \vdots \\ \frac{\partial^2 f}{\partial x_1 \partial x_{n-1}}(y) & \cdots & \frac{\partial^2 f}{\partial x_{n-1} \partial x_{n-1}}(y) & \frac{\partial f}{\partial x_{n-1}}(y) \\ \frac{\partial^2 f}{\partial x_1 \partial x_n}(y) & \cdots & \frac{\partial^2 f}{\partial x_{n-1} \partial x_n}(y) & \frac{\partial f}{\partial x_n}(y) \end{bmatrix}$$

is nonsingular. Indeed, differentiating the identity $(*)$ and evaluating at y, we obtain, for $1 \leq i, j \leq n-1$,

$$\frac{\partial^2 f}{\partial x_j \partial x_i}(y) = -\frac{\partial f}{\partial x_n}(y) \frac{\partial^2 h}{\partial u_j \partial u_i}(y).$$

Since $(\partial f / \partial x_n)(y) \neq 0$ and y is a nondegenerate critical point of h, the claim follows.

By Lemma 11.5.1, the number of critical points of h, and therefore the sum of the Betti numbers of W, is less than or equal to the product

$$\left(\deg \frac{\partial f}{\partial x_1}\right) \cdots \left(\deg \frac{\partial f}{\partial x_{n-1}}\right) (\deg f) = 2d(2d-1)^{n-1}.$$

$\square$

Theorem 11.5.3. *Let $V \subset \mathbb{R}^n$ be an algebraic set defined by equations of degree less than or equal to d. Then the sum of the Betti numbers of V is less than or equal to $d(2d-1)^{n-1}$.*

Proof. By Corollary 9.3.7, there exists an $r \in \mathbb{R}$, $r > 0$, such that $V \cap \overline{B}_n(0,r)$ is a deformation retract of V. It suffices to estimate the sum of the Betti numbers of $V \cap \overline{B}_n(0,r)$. Assume that $V = \mathcal{Z}(f_1, \ldots, f_p)$, and define $\Phi(x, \epsilon) = f_1^2 + \cdots + f_p^2 + \epsilon(\|x\|^2 - r^2)$ for all $x \in \mathbb{R}^n$ and all $\epsilon > 0$. Observe that no critical point of Φ is contained in the set of zeros of Φ. Therefore, by Bertini's theorem 9.6.4, there is an $a \in \mathbb{R}$, $a > 0$, such that, for every $\epsilon \in]0, a[$, the set $W_\epsilon = \{x \in \mathbb{R}^n \mid \Phi(x, \epsilon) = 0\}$ is a nonsingular hypersurface of $\mathbb{R}^n$. Moreover, W_ϵ is the boundary of the compact set $K_\epsilon = \{x \in \mathbb{R}^n \mid \Phi(x, \epsilon) \leq 0\}$. By Proposition 11.5.2, the sum of the Betti numbers of W_ϵ is less than or equal to $2d(2d-1)^{n-1}$.

Claim. *The sum of the Betti numbers of W_ϵ is twice the sum of the Betti numbers of K_ϵ.*

Proof. Let $E_\epsilon = \mathbb{R}^n \setminus K_\epsilon$ and $\overline{E}_\epsilon = E_\epsilon \cup W_\epsilon$. Denote by $\tilde{H}_*$ reduced homology. Since $\tilde{H}_*(\mathbb{R}^n) = 0$, the Mayer-Vietoris exact sequence for reduced homology implies that $\tilde{H}_i(W_\epsilon) \simeq \tilde{H}_i(K_\epsilon) \oplus \tilde{H}_i(\overline{E}_\epsilon)$. Since W_ϵ has a collar in $\overline{E}_\epsilon$, we have $\tilde{H}_i(\overline{E}_\epsilon) \simeq \tilde{H}_i(E_\epsilon)$. From the Alexander duality theorem ([308], p. 296, Theorem 16), we deduce that $\tilde{H}_i(E_\epsilon) \simeq H^{n-i-1}(K_\epsilon)$. Computing the sum of the dimensions of the $\tilde{H}_i(W_\epsilon)$, we obtain

$$c_n = 0 \quad \text{and} \quad b_0 - 1 + \sum_{1 \leq i < n} b_i = c_0 - 1 + \sum_{1 \leq i < n} c_i + \sum_{0 \leq i < n} c_i,$$

where b_i and c_i are the Betti numbers of W_ϵ and K_ϵ, respectively. This proves the claim.

The claim implies that the sum of the Betti numbers of K_ϵ is less than or equal to $d(2d-1)^{n-1}$. Since $V \cap \overline{B}_n(0,r) = \bigcap_{\epsilon \in]0,a[} K_\epsilon$ and all these sets can be triangulated, one has $H_i(V \cap \overline{B}_n(0,r)) = \varprojlim \tilde{H}_i(K_\epsilon)$ ([129], Chap. 10, §3, Theorem 3.1). Hence, the sum of the Betti numbers of $V \cap \overline{B}_n(0,r)$ is less than or equal to $d(2d-1)^{n-1}$. $\square$

It now remains to prove that this upper bound is valid for any real closed field.

Proposition 11.5.4. *Let R be a real closed field and $V \subset R^n$ an algebraic set defined by equations of degree less than or equal to d. Then the sum of the Betti numbers of V is less than or equal to $d(2d-1)^{n-1}$.*

Proof. We first work over $\mathbb{R}_{\text{alg}}$, the smallest real closed field. We identify a system of k polynomials $(f_1, \ldots, f_k)$ in n variables of degree less than or equal to d with the point of $\mathbb{R}_{\text{alg}}^N$, $N = k\binom{n+d}{d}$, whose coordinates are the coefficients of $f_1, \ldots, f_k$. Let

$$X = \{(f_1, \ldots, f_k, x) \in \mathbb{R}_{\text{alg}}^N \times \mathbb{R}_{\text{alg}}^n \mid f_1(x) = \cdots = f_k(x) = 0\},$$

and let $\Pi : X \to \mathbb{R}_{\text{alg}}^N$ be the canonical projection. By Theorem 9.3.2, there exist a finite partition of $\mathbb{R}_{\text{alg}}^N$ into semi-algebraic sets $A_1, \ldots, A_p$, semi-algebraic sets $F_1, \ldots, F_p$ contained in $\mathbb{R}_{\text{alg}}^n$ and semi-algebraic homeomorphisms $\theta_i : \Pi^{-1}(A_i) \to A_i \times F_i$, for $i = 1, \ldots, p$, such that the composition of θ_i with the projection $A_i \times F_i \to A_i$ is $\Pi|_{\Pi^{-1}(A_i)}$. The F_i are, in fact, algebraic subsets of $\mathbb{R}_{\text{alg}}^n$ defined by k equations of degree less than or equal to d. By Theorem 11.5.3, the sum of the Betti numbers of $(F_i)_\mathbb{R}$ is less than or equal to $d(2d-1)^{n-1}$, and, by invariance of the homology groups under extension of real closed field (Proposition 11.7.9), the same bound holds for the sum of the Betti numbers of F_i. Now, let $V \subset R^n$ be defined by k equations $f_1 = \cdots = f_k = 0$ of degree less than or equal to d with coefficients in R. We have $\Pi_R^{-1}(f_1, \ldots, f_k) = \{(f_1, \ldots, f_k)\} \times V$. The point $(f_1, \ldots, f_k) \in R^N$ belongs to some $(A_i)_R$, and the semi-algebraic homeomorphism $(\theta_i)_R$ induces a semi-algebraic homeomorphism from V onto $(F_i)_R$. Again, by Proposition 11.7.9, the sum of the Betti numbers of $(F_i)_R$ is less than or equal to $d(2d-1)^{n-1}$, and the same bound holds for the sum of the Betti numbers of V. □

Remark 11.5.5. a) The bound for the sum of the Betti numbers is mostly used in applications as a bound for b_0, the number of semi-algebraically connected components. For instance, it is used in [35] to obtain lower bounds on the complexity of algorithms.

b) One can also find an upper bound for the sum of the Betti numbers of semi-algebraic sets. It is quite clear that any such estimate necessarily depends not only on the degree of the inequalities, but also upon their number. Using the results of this section, one can prove the following [231]:

If $X \subset R^n$ is a basic closed semi-algebraic subset defined by p polynomial inequalities $f_1 \geq 0, \ldots, f_p \geq 0$ of degree $\leq d$, then the sum of the Betti numbers of X is $\leq \frac{1}{2}(dp+2)(dp+1)^{n-1}$.

Note that another bound for the number of semi-algebraically connected components of X as above can be easily obtained from Proposition 11.5.4. Indeed, X is the projection of the algebraic subset Y of R^{n+p} defined by the equations $f_i(x) = y_i^2$, for $i = 1, \ldots, p$. Hence, if $d \geq 2$, $b_0(X) \leq b_0(Y) \leq d(2d-1)^{n+p-1}$.

11.6 Nonsingular Algebraic Curves in the Real Projective Plane

In this section we present a brief introduction to the problem of describing the topological types of embeddings of nonsingular algebraic curves in $\mathbb{P}_2(\mathbb{R})$ (which is the first part of Hilbert's 16th problem). The problem has been extensively investigated, mainly by topological methods.

We shall only consider the case $R = \mathbb{R}$. All results in this section still hold for an arbitrary real closed field, replacing connected components with semi-algebraically connected components. This can be proved using the same technique as in Section 5.

Let Γ be a nonsingular algebraic curve in $\mathbb{P}_2(\mathbb{R})$. The curve Γ has finitely many connected components, each Nash diffeomorphic to a circle. There are two possible cases for an embedding p of the circle S^1 into $\mathbb{P}_2(\mathbb{R})$:

(i) $\mathbb{P}_2(\mathbb{R}) \setminus p(S^1)$ has two connected components, one homeomorphic to an open disk (the interior) and the other, to a Möbius band (the exterior);

(ii) $\mathbb{P}_2(\mathbb{R}) \setminus p(S^1)$ is connected and homeomorphic to an open disk.

The embedded circle $p(S^1)$ is said to be an *oval* in the first case, and a *pseudo-line* in the second case. The inverse image of an oval by the canonical covering $\pi : S^2 \to \mathbb{P}_2(\mathbb{R})$ has two connected components, while the inverse image of a pseudo-line has only one.

Proposition 11.6.1. *Let Γ be a nonsingular algebraic curve of degree d in $\mathbb{P}_2(\mathbb{R})$.*

(i) If d is even, every connected component of Γ is an oval, and $[\Gamma] = 0$ in $H_1(\mathbb{P}_2(\mathbb{R}), \mathbb{Z}/2)$.

(ii) If d is odd, one of the connected components of Γ is a pseudo-line, while the others are ovals. In this case, $[\Gamma] \neq 0$ in $H_1(\mathbb{P}_2(\mathbb{R}), \mathbb{Z}/2)$.

Proof. Let $F = 0$ be a homogeneous equation of Γ, where F is a form of degree d in 3 variables. Since the complement of a pseudo-line in $\mathbb{P}_2(\mathbb{R})$ is homeomorphic to the affine plane, it is clear that Γ cannot have more than one pseudo-line among its connected components. Now let γ be a path in S^2 transversal to $\pi^{-1}(\Gamma)$ and joining two antipodal points which are not zeros of F. If C is a connected component of Γ, then clearly γ intersects $\pi^{-1}(C)$ in an even (resp. odd) number of points if C is an oval (resp. a pseudo-line). Counting the sign changes of F along γ, we see that Γ has a connected component which is a pseudo-line if and only if F takes opposite signs at antipodal points, i.e. if and only if d is odd. $\square$

Theorem 11.6.2 (Harnack's theorem). *A nonsingular algebraic curve of degree d in $\mathbb{P}_2(\mathbb{R})$ has at most $g(d) + 1$ connected components, where $g(d) = (d-1)(d-2)/2$.*

Proof. We may assume that $d > 2$. It suffices to consider irreducible curves, because the inequality

$$g(d_1) + 1 + g(d_2) + 1 \leq g(d_1 + d_2) + 1$$

holds whenever $d_1 > 1$ or $d_2 > 1$. Suppose that Γ is a nonsingular irreducible curve of degree d with more than $g(d) + 1$ connected components. We shall show that this assumption leads to a contradiction. The curve Γ contains $p = g(d) + 1$ ovals $\Omega_1, \ldots, \Omega_p$ and at least one other connected component. We choose $\frac{1}{2}d(d-1) - 1$ points on Γ. Since $\frac{1}{2}d(d-1) - 1 \geq g(d) + 1$ for

11.6 Nonsingular Algebraic Curves in the Real Projective Plane

$d > 2$, we can choose one point on each of the ovals $\Omega_1, \ldots, \Omega_p$ and the others on another connected component of Γ. There is a curve Δ of degree $d - 2$ passing through these $\frac{1}{2}d(d-1) - 1$ points. The curves Γ and Δ have no common irreducible component, since Γ is irreducible and the degree of Δ is $d - 2$. By Bezout's theorem, the number of intersection points of Γ and Δ, counted with multiplicity, is not greater than $d(d-2)$. If Δ intersects an oval Ω_i in a point with multiplicity 1, then Δ necessarily intersects Ω_i in another point. Indeed, Theorem 9.5.7 implies that the number of intersection points Q of Δ and Ω_i, such that there is an odd number of half-branches of Δ centred at Q and contained in the interior of Ω_i, is even. Hence, the number of intersection points of Γ and Δ, counted with multiplicity, should be at least $\frac{1}{2}d(d-1) - 1 + g(d) + 1 = (d-1)^2$, which exceeds $d(d-2)$. This contradiction implies the theorem. □

Proposition 11.6.3. *The estimate $g(d) + 1$ is the best possible. More precisely, for each d, there exists a nonsingular curve of degree d in $\mathbb{P}_2(\mathbb{R})$, which has exactly $g(d) + 1$ connected components.*

Sketch of Harnack's construction: Choose a line L in $\mathbb{P}_2(\mathbb{R})$. Starting with $d = 2$, we construct, by induction, a nonsingular curve Γ_d of degree d in $\mathbb{P}_2(\mathbb{R})$, with exactly $g(d) + 1$ connected components, and which, moreover, has the following properties:

(i) Γ_d has a connected component C_d intersecting L in d distinct points.

(ii) There is an orientation of L and an orientation of C_d, such that these d distinct points are arranged in the same order on L as on C_d. We denote these points by $a_1, \ldots, a_d$, taken in this order.

(iii) For each $i = 1, \ldots, d-1$, the union of the interval $[a_i, a_{i+1}]$ on L and the segment of C_d bounded by a_i and a_{i+1} form an oval in $\mathbb{P}_2(\mathbb{R})$ (which is, of course, not smooth).

For $d = 2$, we can choose Γ_d to be a nonsingular conic intersecting L in two distinct points. Suppose that we have already constructed Γ_d with $g(d) + 1$ components and having properties (i)-(iii). The curve Γ_{d+1} is then constructed as follows: we choose distinct points $b_1, \ldots, b_{d+1}$ on L such that $a_1, \ldots, a_d, b_1, \ldots, b_{d+1}$ are ordered according to the orientation of L. Choose lines $L_1, \ldots, L_{d+1}$ distinct from L and passing through $b_1, \ldots, b_{d+1}$, respectively. The curve Γ_{d+1} is a small perturbation of the union $\Gamma_d \cup L$. Identifying the curves with their homogeneous equations, we define

$$\Gamma_{d+1} = L\Gamma_d + \epsilon \prod_{i=1}^{d+1} L_i,$$

where $\epsilon \in \mathbb{R}$ is chosen small enough, and its sign is the opposite of the sign of $L\Gamma_d / \prod_{i=1}^{d+1} L_i$ in the interior of the ovals between a_i and a_{i+1}. With this choice of sign, the effect of the perturbation is to shrink these ovals towards their interior. Hence, the perturbation of $C_d \cup L$ has $d - 1$ ovals and an additional connected component C_{d+1} intersecting L in $b_1, \ldots, b_{d+1}$. The curve Γ_{d+1} has properties (i)-(iii). The perturbation of the ovals of Γ_d which do not intersect L is small; therefore, the curve Γ_{d+1} has $g(d) + 1 + d - 1 = g(d+1) + 1$ connected components. Figure 11.6 illustrates the construction of Γ_4. □

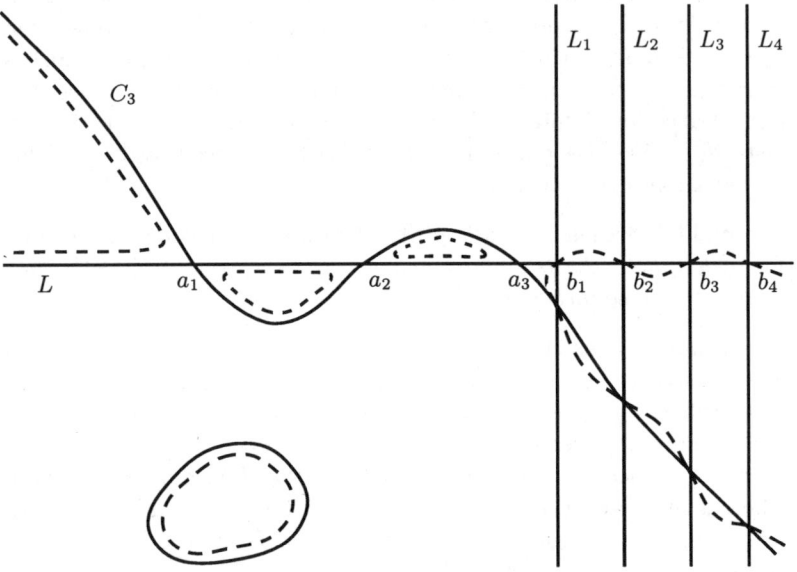

Figure 11.6. Harnack's construction of an M-curve of degree 4

The curves of degree d which have $g(d) + 1$ connected components, are called *M-curves*. Note that, if the complexification of the curve Γ is nonsingular, its genus is precisely $g(d)$.

We complement Harnack's theorem with a few remarks concerning the number of connected components of a projective (not necessarily plane) real algebraic curve. Let $X \subset \mathbb{P}_n(\mathbb{R})$ be a nonsingular irreducible real algebraic curve. It is well known that there exists a unique (up to a biregular isomorphism over $\mathbb{R}$) complex projective nonsingular irreducible curve $X_\mathbb{C}$ defined over $\mathbb{R}$, such that the set of real points of $X_\mathbb{C}$ is biregularly isomorphic to X. We may, therefore, assume without loss of generality that X is just the real part of $X_\mathbb{C}$. The curve X is said to be *dividing* if $X_\mathbb{C} \setminus X$ is disconnected (in which case $X_\mathbb{C} \setminus X$ has precisely 2 connected components). If $s(X)$ is the number of connected components of X and $g(X)$ is the genus of X (that is, by definition, the genus of $X_\mathbb{C}$), then $1 \leq s(X) \leq g(X)$ (resp. $1 \leq s(X) \leq g(X) + 1$ and $s(X) \equiv g(X) + 1 \pmod{2}$), if X is nondividing (resp. dividing). Moreover, given integers g and s satisfying the conditions above, there exists a nondividing (resp. dividing) curve X with $s(X) = s$ and

11.6 Nonsingular Algebraic Curves in the Real Projective Plane

$g(X) = g$. These facts have been known for about 100 years (cf. [139] and the references cited there).

Harnack's theorem gives no information concerning the possible relative positions of the components of a plane curve. This question is the first part of Hilbert's 16th problem. The relative positions of the ovals of a nonsingular curve Γ in the projective plane can be described using the notion of a *nest*. One oval is *nested* in another if it is contained in its interior; the *depth* of an oval Ω of Γ is the number of ovals of Γ containing Ω in their interiors. Bezout's theorem imposes restrictions on the number and position of possible nests. For example, an M-curve of degree 4 with 3 or 4 ovals cannot have a nest; indeed, if such a curve had an oval Ω contained in the interior of an oval Ω', then any line connecting a point in the interior of Ω to a point in the interior of an oval different from Ω and Ω' would intersect the curve in at least 6 points. Hence, the only possible configuration for an M-curve of degree 4 is 4 ovals, each without a nest.

This simple argument does not suffice to solve the problem of relative positions of ovals of M-curves of degree 6. This case was not completely solved until 1971. The M-curves of degree 6 have 11 ovals. The argument based on Bezout's theorem shows that ovals with a depth equal to 2 cannot occur, and also that there is at most one oval containing other ovals in its interior. Harnack's construction (Proposition 11.6.3) gives an M-curve with the following configuration: one oval containing another one in its interior, and the 9 remaining ovals outside, none of which is nested in any other. This configuration is denoted by $1\langle 1 \rangle \coprod 9$. A construction by Hilbert [154] gives a different configuration, denoted $1\langle 9 \rangle \coprod 1$: one oval containing nine ovals in its interior, and the remaining oval outside.

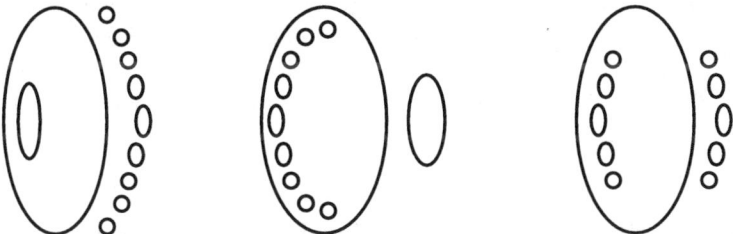

Figure 11.7. The three possible configurations for an M-curve of degree 6

The existence of other configurations for M-curves of degree 6 was doubted for a long time, until Gudkov [140] constructed such a curve with a configuration $1\langle 5 \rangle \coprod 5$. These three are the only possible configurations for M-curves of degree 6. This claim is a consequence of a famous congruence, conjectured by Gudkov and proved by Rokhlin [279]. Before stating this result, we fix some notation. Let Γ be a nonsingular algebraic curve of even degree in $\mathbb{P}_2(\mathbb{R})$. We can choose a homogeneous equation F of Γ, such that

$F(x,y,z) \leq 0$, for every $(x:y:z)$ outside all ovals of Γ. Then we set

$$B_+ = \{(x:y:z) \in \mathbb{P}_2(\mathbb{R}) \mid F(x,y,z) \geq 0\}.$$

Theorem 11.6.4. *Let Γ be an M-curve of even degree $2k$. Then the Euler-Poincaré characteristic $\chi(B_+)$ is congruent to k^2 modulo 8.*

Reference for the proof: [342]. □

The Euler-Poincaré characteristic $\chi(B_+)$ can be computed in the following way. An oval of Γ is said to be *even* (resp. *odd*) if its depth is even (resp. odd). Denote by p (resp. n) the number of even (resp. odd) ovals of Γ. Then $\chi(B_+) = p - n$. For an M-curve of degree 6, $p + n = 11$ and, by Theorem 11.6.4, $p - n \equiv 9 \pmod 8$. The only possibilities for (p,n) are therefore $(10,1)$, $(2,9)$ and $(6,5)$. Hence, the three configurations described above are the only possible ones.

The numbers p and n were first considered by Ragsdale [262], who formulated the conjecture that, for every curve of degree $2k$,

$$p \leq \frac{3k(k-1)}{2} + 1, \quad n \leq \frac{3k(k-1)}{2}.$$

Petrovsky [253] also formulated similar conjectures, and proved the inequalities

$$p - n \leq \frac{3k(k-1)}{2} + 1, \quad n - p \leq \frac{3k(k-1)}{2}.$$

The Ragsdale-Petrovsky conjectures were recently disproven by Itenberg who produced a family of counter-examples such that the difference between p (or n) and $3k(k-1)/2$ is asymptotic to $\frac{1}{8}k^2$ as $k \to \infty$ [176]. The construction of these counter-examples is done by "patchworking algebraic curves" [178]. This is a combinatorial version of Viro's gluing method, which has been successfully applied to the construction of real algebraic varieties with controlled topology [331] and other geometric objects, such as polynomial vector fields with prescribed singularities [177].

11.7 Appendix: Homology of Semi-algebraic Sets over a Real Closed Field

Throughout this section, R denotes a real closed field and Λ a field. The easiest way to define homology groups $H_r(A, \Lambda)$, for a semi-algebraic subset A of R^n, is to use triangulation and define H_r combinatorially.

Definition 11.7.1. *If $K \subset R^n$ is a finite simplicial complex, then $H_r(|K|, \Lambda)$ is, by definition, the simplicial homology group $H_r(K, \Lambda)$, computed combinatorially. If $A \subset R^n$ is a closed and bounded semi-algebraic set and*

11.7 Appendix: Homology of Semi-algebraic Sets over a Real Closed Field

$\Phi : |K| \to A$ a semi-algebraic triangulation of A, then $H_r(A, \Lambda)$ is, by definition, equal to $H_r(|K|, \Lambda)$. If $B \subset R^n$ is an arbitrary semi-algebraic set and A a semi-algebraic deformation retract of B which is closed and bounded (Corollary 9.3.7 and Remark 9.3.8), then $H_r(B, \Lambda)$ is, by definition, equal to $H_r(A, \Lambda)$.

This definition only makes sense if the homology groups defined in this way are independent of the choice of the triangulation and the retraction. This independence, while well-known for $R = \mathbb{R}$, should be proved for an arbitrary real closed field R. This can be done using the Čech cohomology of the space $\widetilde{B}$ (Proposition 7.2.2). We denote Čech cohomology by $\check{H}^*$. Note that if B is a semi-algebraic set, the Čech cohomology groups $\check{H}^r(\widetilde{B}, \Lambda)$ coincide with the cohomology groups $H^r(\widetilde{B}, \Lambda)$ (defined as derived functors of the global sections functor), [80].

We first prove that two homotopic mappings induce the same homomorphism of Čech cohomology groups. This has to be stated in the semi-algebraic framework. Let $[0,1]_R = \{t \in R \mid 0 \leq t \leq 1\}$.

Definition 11.7.2. *Two continuous semi-algebraic mappings $f_0, f_1 : A \to B$ are said to be* semi-algebraically homotopic *if there exists a continuous semi-algebraic mapping $h : A \times [0,1]_R \to B$, such that, for every x in A,*

$$h(x,0) = f_0(x) \quad \text{and} \quad h(x,1) = f_1(x) .$$

Proposition 11.7.3. *Let $f_0, f_1 : A \to B$ be two continuous semi-algebraic mappings that are semi-algebraically homotopic. Then the homomorphisms $\tilde{f}_0^*, \tilde{f}_1^* : \check{H}^*(\widetilde{B}, \Lambda) \to \check{H}^*(\widetilde{A}, \Lambda)$, induced by f_0 and f_1, are equal.*

Sketch of proof: One can follow the proof of Theorem 5.1 in [129], Chap. 9, where the problem is reduced to the case $B = A \times [0,1]_R$, $f_0(x) = (x,0)$ and $f_1(x) = (x,1)$. Lemma 5.6 in [129] (concerning the existence of "stacked coverings") has to be replaced by the following lemma:

Lemma 11.7.4. *Let $(V_j)_{j=1,\ldots,p}$ be a finite open semi-algebraic cover of $A \times [0,1]_R$. There exists a finite open semi-algebraic cover $(U_i)_{i=1,\ldots,q}$ of A, and, for each $i = 1,\ldots,q$, continuous semi-algebraic functions $0 = \varphi_{i,0} < \cdots < \varphi_{i,k} < \cdots < \varphi_{i,r_i} = 1$ from U_i to R, such that, for every j, $1 \leq j \leq p$, there exists a pair (i,k), $1 \leq i \leq q$ and $1 \leq k \leq r_i$, with*

$$\{(x,t) \in A \times [0,1]_R \mid x \in U_i \text{ and } \varphi_{i,k-1}(x) \leq t \leq \varphi_{i,k}(x)\} \subset V_j .$$

Proof. Let α be a point in $\widetilde{A}$. The fibre at α of the constant semi-algebraic family $A \times [0,1]_R$ is $[0,1]_{k(\alpha)}$ (Example 7.4.2), and the fibres $(V_j)_\alpha$ form an open semi-algebraic cover of $[0,1]_{k(\alpha)}$. We can find a finite sequence of elements of $k(\alpha)$, $0 = b_0^\alpha < \cdots < b_k^\alpha < \cdots < b_{r_\alpha}^\alpha = 1$, such that, for every j, $1 \leq j \leq p$, there exists k, $1 \leq k \leq r_\alpha$, with $[b_{k-1}^\alpha, b_k^\alpha] \subset (V_j)_\alpha$. By Corollary

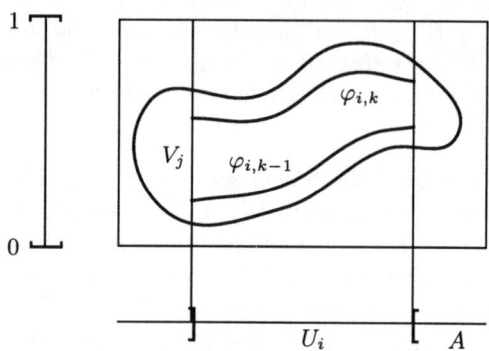

Figure 11.8.

7.3.5, there exist an open semi-algebraic subset U^α of A, with $\alpha \in \widetilde{U^\alpha}$, and continuous semi-algebraic functions $\varphi_1^\alpha, \ldots, \varphi_{r_\alpha-1}^\alpha : U^\alpha \to R$, such that $\varphi_k^\alpha(\alpha) = b_k^\alpha$; take $\varphi_0^\alpha = 0$ and $\varphi_{r_\alpha}^\alpha = 1$. Shrinking U^α, if necessary, we have

$$0 = \varphi_0^\alpha < \cdots < \varphi_k^\alpha < \cdots < \varphi_{r_\alpha}^\alpha = 1,$$

and, for every j, $1 \leq j \leq p$, there exists k, $1 \leq k \leq r_\alpha$, with

$$\{(x,t) \in U^\alpha \times [0,1]_R \mid \varphi_{k-1}^\alpha(x) < t < \varphi_k^\alpha(x)\} \subset V_j,$$

since these properties hold at α. Since the $\widetilde{U^\alpha}$ cover $\widetilde{A}$, and $\widetilde{A}$ is compact, we can extract a finite cover $\widetilde{U}_1, \ldots, \widetilde{U}_q$. Hence, we get a finite cover $U_1, \ldots, U_q$ of A, which satisfies the required properties. □

Remark 11.7.5. One has to preserve the finiteness of the cover in the semi-algebraic case; this is the reason why the genuine rectangles of the "stacked coverings" have to be replaced with "twisted" rectangles, whose horizontal sides are graphs of continuous semi-algebraic functions. Lemma 11.7.4 can also be proved without the real spectrum, by using slicing (cf. Section 2.3).

Theorem 11.7.6. *Let B be a semi-algebraic subset of R^n, A a closed and bounded semi-algebraic set, which is a semi-algebraic deformation retract of B. Let $\Phi : |K| \to A$ be a semi-algebraic triangulation of A. Then the simplicial homology group $H_r(K, \Lambda)$ is isomorphic to the dual space of $\check{H}^r(B, \Lambda)$.*

Proof. Proposition 11.7.3 shows that $\check{H}^r(\widetilde{B}, \Lambda)$ is isomorphic to $\check{H}^r(\widetilde{A}, \Lambda)$, and $\check{H}^r(\widetilde{A}, \Lambda)$ is isomorphic to $\check{H}^r(\widetilde{|K|}, \Lambda)$, since there is a homeomorphism $\widetilde{\Phi} : \widetilde{|K|} \to \widetilde{A}$ (Proposition 7.2.8). Hence, it suffices to prove the duality between $\check{H}^r(\widetilde{|K|}, \Lambda)$ and $H_r(K, \Lambda)$. If σ is a simplex of K, denote by U_σ the "star" of σ in the first barycentric subdivision K' of K, that is,

$$U_\sigma = \bigcup \{(\sigma')^0 \mid \sigma' \text{ simplex of } K', \sigma' \cap \sigma \neq \emptyset\}.$$

11.7 Appendix: Homology of Semi-algebraic Sets over a Real Closed Field

We have $U_\sigma \cap U_\tau = U_{\sigma \cap \tau}$ and, for $p > 0$, $\check{H}^p(\widetilde{U}_\sigma, \Lambda) = 0$, since U_σ has a semi-algebraic deformation retraction to a point. Let $\mathcal{U}$ be the cover of $|K|$ consisting of all U_σ. Then $\check{H}^r(\widetilde{\mathcal{U}}, \Lambda)$ is dual to $H_r(K, \Lambda)$ computed combinatorially, since $\widetilde{\mathcal{U}}$ and K have the same nerve, and, by [137], Chap. 2, §5, Corollary of Theorem 5.4.1, $\check{H}^r(\widetilde{\mathcal{U}}, \Lambda)$ coincides with $\check{H}^r(\widetilde{|K|}, \Lambda)$. □

Corollary 11.7.7. *The group $H_r(B, \Lambda)$ is well defined, i.e. $H_r(B, \Lambda)$ is independent of the choice of the retraction and the triangulation. Moreover, $H_r(B, \Lambda)$ is dual to $\check{H}^r(\widetilde{B}, \Lambda)$.*

Remark 11.7.8. The duality between $H_r(B, \Lambda)$ and $\check{H}^r(\widetilde{B}, \Lambda)$ also shows that homology of semi-algebraic sets over a real closed field is functorial. If $f : B \to C$ is a continuous semi-algebraic mapping, the continuous mapping $\widetilde{f} : \widetilde{B} \to \widetilde{C}$ induces a homomorphism $\widetilde{f}^* : \check{H}^r(\widetilde{C}, \Lambda) \to \check{H}^r(\widetilde{B}, \Lambda)$ between the Čech cohomology groups. Hence, by duality, we obtain a homomorphism $f_* : H_r(B, \Lambda) \to H_r(C, \Lambda)$.

The homology defined above is invariant under extension of the base real closed field.

Proposition 11.7.9. *Let R' be a real closed field extension of R. Then $H_r(B_{R'}, \Lambda) = H_r(B, \Lambda)$, for every semi-algebraic set $B \subset R^n$.*

Proof. Definition 11.7.1 allows one to reduce the problem to the case of a simplicial complex; indeed, the extension to R' of a semi-algebraic triangulation is a semi-algebraic triangulation, and the extension to R' of a semi-algebraic deformation retract is a semi-algebraic deformation retract (cf. Section 5.3). In the case of a simplicial complex, the combinatorial computation is not modified when the base field changes. □

Remark 11.7.10. The homology groups can also be determined combinatorially by using a semi-algebraic cellular decomposition (in particular, the one induced by a stratifying family of polynomials, cf. Proposition 9.1.12). It suffices to take a semi-algebraic triangulation compatible with the cellular decomposition, and then follow the procedure described above.

In Section 4, we also use Borel-Moore homology for locally closed semi-algebraic sets. In order to have such a homology theory for an arbitrary real closed field, we first define the relative homology groups $H_r(A, B; \Lambda)$, where $B \subset A$ are two closed and bounded semi-algebraic sets.

Definition 11.7.11. *Let $\Phi : |K| \to A$ be a semi-algebraic triangulation of A, such that $\Phi^{-1}(B) = |L|$, where L is a subcomplex of K. The relative homology group $H_r(A, B; \Lambda)$ is, by definition, the simplicial relative homology group $H_r(K, L; \Lambda)$, computed combinatorially.*

We have to verify that the homology groups defined in this way do not depend on the triangulation (this is well-known if $R = \mathbb{R}$). The following argument can be used: given two semi-algebraic triangulations $\Phi_1 : |K_1| \to A$ and $\Phi_2 : |K_2| \to A$, there exists a third semi-algebraic triangulation $\Phi_3 : |K_3| \to A$ and homeomorphisms $\theta_1 : |K_3| \to |K_1|$ and $\theta_2 : |K_3| \to |K_2|$, such that $\Phi_i \circ \theta_i = \Phi_3$, for $i = 1, 2$, and each simplex of K_i is the union of images by θ_i of some simplices of K_3 (this is an easy consequence of Theorem 9.2.1).

Note that $H_r(A, \emptyset; \Lambda)$ is the homology group $H_r(A, \Lambda)$ previously defined.

Proposition 11.7.12. *Let $C \subset B \subset A$ be closed and bounded semi-algebraic sets. Then there is a long exact sequence:*

$$\cdots \to H_{r+1}(A, B; \Lambda) \to H_r(B, C; \Lambda) \to H_r(A, C; \Lambda) \to H_r(A, B; \Lambda) \to \cdots$$

Proof. The choice of a semi-algebraic triangulation of A, such that B and C are images of subcomplexes, reduces the proof to the case of finite simplicial complexes, where the argument is of a combinatorial nature. $\square$

The Borel-Moore homology of a locally closed semi-algebraic set S will be defined using relative homology, by representing S as the difference of two closed and bounded semi-algebraic sets. For this, we use the semi-algebraic Alexandrov compactification of S (Definition 2.5.11).

Definition 11.7.13. *Let S be a locally closed semi-algebraic set. The* Borel-Moore *homology groups $H_r^{\mathrm{BM}}(S, \Lambda)$ are defined by*

$$H_r^{\mathrm{BM}}(S, \Lambda) = \begin{cases} H_r(S, \Lambda) & \text{if } S \text{ is closed and bounded,} \\ H_r(\dot{S}, \dot{S} \setminus \eta(S); \Lambda) & \text{otherwise,} \end{cases}$$

where $(\dot{S}, \eta)$ is the semi-algebraic Alexandrov compactification of S.

Proposition 11.7.14. *Let $B \subset A$ be two closed and bounded semi-algebraic sets. Then $H_r^{\mathrm{BM}}(A \setminus B, \Lambda) = H_r(A, B; \Lambda)$.*

Proof. The proposition is obvious if $A \setminus B$ is closed and bounded. Otherwise, let $(\dot{S}, \eta)$ be the semi-algebraic Alexandrov compactification of $S = A \setminus B$, and let $\varphi : A \to \dot{S}$ be the continuous semi-algebraic mapping defined by $\varphi|_S = \eta$ and $\varphi(B) = \dot{S} \setminus \eta(S)$ (Lemma 2.5.10). Assume that A has a semi-algebraic cellular decomposition induced by a stratifying family of polynomials and compatible with B (Proposition 9.1.12). The image by φ of this cellular decomposition induces a semi-algebraic cellular decomposition of $\dot{S}$, and the combinatorial computation gives $H_r(A, B; \Lambda) = H_r(\dot{S}, \dot{S} \setminus \eta(S); \Lambda)$. $\square$

Proposition 11.7.15. *Let S be a locally closed semi-algebraic set and T a closed semi-algebraic subset of S. Then there is a long exact sequence:*

$$\cdots \to H_{r+1}^{\mathrm{BM}}(S \setminus T, \Lambda) \to H_r^{\mathrm{BM}}(T, \Lambda) \to H_r^{\mathrm{BM}}(S, \Lambda) \to H_r^{\mathrm{BM}}(S \setminus T, \Lambda) \to \cdots$$

11.7 Appendix: Homology of Semi-algebraic Sets over a Real Closed Field

Proof. By using the Alexandrov compactification, we may assume $S = A \setminus C$, where $C \subset A$ are closed and bounded semi-algebraic sets. Then we apply Proposition 11.7.12 with $B = C \cup T$. □

11.7.16 Local Homology. In the previous sections, we used local homology $H_*(V, V \setminus a; \Lambda)$. However, the pair $(V, V \setminus a)$ does not fit into the framework of relative homology developed only for pairs of closed and bounded semi-algebraic sets. We shall explain how to define $H_*(V, V \setminus a; \Lambda)$, without constructing a general theory of relative homology. Let S be a locally closed semi-algebraic set and a a point of S. We choose a closed and bounded semi-algebraic neighbourhood A of a in S, and a continuous semi-algebraic mapping $\theta : G \times [0, 1] \to A$ (where $G = A \setminus \text{int}(A)$ and $\text{int}(A)$ is the interior of A in S), such that
  (i) $\theta(x, 1) = x$ for every x in G,
  (ii) $\theta|_{G \times]0,1]}$ is a homeomorphism onto $A \setminus a$,
  (iii) $\theta(x, 0) = a$ for every x in G.

In other words, A is equipped with a conic structure, with base G and vertex a. Such a conic structure can be obtained, for example, from Theorem 9.3.6. We set $H_r(S, S \setminus a; \Lambda) = H_r(A, G; \Lambda)$. Starting from a semi-algebraic triangulation $\Phi : |K| \to G$ and using the conic structure of A, we obtain a semi-algebraic triangulation $\Psi : |L| \to A$, where L is a cone with base K. The combinatorial computation then gives

$$H_0(S, S \setminus a; \Lambda) = \Lambda \quad \text{and} \quad H_{r+1}(S, S \setminus a; \Lambda) = H_r(G, \Lambda) .$$

We have to prove that the local homology groups defined in this way are independent of the choice of A and its conic structure. To show this, we can follow [292], Chap. 5, §32, Theorem 1, and prove that, given another $(A', \theta' : G' \times [0, 1] \to A')$ with the same properties, there is a semi-algebraic homotopy equivalence between G and G'. Moreover, since G is a semi-algebraic deformation retract of $A \setminus a$, we can easily obtain the long exact sequence

$$\cdots \to H_{r+1}(S, S \setminus a; \Lambda) \to H_r(S \setminus a, \Lambda) \to H_r(S, \Lambda) \to H_r(S, S \setminus a; \Lambda) \to \cdots$$

Bibliographic Notes. Theorem 11.1.1 can be found in Thom's paper [321] and Theorem 11.1.2 in [96]. Theorem 11.2.2 for real analytic sets is due to Sullivan [312]. We followed the proof of [146]. Another proof is given in [79]. A systematic study of combinatorial numerical invariants of real algebraic sets is given in [229, 250]. The existence of the fundamental class of a real analytic set was proved by Borel and Haefliger [64], without using triangulations. Proposition 11.3.3 is contained in [128]. The group $H^1_{\text{alg}}(X, \mathbb{Z}/2)$, for a compact nonsingular real algebraic variety X, was investigated systematically by Bochnak and Kucharz in [48]. Some examples of algebraic varieties with $H^1_{\text{alg}} \neq H^1$ are given in [34, 56, 145, 171, 196, 222, 271, 305, 306]. Example

11.3.9 is taken from the survey [58] which also contains further facts concerning the group H^*_{alg}. The group $H^2_{\mathrm{alg}}(X, \mathbb{Z}/2)$ is studied in [30, 54, 319]. A more detailed treatment of the group H^*_{alg} is given in [59]. Lemma 11.4.3 is in [33]. Theorem 11.4.2 is due to Bialynicki-Birula and Rosenlicht [37] for $V = \mathbb{R}^n$ and to Borel [63] for arbitrary nonsingular V. The upper bound for the sum of the Betti numbers given in Theorem 11.5.3 is due to Milnor [231]. Thom gave similar bounds in [323]. Proposition 11.5.2 is contained in [249]. The bound in the case of an arbitrary real closed field can be found in [109]. Harnack's theorem is certainly amongst the oldest results in real algebraic geometry; the paper [149] dates back to 1876. We refer the reader, who is interested in the first part of Hilbert's 16th problem, to [2, 139, 141, 178, 270, 342, 332]. The study of the moduli spaces of real algebraic curves, which is currently an active area of research, is beyond the scope of this book (we refer the reader to [248, 307] and to the references cited therein). We also do not give an account of a number of important works concerning the theory of real algebraic surfaces, for which we suggest [306] and [192].

The homology of semi-algebraic sets over an arbitrary real closed field is studied in [109].

12. Algebraic Vector Bundles

Abstract. In the first section, we define algebraic R-vector bundles over an affine real algebraic variety X; the definition requires that an algebraic vector bundle be algebraically isomorphic to a subbundle of a trivial bundle. There are several justifications for this requirement. One such justification is that it results in an equivalence of the category of algebraic R-bundles with the category of projective modules of finite type over the ring $\mathcal{R}(X)$ of regular functions on X. Section 2 reviews some basic facts concerning the divisor class group of a ring, with some applications to the question of the factoriality of $\mathcal{R}(X)$. In Sections 3–6, we are mainly concerned with vector bundles over compact nonsingular algebraic subsets of $\mathbb{R}^n$. In Section 3, we use the Stone-Weierstrass theorem to compare algebraic and topological vector bundles. In Section 4, we study algebraic line bundles and we explain the relation between the algebraic approximation of $\mathcal{C}^\infty$ hypersurfaces of X and the group $H^1_{\text{alg}}(X, \mathbb{Z}/2)$. Section 5 contains a characterization of those topological vector bundles over X which are isomorphic to algebraic vector bundles, in the case that X is a compact nonsingular algebraic curve or surface. This allows one to compare the algebraic K-theory of $\mathcal{R}(X)$ with the topological K-theory of X. The group $H^{\text{alg}}_*(X, \mathbb{Z}/2)$ plays an important role in this context. In Section 6 we study algebraic $\mathbb{C}$-vector bundles. The results of this section will be particularly useful in Chap. 13. Finally, Section 7 is devoted to Nash and semi-algebraic vector bundles. The main tool here is Efroymson's approximation theorem. We obtain a purely topological characterization of the affine Nash manifolds whose ring of Nash functions is factorial.

In this chapter we assume knowledge of the basic facts concerning topological or $\mathcal{C}^\infty$ vector bundles (cf. [236] or [174]).

12.1 Algebraic Vector Bundles

Throughout this section, X is an affine real algebraic variety over a real closed field R (Definition 3.2.9).

Definition 12.1.1. *A* pre-algebraic R-vector bundle *over X is a triple $\xi = (E, p, X)$, where:*

(i) E is a real algebraic variety (not necessarily affine), and $p : E \to X$, a regular mapping,

(ii) for each $x \in X$, the fibre $p^{-1}(x)$ is a finite dimensional R-vector space,

(iii) there exist a finite covering $(U_i)_{i \in I}$ of X by Zariski open sets and, for each $i \in I$, an integer n and a biregular isomorphism $\varphi_i : U_i \times R^n \to p^{-1}(U_i)$, such that $p \circ \varphi_i$ is the canonical projection of $U_i \times R^n$ onto U_i and, for every $x \in U_i$, the restriction $\{x\} \times R^n \to p^{-1}(x)$ of φ_i is a R-linear isomorphism.

An algebraic section of ξ is a regular mapping $s : X \to E$ such that $p \circ s = \mathrm{Id}_X$. The variety X (resp. E) is called the base space (resp. the total space) of ξ.

Given two pre-algebraic R-vector bundles $\xi = (E, p, X)$ and $\xi' = (E', p', X)$ over X, an algebraic morphism $\psi : \xi \to \xi'$ is a regular mapping $\psi : E \to E'$ such that $p' \circ \psi = p$ and, for every $x \in X$, $\psi_x : p^{-1}(x) \to p'^{-1}(x)$ is R-linear. The bundles ξ and ξ' are algebraically isomorphic (which is denoted by $\xi \simeq_{\mathrm{alg}} \xi'$) if there exist algebraic morphisms $\psi : \xi \to \xi'$ and $\varphi : \xi' \to \xi$ such that $\varphi \circ \psi = \mathrm{Id}_\xi$ and $\psi \circ \varphi = \mathrm{Id}_{\xi'}$.

The *rank* of an R-vector bundle $\xi = (E, p, X)$ is the function from X to $\mathbb{N}$ which assigns to $x \in X$ the dimension of the R-vector space $p^{-1}(x)$. By (iii), the rank is locally constant for the Zariski topology. Hence, if X is connected in the Zariski topology, the rank of a pre-algebraic R-vector bundle over X is constant. We shall always assume in the proofs that this is the case. This involves no loss of generality.

We shall say "vector bundle" for "R-vector bundle". We shall usually mention the field of scalars when considering vector bundles with a field of scalars different from R (mainly $C = R[i]$, cf. Section 12.6).

We denote by ϵ^n_X (or ϵ^n if no confusion is possible) the vector bundle $(X \times R^n, \pi, X)$, where π is the canonical projection. A pre-algebraic vector bundle over X is said to be *algebraically trivial* if it is algebraically isomorphic to ϵ^n_X for some n.

Let $f : Y \to X$ be a regular mapping between affine real algebraic varieties, and let $\xi = (E, p, X)$ be a pre-algebraic vector bundle. The *induced vector bundle* $f^*(\xi) = (E', p', Y)$, where

$$E' = \{(v, y) \in E \times Y \mid p(v) = f(y)\}$$

and $p'(v, y) = y$, is equipped with a canonical structure of pre-algebraic vector bundle. If Y is an algebraic subvariety of X, and f is the inclusion mapping $Y \hookrightarrow X$, the bundle $f^*(\xi)$ is called the *restriction of the bundle ξ to Y* and is denoted by $\xi|_Y$.

Property (iii) of the definition means that, for every pre-algebraic vector bundle ξ over X, there exists a finite covering $(U_i)_{i \in I}$ of X by Zariski open sets such that, for every i, $\xi|_{U_i}$ is algebraically isomorphic to $\epsilon^n_{U_i}$. An algebraic isomorphism $\varphi : \epsilon^n_U \to \xi|_U$ is called a *trivialization of ξ over U*. Given a family of trivializations $(\varphi_i : \epsilon^n_{U_i} \to \xi|_{U_i})_{i \in I}$, where the U_i are Zariski open sets covering X, we obtain a family of regular mappings

$$g_{i,j} : U_i \cap U_j \longrightarrow GL(n, R) \quad \text{for } i, j \in I$$

such that, for every $x \in U_i \cap U_j$ and $v \in R^n$,

$$\varphi_i^{-1} \circ \varphi_j(x, v) = (x, g_{i,j}(x) \cdot v) \,.$$

The $g_{i,j}$ are called the *transition functions* of the bundle ξ (for a given family of trivializations). These mappings satisfy $g_{i,j}(x) \cdot g_{j,k}(x) = g_{i,k}(x)$ for every $x \in U_i \cap U_j \cap U_k$. Conversely, given a covering of X by Zariski open subsets $(U_i)_{i \in I}$ (where we may assume I finite) and a family of regular mappings $g_{i,j} : U_i \cap U_j \to GL(n, R)$, for $i, j \in I$, such that $g_{i,i}(x)$ is the identity matrix for every $x \in U_i$ and $g_{i,j}(x) \cdot g_{j,k}(x) = g_{i,k}(x)$ for every $x \in U_i \cap U_j \cap U_k$, there exists a pre-algebraic vector bundle ξ over X, unique up to an algebraic isomorphism, such that ξ is algebraically trivial over each U_i and the $g_{i,j}$ are the transition functions of ξ. Indeed, the total space E of ξ is obtained by gluing the $U_i \times R^n$ together along the biregular isomorphisms

$$U_j \times R^n \supset (U_i \cap U_j) \times R^n \longrightarrow (U_i \cap U_j) \times R^n \subset U_i \times R^n$$
$$(x, v) \longmapsto (x, g_{i,j}(x) \cdot v) \,,$$

and we take the obvious projection $p : E \to X$.

If ξ and ξ' are two pre-algebraic vector bundles over X, the vector bundles $\xi \oplus \xi'$ (*Whitney sum*), $\xi \otimes \xi'$ (*tensor product*), $\bigwedge^k \xi$ (*exterior powers*) and $\xi^\vee$ (*dual vector bundle*) have a canonical structure of pre-algebraic vector bundle. This can be shown by choosing a family of algebraic trivializations of ξ and ξ' over the same covering of X, and using the transition functions of ξ and ξ'. The transition functions of the vector bundles $\xi \oplus \xi'$, $\xi \otimes \xi'$, $\bigwedge^k \xi$ and $\xi^\vee$ are obtained, respectively, by composing the transition functions of ξ and ξ' with the regular mappings

$$GL(n, R) \times GL(n', R) \xrightarrow{\oplus} GL(n + n', R) \,,$$
$$GL(n, R) \times GL(n', R) \xrightarrow{\otimes} GL(nn', R) \,,$$
$$GL(n, R) \xrightarrow{\bigwedge^k} GL\left(\binom{n}{k}, R\right) \,,$$
$$GL(n, R) \xrightarrow{{}^t(-)^{-1}} GL(n, R) \,.$$

The bundle $\mathrm{Hom}(\xi, \xi')$ is also equipped with a canonical structure of pre-algebraic vector bundle, and $\mathrm{Hom}(\xi, \xi') \simeq_{\mathrm{alg}} \xi^\vee \otimes \xi'$.

Let $\xi = (E, p, X)$ and $\xi' = (E', p', X)$ be two pre-algebraic vector bundles over X. We say that ξ' is a *pre-algebraic vector subbundle* of ξ if E' is contained in E, $p' = p|_{E'}$ and the inclusion mapping $i : E' \hookrightarrow E$ is an algebraic morphism from ξ' into ξ. Then E' is Zariski closed in E.

If $\psi : \eta \to \xi$ is an injective algebraic morphism, the image of ψ is a pre-algebraic vector subbundle of ξ. More generally, assume that ψ has constant rank, i.e. there exists an integer k such that, for every $x \in X$, the linear mapping ψ_x has rank k. Then the vector bundles $\mathrm{Ker}(\psi)$, $\mathrm{Im}(\psi)$ and $\mathrm{Coker}(\psi)$ are equipped with a canonical structure of pre-algebraic vector bundle, so that $\mathrm{Ker}(\psi)$ (resp. $\mathrm{Im}(\psi)$) is a pre-algebraic vector subbundle of η (resp. ξ), and the surjective morphism $\xi \to \mathrm{Coker}(\psi)$ is an algebraic morphism. $\mathrm{Coker}(\psi)$ is said to be a *pre-algebraic quotient vector bundle* of ξ. All these assertions can be proved by checking that the arguments used in [174], Chap. 3, §8 can be adapted to the algebraic framework.

The classical equivalence between vector bundles and locally free sheaves holds for pre-algebraic vector bundles and sheaves of $\mathcal{R}_X$-modules.

Definition 12.1.2. *Let X be an affine real algebraic variety and $\mathcal{R}_X$ its sheaf of regular functions. A locally free algebraic sheaf of finite type over X is a sheaf $\mathcal{F}$ of $\mathcal{R}_X$-modules such that there exist a covering of X by Zariski open subsets $(U_i)_{i \in I}$ and, for every $i \in I$, an integer n such that $\mathcal{F}|_{U_i}$ is isomorphic to $\mathcal{R}_{U_i}^n$.*

Proposition 12.1.3. *Let ξ be an algebraic vector bundle over an affine real algebraic variety X. For every Zariski open subset U of X, denote by $\mathcal{L}_{\mathrm{alg}}(\xi)(U)$ the $\mathcal{R}(U)$-module of algebraic sections of $\xi|_U$. Then $\mathcal{L}_{\mathrm{alg}}(\xi)$ is a sheaf of $\mathcal{R}_X$-modules, and $\mathcal{L}_{\mathrm{alg}}$ is an equivalence of the category of pre-algebraic vector bundles over X with the category of locally free algebraic sheaves of finite type over X.*

Hints for the proof: The arguments used in [295], chapter 6, §1, Theorem 2 remain valid in the case considered in the proposition. □

We now describe some pre-algebraic vector bundles over the grassmannians. We identify $\mathbb{G}_{n,k}(R)$, the grassmannian of k-dimensional vector subspaces of R^n, with the algebraic set of matrices of orthogonal projections (cf. Theorem 3.4.4):

$$\mathbb{G}_{n,k}(R) = \{A \in \mathbb{M}_{n,n}(R) \mid {}^t A = A,\ A^2 = A \text{ and } \mathrm{trace}(A) = k\}.$$

Proposition 12.1.4. *Let*

$$\begin{aligned} E_{n,k} &= \{(A,v) \in \mathbb{G}_{n,k}(R) \times R^n \mid A \cdot v = v\},\\ E_{n,k}^\perp &= \{(A,v) \in \mathbb{G}_{n,k}(R) \times R^n \mid A \cdot v = 0\}. \end{aligned}$$

Let $p_{n,k}$ (resp. $p_{n,k}^\perp$) be the canonical projection of $E_{n,k}$ (resp. $E_{n,k}^\perp$) onto $\mathbb{G}_{n,k}(R)$. Then $\gamma_{n,k} = (E_{n,k}, p_{n,k}, \mathbb{G}_{n,k}(R))$ and $\gamma_{n,k}^\perp = (E_{n,k}^\perp, p_{n,k}^\perp, \mathbb{G}_{n,k}(R))$ are pre-algebraic vector bundles over $\mathbb{G}_{n,k}(R)$ of rank k and $n-k$, respectively. Moreover, $\gamma_{n,k} \oplus \gamma_{n,k}^\perp \simeq_{\mathrm{alg}} \epsilon^n$.

The bundle $\gamma_{n,k}$ is called *the universal vector bundle over $\mathbb{G}_{n,k}(R)$*.

Proof. Clearly, $E_{n,k}$ is an algebraic set. Now we proceed to describe algebraic trivializations of $\gamma_{n,k}$ over Zariski open subsets of $\mathbb{G}_{n,k}(R)$. Let $\sigma = \{\sigma_1, \ldots, \sigma_k\} \subset \{1, \ldots, n\}$, and let V_σ be the k-dimensional vector subspace of R^n generated by $e_{\sigma_1}, \ldots, e_{\sigma_k}$, where $(e_1, \ldots, e_n)$ is the canonical basis of R^n. Denote by U_σ the Zariski open subset of $\mathbb{G}_{n,k}(R)$ consisting of matrices A such that the mapping $v \mapsto A \cdot v$ is injective on V_σ. Choose a linear isomorphism $i : R^k \to V_\sigma$. Then the mapping

$$\psi_\sigma : U_\sigma \times R^k \longrightarrow p_{n,k}^{-1}(U_\sigma),$$

defined by $\psi_\sigma(A, x) = (A, A \cdot i(x))$, is a biregular isomorphism, and is linear on each fibre. The Zariski open sets U_σ cover $\mathbb{G}_{n,k}(R)$, and $\gamma_{n,k}$ is algebraically trivial over each U_σ.

Since $\gamma_{n,k}^\perp$ is isomorphic to the induced bundle $f^*(\gamma_{n,n-k})$, where $f : \mathbb{G}_{n,k}(R) \to \mathbb{G}_{n,n-k}(R)$ is the canonical biregular isomorphism described in Proposition 3.4.5, $\gamma_{n,k}^\perp$ is also a pre-algebraic vector bundle.

Finally, the algebraic isomorphism $\gamma_{n,k} \oplus \gamma_{n,k}^\perp \simeq_{\mathrm{alg}} \epsilon^n$ is defined by

$$((A, v_1), (A, v_2)) \mapsto (A, v_1 + v_2). \qquad \square$$

The category of all pre-algebraic vector bundles is too large to be useful in real algebraic geometry, and we shall soon restrict our attention to a special subcategory. The following example of a pre-algebraic vector bundle which is not generated by global algebraic sections shows one of the drawbacks of the category of all pre-algebraic vector bundles.

Example 12.1.5. Let

$$P = X^2(X - 1)^2 + Y^2 \in R[X, Y].$$

The irreducible polynomial P has only two zeros $c_1 = (0,0)$ and $c_2 = (1,0)$ in R^2. Set $U_i = R^2 \setminus \{c_i\}$. The transition function

$$\begin{aligned} g_{1,2} : U_1 \cap U_2 &\longrightarrow GL(1, R) = R^* \\ (x, y) &\longmapsto P(x, y) \end{aligned}$$

defines a pre-algebraic vector bundle ξ of rank 1 over R^2. Using the algebraic trivializations over U_1 and U_2, a global algebraic section of ξ can be described as a pair (s_1, s_2), where $s_i : U_i \to R$ is a regular function, for $i = 1, 2$, and $s_1 = g_{1,2} s_2$. For $i = 1, 2$, set $s_i = f_i/h_i$, where f_i and h_i are relatively prime polynomials. Then $f_1 h_2 = P f_2 h_1$. Since P does not divide h_2, we have $f_1 = \lambda P f_2$ and $h_2 = \lambda^{-1} h_1$, where $\lambda \in R^*$. This shows that every algebraic global section of the bundle ξ vanishes at c_2. Hence, ξ is not generated by its global algebraic sections. This shows that ξ is not algebraically trivial, although it is topologically trivial. Note also that the sheaf $\mathcal{L}_{\mathrm{alg}}(\xi)$ of algebraic sections of ξ is a locally free algebraic sheaf of finite type that is not generated by its global sections: contrary to the case of algebraically closed fields, here there is no "Theorem A" ([293], Chap. 2, §3, Theorem 2).

A useful property, which holds for every affine variety over an algebraically closed field, is the equivalence between algebraic vector bundles and projective modules of finite type over the ring of regular functions. This equivalence is a particular case of the equivalence between coherent algebraic sheaves and modules of finite type (cf. [293], Chap. 2, §4). If M is a $\mathcal{R}(X)$-module, the sheaf of $\mathcal{R}_X$-modules associated to the presheaf $U \mapsto \mathcal{R}(U) \otimes_{\mathcal{R}(X)} M$ is denoted by $\mathcal{R}_X \otimes_{\mathcal{R}(X)} M$. The same arguments as in [293] show that, if M is a projective module of finite type, then $\mathcal{R}_X \otimes_{\mathcal{R}(X)} M$ is a locally free algebraic sheaf of finite type. Hence, to every projective module M of finite type over $\mathcal{R}(X)$ is associated a pre-algebraic vector bundle ξ over X, defined up to an algebraic isomorphism, such that $\mathcal{L}_{\mathrm{alg}}(\xi) \simeq \mathcal{R}_X \otimes_{\mathcal{R}(X)} M$. This vector bundle ξ is, of course, generated by its global algebraic sections. Hence, the vector bundle of Example 12.1.5 is not associated to a projective module of finite type over $\mathcal{R}(R^2)$. However, the module of its global algebraic sections is projective, and even isomorphic to $\mathcal{R}(R^2)$ by the mapping $(s_1, s_2) \mapsto s_2 = f_2/h_2$, since $h_2 = \lambda^{-1} h_1$ has no zero on R^2.

We shall now introduce the subcategory of the category of pre-algebraic vector bundles consisting of bundles associated to a projective module of finite type over $\mathcal{R}(X)$.

Definition 12.1.6. *A pre-algebraic vector bundle ξ over X is said to be* algebraic *if there exists an injective algebraic morphism from ξ to a trivial bundle ϵ_X^n (i.e. ξ is algebraically isomorphic to a pre-algebraic vector subbundle of a trivial bundle).*

Theorem 12.1.7. *Let $\xi = (E, p, X)$ be a pre-algebraic vector bundle of rank k over X. Then the following properties are equivalent:*

(i) *ξ is algebraic.*

(ii) *For every x in X, there exist global algebraic sections $s_1, \ldots, s_k$ of ξ such that $s_1(x), \ldots, s_k(x)$ generate the fibre $p^{-1}(x)$ as an R-vector space.*

(iii) *There exists a surjective algebraic morphism from a trivial bundle ϵ_X^n onto ξ (i.e. ξ is algebraically isomorphic to a pre-algebraic quotient of a trivial bundle).*

(iv) *There exists a pre-algebraic vector bundle ξ' over X such that $\xi \oplus \xi'$ is algebraically isomorphic to a trivial bundle ϵ_X^n (i.e. ξ is a pre-algebraic direct factor of a trivial bundle).*

(v) *There exists a regular mapping $f : X \to \mathbb{G}_{n,k}(R)$, for some $n \geq k$, such that ξ is algebraically isomorphic to $f^*(\gamma_{n,k})$.*

(vi) *There exists a projective module of finite type M over $\mathcal{R}(X)$ such that $\mathcal{L}_{\mathrm{alg}}(\xi)$ is isomorphic to $\mathcal{R}_X \otimes_{\mathcal{R}(X)} M$.*

Proof. (i) $\Rightarrow$ (v) Let $\varphi : \xi \to \epsilon_X^n$ be an injective algebraic morphism. Let $f : X \to \mathbb{G}_{n,k}(R)$ be defined by $\{x\} \times f(x) = \varphi(p^{-1}(x))$. It suffices to prove that f is regular. To this end, consider a Zariski open subset U of X such that there is an algebraic trivialization $\psi : \epsilon_U^k \to \xi|_U$. For $i = 1, \ldots, k$, let $\varphi_i : U \to R^n$ be the regular mapping defined by $(x, \varphi_i(x)) = \varphi(\psi(x, e_i))$ for every x in U, where $(e_1,\ldots,e_k)$ is the canonical basis of R^k. Since $f(x)$ is generated by the $\varphi_i(x)$, Proposition 3.4.7 applied to the mappings φ_i shows that $f|_U$ is regular.

(v) $\Rightarrow$ (iv) Since $\gamma_{n,k} \oplus \gamma_{n,k}^\perp = \epsilon^n$, we have $f^*(\gamma_{n,k}) \oplus f^*(\gamma_{n,k}^\perp) = \epsilon_X^n$.

(iv) $\Rightarrow$ (vi) We have $\mathcal{L}_{\mathrm{alg}}(\xi) \oplus \mathcal{L}_{\mathrm{alg}}(\xi') \simeq \mathcal{R}_X^n$. Let Π be the composition $\mathcal{R}_X^n \to \mathcal{L}_{\mathrm{alg}}(\xi) \to \mathcal{R}_X^n$. The mapping Π is a projection (i.e. $\Pi^2 = \Pi$), and its image is isomorphic to $\mathcal{L}_{\mathrm{alg}}(\xi)$. By taking global sections, Π induces a projection $\Pi(X) : \mathcal{R}(X)^n \to \mathcal{R}(X)^n$ such that $\mathcal{R}_X \otimes_{\mathcal{R}(X)} \Pi(X) = \Pi$. The image of $\Pi(X)$ is a projective module M of finite type over $\mathcal{R}(X)$, and $\mathcal{R}_X \otimes_{\mathcal{R}(X)} M$ is the image of $\Pi = \mathcal{R}_X \otimes_{\mathcal{R}(X)} \Pi(X)$. Hence, $\mathcal{L}_{\mathrm{alg}}(\xi)$ is isomorphic to $\mathcal{R}_X \otimes_{\mathcal{R}(X)} M$.

(vi) $\Rightarrow$ (ii) If $\mathcal{L}_{\mathrm{alg}}(\xi)$ is isomorphic to $\mathcal{R}_X \otimes_{\mathcal{R}(X)} M$, then it is generated by its global sections. Since $p^{-1}(x) \simeq \mathcal{L}_{\mathrm{alg}}(\xi)_x / \mathfrak{m}_{X,x} \mathcal{L}_{\mathrm{alg}}(\xi)_x$, where $\mathfrak{m}_{X,x}$ is the maximal ideal of $\mathcal{R}_{X,x}$, the fibre $p^{-1}(x)$ is generated, as an R-vector space, by the values at x of global algebraic sections of ξ.

(ii) $\Rightarrow$ (iii) The elements $s_1(x),\ldots,s_k(x)$ form a basis of $p^{-1}(x)$. Let $\varphi : \epsilon_U^k \to \xi|_U$ be an algebraic trivialization of ξ, where U is a Zariski open neighbourhood of x in X. The determinant of $(\varphi^{-1}(s_1(y)),\ldots,\varphi^{-1}(s_k(y)))$ is a regular function of $y \in U$. Hence, there is a Zariski open neighbourhood V of x in X such that, for every $y \in V$, $s_1(y),\ldots,s_k(y)$ is a basis of $p^{-1}(y)$. Since X is covered by finitely many such open subsets V, we can find a finite family of global algebraic sections $s_1,\ldots,s_n$ of ξ, such that $(s_1(x),\ldots,s_n(x))$ generate $p^{-1}(x)$ for every x in X. These global algebraic sections induce a surjective algebraic morphism from the trivial bundle ϵ_X^n to ξ.

(iii) $\Rightarrow$ (i) Given a surjective algebraic morphism $\epsilon_X^n \to \xi$, we obtain, by dualizing, an injective algebraic morphism $\xi^\vee \to (\epsilon_X^n)^\vee \simeq_{\mathrm{alg}} \epsilon_X^n$. Hence, the dual bundle $\xi^\vee$ is algebraic. The preceding arguments imply that $\xi^\vee$ is a pre-algebraic direct factor of the trivial bundle ϵ_X^n. Hence, ξ is also a pre-algebraic direct factor of ϵ_X^n, which implies that ξ is algebraic. $\square$

We now present a few straightforward properties of algebraic vector bundles.

Proposition 12.1.8.

(i) *The vector bundles $\gamma_{n,k}$ and $\gamma_{n,k}^\perp$ over $\mathbb{G}_{n,k}(R)$ are algebraic.*

(ii) *If ξ is an algebraic vector bundle over X, and $f : Y \to X$ is a regular mapping, then $f^*(\xi)$ is algebraic.*

(iii) *If ξ and η are algebraic vector bundles over X, then $\xi \oplus \eta$, $\xi \otimes \eta$, $\xi^\vee$, $\bigwedge^q \xi$ and $\mathrm{Hom}(\xi,\eta)$ are algebraic vector bundles.*

Proof. Statements (i) and (ii) follow from Theorem 12.1.7(v). For (iii), the cases of $\xi \oplus \eta$ and $\xi \otimes \eta$ are straightforward. The case of $\xi^\vee$ follows from the proof of Theorem 12.1.7, (iii) $\Rightarrow$ (i). For $\bigwedge^q \xi$, we can use the canonical surjection $\bigotimes^q \xi \to \bigwedge^q \xi$, the fact that $\bigotimes^q \xi$ is algebraic and property (iii) of Theorem 12.1.7. The case of $\mathrm{Hom}(\xi,\eta)$ follows from the algebraic isomorphism $\mathrm{Hom}(\xi,\eta) \simeq_{\mathrm{alg}} \xi^\vee \otimes \eta$. $\square$

Proposition 12.1.9. *Let $V \subset R^n$ be a nonsingular algebraic set of dimension k. The tangent bundle and the normal bundle of V are algebraic.*

Proof. The tangent (resp. normal) bundle of V is isomorphic to $\tau_V^*(\gamma_{n,k})$ (resp. $\tau_V^*(\gamma_{n,k}^\perp)$), where $\tau_V : V \to \mathbb{G}_{n,k}(R)$ is the regular mapping defined by $\tau_V(x) = T_x(V)$ for $x \in V$ (cf. Corollary 3.4.8). $\square$

Before stating the next theorem, let us recall four properties which characterize the Stiefel-Whitney cohomology classes of an $\mathbb{R}$-vector bundle. To each topological $\mathbb{R}$-vector bundle $\xi = (E, p, B)$ over a paracompact base space B there corresponds a sequence of cohomology classes

$$w_i(\xi) \in H^i(B, \mathbb{Z}/2), \quad i = 0, 1, 2, \ldots,$$

called *the Stiefel-Whitney classes of ξ*, which satisfy the following properties:

(i) The class $w_0(\xi)$ is the unit element $1 \in H^0(B, \mathbb{Z}/2)$, and $w_i(\xi) = 0$ for i greater than the rank of ξ.

(ii) If $f : B_1 \to B$ is a continuous mapping, and ξ is an $\mathbb{R}$-vector bundle over B, then

$$w_i(f^*(\xi)) = f^*(w_i(\xi)) .$$

(iii) (The Whitney product theorem) If ξ and η are $\mathbb{R}$-vector bundles over the same base space, then

$$w_k(\xi \oplus \eta) = \sum_{i=0}^{k} w_i(\xi) \cup w_{k-i}(\eta) ,$$

where the symbol $\cup$ stands for cup product.

(iv) For a nontrivial $\mathbb{R}$-line bundle γ on $\mathbb{P}_1(\mathbb{R})$, the class $w_1(\gamma)$ is nonzero.

The existence and uniqueness of Stiefel-Whitney cohomology classes is established in [236]. Given an $\mathbb{R}$-vector bundle ξ of rank k over B, its *total Stiefel-Whitney class* $w(\xi)$ is the element $\sum_{i=0}^{k} w_i(\xi)$ of the total cohomology ring $H^*(B, \mathbb{Z}/2)$.

Given a C^∞ manifold M, we use the abbreviation $w_i(M)$ for the i-th Stiefel-Whitney class of the tangent bundle of M. This class is called the i-th Stiefel-Whitney class of M.

Theorem 12.1.10. *($R = \mathbb{R}$) If ξ is an algebraic vector bundle on a compact nonsingular affine real algebraic variety X, then the total Stiefel-Whitney class $w(\xi)$ of ξ belongs to $H_{\text{alg}}^*(X, \mathbb{Z}/2)$.*

Proof. By Theorem 12.1.7 (v), we may assume that ξ is isomorphic to $f^*(\gamma_{n,p})$, for some regular mapping $f : X \to \mathbb{G}_{n,p}(\mathbb{R})$. Then $w(\xi) = w(f^*(\gamma_{n,p})) = f^*(w(\gamma_{n,p}))$, and, hence, the conclusion follows in view of Proposition 11.3.3 and Theorem 11.3.4 (ii). $\square$

Observe that, by Proposition 12.1.9 and Theorem 12.1.10, the total Stiefel-Whitney class $w(X) = w(T(X))$ of X is in $H^*_{\text{alg}}(X, \mathbb{Z}/2)$.

The following lemma provides a useful tool to construct algebraic vector bundles.

Lemma 12.1.11. *Let $(U_i)_{i=1,\ldots,q}$ be a Zariski open covering of an affine real algebraic variety X. Let $h_{i,j} : U_j \to \mathbb{M}_{k,k}(R)$ be regular mappings such that $h_{i,j}(U_i \cap U_j) \subset GL(k,R)$, $h_{i,i}(x)$ is the identity matrix and $h_{i,j} \cdot h_{j,\ell} = h_{i,\ell}$ on $U_j \cap U_\ell$ for all $i,j,\ell = 1, \ldots, q$. Let E be the set of $(x, v_1, \ldots, v_q) \in X \times (R^k)^q$ such that $v_i = h_{i,j}(x) \cdot v_j$, for every j such that $x \in U_j$ and every $i = 1, \ldots, q$. Let $p : E \to X$ be defined by $p(x, v_1, \ldots, v_q) = x$.*

Then $\xi = (E, p, X)$ is an algebraic vector bundle, which is trivial over each U_i and has the $h_{i,j}|_{U_i \cap U_j}$ as transition functions.

Note that the mappings $h_{i,j}$ have to be defined on U_j and not only on $U_i \cap U_j$.

Proof. Since $E \cap (U_i \times (R^k)^q)$ is closed in $U_i \times (R^k)^q$ for the Zariski topology, E is a real algebraic variety. The mappings
$$(x, v) \longmapsto (x, h_{1,j}(x) \cdot v, \ldots, h_{q,j}(x) \cdot v)$$
are biregular isomorphisms from $U_j \times R^k$ onto $p^{-1}(U_j)$, for $j = 1, \ldots, q$. Hence, they are algebraic trivializations of ξ. Moreover, ξ is, by construction, a pre-algebraic subbundle of the trivial bundle ϵ_X^{kq}. Therefore, the bundle ξ is algebraic. □

Theorem 12.1.7 allows one to prove the following equivalence of categories.

Proposition 12.1.12. *Given an algebraic vector bundle ξ over X, denote by $\Gamma_{\text{alg}}(\xi)$ the $\mathcal{R}(X)$-module of global algebraic sections of ξ. Then Γ_{alg} is an equivalence of the category of algebraic vector bundles over X with the category of projective modules of finite type over $\mathcal{R}(X)$.*

Proof. By property (vi) of Theorem 12.1.7, it suffices to verify that, if M is a projective module of finite type over $\mathcal{R}(X)$, the module of global sections of $\mathcal{R}_X \otimes_{\mathcal{R}(X)} M$ is isomorphic to M. This can be easily proved, by representing M as a direct factor of a free module of finite type $\mathcal{R}(X)^n$. □

Remark 12.1.13. The usual constructions are preserved by the equivalence of categories Γ_{alg}. If ξ and η are two algebraic vector bundles over X, then $\Gamma_{\text{alg}}(\xi \oplus \eta) = \Gamma_{\text{alg}}(\xi) \oplus \Gamma_{\text{alg}}(\eta)$, $\Gamma_{\text{alg}}(\xi \otimes \eta) = \Gamma_{\text{alg}}(\xi) \otimes \Gamma_{\text{alg}}(\eta)$, $\Gamma_{\text{alg}}(\xi^\vee) = \Gamma_{\text{alg}}(\xi)^\vee$, $\Gamma_{\text{alg}}(\bigwedge^q \xi) = \bigwedge^q \Gamma_{\text{alg}}(\xi)$ and $\Gamma_{\text{alg}}(\text{Hom}(\xi, \eta)) = \text{Hom}(\Gamma_{\text{alg}}(\xi), \Gamma_{\text{alg}}(\eta))$.

Remark 12.1.14. Note that the total space of an algebraic vector bundle over an affine real algebraic variety is itself an affine real algebraic variety. It is shown in [173, 223] that the converse is also valid: a pre-algebraic vector bundle with affine total space is algebraic.

12.2 Algebraic Line Bundles and the Divisor Class Group of the Ring of Regular Functions

The main part of this section consists of a review of some basic facts concerning divisors, for which we refer the reader to [66], [23], Chap. 3, §7 or [150], Chap. 2, §6.

Let A be an arbitrary commutative ring. The set of isomorphism classes of projective modules of rank 1 over A (also called *invertible modules*), equipped with the tensor product, is a commutative group, called the *Picard group of A* and denoted by Pic(A) (cf. [66], Chap. 2, §5, Proposition 7). Note that the inverse of the class of an invertible module M is the class of the dual module $M^\vee$.

If X is an affine real algebraic variety, an *algebraic line bundle* over X is an algebraic vector bundle of rank 1. We denote by $V^1_{\text{alg}}(X)$ the set of isomorphism classes of algebraic line bundles over X.

Proposition 12.2.1. *Let X be an affine real algebraic variety. Then $V^1_{\text{alg}}(X)$, equipped with the tensor product, is a commutative group, isomorphic to the group Pic($\mathcal{R}(X)$) of isomorphism classes of invertible $\mathcal{R}(X)$-modules.*

Proof. This is a straightforward consequence of Proposition 12.1.12. □

Now assume A to be an integral domain and let K be its field of fractions. A *fractional ideal* of A is a sub-A-module M of K such that there exists $b \in K^*$ with $bM \subset A$. A *principal fractional ideal* of A is a fractional ideal of the form bA, for some $b \in K^*$. An *invertible fractional ideal* of A is a fractional ideal M such that M is an A-module of finite type and, for every maximal ideal $\mathfrak{m}$ of A, there exists $b \in K^*$ such that $M_\mathfrak{m} = bA_\mathfrak{m}$. The invertible fractional ideals form a group with respect to product (often called the Cartier group of A). Every invertible fractional ideal is an invertible A-module. The quotient of the group of invertible fractional ideals by the subgroup of principal fractional ideals (called the *group of classes of invertible fractional ideals of A*) is isomorphic to the group Pic(A) (cf. [66], Chap. 2, §5, n. 6 and 7).

We assume now that A is a Krull domain; if A is noetherian, this means that A is integrally closed. A *divisorial fractional ideal* of A is a fractional ideal M of A such that $M = A : (A : M)$ (if N is a sub-A-module of K, then $A : N = \{b \in K \mid bN \subset A\}$). These divisorial fractional ideals form a group with respect to product, which is canonically isomorphic to the free commutative group generated by the height 1 prime ideals of A. The quotient of this group by the subgroup of principal fractional ideals is called the *divisor class group* of A and is denoted by Cl(A) ([66], Chap. 7, §1).

Proposition 12.2.2. *Let A be a Krull domain. Then*

(i) A is a factorial ring if and only if Cl(A) = 0.

(ii) If S is a multiplicative subset of A such that $0 \notin S$, the canonical homomorphism Cl(A) → Cl($S^{-1}A$) is surjective and its kernel is generated

12.2 Algebraic Line Bundles and the Divisor Class Group 307

by the classes of height 1 prime ideals of A having nonempty intersection with S.

References for the proof: For (i), see Definition 1 and Proposition 2 in [66], Chap. 7 §3.

(ii) [66], Chap. 7, §2, Proposition 17. □

Now we compare the Picard group with the divisor class group. An invertible fractional ideal is divisorial. Hence, Pic(A) can be identified with a subgroup of Cl(A). There is no equality in general; however, the two groups coincide in a case that is geometrically interesting. A domain A is said to be *locally factorial* if, for every maximal ideal $\mathfrak{m}$ of A, the localization $A_\mathfrak{m}$ is factorial. A locally factorial noetherian domain is integrally closed and, therefore, a Krull domain.

Proposition 12.2.3. *Let A be a locally factorial noetherian domain. Then every divisorial fractional ideal of A is invertible, and the groups* Pic(A) *and* Cl(A) *are equal.*

Reference for the proof: [66], Chap. 7 §3, Proposition 1. □

Corollary 12.2.4. *Let X be an irreducible affine real algebraic variety. Assume that, for every x in X, the local ring $\mathcal{R}_{X,x}$ is factorial (this is the case if X is nonsingular). Then the group $V^1_{\mathrm{alg}}(X)$ is isomorphic to the divisor class group* Cl($\mathcal{R}(X)$) *of the ring of regular functions $\mathcal{R}(X)$.*

Remark 12.2.5. We now illustrate this result by the explicit construction of an algebraic line bundle whose class in $V^1_{\mathrm{alg}}(X)$ is the image of the class of a given invertible fractional ideal I of $\mathcal{R}(X)$ by the isomorphism Cl($\mathcal{R}(X)$) → $V^1_{\mathrm{alg}}(X)$, where X satisfies the assumptions of Corollary 12.2.4.

There exist a finite covering $(U_i)_{i=1,\ldots,q}$ of X by Zariski open subsets and $f_i \in I$, for $i = 1, \ldots, q$, such that $I\mathcal{R}(U_i) = f_i\mathcal{R}(U_i)$. Let E be the set of $(x, v_1, \ldots, v_q) \in X \times R^q$ such that $v_i = (f_i/f_j)(x)v_j$, for every j such that $x \in U_j$ and every $i = 1, \ldots, q$. Set $p(x, v_1, \ldots, v_q) = x$. Then, by Lemma 12.1.11, $\eta = (E, p, X)$ is an algebraic line bundle. The dual bundle $\xi = \eta^\vee$ is also algebraic. The transition functions of ξ are $g_{ij} = f_j/f_i$. We claim that the class of ξ is the image of the class of I by the isomorphism Cl($\mathcal{R}(X)$) → $V^1_{\mathrm{alg}}(X)$. It suffices to prove that $\Gamma_{\mathrm{alg}}(\xi)$ is isomorphic, as an $\mathcal{R}(X)$-module, to I. An algebraic section of ξ is given by $(s_1, \ldots, s_q)$, with $s_i \in \mathcal{R}(U_i)$ and $s_i f_i = s_j f_j$ for every i, j. If $h = s_1 f_1 = \cdots = s_q f_q$, then $h \in \bigcap_{i=1}^q I\mathcal{R}(U_i) = I$, and the homomorphism assigning h to $(s_1, \ldots, s_q)$ is an isomorphism from $\Gamma_{\mathrm{alg}}(\xi)$ onto I.

Corollary 12.2.6. *Let X be an irreducible affine real algebraic variety. The following properties are equivalent:*

(i) *$\mathcal{R}(X)$ is factorial.*

(ii) *The local ring $\mathcal{R}_{X,x}$ is factorial for every x in X, and every algebraic line bundle over X is algebraically trivial.*

If $V_{\text{alg}}^1(X)$ is nontrivial, it follows that $\mathcal{R}(X)$ is not factorial. This is the case if X is nonorientable.

Proposition 12.2.7. *($R = \mathbb{R}$) Let X be a nonsingular affine real algebraic variety. If X is nonorientable as a C^∞ manifold, then $V_{\text{alg}}^1(X) \neq 0$ and the ring $\mathcal{R}(X)$ is not factorial.*

Proof. By Proposition 12.1.9, the tangent bundle $T(X)$ is algebraic. Hence, by Proposition 12.1.8 (iii), $\bigwedge^d T(X)$ is an algebraic line bundle, where $d = \dim(X)$. Since X is nonorientable, $\bigwedge^d T(X)$ is topologically and, therefore, algebraically nontrivial. Hence $V_{\text{alg}}^1(X) \neq 0$. □

Remark 12.2.8. It would be interesting to know whether Proposition 12.2.7 is still valid if X is an affine real algebraic variety (possibly with singularities) and, at the same time, a nonorientable topological manifold.

12.3 Approximation of Continuous Sections of an Algebraic Vector Bundle by Algebraic Sections

Throughout this section, we assume that $R = \mathbb{R}$ and X *is a compact affine real algebraic variety* (up to a biregular isomorphism, we may assume that X is a compact algebraic subset of some $\mathbb{R}^q$). We need this assumption in order to use the Stone-Weierstrass approximation theorem (8.8.5).

Theorem 12.3.1. *Let $\xi = (E, p, X)$ be an algebraic vector bundle. Let $\sigma : X \to E$ be a continuous section of ξ. Then, for every open neighbourhood U of $\sigma(X)$ in E, there exists an algebraic section $s : X \to E$ of ξ such that $s(X) \subset U$.*

Proof. By Theorem 12.1.7 (iii), there are finitely many global algebraic sections $s_1, \ldots, s_n$ of ξ such that, for every point x in X, $s_1(x), \ldots, s_n(x)$ generate the fibre $p^{-1}(x)$ as an $\mathbb{R}$-vector space. Fix a point $x \in X$ and let k be the rank of ξ. We can find k sections among $s_1, \ldots, s_n$, say $s_1, \ldots, s_k$, such that $s_1(x), \ldots, s_k(x)$ form a basis of $p^{-1}(x)$. Then $s_1(y), \ldots, s_k(y)$ form a basis of $p^{-1}(y)$ for every y in an open neighbourhood V_x of x. The continuous section $\sigma|_{V_x}$ can be represented as $\sigma|_{V_x} = \alpha_{1,x} s_1|_{V_x} + \cdots + \alpha_{k,x} s_k|_{V_x}$, where the $\alpha_{i,x}$ are continuous functions from V_x to $\mathbb{R}$. Using a partition of unity subordinate to a finite subcovering of the covering $(V_x)_{x \in X}$, we obtain

$$\sigma = \alpha_1 s_1 + \cdots + \alpha_n s_n ,$$

where $\alpha_1, \ldots, \alpha_n$ are continuous functions from X to $\mathbb{R}$. By the Stone-Weierstrass theorem, there are regular functions $\beta_1, \ldots, \beta_n$ from X to $\mathbb{R}$ close enough to $\alpha_1, \ldots, \alpha_n$, so that the image of the algebraic section $s = \beta_1 s_1 + \cdots + \beta_n s_n$ is contained in U. □

12.3 Approximation of Continuous Sections by Algebraic Sections

Given a C^∞ manifold Y and a C^∞ vector bundle η over Y, we denote by $\Gamma^\infty(\eta)$ the space of all C^∞ sections of η endowed with the C^∞ topology.

Theorem 12.3.2. *If X is nonsingular and ξ is an algebraic vector bundle over X, the set of algebraic sections of ξ is dense in $\Gamma^\infty(\xi)$.*

Proof. Using the notation of the proof of Theorem 12.3.1 we observe that, if σ is a C^∞ section, the α_i are C^∞ and can be approximated in the C^∞ topology by regular functions. □

We can now compare topological and algebraic vector bundles over compact real algebraic varieties.

If the vector bundles ξ and η are topologically (resp. C^∞) isomorphic, we denote this by $\xi \simeq_{\text{top}} \eta$ (resp. $\xi \simeq_{C^\infty} \eta$). If ξ and η are C^∞ vector bundles, $\xi \simeq_{\text{top}} \eta$ implies $\xi \simeq_{C^\infty} \eta$.

Let $V^1(X)$ be the group of isomorphism classes of topological line bundles over X, with $\otimes$ as group operation.

Theorem 12.3.3. *Let ξ and η be algebraic vector bundles over X. If $\xi \simeq_{\text{top}} \eta$, then $\xi \simeq_{\text{alg}} \eta$. In particular, the canonical group homomorphism $V^1_{\text{alg}}(X) \to V^1(X)$ is injective.*

Proof. Let $\varphi : \xi \to \eta$ be a topological isomorphism. Then φ determines a continuous section σ of the bundle $\text{Hom}(\xi, \eta)$. The image of σ is contained in the open subset $\text{Iso}(\xi, \eta)$ of all isomorphisms $\xi \to \eta$. Since $\text{Hom}(\xi, \eta)$ is an algebraic vector bundle (Proposition 12.1.8 (iii)), there exists an algebraic section of $\text{Hom}(\xi, \eta)$ contained in $\text{Iso}(\xi, \eta)$ (cf. Theorem 12.3.1). Hence $\xi \simeq_{\text{alg}} \eta$. □

Given a commutative ring A, we denote by $\text{Proj}(A)$ the set of isomorphism classes of projective A-modules of finite type.

Corollary 12.3.4. *Let $C^0(X)$ be the ring of continuous functions $X \to \mathbb{R}$. The canonical mapping $\text{Proj}(\mathcal{R}(X)) \to \text{Proj}(C^0(X))$, induced by the inclusion $\mathcal{R}(X) \hookrightarrow C^0(X)$, is injective.*

Proof. We use Theorem 12.3.3, Proposition 12.1.12 and the well-known equivalence of the category of vector bundles over X with the category of projective modules of finite type over $C^0(X)$ ([314]). □

The canonical injection $\text{Proj}(\mathcal{R}(X)) \to \text{Proj}(C^0(X))$ is not always surjective. To study this question we use K-theory. Recall that the group $\widetilde{K}_0(A)$ of a commutative ring A is constructed as follows. Two projective A-modules of finite type M and N are said to be *stably equivalent* if there exist $m, n \in \mathbb{N}$ such that $M \oplus A^m \simeq N \oplus A^n$. The set of classes of stably equivalent projective A-modules of finite type, equipped with the operation induced by $\oplus$, forms a commutative group $\widetilde{K}_0(A)$. A ring homomorphism $f : A \to B$ induces a homomorphism of groups $\widetilde{K}_0(f) : \widetilde{K}_0(A) \to \widetilde{K}_0(B)$, and the assignment of $\widetilde{K}_0(f)$ to f is functorial.

Similarly, one defines the analogous group for vector bundles over X. Two topological vector bundles ξ and η over X are said to be *stably equivalent* if $\xi \oplus \epsilon_X^m \simeq_{\text{top}} \eta \oplus \epsilon_X^n$ for some m and n in $\mathbb{N}$. In particular, ξ is said to be *stably trivial* if $\xi \oplus \epsilon_X^m \simeq_{\text{top}} \epsilon_X^n$, for some m and n in $\mathbb{N}$. The set of classes of stably equivalent topological vector bundles over X, equipped with the operation $\oplus$ induced by the Whitney sum, constitutes a group denoted by $\widetilde{KO}(X)$ (the Grothendieck group of X). We have $\widetilde{KO}(X) \simeq \widetilde{K}_0(\mathcal{C}^0(X))$ (cf. [314]).

By Proposition 12.1.12, replacing in the definition of $\widetilde{KO}(X)$ topological vector bundles with algebraic ones, we obtain a group isomorphic to $\widetilde{K}_0(\mathcal{R}(X))$.

Proposition 12.3.5. *If a topological vector bundle ξ over X is stably equivalent to an algebraic vector bundle, then ξ is topologically isomorphic to an algebraic vector bundle.*

Proof. It suffices to prove that, if ζ and η are two algebraic vector bundles over X such that $\xi \oplus \zeta \simeq_{\text{top}} \eta$ then ξ is topologically isomorphic to an algebraic vector bundle. Let $k : \xi \to \eta$ and $j : \zeta \to \eta$ be $\mathcal{C}^0$ morphisms such that

$$\begin{aligned} (k,j) : \xi \oplus \zeta &\longrightarrow \eta \\ (u,v) &\longmapsto k(u) + j(v) \end{aligned}$$

is a topological isomorphism. Then j determines a continuous section of the algebraic vector bundle $\text{Hom}(\zeta, \eta)$. Approximating this section close enough by an algebraic section (Theorem 12.3.1), we obtain an algebraic morphism $j' : \zeta \to \eta$ such that $(k, j') : \xi \oplus \zeta \to \eta$ is still an isomorphism. Then ξ is topologically isomorphic to the vector bundle $\text{Coker}(j')$, which is algebraic. $\square$

Theorem 12.3.6.
(i) *The canonical homomorphism $\widetilde{K}_0(\mathcal{R}(X)) \to \widetilde{K}_0(\mathcal{C}^0(X))$ is injective.*

(ii) *If every element of $\widetilde{KO}(X)$ is represented by an algebraic vector bundle, then every topological vector bundle over X is topologically isomorphic to an algebraic vector bundle. In particular, if*

$$\widetilde{K}_0(\mathcal{R}(X)) \to \widetilde{K}_0(\mathcal{C}^0(X))$$

is surjective, then

$$\text{Proj}(\mathcal{R}(X)) \to \text{Proj}(\mathcal{C}^0(X))$$

is surjective.

Proof. (i) follows from Corollary 12.3.4, and (ii) from Proposition 12.3.5. $\square$

12.3 Approximation of Continuous Sections by Algebraic Sections 311

If X is an algebraic set, the homomorphism $\widetilde{K}_0(\mathcal{P}(X)) \to \widetilde{K}_0(\mathcal{C}^0(X))$ is not, in general, injective: the counter-example

$$X = \{(y,z) \in \mathbb{R}^2 \mid y^2 + z^2 + z^4 = 1\}$$

is given in [130] (note that X is nonsingular and homeomorphic to S^1).

Example 12.3.7. a) Every topological vector bundle over the sphere S^n is topologically isomorphic to an algebraic vector bundle. Indeed, $\widetilde{KO}(S^n) = 0$ for $n \not\equiv 0, 1, 2, 4 \pmod{8}$ ([174], p. 109), i.e. every topological vector bundle is stably trivial in this case.

For $n \equiv 0, 1, 2, 4 \pmod{8}$, the homomorphism $\widetilde{K}_0(\mathcal{P}(S^n)) \to \widetilde{K}_0(\mathcal{C}^0(S^n))$ is surjective according to [131] (it is also injective according to [316]). This implies the surjectivity of $\widetilde{K}_0(\mathcal{R}(S^n)) \to \widetilde{K}_0(\mathcal{C}^0(S^n))$. The hypothesis of Theorem 12.3.6 (ii) is thus satisfied for the *standard* spheres.

b) We may ask whether every topological vector bundle over an affine real algebraic variety X homeomorphic to S^n is still isomorphic to an algebraic vector bundle. The answer is positive if $n \not\equiv 0, 1, 2, 4 \pmod{8}$, for the same reason as in a). For $n \equiv 0, 1, 2, 4 \pmod{8}$, in which case $\widetilde{KO}(S^n) \simeq \mathbb{Z}/2$ if $n \equiv 1, 2 \pmod{8}$ and $\widetilde{KO}(S^n) \simeq \mathbb{Z}$ if $n \equiv 0, 4 \pmod{8}$, the answer is not always positive (except for $n = 1, 2$; cf. Section 12.5). Indeed, for every positive integer k, there exists a nonsingular algebraic hypersurface $X_{4k} \subset \mathbb{R}^{4k+1}$ (resp. $\Sigma_{2k} \subset \mathbb{R}^{2k+1}$), diffeomorphic to S^{4k} (resp. S^{2k}), such that every algebraic $\mathbb{R}$- (resp. $\mathbb{C}$-)vector bundle over X_{4k} (resp. Σ_{2k}) is stably trivial (a more general result is stated in Theorem 12.6.12). In Example 12.6.9, we shall construct a real algebraic set $\Sigma \subset \mathbb{R}^5$, homeomorphic to S^2, and such that every algebraic $\mathbb{C}$-vector bundle over Σ is trivial.

c) Every topological vector bundle over $\mathbb{P}_n(\mathbb{R})$ is isomorphic to an algebraic vector bundle. Indeed, if X is an affine real algebraic variety homeomorphic to $\mathbb{P}_n(\mathbb{R})$, the following conditions are equivalent:

(i) Every topological vector bundle over X is topologically isomorphic to an algebraic vector bundle.

(ii) There exists a nonorientable algebraic vector bundle over X.

The fact that (ii) implies (i) is a straightforward consequence of Theorem 12.3.6 and the fact that $\widetilde{KO}(X)$ is generated by the class of the unique nontrivial line bundle (cf. [174], p. 223, Theorem 12.7). Condition (ii) is, of course, satisfied by the universal bundle $\gamma_{n,1}$ over $\mathbb{P}_n(\mathbb{R})$. Moreover, if X is nonsingular and homeomorphic to $\mathbb{P}_{2k}(\mathbb{R})$, condition (ii) is satisfied by the tangent bundle $T(X)$. However, for each positive integer k, there exists a nonsingular affine algebraic variety diffeomorphic to $\mathbb{P}_{2k+1}(\mathbb{R})$, for which conditions (i) and (ii) are not satisfied. It suffices to take an algebraic model X of $\mathbb{P}_{2k+1}(\mathbb{R})$ with $H^1_{\text{alg}}(X, \mathbb{Z}/2) = 0$. Such a model exists by Theorem 11.3.7.

Remark 12.3.8. We shall prove in the next section (Example 12.4.15) that compactness of X is essential in Theorem 12.3.3 and Corollary 12.3.4. It is not known whether an algebraic bundle over $\mathbb{R}^n$ is always algebraically trivial (i.e. whether every projective module of finite type over $\mathcal{R}(\mathbb{R}^n)$ is free), except when $n \leq 4$, in which case the answer is affirmative (cf. [36]). Since, by [205], $\widetilde{K}_0(\mathcal{R}(\mathbb{R}^n)) = \widetilde{K}_0(\mathcal{P}(\mathbb{R}^n)) = 0$, every algebraic bundle ξ over $\mathbb{R}^n$ is algebraically stably trivial (there exist $p, q \in \mathbb{N}$ such that $\xi \oplus \epsilon^p \simeq_{\text{alg}} \epsilon^q$).

12.4 Algebraic Approximation of $\mathcal{C}^\infty$ Hypersurfaces

In this section, we are concerned with line bundles and hypersurfaces defined by sections of such bundles. Recall that, if S is a topological space, we denote by $V^1(S)$ the group of isomorphism classes of topological line bundles over S. If S is a $\mathcal{C}^\infty$ manifold, we obtain a canonically isomorphic group by taking $\mathcal{C}^\infty$ isomorphism classes of $\mathcal{C}^\infty$ line bundles.

Given a paracompact topological space S, let

$$w_1 : V^1(S) \longrightarrow H^1(S, \mathbb{Z}/2)$$

be the mapping which assigns to the class of a line bundle ξ over S its first Stiefel-Whitney class $w_1(\xi)$. Recall that, if $(g_{i,j} : U_i \cap U_j \to \mathbb{R}^*)_{i,j \in I}$ is a system of transition functions for ξ, then $w_1(\xi)$ is the cohomology class represented by the 1-cocycle $(c_{i,j} : U_i \cap U_j \to \mathbb{Z}/2)_{i,j \in I}$ defined by $c_{i,j}(x) = 0$ (resp. 1) if $g_{i,j}(x) > 0$ (resp. $g_{i,j}(x) < 0$). It is known and easy to verify that w_1 is a well-defined group isomorphism.

Our goal now is to prove that, for a compact nonsingular real algebraic variety X over $\mathbb{R}$, one has $w_1(V^1_{\text{alg}}(X)) = H^1_{\text{alg}}(X, \mathbb{Z}/2)$. Here we consider $V^1_{\text{alg}}(X)$ as a subgroup of $V^1(X)$ (cf. Theorem 12.3.3). First we need some preparation.

Assume that S is a compact $\mathcal{C}^\infty$ manifold of dimension d. Given a vector bundle ξ over S and a section σ of ξ, we denote by

$$\sigma^{-1}(0) = \{x \in S \mid \sigma(x) = 0\}$$

the zero set of σ. Assume further that ξ is a $\mathcal{C}^\infty$ line bundle. We can choose a $\mathcal{C}^\infty$ section σ of ξ transverse to the zero section (for the notion of transversality, see [160]). The set $\sigma^{-1}(0)$ is then a compact $\mathcal{C}^\infty$ hypersurface of S (or $\sigma^{-1}(0) = \emptyset$). As usual, $[\sigma^{-1}(0)]$ denotes the homology class in $H_{d-1}(S, \mathbb{Z}/2)$ represented by $\sigma^{-1}(0)$. We claim that the homology class $[\sigma^{-1}(0)]$ is independent of the choice of σ.

Let σ_1, σ_2 be two $\mathcal{C}^\infty$ sections of ξ, transverse to the zero section. One sees easily that a small perturbation of σ_2 does not change the class $[\sigma_2^{-1}(0)] \in H_{d-1}(S, \mathbb{Z}/2)$. Hence, we may assume that $\sigma_1^{-1}(0)$ is transverse to $\sigma_2^{-1}(0)$. Since $V^1(S)$ is a $\mathbb{Z}/2$-vector space, the bundle $\xi \otimes \xi$ is trivial. Therefore, if $\xi \otimes \xi = (E, p, S)$, there exists a $\mathcal{C}^\infty$ function $h : E \to \mathbb{R}$ whose restriction to each fibre $p^{-1}(x)$ is a linear isomorphism. Set $f = h \circ (\sigma_1 \otimes \sigma_2) : S \to \mathbb{R}$. We have $f^{-1}(0) = \sigma_1^{-1}(0) \cup \sigma_2^{-1}(0)$, and $[\sigma_1^{-1}(0)] + [\sigma_2^{-1}(0)] = [f^{-1}(0)] = 0$

in $H_{d-1}(S,\mathbb{Z}/2)$, since $f^{-1}(0)$ is the boundary of the compact manifold with corners $f^{-1}([0,+\infty[)$. Hence, $[\sigma_1^{-1}(0)] = [\sigma_2^{-1}(0)]$, which proves the claim.

Denote by
$$\phi : V^1(S) \longrightarrow H_{d-1}(S,\mathbb{Z}/2)$$
the mapping defined by $\phi(\xi) = [\sigma^{-1}(0)]$, where σ is a $\mathcal{C}^\infty$ section of ξ transverse to the zero section (since no confusion is possible, we make no distinction between ξ and its isomorphism class in $V^1(S)$).

Theorem 12.4.1. *With the notation as above,*
$$\phi = D \circ w_1 ,$$
where $D : H^1(S,\mathbb{Z}/2) \to H_{d-1}(S,\mathbb{Z}/2)$ is the Poincaré duality isomorphism. In particular, ϕ is an isomorphism.

Proof. Given a $\mathcal{C}^\infty$ line bundle ξ over S, choose a $\mathcal{C}^\infty$ mapping $f : S \to \mathbb{P}_n(\mathbb{R})$ such that $f^*(\gamma) \simeq_{\mathcal{C}^\infty} \xi$, where $\gamma = \gamma_{n,1}$ is the line bundle over $\mathbb{P}_n(\mathbb{R})$ defined in Proposition 12.1.4. Without loss of generality, we may assume that f is transverse to $\mathbb{P}_{n-1}(\mathbb{R}) \subset \mathbb{P}_n(\mathbb{R})$. Let σ be a $\mathcal{C}^\infty$ section of γ transverse to the zero section and such that $\sigma^{-1}(0) = \mathbb{P}_{n-1}(\mathbb{R})$. Then the section s of $f^*(\gamma)$ defined by $s(x) = (\sigma(f(x)),x)$, for $x \in S$, is transverse to the zero section. It follows from properties (ii) and (iv) of the Stiefel-Whitney classes (recalled in Section 12.1) that
$$w_1(\xi) = w_1(f^*(\gamma)) = f^*(u) ,$$
where $u \in H^1(\mathbb{P}_n(\mathbb{R}),\mathbb{Z}/2)$ is the Poincaré dual of the class represented by $\mathbb{P}_{n-1}(\mathbb{R})$. To finish the proof, it suffices to show that $D(f^*(u)) = [s^{-1}(0)]$. This, however, is an immediate consequence of the following result: if $\varphi : M \to N$ is a $\mathcal{C}^\infty$ mapping between compact $\mathcal{C}^\infty$ manifolds, and φ is transverse to a compact $\mathcal{C}^\infty$ submanifold Y of N, then $[\varphi^{-1}(Y)] = D_M(\varphi^*(D_N^{-1}([Y])))$ (cf. [144]). Since $s^{-1}(0) = f^{-1}(\mathbb{P}_{n-1}(\mathbb{R}))$, the theorem follows. □

Remark 12.4.2. Theorem 12.4.1 shows that every element of $H_{d-1}(S,\mathbb{Z}/2)$ can be represented by a compact $\mathcal{C}^\infty$ hypersurface of S.

Remark 12.4.3. Let Y be a closed $\mathcal{C}^\infty$ hypersurface of S. Using local equations of Y and proceeding in the same way as in Remark 12.2.5, we can construct a $\mathcal{C}^\infty$ line bundle ξ over S and a $\mathcal{C}^\infty$ section s of ξ transverse to the zero section, such that $Y = s^{-1}(0)$. If ξ' is a $\mathcal{C}^\infty$ line bundle over S such that $\phi(\xi') = [Y]$, then Theorem 12.4.1 shows that $\xi \simeq_{\mathcal{C}^\infty} \xi'$. Hence, ξ' has a $\mathcal{C}^\infty$ section s' transverse to the zero section and such that $Y = s'^{-1}(0)$. In particular, if Y is a $\mathcal{C}^\infty$ hypersurface of S, with $[Y] = 0$ in $H_{d-1}(S,\mathbb{Z}/2)$, there exists a $\mathcal{C}^\infty$ function $h : S \to \mathbb{R}$ such that 0 is a regular value of h and $h^{-1}(0) = Y$.

For the remainder of this section, we assume that X *is a compact affine real algebraic variety over* $\mathbb{R}$.

Before showing that $\phi(V^1_{\text{alg}}(X)) = H^{\text{alg}}_{d-1}(X, \mathbb{Z}/2)$ and, therefore, that $w_1(V^1_{\text{alg}}(X)) = H^1_{\text{alg}}(X, \mathbb{Z}/2)$, we need a result which is interesting in its own right.

Proposition 12.4.4. *Assume that X is irreducible and nonsingular of dimension d. Let $\mathfrak{p}_1, \ldots, \mathfrak{p}_q$ be height 1 prime ideals of $\mathcal{R}(X)$, and set $Z_i = \mathcal{Z}_X(\mathfrak{p}_i)$. If $\sum_{i=1}^q n_i[Z_i] = 0$ in $H_{d-1}(X, \mathbb{Z}/2)$, where $n_1, \ldots, n_q$ are positive integers, then the ideal $\mathfrak{p}_1^{n_1} \cdots \mathfrak{p}_q^{n_q}$ is principal.*

Proof. Since X is nonsingular and the $\mathfrak{p}_i$ are of height 1, we can find a finite covering of X by Zariski open sets U_ℓ such that, for every $i = 1, \ldots, q$, there exists $f_{i,\ell} \in \mathfrak{p}_i$ generating $\mathfrak{p}_i \mathcal{R}(U_\ell)$. For each ℓ, we choose semi-algebraic subsets F'_ℓ, F_ℓ, U'_ℓ of X such that F_ℓ and F'_ℓ are closed, U'_ℓ is open, $F'_\ell \subset U'_\ell \subset F_\ell \subset U_\ell$ and the F'_ℓ cover X. Let $\Psi : |K| \to X$ be a semi-algebraic triangulation of X compatible with the F'_ℓ, U'_ℓ, F_ℓ and the Z_i. By abuse of language, we identify a simplex σ of K with its image by Ψ. Let σ be a d-simplex of K. Set

$$S(\sigma) = \bigcup\{\tau \mid \tau \text{ simplex of } K, \ \tau \cap \sigma \neq \emptyset\}.$$

If $\sigma \subset F'_\ell$, then $S(\sigma) \subset F_\ell$. For each d-simplex σ and each $i = 1, \ldots, q$, we can choose $f_{i,\sigma} \in \mathfrak{p}_i$ such that, for every point x in $S(\sigma)$, $f_{i,\sigma}$ generates $\mathfrak{p}_i \mathcal{R}_{X,x}$. It suffices to take $f_{i,\sigma} = f_{i,\ell}$ for some ℓ such that $\sigma \subset F'_\ell$.

Note that if τ is a $(d-1)$-simplex of K contained in $S(\sigma)$, then the sign of $f_{i,\sigma}$ changes when crossing τ if and only if $\tau \subset Z_i$. Indeed, if $\tau \not\subset Z_i$, then $f_{i,\sigma}$ has the same sign on both sides of τ, and, if $\tau \subset Z_i$, then τ necessarily contains some zero, say v, of $f_{i,\sigma}$ such that $d_v f_{i,\sigma} \neq 0$.

Let $\tau_1, \ldots, \tau_s$ be the $(d-1)$-simplices of K contained in the union of all Z_i such that n_i is odd. It follows from Proposition 11.3.1 that the chain $\tau_1 + \cdots + \tau_s$ is a cycle. Moreover, since $\sum_{i=1}^q n_i[Z_i] = 0 \in H_{d-1}(X, \mathbb{Z}/2)$, this cycle is a boundary. Let $\sigma_1, \ldots, \sigma_t$ be d-simplices of K such that $\tau_1 + \cdots + \tau_s$ is the boundary of the chain $\sigma_1 + \cdots + \sigma_t$. Set $f_\sigma = \pm f_{1,\sigma}^{n_1} \cdots f_{q,\sigma}^{n_q}$, where the sign of f_σ is chosen in such a way that f_σ is positive on σ if and only if σ occurs among $\sigma_1, \ldots, \sigma_t$. With this choice, we have, by the preceding remark, $f_\sigma / f_{\sigma'} > 0$ on $S(\sigma) \cap S(\sigma')$. Using the f_σ we shall construct a generator of the ideal $I = \mathfrak{p}_1^{n_1} \cdots \mathfrak{p}_q^{n_q}$.

For d-simplices σ and σ', we denote $f_\sigma / f_{\sigma'}$ by $h_{\sigma\sigma'}$. Note that, for $x \in \sigma'$, we have $h_{\sigma\sigma'} \in \mathcal{R}_{X,x}$ since $f_{\sigma'}$ generates $I\mathcal{R}_{X,x}$. Let ϵ be a positive real number such that, for all d-simplices σ and σ' and every $x \in \sigma'$, we have $r|h_{\sigma\sigma'}(x)| \leq 1/\epsilon$, where r is the total number of d-simplices of K. Applying the Stone-Weierstrass theorem, we can choose, for each d-simplex σ, a regular function $\varphi_\sigma \in \mathcal{R}(X)$ such that $\varphi_\sigma > 0$ on X, $\varphi_\sigma > 1$ on σ and $\varphi_\sigma < \epsilon$ outside $S(\sigma)$. Let $f = \sum_\sigma \varphi_\sigma f_\sigma$. We claim that, for every point x in X, f generates $I\mathcal{R}_{X,x}$. Indeed, let σ' be a d-simplex such that $x \in \sigma'$. Let

12.4 Algebraic Approximation of C^∞ Hypersurfaces

$$L_1 = \{\sigma \mid \sigma \text{ is a } d\text{-simplex of } K, \ x \in S(\sigma)\}$$
$$L_2 = \{\sigma \mid \sigma \text{ is a } d\text{-simplex of } K, \ x \notin S(\sigma)\}.$$

Then
$$f(x) = f_{\sigma'}(x)(\sum_{\sigma \in L_1} \varphi_\sigma(x) h_{\sigma\sigma'}(x) + \sum_{\sigma \in L_2} \varphi_\sigma(x) h_{\sigma\sigma'}(x)).$$

If $\sigma \in L_1$, then $h_{\sigma\sigma'}(x)$ is positive. Hence, $\sum_{\sigma \in L_1} \varphi_\sigma(x) h_{\sigma\sigma'}(x) > \varphi_{\sigma'}(x) > 1$. According to the choice of ϵ, we have $\sum_{\sigma \in L_2} \varphi_\sigma(x)|h_{\sigma\sigma'}(x)| < 1$. Therefore $f/f_{\sigma'}$ is invertible in $\mathcal{R}_{X,x}$, which proves the claim. Now, if $g \in I$, then $g/f \in \mathcal{R}_{X,x}$ for every x in X. Hence, by Proposition 3.2.3, $g/f \in \mathcal{R}(X)$. □

We say that *the sign of a function* $f : X \to \mathbb{R}$ *changes at* $x \in X$ if, for every neighbourhood U of x, there exist points y_1 and y_2 in U such that $f(y_1)f(y_2) < 0$. Recall that, if Z is a semi-algebraic set and k a nonnegative integer, $Z^{(k)}$ denotes the subset of $x \in Z$ such that $\dim(Z_x) = k$.

Corollary 12.4.5. *Let X be a compact nonsingular affine real algebraic variety of dimension d. Let Z be a Zariski closed subset of X of dimension $d - 1$. Then $[Z] = 0$ in $H_{d-1}(X, \mathbb{Z}/2)$ if and only if there exists a function $f \in \mathcal{R}(X)$ such that $Z^{(d-1)}$ is equal to the set of points of X where the sign of f changes.*

Proof. If $Z^{(d-1)}$ is the set of points where the sign of $f \in \mathcal{R}(X)$ changes, then $[Z] = 0$ in $H_{d-1}(X, \mathbb{Z}/2)$. Indeed, take a semi-algebraic triangulation of X compatible with $f^{-1}(0)$. Denote by $\tau_1, \ldots, \tau_k$ (resp. $\sigma_1, \ldots, \sigma_\ell$) the $(d-1)$-simplices (resp. d-simplices) contained in $Z^{(d-1)}$ (resp. $f^{-1}([0, +\infty[)$). Then the cycle $\tau_1 + \cdots + \tau_k$ representing $[Z]$ is the boundary of the chain $\sigma_1 + \cdots + \sigma_\ell$.

Conversely, assume $[Z] = 0$ in $H_{d-1}(X, \mathbb{Z}/2)$. Without loss of generality, we may assume that all irreducible components of Z are of dimension $d-1$, since removing irreducible components of dimension smaller than $d-1$ changes neither $[Z]$ nor $Z^{(d-1)}$. By Proposition 12.4.4, the ideal $\mathcal{I}_{\mathcal{R}(X)}(Z)$ is then principal. Let $f \in \mathcal{R}(X)$ be a generator of $\mathcal{I}_{\mathcal{R}(X)}(Z)$. The set of points where the sign of f changes is closed in Z, contains the nonsingular points of Z and is of local dimension $d-1$ at each point (cf. Theorem 4.5.1 (iv) ⇒ (v)). This set is therefore precisely $Z^{(d-1)}$. □

Theorem 12.4.6. *Let X be a compact nonsingular affine real algebraic variety of dimension d. Given an element α of $H_{d-1}(X, \mathbb{Z}/2)$, the following properties are equivalent:*

(i) *α is the image by $\phi : V^1(X) \to H_{d-1}(X, \mathbb{Z}/2)$ of a class represented by an algebraic line bundle.*

(ii) *There exists a nonsingular algebraic hypersurface $Y \subset X$ such that $\alpha = [Y]$.*

(iii) *$\alpha \in H^{\text{alg}}_{d-1}(X, \mathbb{Z}/2)$.*

In particular, ϕ induces an isomorphism

$$V^1_{\text{alg}}(X) \longrightarrow H^{\text{alg}}_{d-1}(X, \mathbb{Z}/2),$$

and $w_1(V^1_{\text{alg}}(X)) = H^1_{\text{alg}}(X, \mathbb{Z}/2)$.

Proof. We may assume that X is irreducible. In what follows, we often make no distinction between a line bundle and its isomorphism class in $V^1(X)$.

(i) $\Rightarrow$ (ii). Let ξ be an algebraic line bundle over X such that $\phi(\xi) = \alpha$. There is an algebraic section s of ξ transverse to the zero section. Indeed, we can first choose any $\mathcal{C}^\infty$ section of ξ transverse to the zero section and then approximate it close enough by an algebraic section (cf. Theorem 12.3.2). Then $Y = s^{-1}(0)$ is a nonsingular algebraic hypersurface of X, and $[Y] = \alpha$ by definition of ϕ.

(ii) $\Rightarrow$ (iii) is obvious.

(iii) $\Rightarrow$ (i). We shall define a homomorphism

$$\psi : H^{\text{alg}}_{d-1}(X, \mathbb{Z}/2) \to V^1_{\text{alg}}(X)$$

such that $\psi \circ \phi(\xi) = \xi$ for each ξ in $V^1_{\text{alg}}(X)$. If $\alpha \in H^{\text{alg}}_{d-1}(X, \mathbb{Z}/2)$, there exists an algebraic hypersurface Y of X such that $\alpha = [Y]$. Let $Z_1, \ldots, Z_q$ be the irreducible components of dimension $d-1$ of Y, and let $\mathfrak{p}_i = \mathcal{I}_{\mathcal{R}(X)}(Z_i)$, for $i = 1, \ldots, q$. We have $[Y] = [Z_1] + \cdots + [Z_q]$. The ideal $\mathfrak{p}_1 \cdots \mathfrak{p}_q$ represents an element of the divisor class group $\text{Cl}(\mathcal{R}(X))$. The image of this element by the canonical isomorphism

$$\text{Cl}(\mathcal{R}(X)) \to V^1_{\text{alg}}(X)$$

(cf. Corollary 12.2.4 and Remark 12.2.5) is, by definition, the line bundle $\psi(\alpha)$. First, we have to verify that ψ is well-defined. Assume that $[Y] = [Y']$, where Y' is another algebraic hypersurface of X, and let $Z'_1, \ldots, Z'_r$ be the irreducible components of dimension $d-1$ of Y'. Set $\mathfrak{p}'_j = \mathcal{I}_{\mathcal{R}(X)}(Z'_j)$ for $j = 1, \ldots, r$. Since $[Y] + [Y'] = 0$ in $H_{d-1}(X, \mathbb{Z}/2)$, Proposition 12.4.4 implies that the ideal $\mathfrak{p}_1 \cdots \mathfrak{p}_q \mathfrak{p}'_1 \cdots \mathfrak{p}'_r$ is principal. Hence, $\mathfrak{p}_1 \cdots \mathfrak{p}_q$ and $\mathfrak{p}'_1 \cdots \mathfrak{p}'_r$ both represent the same class in $\text{Cl}(\mathcal{R}(X))$.

Now let ξ be an algebraic line bundle over X. Let s be an algebraic section of ξ transverse to the zero section and such that $\phi(\xi) = [s^{-1}(0)]$. If $g_{ij} \in \mathcal{R}(U_i \cap U_j)$ are transition functions of ξ corresponding to an appropriate Zariski open covering $(U_i)_{i=1,\ldots,q}$ of X, then s is given by a system $(s_1, \ldots, s_q)$ with $s_i \in \mathcal{R}(U_i)$ and $s_i = g_{ij} s_j$ on $U_i \cap U_j$. All irreducible components $Z_1, \ldots, Z_r$ of $s^{-1}(0)$ are nonsingular of dimension $d-1$. If $\mathfrak{q}_\ell = \mathcal{I}_{\mathcal{R}(X)}(Z_\ell)$ and $I = \mathfrak{q}_1 \cdots \mathfrak{q}_r$, then $I\mathcal{R}(U_i) = s_i \mathcal{R}(U_i)$ and, by Remark 12.2.5, the image of the class of I by the canonical isomorphism $\text{Cl}(\mathcal{R}(X)) \to V^1_{\text{alg}}(X)$ is the bundle with transition functions $h_{ij} = s_j/s_i = g_{ji}$, i.e. the dual bundle $\xi^\vee$. Therefore, the image of the class of ξ by $\psi \circ \phi$ is the class of $\xi^\vee$. Since $\xi \otimes \xi^\vee \simeq_{\text{alg}} \epsilon^1_X$, these classes are equal in $V^1_{\text{alg}}(X)$. This shows that $\psi \circ \phi(\xi) = \xi$.

In order to prove that $\phi|_{V^1_{\text{alg}}(X)}$ is an isomorphism onto $H^{\text{alg}}_{d-1}(X, \mathbb{Z}/2)$ whose inverse is ψ, it only remains to verify that ψ is injective. If $\psi([Y]) = 0$, then the ideal $\mathfrak{p}_1 \cdots \mathfrak{p}_q$ is principal (with the same notation as above). Let $f \in \mathcal{R}(V)$ be a generator of $\mathfrak{p}_1 \cdots \mathfrak{p}_q$. Choose a semi-algebraic triangulation $\Phi : |K| \to X$ compatible with Y. We identify the simplices of K with their images by Φ. Let $\tau_1, \ldots, \tau_s$ be the $(d-1)$-simplices contained in Y, and $\sigma_1, \ldots, \sigma_t$, the d-simplices where f is nonnegative. Then the cycle $\tau_1 + \cdots + \tau_s$ is the boundary of the chain $\sigma_1 + \cdots + \sigma_t$. Since $[Y]$ is the class of $\tau_1 + \cdots + \tau_s$ in $H_{d-1}^{\text{alg}}(X, \mathbb{Z}/2)$, we have $[Y] = 0$. This shows that ψ is injective.

Finally, since $w_1 = D^{-1} \circ \phi$ (cf. Theorem 12.4.1), it follows that $w_1(V_{\text{alg}}^1(X)) = H_{\text{alg}}^1(X, \mathbb{Z}/2)$. This completes the proof. □

Remark 12.4.7. Let $f : X \to Y$ be a regular mapping between compact nonsingular affine real algebraic varieties. Theorem 11.3.4(ii) implies, in particular, that $f^*(H_{\text{alg}}^1(Y, \mathbb{Z}/2)) \subset H_{\text{alg}}^1(X, \mathbb{Z}/2)$. This can now be seen independently, by observing that, for ξ in $V_{\text{alg}}^1(Y)$, the induced bundle $f^*(\xi)$ is in $V_{\text{alg}}^1(X)$, and using the isomorphism $V_{\text{alg}}^1(X) \to H_{\text{alg}}^1(X, \mathbb{Z}/2)$ defined by w_1 (cf. Theorem 12.4.6).

It should be mentioned that the implication (iii) ⇒ (ii) in Theorem 12.4.6 is no longer valid, in general, in codimension > 1. The paper [319] contains an example of a compact nonsingular affine real algebraic variety X of dimension 9, for which a homology class in $H_7^{\text{alg}}(X, \mathbb{Z}/2)$ is not representable by a compact $\mathcal{C}^\infty$ submanifold of X.

Corollary 12.4.8. *If X is a compact nonsingular nonorientable affine real algebraic variety of dimension d, then $H_{d-1}^{\text{alg}}(X, \mathbb{Z}/2) \neq 0$.*

Proof. Follows from Proposition 12.2.7 and Theorem 12.4.6. □

Observe that Corollary 12.4.8 follows also from Theorem 12.1.10

By applying Theorem 12.4.6, the results stated in Theorems 11.3.7 and 11.3.8 can be interpreted in terms of V_{alg}^1. For example, one deduces immediately the following.

Theorem 12.4.9. *Given a compact connected $\mathcal{C}^\infty$ manifold M of dimension ≥ 2, the following conditions are equivalent:*
  (i) *M has an algebraic model X with $V_{\text{alg}}^1(X) = 0$ and $\mathcal{R}(X)$ factorial.*
  (ii) *M is orientable.*

This result shows, in particular, that there exist nonsingular compact affine real algebraic varieties with topological vector bundles which are not isomorphic to algebraic vector bundles. However, according to the paper [33], for every compact $\mathcal{C}^\infty$ manifold M, there exists an algebraic model X of M such that every topological vector bundle over X is topologically isomorphic to an algebraic vector bundle.

We now prove the main result of this section, concerning algebraic approximation of compact $\mathcal{C}^\infty$ hypersurfaces of a nonsingular affine real algebraic variety.

Definition 12.4.10. *A compact C^∞ submanifold Y of a nonsingular affine real algebraic variety X is said to have an* algebraic approximation *in X if, for every open neighbourhood Ω of the inclusion mapping $Y \hookrightarrow X$ in $C^\infty(Y,X)$ (with respect to the C^∞ topology), there exists $h \in \Omega$ such that $h(Y)$ is a nonsingular Zariski closed subset of X.*

Recall that, given a C^∞ manifold M, a C^∞ mapping $F: M \times [0,1] \to M$ is said to be a *diffeotopy* of M if, for every $t \in [0,1]$, the mapping $F_t : M \to M$ defined by $F_t(x) = F(x,t)$ is a diffeomorphism of M and F_0 is the identity mapping. Two C^∞ submanifolds N_0 and N_1 of M are said to be *diffeotopic* if there exists a diffeotopy F of M, such that $N_1 = F_1(N_0)$.

Note that, if Ω in Definition 12.4.10 is small enough, every mapping h in Ω is an embedding and the submanifolds Y and $h(Y)$ are diffeotopic in X (cf. [160], pp. 38 and 180).

Theorem 12.4.11. *Let X be a compact nonsingular affine real algebraic variety of dimension d. Given a compact C^∞ hypersurface Y of X, the following conditions are equivalent:*
  (i) *The homology class $[Y]$ represented by Y belongs to $H_{d-1}^{\mathrm{alg}}(X, \mathbb{Z}/2)$.*
  (ii) *Y has an algebraic approximation in X.*
  (iii) *There exists a diffeotopy F of X, with F_1 arbitrarily close to the identity, such that $F_1(Y)$ is a nonsingular algebraic hypersurface of X.*

Proof. (i) $\Rightarrow$ (iii). If $[Y] \in H_{d-1}^{\mathrm{alg}}(X, \mathbb{Z}/2)$, there exists an algebraic line bundle ξ over X such that $\phi(\xi) = [Y]$ (cf. Theorem 12.4.6). According to Remark 12.4.3, there exists a C^∞ section σ of ξ transverse to the zero section and such that $Y = \sigma^{-1}(0)$. By Theorem 12.3.2, we can choose an algebraic section s of ξ arbitrarily close to σ with respect to the C^∞ topology. If s is close enough to σ, the set $s^{-1}(0)$ is a nonsingular algebraic hypersurface of X, and there exists a diffeotopy F of X, with F_1 arbitrarily close to the identity, such that $F_1(Y) = s^{-1}(0)$.

(ii) $\Rightarrow$ (iii) is a matter of differential topology for which we refer the reader to [160], Theorem 1.6, p. 181.

The implications (iii) $\Rightarrow$ (i) and (iii) $\Rightarrow$ (ii) are obvious. $\square$

We shall now apply Theorem 12.4.11 to the problem of constructing algebraic varieties with $H_{\mathrm{alg}}^1 \neq H^1$ (cf. Section 11.3).

Example 12.4.12. Let us consider again the nonsingular algebraic set X constructed in Example 11.3.9 (we use the notation of this example). We shall indicate a different way of showing that, in the case when X is diffeomorphic to $S^n \times S^1$, the class $[D]$ represented by the hypersurface D is not in $H_n^{\mathrm{alg}}(X, \mathbb{Z}/2)$. Indeed, if $[D]$ is in $H_n^{\mathrm{alg}}(X, \mathbb{Z}/2)$, then, by Theorem 12.4.11, there exists a diffeomorphism $g : X \to X$, which can be chosen arbitrarily close to the identity, such that $g(D)$ is a nonsingular algebraic hypersurface of X. If g is taken sufficiently close to the identity, the projection $\pi : W \to C$ maps $g(D)$ bijectively onto the connected component C_1 of C. Since $\pi|_{g(D)} : g(D) \to C$ is then injective and regular, and $\dim(C \setminus C_1) = \dim(C_2) = n$, one gets a contradiction to Lemma 11.4.3. Hence $[D] \notin H_n^{\mathrm{alg}}(X, \mathbb{Z}/2)$.

Example 12.4.13. Let X be a compact nonsingular affine real algebraic variety of dimension $d \geq 2$, and let $\sigma : Y = E(X, x) \to X$ be the blowing up of X with center $x \in X$. As a C^∞ manifold, Y is diffeomorphic to the connected sum of X and $\mathbb{P}_d(\mathbb{R})$. Let $\beta(X)$ denote $\dim_{\mathbb{Z}/2} H_{\mathrm{alg}}^1(X, \mathbb{Z}/2)$. We claim that $\beta(Y) = \beta(X) + 1$.

Since $\sigma^* : H^1(X, \mathbb{Z}/2) \to H^1(Y, \mathbb{Z}/2)$ is injective and maps $H_{\mathrm{alg}}^1(X, \mathbb{Z}/2)$ into $H_{\mathrm{alg}}^1(Y, \mathbb{Z}/2)$, and the cohomology class corresponding to $[\sigma^{-1}(x)]$ is in $H_{\mathrm{alg}}^1(Y, \mathbb{Z}/2) \setminus \sigma^*(H^1(X, \mathbb{Z}/2))$, it follows that $\beta(Y) \geq \beta(X) + 1$. To prove the claim, it suffices to show that, if $D \subset X$ is a C^∞ compact hypersurface with $[D] \notin H_{d-1}^{\mathrm{alg}}(X, \mathbb{Z}/2)$ and $x \notin D$, the class $[\sigma^{-1}(D)]$ represented by the hypersurface $\sigma^{-1}(D)$ is not in $H_{d-1}^{\mathrm{alg}}(Y, \mathbb{Z}/2)$. If $[\sigma^{-1}(D)]$ were in $H_{d-1}^{\mathrm{alg}}(Y, \mathbb{Z}/2)$, we could find, by Theorem 12.4.11, a nonsingular real algebraic hypersurface Z of Y so close to $\sigma^{-1}(D)$ that $Z \cap \sigma^{-1}(x) = \emptyset$ and $[\sigma(Z)] = [D]$ in $H_{d-1}(X, \mathbb{Z}/2)$. Since the Zariski closure of $\sigma(Z)$ is contained in $\sigma(Z) \cup \{x\}$, that would imply $[D] \in H_{d-1}^{\mathrm{alg}}(X, \mathbb{Z}/2)$. Hence $\beta(Y) = \beta(X) + 1$.

We now return to the question of the factoriality of $\mathcal{R}(X)$. The isomorphism
$$H_{\mathrm{alg}}^1(X, \mathbb{Z}/2) \simeq H_{d-1}^{\mathrm{alg}}(X, \mathbb{Z}/2)$$
allows one to complete the statement of Corollary 12.2.6.

Proposition 12.4.14. *Given a compact irreducible nonsingular affine real algebraic variety X of dimension d, the following properties are equivalent:*
  (i) $\mathcal{R}(X)$ *is factorial.*
  (ii) $H_{d-1}^{\mathrm{alg}}(X, \mathbb{Z}/2) = 0$.

To conclude this section, we shall show that Theorem 12.3.3, Corollary 12.3.4 and Proposition 12.4.14 are no longer valid for noncompact algebraic varieties.

Example 12.4.15. There exists a nonsingular irreducible affine real algebraic surface V such that
  (i) V has two connected components, each diffeomorphic to $\mathbb{R}^2$ (hence, $H^1(V, \mathbb{Z}/2) = H_1(V, \mathbb{Z}/2) = 0$),
  (ii) $\mathrm{Cl}(\mathcal{R}(V)) \neq 0$ (hence, $\mathcal{R}(V)$ is not factorial and $V_{\mathrm{alg}}^1(V) \neq 0$).

To construct V, we proceed as follows. Assume that there exist a nonsingular affine real algebraic variety X and a C^∞ diffeomorphism $\varphi : S^1 \times S^1 \to X$, such that $Z_1 = \varphi(S^1 \times \{x_1\})$, $Z_2 = \varphi(\{x_2\} \times S^1)$ and $Z = \varphi(\{x_3, x_4\} \times S^1)$ are irreducible nonsingular algebraic curves in X (where $x_1, \ldots, x_4$ are distinct points of S^1). Define V to be the Zariski open subset $X \setminus (Z_1 \cup Z)$.

Clearly, both connected components of V are diffeomorphic to $\mathbb{R}^2$. We shall prove that V also has property (ii).

Set $\mathfrak{p}_i = \mathcal{I}_{\mathcal{R}(X)}(Z_i)$ for $i = 1, 2$, $\mathfrak{p} = \mathcal{I}_{\mathcal{R}(X)}(Z)$. The group $H_1^{\text{alg}}(X, \mathbb{Z}/2)$ is isomorphic to $\mathbb{Z}/2 \times \mathbb{Z}/2$ and generated by $[Z_1]$ and $[Z_2]$. It follows, using the isomorphism $H_1^{\text{alg}}(X, \mathbb{Z}/2) \to V_{\text{alg}}^1(X) \to \text{Cl}(\mathcal{R}(X))$, that the group $\text{Cl}(\mathcal{R}(X))$ is isomorphic to $\mathbb{Z}/2 \times \mathbb{Z}/2$ and generated by the classes of $\mathfrak{p}_1$ and $\mathfrak{p}_2$. Moreover, since $[Z] = 0$, the class of $\mathfrak{p}$ is null. Recall that $\mathcal{R}(V)$ is the ring of fractions of $\mathcal{R}(X)$ for the multiplicative subset

$$S = \{f \in \mathcal{R}(X) \mid f^{-1}(0) \subset Z_1 \cup Z\}.$$

It follows, by Proposition 12.2.2, that the canonical homomorphism

$$\text{Cl}(\mathcal{R}(X)) \to \text{Cl}(\mathcal{R}(V))$$

is surjective and its kernel is generated by the classes of height 1 real prime ideals of $\mathcal{R}(X)$ having nonempty intersection with S. The only such ideals are $\mathfrak{p}_1$ and $\mathfrak{p}$. The kernel of the homomorphism $\text{Cl}(\mathcal{R}(X)) \to \text{Cl}(\mathcal{R}(V))$ is thus generated by the class of $\mathfrak{p}_1$, and therefore $\text{Cl}(\mathcal{R}(V))$ is isomorphic to $\mathbb{Z}/2$.

It remains to construct X with the required properties. First, we choose a $\mathcal{C}^\infty$ embedding $\varphi' : S^1 \times S^1 \to S^3$ such that $Z_1 = \varphi'(S^1 \times \{x_1\})$, $Z_2 = \varphi'(\{x_2\} \times S^1)$ and $Z = \varphi'(\{x_3, x_4\} \times S^1)$ are nonsingular irreducible algebraic curves in S^3. Let X' denote the $\mathcal{C}^\infty$ surface $\varphi'(S^1 \times S^1)$. There exists a $\mathcal{C}^\infty$ function $h' : S^3 \to \mathbb{R}$ such that 0 is a regular value of h' and $h'^{-1}(0) = X'$ (cf. Remark 12.4.3). Let $P_1, \ldots, P_k$ be regular functions on S^3 generating $\mathcal{I}_{\mathcal{R}(S^3)}(Z_1 \cup Z_2 \cup Z)$. Since Z_1, Z_2 and Z are nonsingular and intersect each other transversally, there are $\mathcal{C}^\infty$ functions $h'_1, \ldots, h'_k$ from S^3 to $\mathbb{R}$ such that $h' = \sum_{j=1}^k h'_j P_j$. Using the Stone-Weierstrass theorem, we can approximate each h'_j by a regular function h_j. Choosing h_j close enough to h'_j and setting $h = \sum_{j=1}^k h_j P_j$, we obtain a nonsingular algebraic surface $X = h^{-1}(0)$, diffeomorphic to X' and containing Z_1, Z_2 and Z.

12.5 Vector Bundles over Algebraic Curves and Surfaces

In this section we shall characterize those topological vector bundles over X which are isomorphic to algebraic vector bundles, in the case where X *is a compact nonsingular affine real algebraic curve or surface over* $\mathbb{R}$. We shall also state, without proof, a similar result concerning varieties of dimension 3.

Theorem 12.5.1. *Let X be a compact nonsingular affine real algebraic curve. Every topological vector bundle of constant rank over X is topologically isomorphic to an algebraic vector bundle.*

Proof. Since X is of dimension 1, every topological vector bundle ξ of rank $n+1$ over X is isomorphic to a Whitney sum $\xi_1 \oplus \epsilon_X^n$, where ϵ_X^n is the rank n trivial bundle and ξ_1 a line bundle. Hence, it suffices to prove the assertion for topological line bundles. Since the group $V^1(X)$ of isomorphism classes of line bundles over X is isomorphic to $H_0(X, \mathbb{Z}/2)$, and $H_0(X, \mathbb{Z}/2) = H_0^{\text{alg}}(X, \mathbb{Z}/2)$, Theorems 12.3.3 and 12.4.6 imply that the canonical homomorphism $V_{\text{alg}}^1(X) \to V^1(X)$ is an isomorphism. □

Corollary 12.5.2. *Let X be a compact connected nonsingular affine real algebraic curve. Then $\widetilde{K}_0(\mathcal{R}(X))$ is isomorphic to $\mathbb{Z}/2$.*

Proof. By Theorem 12.5.1, the canonical homomorphism $\widetilde{K}_0(\mathcal{R}(X)) \to \widetilde{K}_0(\mathcal{C}^0(X)) \simeq \widetilde{KO}(X)$ is an isomorphism. Since X is homeomorphic to a circle, $\widetilde{KO}(X)$ is isomorphic to $\mathbb{Z}/2$. □

For the study of vector bundles over algebraic surfaces, we shall use Stiefel-Whitney classes. Besides those properties already recalled in Section 12.1, we shall need the result that a topological vector bundle ξ is orientable if and only if $w_1(\xi) = 0$ (cf. [236] p. 148).

Theorem 12.5.3. *Let X be a compact nonsingular affine real algebraic surface. Let ξ be a topological vector bundle of constant rank over X. Then ξ is topologically isomorphic to an algebraic vector bundle over X if and only if $w_1(\xi) \in H_{\text{alg}}^1(X, \mathbb{Z}/2)$.*

Corollary 12.5.4. *Let X be a compact connected nonsingular affine real algebraic surface. Then every topological vector bundle over X is topologically isomorphic to an algebraic vector bundle if and only if $H_1(X, \mathbb{Z}/2) = H_1^{\text{alg}}(X, \mathbb{Z}/2)$.*

Proof. Follows from Theorems 12.4.6 and 12.5.3. □

Before giving the proof of Theorem 12.5.3, we need some preparation.

Lemma 12.5.5. *Let X be a compact nonsingular affine real algebraic variety, and Y, a nonsingular Zariski closed subset of X. Every $\mathcal{C}^\infty$ function $g : X \to \mathbb{R}$ vanishing on Y can be approximated in the $\mathcal{C}^\infty$ topology by regular functions $X \to \mathbb{R}$ also vanishing on Y.*

Proof. Let $h_1, \ldots, h_p$ be generators of the ideal $\mathcal{I}_{\mathcal{R}(X)}(Y)$. Since Y is nonsingular, we can represent the germ of g at a point $x \in X$ in the form $g_x = \lambda_{1,x} h_{1,x} + \cdots + \lambda_{p,x} h_{p,x}$, where the $\lambda_{i,x}$ are germs of $\mathcal{C}^\infty$ functions at x. Using a partition of unity, this allows us to represent g globally as $g = \lambda_1 h_1 + \cdots + \lambda_p h_p$, where $\lambda_i \in \mathcal{C}^\infty(X)$. Then it suffices to apply the Stone-Weierstrass theorem to the functions λ_i. □

Lemma 12.5.6. *Let X and Y be as in Lemma 12.5.5 and let k be the codimension of Y in X. If the normal bundle of Y in X is topologically trivial, there exist regular functions $f_1, \ldots, f_k \in \mathcal{R}(X)$ such that 0 is a regular value of $f = (f_1, \ldots, f_k) : X \to \mathbb{R}^k$ and $f^{-1}(0) = Y \cup Y'$, where Y' is a nonsingular Zariski closed subset of codimension k of X disjoint from Y (or $Y' = \emptyset$).*

Proof. Using a tubular neighbourhood of Y in X and the fact that the normal bundle of Y in X is trivial, we can choose $\mathcal{C}^\infty$ functions $g_1, \ldots, g_k$ defined in a neighbourhood U of Y in X, such that 0 is a regular value of $g = (g_1, \ldots, g_k) : U \to \mathbb{R}^k$ and $g^{-1}(0) = Y$. Then we take $\mathcal{C}^\infty$ functions $\bar{g}_i : X \to \mathbb{R}$ such that $\bar{g}_i = g_i$ in a neighbourhood of Y and such that 0 is still a regular value of $\bar{g} = (\bar{g}_1, \ldots, \bar{g}_k) : X \to \mathbb{R}^k$. By Lemma 12.5.5, there exists regular functions $f_1, \ldots, f_k$ on X such that 0 is a regular value of $f = (f_1, \ldots, f_k) : X \to \mathbb{R}^k$ and $f|_Y = 0$. The required property of $Y' = f^{-1}(0) \setminus Y$ follows from Proposition 3.3.17. □

Proposition 12.5.7. *Let X be a compact nonsingular affine real algebraic variety. Let $Y \subset X$ be a codimension 2 nonsingular Zariski closed subset of X, and assume that the normal bundle of Y in X is trivial. Then there exist an orientable algebraic vector bundle ξ of rank 2 over X and an algebraic section s of ξ transverse to the zero section, such that $Y = s^{-1}(0)$.*

Proof. By Lemma 12.5.6, there exists a regular mapping $f = (f_1, f_2) : X \to \mathbb{R}^2$ such that 0 is a regular value of f and $f^{-1}(0) = Y \cup Y'$, where Y' is a Zariski closed subset of X disjoint from Y. We choose two functions ψ_1 and ψ_2 in $\mathcal{R}(X)$ such that $Y = \psi_1^{-1}(0)$ and $Y' = \psi_2^{-1}(0)$. Since the ideal $\mathcal{I}_{\mathcal{R}(X)}(Y \cup Y')$ consisting of regular functions vanishing on $Y \cup Y'$ is generated by f_1 and f_2, there exist $h_1, h_2 \in \mathcal{R}(X)$ such that $\psi_1 \psi_2 = h_1 f_1 + h_2 f_2$. Consider the regular mappings $g_{2,1} : U_1 = X \setminus Y \to M_{2,2}(\mathbb{R})$ and $g_{1,2} : U_2 = X \setminus Y' \to M_{2,2}(\mathbb{R})$ defined by

$$g_{2,1} = \begin{bmatrix} f_1 \psi_2 \psi_1^{-1} & -h_2 \psi_1^{-2} \\ f_2 \psi_2 \psi_1^{-1} & h_1 \psi_1^{-2} \end{bmatrix}, \quad g_{1,2} = \begin{bmatrix} h_1 \psi_2^{-2} & h_2 \psi_2^{-2} \\ -f_2 \psi_1 \psi_2^{-1} & f_1 \psi_1 \psi_2^{-1} \end{bmatrix}.$$

For every $x \in U_1 \cap U_2$, one has $g_{1,2}(x) = (g_{2,1}(x))^{-1}$. Lemma 12.1.11 implies that the bundle $\xi = (E, p, X)$, where

$$E = \{(x, v_1, v_2) \in X \times \mathbb{R}^2 \times \mathbb{R}^2 \mid v_1 = g_{1,2}(x) \cdot v_2 \text{ if } x \in U_2$$
$$\text{and } v_2 = g_{2,1}(x) \cdot v_1 \text{ if } x \in U_1\}$$

and $p(x, v_1, v_2) = x$, is an algebraic vector bundle. Since $\det(g_{1,2}) = \psi_1^2 / \psi_2^2$ is positive on $U_1 \cap U_2$, the bundle ξ is orientable. The algebraic section s of ξ defined by

$$s = (\mathrm{Id}_X, (\psi_1, 0), (f_1 \psi_2, f_2 \psi_2))$$

is transverse to the zero section of ξ, and $Y = s^{-1}(0)$. □

12.5 Vector Bundles over Algebraic Curves and Surfaces 323

Corollary 12.5.8. *Let X and Y be as in Proposition 12.5.7. There exists an orientable algebraic vector bundle ξ of rank 2 over X, such that $w_2(\xi) \in H^2(X, \mathbb{Z}/2)$ corresponds, by Poincaré duality, to the homology class $[Y] \in H_{d-2}(X, \mathbb{Z}/2)$ (where $d = \dim(X)$).*

Proof. If ξ is an orientable $\mathcal{C}^\infty$ vector bundle of rank 2 over X and s is a $\mathcal{C}^\infty$ section transverse to the zero section of ξ, then $w_2(\xi)$ corresponds, by Poincaré duality, to $[s^{-1}(0)] \in H_{d-2}(X, \mathbb{Z}/2)$ ([299], p. 1006, Lemma 3). The existence of ξ follows from Proposition 12.5.7. □

Remark 12.5.9. In the case of surfaces, the set Y is finite and Corollary 12.5.8 can be proved using the following facts. Let X be a connected compact $\mathcal{C}^\infty$ surface. Then:

(i) If X is nonorientable, there exist exactly two nonisomorphic orientable $\mathcal{C}^\infty$ vector bundles of rank 2 over X. The nontrivial one is characterized by the existence of a $\mathcal{C}^\infty$ section transverse to the zero section and having an odd number of zeros.

(ii) If X is orientable, the orientable $\mathcal{C}^\infty$ vector bundles ξ of rank 2 are classified by nonnegative integers $|\langle e(\xi), \mu \rangle|$, where $e(\xi) \in H^2(X, \mathbb{Z})$ is the Euler class of ξ ([236], p. 98) and $\mu \in H_2(X, \mathbb{Z})$ is a generator of $H_2(X, \mathbb{Z}) \simeq \mathbb{Z}$. Moreover, $|\langle e(\xi), \mu \rangle|$ is the smallest possible number of zeros of a $\mathcal{C}^\infty$ section of ξ transverse to the zero section.

In both cases, $w_2(\xi)$ corresponds to $[s^{-1}(0)]$ by Poincaré duality.

Proof of Theorem 12.5.3. If ξ is an algebraic vector bundle over X, then $w_1(\xi)$ is in $H^1_{\text{alg}}(X, \mathbb{Z}/2)$ (cf. Theorem 12.1.10 or Remark 12.4.7). Now let ξ be a topological vector bundle over X such that $w_1(\xi) \in H^1_{\text{alg}}(X, \mathbb{Z}/2)$. By Theorem 12.4.6, there exists an algebraic line bundle η_1 over X such that $w_1(\xi) = w_1(\eta_1)$. Consider the class $w_2(\xi) + w_1(\xi) \cup w_1(\xi)$ in $H^2(X, \mathbb{Z}/2)$. Its image in $H_0(X, \mathbb{Z}/2)$ by Poincaré duality is represented by a finite set of points Y. Applying Corollary 12.5.8, we obtain an orientable algebraic vector bundle η_2 of rank 2 over X such that $w_2(\eta_2) = w_2(\xi) + w_1(\xi) \cup w_1(\xi)$. Since η_2 is orientable, $w_1(\eta_2) = 0$. By the Whitney product formula, $w_1(\xi \oplus \eta_1 \oplus \eta_2) = 0$ and $w_2(\xi \oplus \eta_1 \oplus \eta_2) = 0$. Since X is of dimension 2, we have $\xi \oplus \eta_1 \oplus \eta_2 \simeq_{\text{top}} \zeta \oplus \epsilon$, where ϵ is a trivial bundle, and ζ, a topological vector bundle of rank 2. In order to prove that ξ is isomorphic to an algebraic vector bundle, it suffices, by Proposition 12.3.5, to prove that ζ is stably isomorphic to an algebraic vector bundle. In fact, we claim that ζ is even stably trivial. This claim follows from the fact that $w_1(\zeta) = w_1(\zeta \oplus \epsilon) = 0$, $w_2(\zeta) = w_2(\zeta \oplus \epsilon) = 0$, and the following purely topological lemma.

Lemma 12.5.10. *Let ζ be a topological vector bundle of rank 2 over a compact $\mathcal{C}^\infty$ manifold X of dimension 2. If $w_1(\zeta) = 0$ and $w_2(\zeta) = 0$, then ζ is stably trivial.*

Proof. The bundle ζ is orientable since $w_1(\zeta) = 0$. We choose an orientation of ζ and denote by $e(\zeta) \in H^2(X, \mathbb{Z})$ the Euler class of this oriented bundle

([236], p. 98). Since $w_2(\zeta) = 0$, $e(\zeta) = 2u$ for some $u \in H^2(X, \mathbb{Z})$ ([236], p. 99). We choose a continuous mapping $f : X \to S^2$ such that $f^*(v) = u$, where v is a generator of $H^2(S^2, \mathbb{Z}) = \mathbb{Z}$. We also choose an orientation of the tangent bundle $T(S^2)$ such that $e(T(S^2)) = 2v$. Then $e(\zeta) = 2u = f^*(2v) = e(f^*(T(S^2)))$. Since ζ and $f^*(T(S^2))$ are two oriented vector bundles of rank 2 with the same Euler class, they are isomorphic. Finally, $f^*(T(S^2))$ is stably trivial since $T(S^2)$ is stably trivial. □□

Corollary 12.5.11. *Let X be a compact connected nonsingular affine real algebraic surface. There is an exact sequence*

$$0 \to \widetilde{K}_0(\mathcal{R}(X)) \to \widetilde{KO}(X) \xrightarrow{\rho} H^1(X, \mathbb{Z}/2)/H^1_{\mathrm{alg}}(X, \mathbb{Z}/2) \to 0 \ .$$

Proof. We have to define the homomorphism ρ. An element α of $\widetilde{KO}(X)$ is the class of some vector bundle ξ. Then $\rho(\alpha)$ is the class of $w_1(\xi) \in H^1(X, \mathbb{Z}/2)$ modulo $H^1_{\mathrm{alg}}(X, \mathbb{Z}/2)$. The mapping ρ is a well-defined epimorphism. By Theorem 12.3.6, the canonical homomorphism $\widetilde{K}_0(\mathcal{R}(X)) \to \widetilde{K}_0(\mathcal{C}^0(X)) \simeq \widetilde{KO}(X)$ is injective. Finally, Theorem 12.5.3 implies that the kernel of ρ is equal to the image of the homomorphism $\widetilde{K}_0(\mathcal{R}(X)) \to \widetilde{KO}(X)$. □

In the next corollary, we shall use invariants which are defined for an arbitrary compact nonsingular affine real algebraic variety X over $\mathbb{R}$. Set

$$\beta_k(X) = \dim_{\mathbb{Z}/2} H^k_{\mathrm{alg}}(X, \mathbb{Z}/2) ,$$
$$\delta(X) = \dim_{\mathbb{Z}/2} \{v \in H^1_{\mathrm{alg}}(X, \mathbb{Z}/2) \mid v \cup v = 0\} .$$

The invariants $\beta_k(X)$ and $\delta(X)$ completely determine the structure of the group $\widetilde{K}_0(\mathcal{R}(X))$, for X of dimension ≤ 3. They will also be useful for the study of the set $\mathcal{R}(X, S^1)$ of regular mappings into S^1 (cf. Section 13.3).

Corollary 12.5.12.
(i) *Let X be a compact connected affine nonsingular real algebraic surface. Then*

$$\widetilde{K}_0(\mathcal{R}(X)) \simeq (\mathbb{Z}/4)^{\beta_1(X) - \delta(X)} \oplus (\mathbb{Z}/2)^{1 - \beta_1(X) + 2\delta(X)}$$

(ii) *Let M be a compact connected C^∞ surface of genus g. The number μ_M of isomorphism classes of groups $\widetilde{K}_0(\mathcal{R}(X))$, as X runs through all algebraic models of M, is as follows*

$$\mu_M = \begin{cases} 2g+1 & \text{if } M \text{ is orientable,} \\ g & \text{if } M \text{ is nonorientable and } g \text{ is odd,} \\ 2g-2 & \text{if } M \text{ is nonorientable and } g \text{ is even.} \end{cases}$$

Proof. (i) We consider the multiplicative group of units of the ring $H^*_{\text{alg}}(X, \mathbb{Z}/2)$, that is, the group

$$L_{\text{alg}}(X) = 1 \oplus H^1_{\text{alg}}(X, \mathbb{Z}/2) \oplus H^2_{\text{alg}}(X, \mathbb{Z}/2)$$

with cup product as group operation. Define a mapping

$$\psi : \widetilde{K}_0(\mathcal{R}(X)) \to L_{\text{alg}}(X) ,$$

which assigns to the element of $\widetilde{K}_0(\mathcal{R}(X))$ represented by an algebraic vector bundle ξ the total Stiefel-Whitney class $w(\xi) = 1 + w_1(\xi) + w_2(\xi)$ of ξ. Clearly, ψ is a well-defined homomorphism (cf. Theorem 12.5.3). As ψ is injective by Lemma 12.5.10, and surjective by Theorem 12.4.6 and Corollary 12.5.8, it follows that ψ is an isomorphism. Since all elements of $L_{\text{alg}}(X)$ different from 1 are of order 2 or 4, the conclusion (i) of the corollary follows by counting.

(ii) follows from (i) and Theorem 11.3.8. □

Finally, we mention without proof (cf. [49] for further details) that a topological vector bundle ξ over a compact connected nonsingular affine real algebraic variety X of dimension 3 is isomorphic to an algebraic vector bundle if and only if its total Stiefel-Whitney class $w(\xi)$ is in $H^{\text{alg}}_*(X, \mathbb{Z}/2)$. This implies the following.

Theorem 12.5.13. *Let X be a compact connected affine nonsingular real algebraic variety of dimension 3. Then*

$$\widetilde{K}_0(\mathcal{R}(X)) \simeq (\mathbb{Z}/4)^{\beta_1(X) - \delta(X)} \oplus (\mathbb{Z}/2)^{\beta_2(X) - \beta_1(X) + 2\delta(X)} .$$

As we have seen, if $\dim(X) \leq 3$, the group $\widetilde{K}_0(\mathcal{R}(X))$ is determined by $H^*_{\text{alg}}(X, \mathbb{Z}/2)$. This is no longer true for varieties of dimension greater than 3. There exists an algebraic model X of the 4-sphere S^4 such that $\widetilde{K}_0(\mathcal{R}(X)) = 0$, while $\widetilde{KO}(X) \simeq \widetilde{KO}(S^4) \simeq \widetilde{K}_0(\mathcal{R}(S^4)) \simeq \mathbb{Z}$ and, of course, $H^{\text{alg}}_*(X, \mathbb{Z}/2) = H_*(X, \mathbb{Z}/2)$ (cf. Theorem 12.6.12 below).

12.6 Algebraic $\mathbb{C}$-vector Bundles

In this section we study algebraic $\mathbb{C}$-vector bundles over an affine real algebraic variety X over $\mathbb{R}$. These bundles, in turn, will be used in Chap. 13 to investigate the set $\mathcal{R}(X, S^{2k})$ of regular mappings into even dimensional spheres.

We regard $\mathbb{C}$ as a real algebraic variety by identifying it with $\mathbb{R}^2$, and we denote by $\mathcal{R}(X, \mathbb{C})$ the $\mathbb{C}$-algebra of regular functions from X into $\mathbb{C}$. Similarly, we denote by $\mathcal{P}(X, \mathbb{C})$ (resp. $\mathcal{C}^0(X, \mathbb{C})$) the $\mathbb{C}$-algebra of polynomial (resp. continuous) functions from X into $\mathbb{C}$. Clearly, one has $\mathcal{R}(X, \mathbb{C}) \simeq \mathcal{R}(X) \otimes_{\mathbb{R}} \mathbb{C}$, etc.

326   12. Algebraic Vector Bundles

By copying the definitions and most of the results of Sections 1–3 of this chapter concerning algebraic $\mathbb{R}$-vector bundles, we can construct the theory of algebraic $\mathbb{C}$-vector bundles. In this theory we have to replace the use of the ring $\mathcal{R}(X)$ by that of $\mathbb{R}(X,\mathbb{C})$. In particular, the category of algebraic $\mathbb{C}$-vector bundles over X is equivalent to the category of projective modules of finite type over $\mathcal{R}(X,\mathbb{C})$. The grassmannian $\mathbb{G}_{n,k}(\mathbb{R})$ used in Section 12.1 should be replaced by the complex grassmannian $\mathbb{G}_{n,k}(\mathbb{C})$, but endowed with its canonical underlying affine real algebraic structure (cf. Proposition 3.4.6). Similarly, in this context, the universal $\mathbb{C}$-vector bundle $\gamma_{n,k}(\mathbb{C})$ over $\mathbb{G}_{n,k}(\mathbb{C})$ replaces the bundle $\gamma_{n,k}$ defined in Proposition 12.1.4. The proof that $\gamma_{n,k}(\mathbb{C})$ is algebraic is completely analogous to that given in Proposition 12.1.8.

It should be stressed that the total space of an algebraic $\mathbb{C}$-vector bundle over X is an affine real algebraic variety, and an algebraic section of such a bundle is, in particular, a regular mapping between affine *real* algebraic varieties. All results concerning the density of the set of algebraic sections stated in Section 12.3 remain valid, with the same proofs, for algebraic sections of algebraic $\mathbb{C}$-vector bundles.

The group $\widetilde{KO}(X)$ of classes of stably equivalent topological $\mathbb{R}$-vector bundles is replaced by the analogously defined group $\widetilde{K}(X)$ of classes of stably equivalent topological $\mathbb{C}$-vector bundles. For X compact, one has also the canonical injective homomorphism

$$\widetilde{K}_0(\mathcal{R}(X,\mathbb{C})) \longrightarrow \widetilde{K}_0(\mathcal{C}^0(X,\mathbb{C})) \simeq \widetilde{K}(X).$$

Example 12.6.1. Every topological $\mathbb{C}$-vector bundle over $S^n \subset \mathbb{R}^{n+1}$ (the standard n-sphere) is isomorphic to an algebraic $\mathbb{C}$-vector bundle.

Indeed, for odd dimensional spheres, the group $\widetilde{K}(S^{2k+1}) = 0$ (cf. [174], p. 109). In other words, every topological $\mathbb{C}$-vector bundle ξ over S^{2k+1} is stably trivial, which implies, as in Proposition 12.3.5, that ξ is isomorphic to an algebraic $\mathbb{C}$-bundle. Since the group $\widetilde{K}(X)$ is a topological invariant, the same conclusion is valid for every affine real algebraic variety homeomorphic to S^{2k+1}.

For even dimensional spheres, the canonical homomorphism

$$\widetilde{K}_0(\mathcal{P}(S^{2k},\mathbb{C})) \to \widetilde{K}_0(\mathcal{R}(S^{2k},\mathbb{C})) \to \widetilde{K}_0(\mathcal{C}^0(S^{2k},\mathbb{C})) \simeq \widetilde{K}(S^{2k}) \simeq \mathbb{Z}$$

is an isomorphism (cf. [131, 316]). This means that every topological $\mathbb{C}$-vector bundle ξ over S^{2k} is stably equivalent to an algebraic one. This implies, again as in Proposition 12.3.5, that ξ is, in fact, isomorphic to an algebraic $\mathbb{C}$-vector bundle.

On the other hand, we shall see in Example 12.6.9 that there exists an affine real algebraic surface, homeomorphic to S^2, on which every algebraic $\mathbb{C}$-vector bundle is trivial.

We denote by $V_\mathbb{C}^1(X)$ the group of isomorphism classes of topological $\mathbb{C}$-line bundles over X, and by $V_{\mathbb{C}-\mathrm{alg}}^1(X)$ the analogous group for algebraic

ℂ-line bundles (the group operation is defined by tensor product). If X is compact, one proves, in the same way as in Theorem 12.3.3, that the canonical homomorphism $V^1_{\mathbb{C}-\text{alg}}(X) \to V^1_{\mathbb{C}}(X)$ is injective. We shall thus regard $V^1_{\mathbb{C}-\text{alg}}(X)$ as a subgroup of $V^1_{\mathbb{C}}(X)$.

Recall that the map $V^1_{\mathbb{C}}(X) \to H^2(X, \mathbb{Z})$, which assigns to the class of a line bundle ξ its first Chern class $c_1(\xi)$, is an isomorphism (cf. [161], p. 49).

If X is a nonsingular affine real algebraic variety, the group $V^1_{\mathbb{C}-\text{alg}}(X)$ is isomorphic to the divisor class group $\text{Cl}(\mathcal{R}(X, \mathbb{C}))$ of $\mathcal{R}(X, \mathbb{C})$ (cf. Corollary 12.2.4). The next result is an immediate consequence of the facts recalled above.

Proposition 12.6.2. *Let X be an affine irreducible nonsingular real algebraic variety.*
 (i) *The ring $\mathcal{R}(X, \mathbb{C})$ is factorial if and only if $V^1_{\mathbb{C}-\text{alg}}(X) = 0$.*
 (ii) *If X is compact and $H^2(X, \mathbb{Z}) = 0$, the ring $\mathcal{R}(X, \mathbb{C})$ is factorial.*

We shall now study algebraic ℂ-vector bundles over the product $X \times S^1$. First we need some preparation.

Lemma 12.6.3. *Let $X \subset \mathbb{R}^n$ be a nonsingular algebraic set. There exists an algebraic set $X' \subset \mathbb{R}^{n+1}$, biregularly isomorphic to X, such that the Zariski closure $X'_{\mathbb{C}}$ of X' in $\mathbb{C}^{n+1}$ is a nonsingular complex algebraic set.*

Proof. Let $\mathcal{I}_{\mathcal{P}(\mathbb{R}^n)}(X) = (f_1, \ldots, f_k)$, and let d be the dimension of X. Denote by g the sum of squares of the $d \times d$ minors of the jacobian matrix $\left[\frac{\partial f_i}{\partial x_j}\right]$, $i = 1, \ldots, k$, $j = 1, \ldots, n$. Then we take

$$X' = \{(x, y) \in \mathbb{R}^{n+1} \mid f_1(x) = \cdots = f_k(x) = 0, \ yg(x) = 1\}.$$

□

Lemma 12.6.4. *Let $X \subset \mathbb{R}^n$ be a real algebraic set such that its Zariski closure $X_{\mathbb{C}}$ in $\mathbb{C}^n$ is a nonsingular complex algebraic set. Set $A = \mathcal{P}(X, \mathbb{C})$.*
 (i) *If S is a multiplicative subset of A not containing 0, the canonical homomorphism $\widetilde{K}_0(A) \to \widetilde{K}_0(S^{-1}A)$ is surjective.*
 (ii) *The canonical homomorphisms $A \to A[T] \to A[T, T^{-1}]$ induce isomorphisms $\widetilde{K}_0(A) \to \widetilde{K}_0(A[T]) \to \widetilde{K}_0(A[T, T^{-1}])$.*

References for the proof: The ring A is regular in the sense of [23]. Hence, (i) follows from Theorem 6.5, p. 499 in [23], and (ii) from Theorem 3.1, p. 636.
□

Theorem 12.6.5. *Let X be an affine nonsingular real algebraic variety. The projection $\pi : X \times S^1 \to X$ induces an isomorphism*

$$\widetilde{K}_0(\mathcal{R}(X, \mathbb{C})) \to \widetilde{K}_0(\mathcal{R}(X \times S^1, \mathbb{C})).$$

Proof. Without loss of generality, we may assume that X is an irreducible algebraic subset of $\mathbb{R}^n$, and that its complexification $X_\mathbb{C}$ is also nonsingular (cf. Lemma 12.6.3).

The important point of the proof is the observation that $\mathcal{P}(X \times S^1, \mathbb{C})$, as a $\mathcal{P}(X, \mathbb{C})$-algebra, is isomorphic to $\mathcal{P}(X, \mathbb{C})[T, T^{-1}]$. Since

$$\mathcal{P}(X \times S^1, \mathbb{C}) \simeq \mathcal{P}(X \times S^1) \otimes_\mathbb{R} \mathbb{C} \simeq \mathcal{P}(X) \otimes_\mathbb{R} \mathcal{P}(S^1) \otimes_\mathbb{R} \mathbb{C}$$
$$\simeq \mathcal{P}(X, \mathbb{C}) \otimes_\mathbb{C} \mathcal{P}(S^1, \mathbb{C}),$$

this is a consequence of the fact that the $\mathbb{C}$-algebras $\mathcal{P}(S^1, \mathbb{C})$ and $\mathbb{C}[T, T^{-1}]$ are isomorphic. Indeed, since $\mathcal{P}(S^1, \mathbb{C}) = \mathbb{C}[U, V]/(U^2 + V^2 - 1)$, the isomorphism is obtained by substituting $T = U + iV$, $T^{-1} = U - iV$.

Choose a point $a \in S^1$ and denote by $i : X \to X \times S^1$ the injection defined by $i(x) = (x, a)$. Now consider the following commutative diagram

$$\begin{array}{ccccc}
\widetilde{K}_0(\mathcal{P}(X, \mathbb{C})) & \xrightarrow{\widetilde{K}_0(\pi^*)} & \widetilde{K}_0(\mathcal{P}(X \times S^1, \mathbb{C})) & \xrightarrow{\widetilde{K}_0(i^*)} & \widetilde{K}_0(\mathcal{P}(X, \mathbb{C})) \\
\alpha \downarrow & & \beta \downarrow & & \alpha \downarrow \\
\widetilde{K}_0(\mathcal{R}(X, \mathbb{C})) & \xrightarrow{\widetilde{K}_0(\pi^*)} & \widetilde{K}_0(\mathcal{R}(X \times S^1, \mathbb{C})) & \xrightarrow{\widetilde{K}_0(i^*)} & \widetilde{K}_0(\mathcal{R}(X, \mathbb{C})),
\end{array}$$

where π^* and i^* are the homomorphisms induced by π and i, and α and β are given by the canonical homomorphisms $\mathcal{P}(X, \mathbb{C}) \hookrightarrow \mathcal{R}(X, \mathbb{C})$ and $\mathcal{P}(X \times S^1, \mathbb{C}) \hookrightarrow \mathcal{R}(X \times S^1, \mathbb{C})$. By Lemma 12.6.4 (ii), the homomorphism

$$\widetilde{K}_0(\mathcal{P}(X, \mathbb{C})) \to \widetilde{K}_0(\mathcal{P}(X \times S^1, \mathbb{C})) \simeq \widetilde{K}_0(\mathcal{P}(X, \mathbb{C})[T, T^{-1}])$$

induced by π^* is an isomorphism. By Lemma 12.6.4 (i) applied to

$$\mathcal{P}(X \times S^1, \mathbb{C}) \hookrightarrow \mathcal{R}(X \times S^1, \mathbb{C}) \simeq S^{-1}\mathcal{P}(X \times S^1, \mathbb{C})$$

(with $S = \{f \in \mathcal{P}(X \times S^1) \mid f^{-1}(0) = \emptyset\}$), the homomorphism β is surjective. It follows that $\widetilde{K}_0(\pi^*) : \widetilde{K}_0(\mathcal{R}(X, \mathbb{C})) \to \widetilde{K}_0(\mathcal{R}(X \times S^1, \mathbb{C}))$ is surjective. Finally, since $\pi \circ i = \mathrm{Id}_X$, the composition $\widetilde{K}_0(i^*) \circ \widetilde{K}_0(\pi^*)$ is the identity on $\widetilde{K}_0(\mathcal{R}(X, \mathbb{C}))$. Hence, $\widetilde{K}_0(\pi^*)$ is an isomorphism from $\widetilde{K}_0(\mathcal{R}(X, \mathbb{C}))$ onto $\widetilde{K}_0(\mathcal{R}(X \times S^1, \mathbb{C}))$. $\square$

Corollary 12.6.6. *Denote by T^n the product $S^1 \times \cdots \times S^1$ (with n factors). Then $\widetilde{K}_0(\mathcal{R}(T^n, \mathbb{C})) = 0$.*

Corollary 12.6.7. *Every algebraic $\mathbb{C}$-vector bundle over $S^1 \times S^1$ is trivial.*

Proof. As observed in the proof of Theorem 12.6.5, the ring $\mathcal{P}(S^1 \times S^1, \mathbb{C})$ is isomorphic to $\mathbb{C}[T, U, T^{-1}, U^{-1}]$. Hence, $\mathcal{R}(S^1 \times S^1, \mathbb{C})$ is factorial, and, by Proposition 12.6.2 (i), $V^1_{\mathbb{C}-\mathrm{alg}}(S^1 \times S^1) = 0$. For surfaces, this implies that every algebraic $\mathbb{C}$-vector bundle is trivial. $\square$

Remark 12.6.8. Every topological (resp. algebraic) $\mathbb{C}$-line bundle over $S^1 \times S^1$ can be considered, in a natural way, as an orientable topological (resp. algebraic) $\mathbb{R}$-vector bundle of rank 2. Every orientable topological $\mathbb{R}$-vector bundle of rank 2 over $S^1 \times S^1$ is obtained in this way. Theorem 12.5.3 and Corollary 12.6.7 show that the analogous statement is false for algebraic vector bundles.

Example 12.6.9. (i) There exists an algebraic set $X \subset \mathbb{R}^5$, homeomorphic to S^2, such that every algebraic $\mathbb{C}$-vector bundle over X is trivial. Let $a \in S^1$, and let $Y = (\{a\} \times S^1) \cup (S^1 \times \{a\}) \subset S^1 \times S^1$. Let $X = S^1 \times S^1/Y$ be the algebraic set obtained by blowing down Y to a point y (Proposition 3.5.6) (observe that X is an algebraic Alexandrov compactification of $\mathbb{R}^2$ which is not biregularly isomorphic to the "standard" compactification S^2). Denote by $\Phi : S^1 \times S^1 \to X$ the regular mapping defining the blowing down, i.e. $\Phi(Y) = \{y\}$ and $\Phi|_{S^1 \times S^1 \setminus Y}$ is a biregular isomorphism onto $X \setminus \{y\}$. It is clear that Φ induces an isomorphism $\Phi^* : H^2(X, \mathbb{Z}) \to H^2(S^1 \times S^1, \mathbb{Z})$. Let ξ be an algebraic $\mathbb{C}$-vector bundle over X, and $c_1(\xi) \in H^2(X, \mathbb{Z})$, its first Chern class. Since, by Corollary 12.6.7, the bundle $\Phi^*(\xi)$ is trivial, one has $\Phi^*(c_1(\xi)) = c_1(\Phi^*(\xi)) = 0$. Therefore $c_1(\xi) = 0$ and, since X is of dimension 2, ξ is trivial. For basic facts concerning Chern classes, we refer the reader to [236].

(ii) Using more sophisticated methods, one can prove the following (cf. [55]): Given a compact connected orientable $\mathcal{C}^\infty$ surface M and a nonnegative integer k, there exists an algebraic model X of M such that

$$V^1_{\mathbb{C}\text{-alg}}(X) = kV^1_{\mathbb{C}}(X).$$

$\square$

Given a $\mathcal{C}^\infty$ submanifold Z of a compact $\mathcal{C}^\infty$ manifold Y, $\dim Y = 2\dim Z$, we denote by $\#_2(Z, Z; Y)$ the modulo 2 self-intersection number of Z in Y (cf. [160], p. 132-133).

Theorem 12.6.10. *Let X be a compact connected nonorientable nonsingular affine real algebraic surface containing a nonsingular algebraic curve $\Gamma \subset X$, such that $\#_2(\Gamma, \Gamma; X) = 1$. Then every topological $\mathbb{C}$-vector bundle over X is topologically isomorphic to an algebraic $\mathbb{C}$-vector bundle.*

Proof. It suffices to show that $V^1_{\mathbb{C}\text{-alg}}(X) = V^1_{\mathbb{C}}(X)$. Moreover, since $H^2(X, \mathbb{Z}) = \mathbb{Z}/2$ and $V^1_{\mathbb{C}}(X)$ is isomorphic to $H^2(X, \mathbb{Z})$, it suffices to show that there exists a nontrivial algebraic $\mathbb{C}$-line bundle over X. Let ξ be the algebraic $\mathbb{R}$-line bundle corresponding to the homology class $[\Gamma] \in H_1^{\text{alg}}(X, \mathbb{Z}/2)$ (cf. Theorem 12.4.6). Let Γ' be a $\mathcal{C}^\infty$ curve in X, diffeotopic and transverse to Γ. Choose $\mathcal{C}^\infty$ sections σ and σ' of ξ, transverse to the zero section of ξ, such that $\sigma^{-1}(0) = \Gamma$ and $\sigma'^{-1}(0) = \Gamma'$ (cf. Remark 12.4.3). Then (σ, σ') is a $\mathcal{C}^\infty$ section of $\xi \oplus \xi$ which is transverse to the zero section and satisfies $(\sigma, \sigma')^{-1}(0) = \Gamma \cap \Gamma'$. By assumption, the number of points in $\Gamma \cap \Gamma'$ is odd.

This number, taken modulo 2, is the Euler number of $\xi \oplus \xi$ (i.e. the mod 2 self-intersection number of X in the total space of $\xi \oplus \xi$). In particular, $\xi \oplus \xi$ is topologically nontrivial. Since the algebraic $\mathbb{C}$-line bundle $\xi \otimes_{\mathbb{R}} \mathbb{C}$, regarded as an $\mathbb{R}$-vector bundle of rank 2, is isomorphic to $\xi \oplus \xi$, the bundle $\xi \otimes_{\mathbb{R}} \mathbb{C}$ is nontrivial. $\square$

Corollary 12.6.11. *Let X be a compact connected nonorientable nonsingular affine real algebraic surface, satisfying either of the following conditions:*
  (i) $H_1(X, \mathbb{Z}/2) = H_1^{\mathrm{alg}}(X, \mathbb{Z}/2)$.
  (ii) *The genus of X is odd.*
Then every topological $\mathbb{C}$-vector bundle is topologically isomorphic to an algebraic $\mathbb{C}$-vector bundle.

Proof. (i) By Theorem 12.4.11, there exists a nonsingular algebraic curve Γ in X whose tubular neighbourhood is a Möbius band. It follows that $\#_2(\Gamma, \Gamma; X) = 1$.
  (ii) Let g be the genus of X. We can choose g disjoint $\mathcal{C}^\infty$ connected curves $\Gamma_1, \ldots, \Gamma_g$ contained in X, each Γ_i having a Möbius band as a tubular neighbourhood, so that $M \backslash \bigcup_{i=1}^g \Gamma_i$ is orientable (cf. [160], p. 206). Let $\Gamma = \bigcup_{i=1}^g \Gamma_i$ and note that $[\Gamma] \in H_1(X, \mathbb{Z}/2)$ corresponds to $w_1(T(X)) \in H_{\mathrm{alg}}^1(X, \mathbb{Z}/2)$ by Poincaré duality (where $T(X)$ is the tangent bundle of X). Hence, by Theorem 12.4.6, $[\Gamma] \in H_1^{\mathrm{alg}}(X, \mathbb{Z}/2)$, and we can assume that Γ is a nonsingular algebraic curve, using Theorem 12.4.11. Since $\#_2(\Gamma, \Gamma; X) \equiv g \pmod{2}$ and g is odd, $\#_2(\Gamma, \Gamma; X) = 1$. $\square$

On the other hand, each compact connected nonorientable surface of *even* genus has an algebraic model X such that every algebraic $\mathbb{C}$-vector bundle over X is trivial (cf. [47]).

Much more is known about algebraic $\mathbb{C}$-vector bundles over real algebraic varieties than was presented above. They were investigated systematically in a series of papers listed in the bibliographic notes at the end of this chapter. Here is a sample of results given without proofs, for which we refer the reader to the original papers. Some of these results will be needed in the next chapter.

Theorem 12.6.12. *Let M be a compact $\mathcal{C}^\infty$ hypersurface of $\mathbb{P}_n(\mathbb{R})$. There exists a $\mathcal{C}^\infty$ diffeomorphism $h : \mathbb{P}_n(\mathbb{R}) \to \mathbb{P}_n(\mathbb{R})$, arbitrarily close, in the $\mathcal{C}^\infty$ topology, to the identity mapping, such that $h(M) = X$ is a nonsingular algebraic hypersurface of $\mathbb{P}_n(\mathbb{R})$ and the groups $\widetilde{K}_0(\mathcal{R}(X))$ and $\widetilde{K}_0(\mathcal{R}(X, \mathbb{C}))$ are finite. If, moreover, $H^{\mathrm{even}}(M, \mathbb{Z})$ is torsion free, then $\widetilde{K}_0(\mathcal{R}(X, \mathbb{C})) = 0$ and $\widetilde{K}_0(\mathcal{R}(X)) = (\mathbb{Z}/2)^l$ for some nonnegative integer l.*

Reference for the proof: cf. [42]. $\square$

The conclusion of Theorem 12.6.12 is not valid for an arbitrary manifold M.

Theorem 12.6.13. *Let M be a compact connected orientable C^∞ manifold of dimension divisible by 4. Assume that the signature of M is different from zero. Then, for every algebraic model X of M, the groups $\widetilde{K}_0(\mathcal{R}(X))$ and $\widetilde{K}_0(\mathcal{R}(X,\mathbb{C}))$ are infinite.*

Reference for the proof: cf. [42, 47]. □

We have seen in Corollary 12.6.6 that, for the standard circle S^1 and every positive integer n, one has $\widetilde{K}_0(\mathcal{R}((S^1)^n,\mathbb{C})) = 0$. However, S^1 is not the only curve having this property. In fact, that is the case for "most" algebraic curves (cf. Theorem 12.6.14 (ii)). The next two results are formulated for projective real algebraic curves. Nevertheless, we are not leaving the framework of affine varieties. Recall that, in the real case, every projective variety is affine (cf. Theorem 3.4.4). In the statement below, given a projective nonsingular real algebraic curve Γ, we denote by $\Gamma_\mathbb{C}$ the unique (up to isomorphism) complex projective nonsingular curve defined over $\mathbb{R}$ and having the real part biregularly isomorphic to Γ.

Let Γ^n denote the product of n copies of Γ.

Theorem 12.6.14. *Let Γ be a connected nonsingular projective real algebraic curve and let n be a positive integer. Then $\widetilde{K}_0(\mathcal{R}(\Gamma^n,\mathbb{C})) = 0$ in each of the following cases:*

(i) *$\Gamma_\mathbb{C}$ is an elliptic curve without complex multiplication.*

(ii) *The endomorphism ring of the Jacobian variety of $\Gamma_\mathbb{C}$ is isomorphic to $\mathbb{Z}$.*

(iii) *Γ is a dividing curve.*

Reference for the proof: [53, 55, 57]. □

On the other hand, there exist curves Γ for which $\widetilde{K}_0(\mathcal{R}(\Gamma^n,\mathbb{C})) = \widetilde{K}_0(\mathcal{C}^0(\Gamma^n,\mathbb{C}))$, for all $n > 0$.

Theorem 12.6.15. *Let n be a positive integer. The set of (isomorphism classes of) connected nonsingular projective real algebraic curves Γ of genus 1 (resp. genus 2) such that*

$$\widetilde{K}_0(\mathcal{R}(\Gamma^n,\mathbb{C})) = \widetilde{K}_0(\mathcal{C}^0(\Gamma^n,\mathbb{C}))$$

is countably infinite (resp. uncountable).

Reference for the proof: [53, 55, 62]. □

12.7 Nash Vector Bundles and Semi-algebraic Vector Bundles

Throughout this section, we shall work over an arbitrary real closed field R.

The definition of semi-algebraic vector bundles is straightforward. Abstract semi-algebraic spaces will not be considered. We shall use atlases of local trivializations.

Definition 12.7.1. *Let $M \subset R^n$ be a semi-algebraic set. Let $\xi = (E, p, M)$ be an R-vector bundle of rank k over M. A family of local trivializations $(U_i, \varphi_i : U_i \times R^k \to p^{-1}(U_i))_{i \in I}$ of ξ is said to be a* semi-algebraic atlas *of ξ if $(U_i)_{i \in I}$ is a finite open semi-algebraic covering of M and the mappings $\varphi_i^{-1} \circ \varphi_j|_{(U_i \cap U_j) \times R^k}$ are continuous semi-algebraic, for every pair $(i, j) \in I \times I$. Two semi-algebraic atlases are equivalent if their union is still a semi-algebraic atlas. A* semi-algebraic vector bundle *is a vector bundle $\xi = (E, p, M)$ equipped with an equivalence class of semi-algebraic atlases.*

Let $(\xi, (U_i, \varphi_i)_{i \in I})$ and $(\xi', (U'_j, \varphi'_j)_{j \in J})$ be two semi-algebraic vector bundles over M. A morphism $\psi : \xi \to \xi'$ of vector bundles is said to be a semi-algebraic morphism *if the mappings $(\varphi'_j)^{-1} \circ \psi \circ \varphi_i|_{(U_i \cap U'_j) \times R^k}$ are continuous semi-algebraic, for every pair $(i, j) \in I \times J$. A section s of ξ is said to be a* semi-algebraic section *if the mappings $\varphi_i^{-1} \circ s|_{U_i} : U_i \to U_i \times R^k$ are continuous semi-algebraic, for every $i \in I$.*

We allow vector bundles to have possibly different ranks over different semi-algebraically connected components of M.

By abuse of notation, we denote by $\xi = (E, p, M)$ a semi-algebraic vector bundle, without specifying the atlas defining its structure. We now give a list of properties of semi-algebraic vector bundles, whose verification is straightforward.

A semi-algebraic atlas is determined by a family of continuous semi-algebraic transition functions:

$$g_{i,j} : U_i \cap U_j \longrightarrow \mathrm{GL}(k, R),$$

where $(U_i)_{i \in I}$ is a finite open semi-algebraic covering of M. If ξ and ξ' are two semi-algebraic vector bundles over M, the bundles $\xi \oplus \xi'$, $\xi \otimes \xi'$, $\xi^\vee$, $\wedge^l \xi$ and $\mathrm{Hom}(\xi, \xi')$ have a canonical structure of semi-algebraic vector bundle. If $f : N \to M$ is a continuous semi-algebraic mapping between two semi-algebraic sets and ξ is a semi-algebraic vector bundle over M, the induced vector bundle $f^*(\zeta)$ has a canonical semi-algebraic structure.

Now let us consider the correspondence between vector bundles and sheaves of locally free modules of finite type. The finiteness condition imposed on atlases leads to the consideration of sheaves over the quasi-compact space $\widetilde{M}$ instead of M (Proposition 7.2.2). Otherwise, we could not even speak about sheaves of continuous semi-algebraic functions (Remark 7.3.3). Let M be a semi-algebraic set, and ξ, a semi-algebraic vector bundle over M. Denote by $\Gamma_{\mathrm{s.a.}}(\xi)$ the set of semi-algebraic sections of ξ. Denote by $\widetilde{\mathcal{L}}_{\mathrm{s.a.}}(\xi)$ the sheaf over $\widetilde{M}$ whose sections over an open subset $\widetilde{U}$ are $\Gamma_{\mathrm{s.a.}}(\xi|_U)$. Then $\widetilde{\mathcal{L}}_{\mathrm{s.a.}}(\xi)$ is a sheaf of $\widetilde{\mathcal{S}^0}$-modules (Proposition 7.3.2). Since $\widetilde{M}$ is quasi-compact, we obtain the following equivalence.

12.7 Nash Vector Bundles and Semi-algebraic Vector Bundles 333

Proposition 12.7.2. *The correspondence $\xi \mapsto \widetilde{\mathcal{L}}_{\text{s.a.}}(\xi)$ is an equivalence of the category of semi-algebraic vector bundles over M with the category of locally free sheaves of $\widetilde{\mathcal{S}^0}$-modules of finite type over $\widetilde{M}$.*

To study semi-algebraic vector bundles, we need a continuous semi-algebraic partition of unity subordinate to a finite open semi-algebraic covering.

Lemma 12.7.3. *Let M be a semi-algebraic set, and $(U_i)_{i=1,\ldots,\ell}$, a finite open semi-algebraic covering of M. There exist continuous semi-algebraic functions $\lambda_i : M \to R$ such that $\sum_{i=1}^{\ell} \lambda_i = 1$ and, for every $i = 1,\ldots,\ell$, $0 \leq \lambda_i \leq 1$ and the closure of $\{x \in M | \lambda_i(x) > 0\}$ is contained in U_i.*

Proof. Let $h_i(x) = \text{dist}(x, M \setminus U_i)$ be the distance between $x \in M$ and $M \setminus U_i$. Define $V_i = \{x \in M \mid h_i(x) > \max(h_1(x), \ldots, h_\ell(x))/2\}$. The V_i form an open semi-algebraic covering of M, and the closure $\overline{V}_i$ of V_i in M is contained in U_i (indeed, for $x \in \overline{V}_i$ we have $h_i(x) \geq \max(h_1(x), \ldots, h_\ell(x))/2$ and thus $h_i(x) > 0$). Let $g_i(x) = \text{dist}(x, M \setminus V_i)$ and define $\lambda_i = g_i / \sum_{j=1}^{\ell} g_j$. □

Proposition 12.7.4. *Let $\xi = (E, p, M)$ be a semi-algebraic vector bundle. There exist a finite number of semi-algebraic sections $s_1, \ldots, s_p$ of ξ such that, for every $x \in M$, the fibre $\xi_x = p^{-1}(x)$ is generated by $s_1(x), \ldots, s_p(x)$ as an R-vector space.*

Proof. Let $(U_i, \varphi_i)_{i=1,\ldots,\ell}$ be a semi-algebraic atlas of ξ, with $\varphi_i : U_i \times R^k \to p^{-1}(U_i)$. Let $(\lambda_i)_{i=1,\ldots,\ell}$ be a semi-algebraic partition of unity subordinate to the covering $(U_i)_{i=1,\ldots,\ell}$. For $1 \leq i \leq \ell$ and $1 \leq j \leq k$, let $s_{i,j}$ be the section of ξ which is zero outside U_i and defined by $s_{i,j}(x) = \varphi_i(x, \lambda_i(x)e_j)$ for $x \in U_i$ (where e_j is the jth vector of the canonical basis of R^k). The sections $s_{i,j}$ are continuous and semi-algebraic, and their values generate the fibre of ξ as an R-vector space at every point of M. □

Corollary 12.7.5. *Let ξ be a semi-algebraic vector bundle of rank k over a semi-algebraic set M. Then:*
   (i) There exists a semi-algebraic vector bundle ξ' over M such that $\xi \oplus \xi'$ is semi-algebraically isomorphic to a trivial bundle ϵ_M^n.
   (ii) There exists a continuous semi-algebraic mapping $f : M \to \mathbb{G}_{n,k}(R)$, for some positive integer n, such that ξ is semi-algebraically isomorphic to $f^*(\gamma_{n,k})$.
   (iii) There exists a projective module of finite type P over $\mathcal{S}^0(M)$ such that $\widetilde{\mathcal{L}}_{\text{s.a.}}(\xi)$ is isomorphic to the sheaf $\widetilde{\mathcal{S}^0} \otimes_{\mathcal{S}^0(M)} P$.

Proof. The proof of Theorem 12.1.7 shows that Proposition 12.7.4 implies the three properties listed above. The necessary modifications are straightforward. □

Corollary 12.7.6. *The functor $\xi \mapsto \Gamma_{\text{s.a.}}(\xi)$ is an equivalence of the category of semi-algebraic vector bundles over M with the category of projective modules of finite type over $\mathcal{S}^0(M)$.*

Proof. Analogous to that of Proposition 12.1.12. □

Proposition 12.7.7. *Let ξ be a semi-algebraic vector bundle over a semi-algebraic set M. Let $f, g : N \to M$ be continuous semi-algebraic mappings that are semi-algebraically homotopic (Definition 11.7.2). The vector bundles $f^*(\xi)$ and $g^*(\xi)$ are then semi-algebraically isomorphic.*

Proof. We can copy the proof of Theorem 4.3 in [174], Chap. 3, with the following change: in Lemma 4.1 of [174], we replace the products $A \times [a, b]$ by the sets $\{(x, t) \mid x \in A \text{ and } \varphi(x) \le t \le \psi(x)\}$, where φ and ψ are two continuous semi-algebraic functions from A to R with $\varphi < \psi$, and we use Lemma 11.7.4. □

Proposition 12.7.8. *Let M be a semi-algebraic subset of $\mathbb{R}^n$. Every topological vector bundle over M is isomorphic to a semi-algebraic vector bundle, and two vector bundles are semi-algebraically isomorphic if and only if they are topologically isomorphic.*

Proof. A semi-algebraic set has a semi-algebraic deformation retract which is compact (cf. Corollary 9.3.7 and Remark 9.3.8). Hence, by Proposition 12.7.7, it suffices to prove the statement when M is compact. Observe that every continuous mapping from M into a Nash submanifold of $\mathbb{R}^p$, in particular into $\mathbb{G}_{n,k}(\mathbb{R})$, is homotopic to a continuous semi-algebraic mapping (this follows from the Stone-Weierstrass theorem and the existence of Nash tubular neighbourhoods). It follows, in the usual way, that every topological vector bundle over M is isomorphic to a semi-algebraic vector bundle. Now, let ξ and ξ' be semi-algebraic vector bundles over M which are topologically isomorphic. The semi-algebraic vector bundle $\text{Hom}(\xi, \xi')$ has a continuous section σ with values in the open subset $\text{Iso}(\xi, \xi')$ of isomorphisms $\xi \to \xi'$. Using Proposition 12.7.4 and the proof of Theorem 12.3.1, we can approximate σ by semi-algebraic sections and, in turn, obtain a semi-algebraic isomorphism between ξ and ξ'. □

Corollary 12.7.9. *The canonical mapping $\text{Proj}(\mathcal{S}^0(M)) \to \text{Proj}(\mathcal{C}^0(M))$, induced by the inclusion $\mathcal{S}^0(M) \hookrightarrow \mathcal{C}^0(M)$, is bijective.*

We now consider Nash vector bundles. Let $M \subset R^n$ be a Nash submanifold. We define a *pre-Nash vector bundle* over M as a vector bundle equipped with an equivalence class of Nash atlases, imitating Definition 12.7.1. The properties of semi-algebraic vector bundles listed after Definition 12.7.1 (including Proposition 12.7.2) remain valid, with obvious modifications, for pre-Nash vector bundles. The following example shows that the analogue of Proposition 12.7.4 does not hold for pre-Nash bundles.

Example 12.7.10. We shall construct a pre-Nash $\mathbb{R}$-line bundle ξ over $\mathbb{R}$ which is not generated by its global Nash sections. The bundle ξ is determined by the transition function

$$g_{1,2} : U_1 \cap U_2 \longrightarrow \mathbb{R}^*$$
$$x \longmapsto (2 + \sqrt{1 - x^2})/\sqrt{3 + x^2},$$

where $U_1 = \,]-\infty, 1[$ and $U_2 = \,]-1, +\infty[$. Note that $g_{1,2}^{-1}(x) = (2 - \sqrt{1 - x^2})/\sqrt{3 + x^2}$. Consider a Nash section of ξ, given by two Nash functions $s_1 : U_1 \to \mathbb{R}$ and $s_2 : U_2 \to \mathbb{R}$ such that $s_1 = g_{1,2} s_2$. The Nash functions s_i, for $i = 1, 2$, extend to algebraic-analytic functions (still denoted by s_i) in neighbourhoods V_i of U_i in $\mathbb{C}$. We now study the analytic continuation of the function s_1 along the following closed path in $V_1 \cup V_2$, starting at 0, going around 1, then going around -1 and coming back to 0:

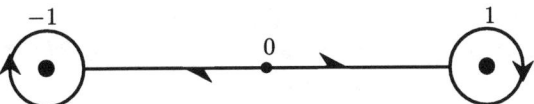

Figure 12.1.

As we go around 1, $s_1 = g_{1,2} s_2$ is changed into $s_1 g_{1,2}^{-2}$, since s_2 is defined in a neighbourhood of 1 and $g_{1,2}$ is changed into $g_{1,2}^{-1}$. As we go around -1, $s_1 g_{1,2}^{-2}$ is changed into $s_1 g_{1,2}^2$. The function we get by analytic continuation after one loop is $s_1 g_{1,2}^2$. Hence, after m loops, we obtain the function $s_1 g_{1,2}^{2m}$. If s_1 is not identically zero, this contradicts the fact that that the Riemann surface of the algebraic function s_1 has finitely many sheets. It follows that s_1 and s_2 are identically zero. The only global Nash section of ξ is the zero section.

The above example motivates the following definition.

Definition 12.7.11. *Let M be a Nash manifold. A pre-Nash vector bundle ξ over M is said to be a* Nash vector bundle *if there exists an injective Nash morphism from ξ into a trivial vector bundle.*

For Nash vector bundles, we formulate a version of Theorem 12.1.7.

Proposition 12.7.12. *Let M be a Nash manifold and ξ a pre-Nash vector bundle of rank k over M. Then the following properties are equivalent:*

(i) *The bundle ξ is Nash.*

(ii) *There exist global Nash sections $s_1, \ldots, s_p$ of ξ such that, for every x in M, the fibre ξ_x is generated by $s_1(x), \ldots, s_p(x)$ as an R-vector space.*

(iii) *There exists a Nash mapping $f : M \to \mathbb{G}_{n,k}(R)$ such that ξ is Nash isomorphic to $f^*(\gamma_{n,k})$.*

(iv) *There exists a projective module P of finite type over $\mathcal{N}(M)$ such that $\tilde{\mathcal{L}}_{\text{Nash}}(\xi)$ is isomorphic to the sheaf $\tilde{\mathcal{N}} \otimes_{\mathcal{N}(M)} P$.*

Proof. Analogous to the proof of Theorem 12.1.7. □

Corollary 12.7.13. *The functor $\xi \mapsto \Gamma_{\text{Nash}}(\xi)$ associating to ξ the module of its global Nash sections is an equivalence of the category of Nash vector bundles over M with the category of projective modules of finite type over $\mathcal{N}(M)$.*

Note that, if $\xi = (E, p, M)$ is a Nash vector bundle, E has a structure of affine Nash manifold (cf. Proposition 12.7.12 (iii)). If M is a Nash submanifold of R^n, the tangent bundle and normal bundle of M are Nash bundles.

Now we shall consider the approximation of sections by Nash sections.

Proposition 12.7.14. *Let $\xi = (E, p, M)$ be a Nash vector bundle over a Nash manifold M.*

(i) *Let $\sigma : M \to E$ be a continuous semi-algebraic section of ξ. For every open semi-algebraic subset U of E containing $\sigma(M)$, there exists a Nash section $s : M \to E$ of ξ such that $s(M) \subset U$.*

(ii) *If $R = \mathbb{R}$, M is compact and $\sigma : M \to E$ is a C^∞ section, then σ can be approximated by Nash sections (in the C^∞ topology).*

Proof. We proceed as in Theorem 12.3.1, using Proposition 12.7.12 (ii), continuous semi-algebraic partitions of unity (Lemma 12.7.3) and the Efroymson approximation theorem 8.8.4 for assertion (i), and the Stone-Weierstrass theorem for assertion (ii). □

Theorem 12.7.15. *Let M be a Nash manifold. Every semi-algebraic vector bundle over M is semi-algebraically isomorphic to a Nash vector bundle. If ξ and ξ' are Nash vector bundles over M, then they are Nash isomorphic if and only if they are semi-algebraically isomorphic.*

Proof. Using tubular neighbourhoods and the Efroymson approximation theorem, it follows that every continuous semi-algebraic mapping $f : M \to \mathbb{G}_{n,k}(R)$ can be approximated by Nash mappings. In particular, f is semi-algebraically homotopic to a Nash mapping. The remainder of the proof is a repetition of the method used in the proof of Proposition 12.7.8. □

Corollary 12.7.16. *The canonical mapping $\text{Proj}(\mathcal{N}(M)) \to \text{Proj}(\mathcal{S}^0(M))$, induced by the inclusion $\mathcal{N}(M) \hookrightarrow \mathcal{S}^0(M)$, is bijective.*

Corollary 12.7.17. *Let M be a Nash submanifold of $\mathbb{R}^n$. Then every topological vector bundle over M is topologically isomorphic to a Nash vector bundle. If ξ and ξ' are Nash vector bundles over M, then they are Nash isomorphic if and only if they are topologically isomorphic.*

Proof. Follows from Theorem 12.7.15 and Proposition 12.7.8. □

12.7 Nash Vector Bundles and Semi-algebraic Vector Bundles 337

Corollary 12.7.18. *Given a Nash submanifold M of $\mathbb{R}^n$, the canonical mapping $\mathrm{Proj}(\mathcal{N}(M)) \to \mathrm{Proj}(\mathcal{C}^0(M))$, induced by the inclusion $\mathcal{N}(M) \hookrightarrow \mathcal{C}^0(M)$, is bijective.*

Corollary 12.7.19. *Let M be a compact Nash submanifold of $\mathbb{R}^n$, and Y, a compact $\mathcal{C}^\infty$ hypersurface of M. Then there exists a $\mathcal{C}^\infty$ diffeomorphism of M, arbitrarily close to the identity, sending Y onto a Nash submanifold of M.*

Proof. We modify the proof of Theorem 12.4.11 (i) $\Rightarrow$ (iii) using Corollary 12.7.17 and Proposition 12.7.14 (ii). $\square$

Corollary 12.7.20. *Let M be a connected Nash submanifold of $\mathbb{R}^n$. The ring $\mathcal{N}(M)$ of Nash functions on M is factorial if and only if $H^1(M, \mathbb{Z}/2) = 0$.*

Proof. The ring $\mathcal{N}(M)$ is locally factorial (Proposition 8.7.15). Hence, the divisor class group $\mathrm{Cl}(\mathcal{N}(M))$ is isomorphic to the group $\mathrm{Pic}(\mathcal{N}(M))$ of isomorphism classes of invertible $\mathcal{N}(M)$-modules (Proposition 12.2.3). By Corollary 12.7.13, the group $\mathrm{Pic}(\mathcal{N}(M))$ is isomorphic to the group $V^1_{\mathrm{Nash}}(M)$ of isomorphism classes of Nash line bundles over M. By Corollary 12.7.17, the group $V^1_{\mathrm{Nash}}(M)$ is isomorphic to the group $V^1(M)$, and the latter group is isomorphic to $H^1(M, \mathbb{Z}/2)$. Thus the groups $\mathrm{Cl}(\mathcal{N}(M))$ and $H^1(M, \mathbb{Z}/2)$ are isomorphic, and the corollary follows. $\square$

Bibliographic Notes. Algebraic vector bundles over real algebraic varieties were first studied by Benedetti and Tognoli in [33]. This paper also contains most of the results described in Section 1, as well as Theorems 12.3.2 and 12.3.3 (it should be mentioned, however, that the claim in [33], that every topological vector bundle on an algebraic variety homeomorphic to S^n is isomorphic to an algebraic one, is false; cf. Example 12.6.9 and Theorem 12.6.12). In some papers, algebraic vector bundles, in the sense of Definition 12.1.6, are called "strongly algebraic". Example 12.1.5 can be found in [326]. Proposition 12.4.4 and Example 12.4.15 can be found in [298]. Theorem 12.4.11 is contained in several papers, including [34]. The results of Section 5 are taken from [48, 49, 197]. The results proved in Section 6 are contained in [45, 46]; the references for those theorems which are cited without proof are given in that section. Theorem 12.6.5 in the polynomial case can be found in [212] and Example 12.7.10 in [170]. Corollary 12.7.20 is proved in [40] (sufficient condition) and [297] (necessary condition). The fact that, for a compact algebraic variety X over $\mathbb{R}$, the homomorphism $\widetilde{K}_0(\mathcal{R}(X)) \to \widetilde{K}_0(\mathcal{C}^0(X))$ is injective is contained in [130]. The isomorphism $\widetilde{K}_0(\mathcal{P}(S^n)) \to \widetilde{K}_0(\mathcal{C}^0(S^n))$ is proved in [131] and [316] (cf. Example 12.3.7). Corollary 12.7.18 is related to the results of [213] and [315] concerning the smallest noetherian rings $\mathcal{A}(X) \subset \mathcal{C}^0(X)$ such that $\mathrm{Proj}(\mathcal{A}(X)) \to \mathrm{Proj}(\mathcal{C}^0(X))$ is bijective.

A systematic treatment of algebraic vector bundles over real algebraic varieties is given in a series of papers: [42, 45, 46, 47, 48, 49, 50, 53, 54, 55, 56, 62, 171, 197].

13. Polynomial or Regular Mappings with Values in Spheres

Abstract. Given a real algebraic set X, we compare the set of polynomial or regular mappings $X \to S^k$ with the corresponding set of continuous or smooth mappings. The results concerning this comparison are very diverse. Section 1 deals with the existence of nonconstant polynomial mappings from S^n into S^k, using mainly the theory of quadratic forms. We prove Wood's theorem, which states that if n is a power of 2 and $k < n$, then every polynomial mapping $S^n \to S^k$ is constant. The Hopf forms are the best known polynomial mappings $S^n \to S^k$. In Section 2, we study the geometry of Hopf forms, which, in turn, is useful for investigating the existence of such forms. Section 3 contains results concerning the set of regular mappings with values in S^1, S^2 or S^4. The choice of these particular spheres is related to the fact that S^1, S^2 and S^4 are biregularly isomorphic, respectively, to the real, complex and quaternionic projective lines. The theory of algebraic vector bundles developed in the previous chapter plays a crucial role. For example, from the fact that every topological ($\mathbb{R}$, $\mathbb{C}$ or $\mathbb{H}$-) line bundle over S^n is isomorphic to an algebraic one, we deduce that $\mathcal{R}(S^n, S^k)$ is dense in $\mathcal{C}^\infty(S^n, S^k)$ for $k = 1, 2, 4$. In Section 4, we study the subset of those homotopy classes of mappings $X \to S^k$ which are represented by regular mappings. We obtain interesting results especially when k is odd. For example, we show that every element of $2\,\pi_n(S^k) \subset \pi_n(S^k)$ can be represented by a regular mapping (when, in addition, $n < 2k - 1$). Finally, the last section contains the characterization of the n-tuples $q_1, \ldots, q_n$ of positive integers such that every regular (resp. polynomial) mapping $S^{q_1} \times \cdots \times S^{q_n} \to S^{q_1 + \cdots + q_n}$ is homotopic to a constant. For this we use some concepts from K-theory.

Throughout this chapter, we work over the field $\mathbb{R}$ of real numbers (but almost all results of the first two sections remain valid over an arbitrary real closed field). *The algebraic sets and affine real algebraic varieties considered in this chapter are thus over* $\mathbb{R}$.

13.1 Polynomial Mappings from S^n into S^k

In this section we investigate the existence of nonconstant polynomial mappings from S^n into S^k. This question is, of course, interesting only if $k < n$, since the canonical inclusion $S^n \hookrightarrow S^k$ is a nonconstant polynomial mapping for $k \geq n$. We begin with two observations. First, we note that the crucial case is the case $k = n - 1$.

Lemma 13.1.1. *There exists a nonconstant polynomial mapping from S^n into S^k (with $k < n$) if and only if, for each $i = 0, 1, \ldots, n - k - 1$, there exists a nonconstant polynomial mapping from S^{n-i} into S^{n-i-1}.*

Proof. We first observe that if $f : S^n \to S^p$ and $g : S^p \to S^q$ are two nonconstant polynomial mappings, then we can find an orthogonal transformation $\beta : S^p \to S^p$ such that $g \circ \beta \circ f$ is nonconstant. The lemma follows using, for the necessary condition, the above remark concerning the inclusions of spheres. $\square$

A mapping $h : S^n \to S^k$ is called a *form* (of degree d) if there exist homogeneous polynomials $H_1, \ldots, H_{k+1} : \mathbb{R}^{n+1} \to \mathbb{R}$ of degree d such that $h(x) = (H_1(x), \ldots, H_{k+1}(x))$ for every $x \in S^n$. The question of the existence of nonconstant polynomial mappings $S^n \to S^k$ is equivalent to the analogous question for forms.

Lemma 13.1.2. *If there exists a nonconstant polynomial mapping $S^n \to S^k$, then there exists a nonconstant form $S^n \to S^k$.*

Proof. Let $q : S^n \to S^n$ be the quadratic form given by

$$q(x) = (x_1^2 - x_2^2 - \cdots - x_{n+1}^2, 2x_1x_2, \ldots, 2x_1x_{n+1}).$$

The form q induces a diffeomorphism from an open neighbourhood of the point $(1, 0, \ldots, 0)$ in S^n onto another open neighbourhood of the same point in S^n. If $f : S^n \to S^k$ is a nonconstant polynomial mapping, $f \circ q : S^n \to S^k$ is also nonconstant. Moreover, each monomial occuring in $f \circ q$ is of even degree. Multiplying each of these monomials by an appropriate power of $\|x\|^2 = x_1^2 + x_2^2 + \cdots + x_{n+1}^2$ (which does not change the value of the mapping on S^n), we obtain a nonconstant form from S^n into S^k. $\square$

Note that every form $S^n \to S^k$ is necessarily of even degree if $n > k$. This follows from the well-known theorem stating that, if $n > k$, there is no continuous mapping $h : S^n \to S^k$ satisfying $h(-x) = -h(x)$ for every $x \in S^n$ (cf. [243], Theorem 68.5, p. 404).

To construct forms $S^n \to S^k$ of degree 2, we use the classical Hopf-Whitehead construction.

Definition 13.1.3. *A bilinear form $F : \mathbb{R}^p \times \mathbb{R}^q \to \mathbb{R}^k$ is said to be* normed *if it satisfies $\|F(x,y)\| = \|x\|\|y\|$, for every $(x, y) \in \mathbb{R}^p \times \mathbb{R}^q$. The* Hopf form *associated to a normed bilinear form $F : \mathbb{R}^p \times \mathbb{R}^q \to \mathbb{R}^k$ is the form $\varphi : S^{p+q-1} \to S^k$ of degree 2 defined by*

$$\varphi(x, y) = (\|x\|^2 - \|y\|^2, 2F(x, y)).$$

Example 13.1.4. The multiplication of complex numbers, quaternions and Cayley numbers provide examples of normed bilinear forms $\mathbb{R}^k \times \mathbb{R}^k \to \mathbb{R}^k$ for $k = 2, 4, 8$, respectively. The associated Hopf forms are the classical Hopf fibrations $S^{2k-1} \to S^k$ for $k = 2, 4, 8$ (cf. [233], p. 102 and Example 13.2.5 below).

A Hopf form is, of course, nonconstant. Thus, the existence of a normed bilinear form $\mathbb{R}^p \times \mathbb{R}^q \to \mathbb{R}^k$ (with $p + q = n + 1$) implies the existence of a nonconstant polynomial mapping $S^n \to S^k$. Note that the existence of a normed bilinear form $\mathbb{R}^p \times \mathbb{R}^q \to \mathbb{R}^k$ is equivalent to the existence of a formula for the product of sums of squares of the type:

$$(x_1^2 + \cdots + x_p^2)(y_1^2 + \cdots + y_q^2) = (\phi_1(x,y))^2 + \cdots + (\phi_k(x,y))^2,$$

where $\phi_1, \ldots, \phi_k$ are bilinear forms in (x,y) with coefficients in $\mathbb{R}$.

Notation 13.1.5. (i) *Let p and q be two positive integers. The smallest positive integer k such that there exists a normed bilinear form from $\mathbb{R}^p \times \mathbb{R}^q \to \mathbb{R}^k$ is denoted by $p * q$.*

(ii) *For every positive integer k represented as $k = 2^{4a+b}d$, where $a, b, d \in \mathbb{N}$, d odd and $0 \leq b \leq 3$, denote $\rho(k) = 8a + 2^b$. The function ρ is called the Hurwitz-Radon function.*

Theorem 13.1.6 (Hurwitz-Radon Theorem). *For every positive integer k, one has $r * k = k$ if and only if $r \leq \rho(k)$.*

Reference for the proof: [204], Theorem 5.11, p. 137.

This result, together with the trivial fact that a normed bilinear form $\mathbb{R}^r \times \mathbb{R}^k \to \mathbb{R}^k$ induces a normed bilinear form $\mathbb{R}^r \times \mathbb{R}^k \to \mathbb{R}^{k+q}$ for every $q \in \mathbb{N}$, provides a sufficient condition for the existence of Hopf forms.

Corollary 13.1.7. *For every $r, k, q \in \mathbb{N}$ with $0 < k$ and $0 < r \leq \rho(k)$, there exists a Hopf form from S^{r+k-1} into S^{k+q}.*

Corollary 13.1.8. *If $n = 2^c d$ with $0 \leq c \leq 3$, d odd and $d \neq 1$, there exists a Hopf form from S^n into S^{n-1}.*

Proof. We have to find r, k and q satisfying the assumption of Corollary 13.1.7 and such that $r + k - 1 = n$ and $k + q = n - 1$. This amounts to choosing k with $0 < k < n$ and $n - k + 1 \leq \rho(k)$. We take $k = n - 2^c = 2^c(d-1)$. Since $d - 1$ is even, we have $\rho(k) \geq 2$ if $c = 0$, $\rho(k) \geq 4$ if $c = 1$, $\rho(k) \geq 8$ if $c = 2$ and $\rho(k) \geq 9$ if $c = 3$. $\square$

This method is limited to the case $0 \leq c \leq 3$, since $\rho(2^5) = 10$ and $2^4 + 1 > 10$. We shall show in Example 13.2.16 c) that there is no Hopf form $S^n \to S^{n-1}$ for $n = 2^4 \cdot 3$.

The following theorem settles the case where n is a power of 2.

Theorem 13.1.9. *If n is a power of 2 and $k < n$, every polynomial mapping from S^n into S^k is constant.*

Proof. By Lemmas 13.1.1 and 13.1.2, it suffices to prove that every form $S^n \to S^{n-1}$ is constant if n is a power of 2. Let $h : S^n \to S^{n-1}$ be a form of degree d given by homogeneous polynomials $H_1, \ldots, H_n$ in the variables $x_1, \ldots, x_{n+1}$. These polynomials satisfy

$$\|H\|^2 = H_1^2 + \cdots + H_n^2 = (x_1^2 + \cdots + x_{n+1}^2)^d = \|x\|^{2d}.$$

Assume that h is a nonconstant form. Then $d > 0$. We may assume that $H_1, \ldots, H_n$ are not all divisible by $\|x\|^2$. Represent H as

$$H(x_1, \ldots, x_{n+1}) = P(x_1, \ldots, x_n, x_{n+1}^2) + x_{n+1}Q(x_1, \ldots, x_n, x_{n+1}^2),$$

where $P(x_1, \ldots, x_n, u)$ and $Q(x_1, \ldots, x_n, u)$ are systems of n polynomials. It follows from the equality $\|H\|^2 = \|x\|^{2d}$ that

(1) $$\|P\|^2 + u\|Q\|^2 = (x_1^2 + \cdots + x_n^2 + u)^d,$$

(2) $$P \cdot Q = \sum_{i=1}^{n} P_i Q_i = 0.$$

Now substitute $u = -(x_1^2 + \cdots + x_n^2)$ in equality (1). It follows that

$$\|F\|^2 - (x_1^2 + \cdots + x_n^2)\|G\|^2 = 0,$$

where $F(x_1, \ldots, x_n) = P(x_1, \ldots, x_n, -(x_1^2 + \cdots + x_n^2))$ and $G(x_1, \ldots, x_n) = Q(x_1, \ldots, x_n, -(x_1^2 + \cdots + x_n^2))$. One has $G \neq 0$ since otherwise $F = G = 0$, which would imply that $u + x_1^2 + \cdots + x_n^2$ divides all components of P and Q, and, hence, that $\|x\|^2$ divides $H_1, \ldots, H_n$. We can thus write $x_1^2 + \cdots + x_n^2 = \|F\|^2/\|G\|^2$ in $\mathbb{R}(x_1, \ldots, x_n)$.

Equality (2) implies $F \cdot G = 0$. Hence

$$x_1^2 + \cdots + x_n^2 + x_{n+1}^2 = \|F + x_{n+1}G\|^2/\|G\|^2.$$

Setting $V = (F + x_{n+1}G)/\|G\|^2$, we obtain

$$x_1^2 + \cdots + x_n^2 + x_{n+1}^2 = \|V\|^2\|G\|^2.$$

Since n is a power of 2 and $\|V\|^2$ and $\|G\|^2$ are sums of n squares in the field $\mathbb{R}(x_1, \ldots, x_{n+1})$, their product $\|V\|^2\|G\|^2$ would also be the sum of n squares in $\mathbb{R}(x_1, \ldots, x_{n+1})$ (Corollary 6.4.13). However, $x_1^2 + \cdots + x_{n+1}^2$ is not a sum of n squares in $\mathbb{R}(x_1, \ldots, x_{n+1})$ (Corollary 6.4.8). Thus the assumption that h is nonconstant leads to a contradiction. $\square$

Corollary 13.1.8 and Theorem 13.1.9 thus solve the problem of the existence of nonconstant polynomial mappings from S^n into S^k for every n not exceeding $47 = 2^4 \cdot 3 - 1$

13.1 Polynomial Mappings from S^n into S^k

Corollary 13.1.10. *Let n and k be positive integers, with $n > k$. Consider the following two properties:*
  (i) *There exists a nonconstant polynomial mapping from S^n into S^k.*
  (ii) *There exists an integer r with $2^r \leq k < n < 2^{r+1}$.*
Then (i) $\Rightarrow$ (ii). Furthermore, (ii) $\Rightarrow$ (i) if $n \leq 47$.

The question whether (ii) $\Rightarrow$ (i) remains open. The table below allows one to grasp better the situation. The sign $+$ (resp. $-$) denotes the existence (resp. nonexistence) of a nonconstant polynomial mapping from S^n into S^k, and the sign ? denotes the unknown cases.

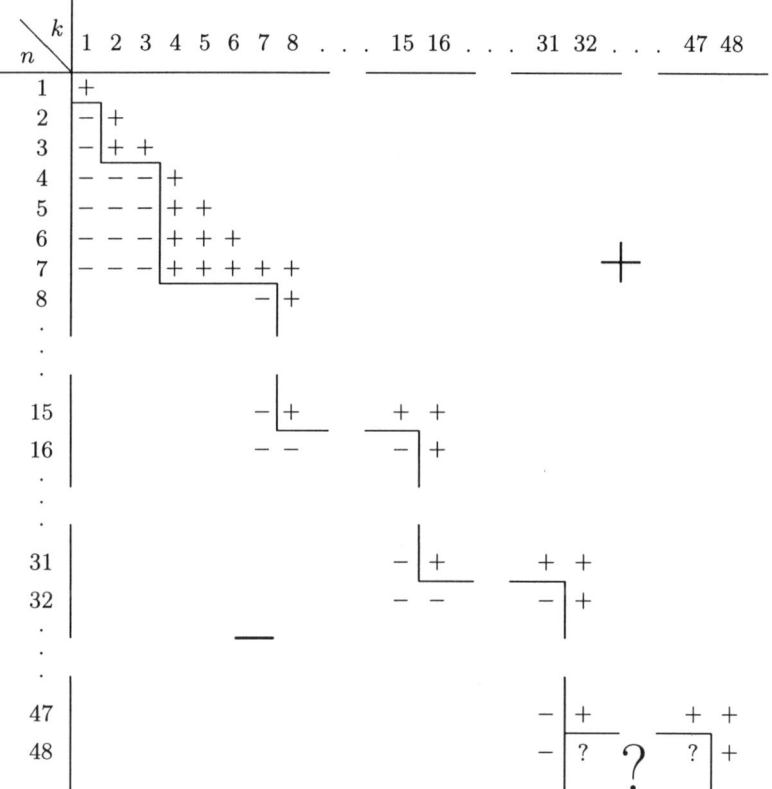

Figure 13.1.

The case $S^{48} \to S^{47}$ is thus the first that remains open. We shall show, in Example 13.2.16 c), that there is no nonconstant form $S^{48} \to S^{47}$ of degree 2. The next result relates Hopf forms to general forms of degree 2.

Theorem 13.1.11. *Let $q : S^n \to S^k$ be a nonconstant form of degree 2. There exist orthogonal transformations $\beta_1 : S^n \to S^n$ and $\beta_2 : S^k \to S^k$ and a Hopf form $\varphi : S^n \to S^k$ such that $\beta_2 \circ q \circ \beta_1$ and φ are homotopic.*

Proof. The form q of degree 2 is induced by a system $Q = (Q_1, \ldots, Q_{k+1})$ of $k+1$ homogeneous polynomials of degree 2 such that, for every $x \in \mathbb{R}^{n+1}$,

$$Q_1(x)^2 + \cdots + Q_{k+1}(x)^2 = \|Q(x)\|^2 = \|x\|^4 \ .$$

Hence, the coefficient of x_i^2 (for $i = 1, \ldots, n+1$) in Q is a vector of $\mathbb{R}^{k+1}$ of length 1. Let $\beta_2 : \mathbb{R}^{k+1} \to \mathbb{R}^{k+1}$ be an orthogonal mapping sending the coefficient of x_1^2 in Q to the vector $(1, 0, \ldots, 0)$. Composing with β_2, we can assume that the first coordinate of Q can be written as

$$Q_1(x) = x_1^2 + x_1 L(x_2, \ldots, x_{n+1}) + C(x_2, \ldots, x_{n+1}) \ ,$$

where L and C are linear and quadratic forms, respectively, and that $Q_2, \ldots, Q_{n+1}$ contain no term involving x_1^2. Comparing the coefficients of x_1^3 in the equality $\|Q(x)\|^2 = \|x\|^4$, we obtain $L = 0$. Composing Q with an appropriate orthogonal transformation $\beta_1 : \mathbb{R}^{n+1} \to \mathbb{R}^{n+1}$ which is the identity on the x_1-axis, we may moreover assume that C is a diagonal form in $x_2, \ldots, x_{n+1}$. The first coordinate of Q is thus of the form

$$Q_1(x) = x_1^2 + \lambda_2 x_2^2 + \cdots + \lambda_{n+1} x_{n+1}^2 \ ,$$

where $|\lambda_i| \leq 1$ for $i = 2, \ldots, n+1$. The λ_i are not all equal to 1, since this would imply $Q_i = 0$ for $i > 1$ and q would be constant. We change the name of the variables, so that Q_1 is written as

$$Q_1(y, z) = y_1^2 + \cdots + y_r^2 + \mu_1 z_1^2 + \cdots + \mu_s z_s^2 = \|y\|^2 + K(z) \ ,$$

with $0 < r$, $0 < s$, $r + s = n + 1$ and $-1 \leq \mu_i < 1$ for $i = 1, \ldots, s$.

The remaining k coordinates of Q can then be written as the vector $A(y) + B(z) + 2G(y, z)$, where A and B are quadratic in y and z, respectively, and G is bilinear in y and z. The equality $\|Q(y, z)\|^2 = (\|y\|^2 + \|z\|^2)^2$ implies $A = 0$ and the following equalities:

(1) $\qquad\qquad\qquad \|B(z)\|^2 + K(z)^2 = \|z\|^4 \ ,$

(2) $\qquad\qquad\qquad B(z) \cdot G(y, z) = 0 \ ,$

(3) $\qquad\qquad\qquad 2\|G(y, z)\|^2 + \|y\|^2 K(z) = \|y\|^2 \|z\|^2 \ .$

We shall now construct a homotopy transforming q into a Hopf form. It suffices to construct such a homotopy in $\mathbb{R}^{k+1} \setminus \{0\}$. We first transform Q_1 into $\|y\|^2 - \|z\|^2$ and B into 0 by setting

$$H(t, y, z) = (\|y\|^2 + (1-t)K(z) - t\|z\|^2, (1-t)B(z) + 2G(y, z)) \ .$$

We have $H(0, y, z) = Q(y, z)$. We claim that $H(t, y, z) \in \mathbb{R}^{k+1} \setminus \{0\}$ if $t \in [0, 1]$ and $(y, z) \in S^n$. Suppose $H(t, y, z) = 0$. By (2), $B(z)$ and $G(y, z)$ are orthogonal, and therefore $(1-t)B(z) = G(y, z) = 0$. If $t \neq 1$, then $B(z) = 0$. Hence, equality (1) and the inequalities $\mu_i < 1$ imply $z = 0$, from which it follows that $\|y\|^2 = 1$ and $H(t, y, z) = (1, 0, \ldots, 0)$. If $t = 1$, then $G(y, z) = 0$,

(3) and the inequalities $\mu_i < 1$ imply that $y = 0$ or $z = 0$, from which it follows that $H(t, y, z) = (\pm 1, 0, \ldots, 0)$. Both cases lead to a contradiction, which proves the claim.

It remains to transform $H(1, y, z) = (\|y\|^2 - \|z\|^2, 2G(y, z))$ into a Hopf form. For this, it suffices to transform G into a normed bilinear form. Note that equality (3) can be rewritten as

$$2\|G(y,z)\|^2 = \|y\|^2((1-\mu_1)z_1^2 + \cdots + (1-\mu_s)z_s^2).$$

We now construct $F(t, y, z)$ by replacing, in $G(y, z)$, the variable z_i with $z_i\sqrt{2}/\sqrt{2 - t(1 + \mu_i)}$, for $i = 1, \ldots, s$. We have obviously $F(0, y, z) = G(y, z)$. Observe that

$$\|F(t,y,z)\|^2 = \|y\|^2 \sum_{i=1}^{s}(1-\mu_i)z_i^2/(2-t(1+\mu_i)),$$

which shows that $F(0, y, z)$ is a normed bilinear form. Moreover, for every $t \in [0, 1]$, $F(t, y, z)$ vanishes only if $y = 0$ or $z = 0$. Hence, $(\|y\|^2 - \|z\|^2, 2F(t, y, z))$ does not vanish for $t \in [0, 1]$ and $(y, z) \in S^n$. It follows that the Hopf form $\varphi(y, z) = (\|y\|^2 - \|z\|^2, 2F(1, y, z))$ is homotopic to q in $\mathbb{R}^{k+1} \setminus \{0\}$ and, hence, in S^k. This completes the proof of the theorem. □

A criterion for a nonconstant form $q : S^n \to S^k$ of degree 2 to be a Hopf form (up to orthogonal transformations) is given in [345].

Remark 13.1.12. The previous theorem is certainly not valid for forms of arbitrary degree. For example, consider the case $S^{31} \to S^{16}$. Corollary 13.1.7 ensures the existence of Hopf forms $S^{31} \to S^{24}$ and $S^{24} \to S^{16}$. Hence, there exists a nonconstant form $S^{31} \to S^{16}$ of degree 4. However, there is no Hopf form (and, by the previous theorem, no nonconstant form of degree 2) from S^{31} into S^{16}. Indeed, the easy observation that $r * s \geq \max(r, s)$ shows that such a form could only originate from a normed bilinear form $\mathbb{R}^{16} \times \mathbb{R}^{16} \to \mathbb{R}^{16}$. But, since $\rho(16) = 9$, the Hurwitz-Radon theorem implies that such a normed bilinear form does not exist.

The results concerning the existence of nonconstant polynomial mappings contained in this section are valid only for *standard spheres*. They are not valid, in general, for algebraic sets biregularly isomorphic to spheres. Indeed, the technique involving quadratic forms used above cannot be applied if the algebraic set is not given by a quadratic equation. We shall see now what can happen with "false" spheres.

Lemma 13.1.13. *For every algebraic set $M \subset \mathbb{R}^p$ containing more than one point and for every positive integer k, there exists a nonconstant regular mapping $M \to S^k$.*

Proof. Choose a linear mapping $\mathbb{R}^p \to \mathbb{R}^k$ which is not constant on M, and compose it with the inverse of the stereographic projection $\mathbb{R}^k \to S^k$. □

Lemma 13.1.14. *Let M and X be two algebraic sets, and $f : M \to X$, a regular mapping. There exist an algebraic set M', a biregular isomorphism $g : M \to M'$ and a polynomial mapping $h : M' \to X$ such that $f = h \circ g$.*

Proof. Take the graph of f for M', $g : M \to M'$ defined by $g(u) = (u, f(u))$ and $h : M' \to X$ defined by $h(u, x) = x$. □

Proposition 13.1.15. *Let n and k be positive integers with $k \leq n/2$. Then there exists an algebraic set Σ^n, biregularly isomorphic to S^n, such that*
  (i) *there exists a nonconstant polynomial mapping $\Sigma^n \to S^k$;*
  (ii) *for every polynomial mapping $f : S^n \to \Sigma^n$, one has $\dim(f(S^n)) < n$.*
In particular, every polynomial mapping $S^n \to \Sigma^n$ has topological degree 0.

Proof. Statement (i) follows from Lemmas 13.1.13 and 13.1.14, and (ii) is a consequence of (i). Indeed, if $g : \Sigma^n \to S^k$ is a nonconstant polynomial mapping, and $f : S^n \to \Sigma^n$, a polynomial mapping with $\dim(f(S^n)) = n$, then $g \circ f : S^n \to S^k$ is nonconstant. Since $k \leq n/2$, this contradicts Corollary 13.1.10. □

Remark 13.1.16. We shall prove in Section 3 that, for $n = 2$ or 4 and Σ^n as above, the set $\mathcal{R}(S^n, \Sigma^n)$ of regular mappings from S^n into Σ^n is dense in the set of $\mathcal{C}^\infty$ mappings. The situations for regular and polynomial mappings are thus very different.

13.2 Hopf Forms and Nonsingular Bilinear Forms

In this section, we study Hopf forms in a more detailed way, and obtain necessary conditions for their existence. This study will use the class of nonsingular bilinear forms, which is larger than the class of normed bilinear forms.

Definition 13.2.1. *A bilinear form $f : \mathbb{R}^r \times \mathbb{R}^s \to \mathbb{R}^k$ is said to be nonsingular if $f(x, y) = 0$ implies $x = 0$ or $y = 0$.*

If r and s are positive integers, $r \# s$ denotes the smallest positive integer k such that there exists a nonsingular bilinear form $\mathbb{R}^r \times \mathbb{R}^s \to \mathbb{R}^k$.

Proposition 13.2.2. *For all positive integers r and s, one has*
  (i) $\max(r, s) \leq r \# s \leq r * s$,
  (ii) $r \# s \leq r + s - 1$.

Proof. (i) is obvious. For (ii), identify $\mathbb{R}^n$ with the space of coefficients of polynomials in one variable of degree at most $n - 1$. With this identification, the product of polynomials defines a nonsingular bilinear form $\mathbb{R}^r \times \mathbb{R}^s \to \mathbb{R}^{r+s-1}$. □

Using various topological methods one obtains the following results concerning $r \# s$.

13.2 Hopf Forms and Nonsingular Bilinear Forms 347

Theorem 13.2.3 (Stiefel-Hopf condition). *If $r\#s \leq k$, the binomial coefficient $\binom{k}{i}$ is even, for every i such that $k - r < i < s$.*

Reference for the proof: [180], p. 141. The result is obtained by "projectivizing" a nonsingular bilinear form $\mathbb{R}^r \times \mathbb{R}^s \to \mathbb{R}^k$ into a mapping $\mathbb{P}_{r-1}(\mathbb{R}) \times \mathbb{P}_{s-1}(\mathbb{R}) \to \mathbb{P}_{k-1}(\mathbb{R})$, and considering the induced homomorphism between the corresponding cohomology rings with coefficients in $\mathbb{Z}/2$. □

Theorem 13.2.4. *Given positive integers r and k, the following properties are equivalent:*
  (i) $r\#k = k$.
  (ii) $r * k = k$.
  (iii) $r \leq \rho(k)$.

Reference for the proof: The implication (i) ⇒ (ii) is proved in [296], Corollary 2.7. The equivalence (ii) ⇔ (iii) is just the Hurwitz-Radon theorem 13.1.6. □

Studying the geometry of Hopf forms, we shall exhibit a family of nonsingular bilinear forms "hidden" behind a Hopf form. We follow rather closely K.Y. Lam's paper [203]. The first step consists in generalizing a property of the Hopf fibrations (Example 13.1.4).

Example 13.2.5. The fibration $\varphi : S^3 \to S^2$, associated to multiplication of complex numbers, is defined by

$$\varphi(z_1, z_2) = (|z_1|^2 - |z_2|^2, 2z_1 z_2), \text{ for } (z_1, z_2) \in S^3 \subset \mathbb{C} \times \mathbb{C} \simeq \mathbb{R}^4.$$

One has $\varphi(z_1, z_2) = \varphi(z'_1, z'_2)$ if and only if there exists $t \in \mathbb{R}$ such that $z'_1 = e^{it} z_1$ and $z'_2 = e^{-it} z_2$. The fibres of $\varphi : S^3 \to S^2$ are thus great circles of S^3.

Lemma 13.2.6. *For every normed bilinear form $f : \mathbb{R}^2 \times \mathbb{R}^2 \to \mathbb{R}^3$, the image of f is a vector subspace of dimension 2 of $\mathbb{R}^3$. Furthermore, there exist an isometry $\beta : \mathbb{R}^2 \to \mathbb{R}^2$ and an isometry $\mu : \text{Im}(f) \to \mathbb{R}^2$ such that the composition:*

$$g : \mathbb{R}^2 \times \mathbb{R}^2 \xrightarrow{\beta \times \text{Id}} \mathbb{R}^2 \times \mathbb{R}^2 \xrightarrow{f} \text{Im}(f) \xrightarrow{\mu} \mathbb{R}^2$$

coincides with multiplication in $\mathbb{C} \simeq \mathbb{R}^2$:

$$g((x_1, x_2), (y_1, y_2)) = (x_1 y_1 - x_2 y_2, \; x_1 y_2 + x_2 y_1).$$

Proof. Denote by e_1, e_2 the canonical basis of $\mathbb{R}^2$, and, for $y \in \mathbb{R}^2$, set $\widehat{e}_i(y) = f(e_i, y)$. The $\widehat{e}_i$ are isometric embeddings $\mathbb{R}^2 \to \mathbb{R}^3$, and, composing with an orthogonal transformation of $\mathbb{R}^3$, we may assume that

$$\begin{aligned} \widehat{e}_1((y_1, y_2)) &= (y_1, y_2, 0), \\ \widehat{e}_2((y_1, y_2)) &= (ay_1 + by_2, a'y_1 + b'y_2, a''y_1 + b''y_2). \end{aligned}$$

Using the fact that f is normed, one sees that the two vectors $\widehat{e}_i((y_1, y_2))$, $i = 1, 2$, are orthogonal and $\|\widehat{e}_2((y_1, y_2))\|^2 = y_1^2 + y_2^2$ for every $(y_1, y_2) \in \mathbb{R}^2$. It follows that $a = b' = 0$, $b = -a'$, $b^2 = 1$ and $a'' = b'' = 0$. Composing, if necessary, with the symmetry sending e_2 to $-e_2$, we may assume that $b = -1$. The lemma follows. □

Remark 13.2.7. The previous property is not satisfied, in general, by a nonsingular bilinear form. A counter-example is provided by $f : \mathbb{R}^2 \times \mathbb{R}^2 \to \mathbb{R}^3$ defined by

$$f((x_1, x_2), (y_1, y_2)) = (x_1y_1, x_1y_2 + x_2y_1, x_2y_2).$$

Also, Lemma 13.2.6 does not hold with $\mathbb{R}^3$ replaced by $\mathbb{R}^4$. A counter-example is obtained by considering the restriction of multiplication of quaternions $\mathbb{R}^4 \times \mathbb{R}^4 \to \mathbb{R}^4$ to $W_1 \times W_2$, where W_1 (resp. W_2) is the $\mathbb{R}$-vector space generated by 1 and i (resp. i and k). The image of $W_1 \times W_2$ generates $\mathbb{R}^4$ over $\mathbb{R}$.

Theorem 13.2.8. *Let $f : \mathbb{R}^r \times \mathbb{R}^s \to \mathbb{R}^k$ be a normed bilinear form and $\varphi : S^{r+s-1} \to S^k$ the Hopf form associated to f. For every point $z \in S^k$, there exists a vector subspace W of $\mathbb{R}^r \times \mathbb{R}^s$ such that $\varphi^{-1}(z) = W \cap S^{r+s-1}$. In particular, $\varphi^{-1}(z)$ is a subsphere of S^{r+s-1}.*

Proof. The case $z = (\pm 1, 0, \ldots, 0)$ is easy since $\varphi^{-1}(1, 0, \ldots, 0) = (\mathbb{R}^r \times \{0\}) \cap S^{r+s-1}$ and $\varphi^{-1}(-1, 0, \ldots, 0) = (\{0\} \times \mathbb{R}^s) \cap S^{r+s-1}$. Assume that $z \neq (\pm 1, 0, \ldots, 0)$. Assume also that $\varphi^{-1}(z)$ neither is empty nor consists of two antipodal points (these are cases corresponding to, respectively, $\dim(W) = 0$ and $\dim(W) = 1$). Consider (x, y) and (x', y') in $\varphi^{-1}(z)$ with $(x', y') \neq \pm(x, y)$. Then x and x' are not colinear. Indeed, since $\|x\|^2 + \|y\|^2 = \|x'\|^2 + \|y'\|^2 = 1$ and $\|x\|^2 - \|y\|^2 = \|x'\|^2 - \|y'\|^2$, we have $\|x\| = \|x'\|$. If x and x' were colinear, we would have $x' = \pm x$. Since f is nonsingular and $f(x, y) = f(x', y')$, it would follow that $(x', y') = \pm(x, y)$. Hence, x and x' generate a vector subspace W_1 of $\mathbb{R}^r$ of dimension 2. In the same way, y and y' generate a vector subspace W_2 of $\mathbb{R}^s$ of dimension 2. The image of $W_1 \times W_2$ by f is contained in the subspace V of $\mathbb{R}^k$ generated by $f(x, y), f(x, y'), f(x', y)$ and $f(x', y')$. Since $f(x, y) = f(x', y')$, the dimension of the subspace V is at most 3. Lemma 13.2.6 implies that, in fact, V is of dimension 2 and the normed bilinear form $\overline{f} : W_1 \times W_2 \to V$ induced by f is multiplication of complex numbers composed with isometries. We already mentioned (Example 13.2.5) that the fibres of the Hopf fibration $S^3 \to S^2$ associated to multiplication of complex numbers are great circles. Hence, the great circle of S^{r+s-1} passing through (x, y) and (x', y') is entirely contained in $\varphi^{-1}(z)$. From this property we deduce that $\varphi^{-1}(z)$ is the intersection of S^{r+s-1} with a vector subspace of $\mathbb{R}^{r+s}$. □

We shall show now how each Hopf form produces a family of nonsingular bilinear forms.

13.2 Hopf Forms and Nonsingular Bilinear Forms

Let $f, \varphi : S^{r+s-1} \to S^k$, z and W be as in Theorem 13.2.8, with z in the image of φ. Let $W^\perp$ be the subspace of $\mathbb{R}^r \times \mathbb{R}^s$ orthogonal to W. Define the mapping
$$g_z : W \times W^\perp \to T_z(S^k)$$
by $g_z(\lambda w, t) = \lambda\, d_w\varphi(t)$, for $w \in \varphi^{-1}(z) = W \cap S^{r+s-1}$, $\lambda \in \mathbb{R}$, $\lambda \geq 0$, where $d_w\varphi : T_w(S^{r+s-1}) \to T_z(S^k)$ is the derivative of φ at w.

Theorem 13.2.9. *The mapping g_z is a nonsingular bilinear form.*

Proof. Since a vector orthogonal to W is tangent to S^{r+s-1} at every point of $W \cap S^{r+s-1}$, the mapping g_z is well-defined. Setting $w = (x,y) \in W \subset \mathbb{R}^r \times \mathbb{R}^s$ and $t = (u,v) \in W^\perp \subset \mathbb{R}^r \times \mathbb{R}^s$, we can define g_z by the following explicit formula
$$g_z((x,y),(u,v)) = (2x \cdot u - 2y \cdot v, 2f(x,v) + 2f(u,y)),$$
from which the bilinearity of g_z is obvious.

It remains to prove that g_z is nonsingular. Assume $(x,y) \neq (0,0)$ and $(u,v) \neq (0,0)$ but $g_z((x,y),(u,v)) = 0$. We can assume that $(x,y) \in \varphi^{-1}(z)$, with $d_{(x,y)}\varphi(u,v) = 0$. Let W_1 (resp. W_2) be the vector subspace of $\mathbb{R}^r$ (resp. $\mathbb{R}^s$) generated by x and u (resp. y and v). Let V be the vector subspace of $\mathbb{R}^k$ generated by $f(W_1 \times W_2)$, and $\overline{f} : W_1 \times W_2 \to V$, the restriction of f to $W_1 \times W_2$.

If $\dim(W_1) = \dim(W_2) = 2$, then, since $f(x,v) + f(u,y) = 0$, we can proceed as in Theorem 13.2.8, and we obtain $\dim(V) = 2$ and $\overline{f}$ is multiplication of complex numbers composed with isometries. The Hopf fibration $\overline{\varphi} : S^3 \to S^2$ associated to $\overline{f}$ is a submersion, and, since $d_{(x,y)}\overline{\varphi}(u,v) = 0$, the vector (u,v) is tangent to the great circle $\overline{\varphi}^{-1}(z) \subset \varphi^{-1}(z)$. This implies that $(u,v) \in W$, contradicting the choice of (u,v).

If $x = 0$, then $z = (-1,0,\ldots,0)$, $W = \{0\} \times \mathbb{R}^s$, so that $v = 0$ since $(u,v) \in W^\perp$. The condition $g_z((x,y),(u,v)) = 0$ becomes $f(u,y) = 0$, and the fact that f is nonsingular implies $u = 0$, a contradiction again. We argue similarly if $y = 0$.

It remains to consider the case $x \neq 0$, $y \neq 0$ and $\dim(W_1) = 1$ or $\dim(W_2) = 1$. Composing with isometries, we may assume that $\overline{f}$ is either multiplication $\mathbb{R} \times \mathbb{R} \to \mathbb{R}$ or of one of the following two forms:

$$\begin{array}{ll}
\mathbb{R} \times \mathbb{R}^2 \to \mathbb{R}^2 & \mathbb{R}^2 \times \mathbb{R} \to \mathbb{R}^2 \\
(a,(b_1,b_2)) \mapsto (ab_1, ab_2) & ((a_1,a_2),b) \mapsto (a_1 b, a_2 b).
\end{array}$$

In each case, a straightforward computation of the associated Hopf form $\overline{\varphi}$ proves that $d_{(x,y)}\overline{\varphi}$ is an isomorphism. Hence, $d_{(x,y)}\overline{\varphi}(u,v) = 0$ implies $(u,v) = 0$. $\square$

In the next corollary we use the notation of Theorems 13.2.8 and 13.2.9.

Corollary 13.2.10. (i) *For every $w \in \varphi^{-1}(z)$, the space $W \cap (\mathbb{R}w)^\perp$ is the kernel of the derivative $d_w\varphi$.*

(ii) *If $\dim(W^\perp) = q$, the Hopf form $\varphi : S^{r+s-1} \to S^k$ has rank q at every point $w \in \varphi^{-1}(z)$.*

(iii) *For every point z in the image of φ, there exists a nonsingular bilinear form $g_z : \mathbb{R}^{r+s-q} \times \mathbb{R}^q \to \mathbb{R}^k$, where q is the rank of φ at some (and, thus, every) point of $\varphi^{-1}(z)$.*

Theorem 13.2.11. (i) *If there exists a surjective Hopf form $\varphi : S^n \to S^k$, then $n \geq k$ and $\rho(k) \geq n - k + 1$.*

(ii) *If there exists a form of degree 2 from S^n into S^k representing a nontrivial element of the homotopy group $\pi_n(S^k)$, then $n \geq k$ and $\rho(k) \geq n - k + 1$.*

Proof. (i) If φ is surjective, its maximal rank is equal to k (by Sard's theorem). By Corollary 13.2.10, there exists a nonsingular bilinear form from $\mathbb{R}^{n+1-k} \times \mathbb{R}^k$ into $\mathbb{R}^k$, and $(n + 1 - k)\#k = k$. We then apply Theorem 13.2.4.

(ii) Follows from (i), using Theorem 13.1.11 and the fact that a homotopically nontrivial mapping into S^k is necessarily surjective. □

Example 13.2.12. If k is odd and $k < n$, every form of degree 2 from S^n into S^k is homotopic to a constant mapping.

Remark 13.2.13. The condition $n \geq k$ and $\rho(k) \geq n - k + 1$ is not sufficient to insure the existence of a form of degree 2 from S^n into S^k homotopically nontrivial. For example, if k is even, every form $\alpha : S^k \to S^k$ of even degree is homotopically trivial since $\alpha(x) = \alpha(-x)$. The existence of a polynomial mapping from S^k into S^k, with k even, of *topological* degree different from $-1, 0$ or 1 is therefore unlikely and, at least, unknown. On the other hand, the Hopf form $S^{k+\rho(k)-1} \to S^k$ associated to the normed bilinear form $\mathbb{R}^k \times \mathbb{R}^{\rho(k)} \to \mathbb{R}^k$ given by the Hurwitz-Radon construction is homotopically nontrivial (K.Y. Lam informed us that this fact is implicitly contained in [24]).

Having in mind other applications of Corollary 13.2.10, it is useful to compare the maximal rank of a normed bilinear form and the maximal rank of the associated Hopf form.

Lemma 13.2.14. *Let $f : \mathbb{R}^r \times \mathbb{R}^s \to \mathbb{R}^k$ be a normed bilinear form, and $\varphi : S^{r+s-1} \to S^k$, the Hopf form associated to f. The maximal rank of φ is greater than or equal to the maximal rank of f.*

Proof. The maximal rank of f is certainly reached at a point $(x_0, y_0) \in \mathbb{R}^r \times \mathbb{R}^s$ such that $x_0 \neq 0$ and $y_0 \neq 0$, and, by bilinearity, we can assume that $\|x_0\|^2 = \|y_0\|^2 = 1/2$. Denote by $S_0^{r-1} \times S_0^{s-1}$ the nonsingular algebraic hypersurface of S^{r+s-1} defined by $\|x\|^2 = \|y\|^2 = 1/2$, and let S_1^{k-1} be

13.2 Hopf Forms and Nonsingular Bilinear Forms   351

the equator of S^k, consisting of those points in S^k whose first coordinate vanishes. Clearly $(x_0, y_0) \in S_0^{r-1} \times S_0^{s-1}$. Let $\varphi_0 : S_0^{r-1} \times S_0^{s-1} \to S_1^{k-1}$ be the restriction of φ. At every point (x, y) in $S_0^{r-1} \times S_0^{s-1}$, the vector $(-x, y)$ is tangent to S^{r+s-1} and normal to $S_0^{r-1} \times S_0^{s-1}$. Moreover,

$$d_{(x,y)}\varphi(-x, y) = (2x \cdot (-x) - 2y \cdot y, \ 2f(x,y) + 2f(-x,y)) = (-2, 0),$$

which is a vector normal to S_1^{k-1} in $\mathbb{R}^{k+1}$. We thus have $\operatorname{rank}(d_{(x,y)}\varphi) = 1 + \operatorname{rank}(d_{(x,y)}\varphi_0)$ for every $(x,y) \in S_0^{r-1} \times S_0^{s-1}$. Since $d_{(x,y)}f(x,y) = 2f(x,y)$ is normal to $T_{2f(x,y)}(S^{k-1})$, $d_{(x,y)}f(-x,y) = 0$ and $\varphi_0(x,y) = (0, 2f(x,y))$, we have $\operatorname{rank}(d_{(x,y)}f) = 1 + \operatorname{rank}(d_{(x,y)}\varphi_0)$. From these two equalities, it follows that $\operatorname{rank}(d_{(x,y)}f) = \operatorname{rank}(d_{(x,y)}\varphi)$ for every $(x,y) \in S_0^{r-1} \times S_0^{s-1}$, which proves the lemma. □

Theorem 13.2.15. *Let $f : \mathbb{R}^r \times \mathbb{R}^s \to \mathbb{R}^k$ be a normed bilinear form. Then*
  (i) *The maximal rank of f is greater than or equal to $r\#s$.*
  (ii) *There exists at least one nonsingular bilinear form $\mathbb{R}^{r+s-q} \times \mathbb{R}^q \to \mathbb{R}^k$ with $r\#s \leq q \leq k$.*

Proof. (i) Let V be a vector subspace of $\mathbb{R}^k$ of maximal dimension satisfying the property $V \cap f(\mathbb{R}^r \times \mathbb{R}^s) = \{0\}$, and let $\Pi : \mathbb{R}^k \to V^\perp$ be the orthogonal projection. It is clear that $\Pi \circ f$ is a nonsingular bilinear form. Moreover, the maximality of V implies that $\Pi \circ f$ is surjective. Hence, the maximal rank of f, which is not smaller than the rank of $\Pi \circ f$, is not smaller than the dimension of $V^\perp$, and the latter is not smaller than $r\#s$.
  (ii) follows from (i), using Lemma 13.2.14 and Corollary 13.2.10 (iii). □

Example 13.2.16. a) If $r\#s = r*s$, then $r + s - (r*s) \leq \rho(r*s)$. Indeed, it suffices to apply Theorem 13.2.15 (ii) to a normed bilinear form from $\mathbb{R}^r \times \mathbb{R}^s$ into $\mathbb{R}^{r*s}$, and the claim follows from Theorem 13.2.4. In particular, using Proposition 13.2.2 (ii) we have that $r*s = r + s - 1$ if $r\#s = r*s$ and $r\#s$ is odd.

  b) The values of $r*s$ for $1 \leq r \leq 9$ are known and one has then $r\#s = r*s$ [203]. The first unknown case is $10*11$. Using topological methods, one can prove that $10\#11 = 17$ [201]. On the other hand, if $10*11$ were 19 or less, Theorem 13.2.15 (ii) would provide a nonsingular bilinear form $\mathbb{R}^{21-q} \times \mathbb{R}^q \to \mathbb{R}^{19}$ with $17 \leq q \leq 19$. This, however, is excluded by the Stiefel-Hopf condition (Theorem 13.2.3). We thus have $10*11 \geq 20$, and this gives the first example for which $r*s > r\#s$.

  c) There is no nonconstant form of degree 2 from S^{48} into S^{47} (which, of course, does not solve the analogous problem for polynomial mappings). It suffices to observe that, by Theorem 13.1.11, there is no normed bilinear form $\mathbb{R}^r \times \mathbb{R}^{49-r} \to \mathbb{R}^{47}$. Indeed, topological considerations (Stiefel-Whitney classes, Stiefel-Hopf condition, etc...) show that the only possible value for r is 17. For instance, one can verify that $\binom{47}{i}$ is odd for $i \geq 32$, and, hence, the

Stiefel-Hopf condition implies the nonexistence of nonsingular bilinear forms $\mathbb{R}^r \times \mathbb{R}^{49-r} \to \mathbb{R}^{47}$ for $r \leq 16$, etc. Assume therefore that there is a normed bilinear form $\mathbb{R}^{17} \times \mathbb{R}^{32} \to \mathbb{R}^{47}$. By Theorem 13.2.4, one has $17\#32 > 32$. By Theorem 13.2.15 (ii) there would be a nonsingular bilinear form $\mathbb{R}^{49-q} \times \mathbb{R}^q \to \mathbb{R}^{47}$ with $32 < q \leq 47$. We have just seen that no such form exists.

Remark 13.2.17. Given a positive integer m (resp. n), it is possible to determine explicitly the least (resp. greatest) value n (resp. m) for which there exists a nonconstant form of degree 2 from S^m into S^n (cf [346]).

13.3 Approximation of Mappings with Values in S^1, S^2 or S^4 by Regular Mappings

The special role played by S^1, S^2 and S^4 is due to the fact that these spheres are biregularly isomorphic to, respectively, the projective spaces $\mathbb{P}_1(\mathbb{R})$, $\mathbb{P}_1(\mathbb{C})$ and $\mathbb{P}_1(\mathbb{H})$ (where $\mathbb{H}$ stands for the field of quaternions). The last two spaces, that is, $\mathbb{P}_1(\mathbb{C})$ and $\mathbb{P}_1(\mathbb{H})$ are considered here as real algebraic varieties, endowed with their natural underlying real algebraic structure. The isomorphism between S^1 (resp. S^2) and $\mathbb{P}_1(\mathbb{R})$ (resp. $\mathbb{P}_1(\mathbb{C})$) is described in Proposition 3.5.2. We shall not treat explictly the case of the quaternions. The noncommutativity of $\mathbb{H}$ makes the description of the isomorphism $S^4 \to \mathbb{P}_1(\mathbb{H})$ more involved, but it presents no essential difficulties. The theory of algebraic $\mathbb{H}$-vector bundles, which is needed below, can be developed along the same lines as in the case of $\mathbb{R}$- or $\mathbb{C}$-vector bundles (cf. Chap. 12). In particular, a continuous or $\mathcal{C}^\infty$ section of an algebraic $\mathbb{H}$-vector bundle over a compact variety can be approximated by algebraic sections.

Throughout this section, $\mathbb{F}$ denotes one of the fields $\mathbb{R}$, $\mathbb{C}$ or $\mathbb{H}$. The grassmannians $\mathbb{G}_{m,p}(\mathbb{F})$ are always considered as real algebraic varieties. As usual, $\gamma_{m,p}(\mathbb{F})$ denotes the universal $\mathbb{F}$-vector bundle over $\mathbb{G}_{m,p}(\mathbb{F})$.

Given affine real algebraic varieties X and Y, X compact, we regard the set $\mathcal{R}(X,Y)$ of regular mappings from X to Y as a subspace of the space $\mathcal{C}^0(X,Y)$ of all continuous mappings from X to Y, endowed with the $\mathcal{C}^0$ topology. If, moreover, both X and Y are nonsingular, $\mathcal{R}(X,Y)$ is regarded as a subspace of the space $\mathcal{C}^\infty(X,Y)$ of all $\mathcal{C}^\infty$ mappings from X to Y, endowed with the $\mathcal{C}^\infty$ topology (cf. [160], Chap. 2).

Theorem 13.3.1. *Let X be a compact affine real algebraic (resp. nonsingular) variety. Given a continuous (resp. $\mathcal{C}^\infty$) mapping $f : X \to \mathbb{G}_{m,p}(\mathbb{F})$, the following properties are equivalent:*

(i) The induced $\mathbb{F}$-vector bundle $f^*(\gamma_{m,p}(\mathbb{F}))$ is topologically isomorphic to an algebraic $\mathbb{F}$-vector bundle.

(ii) The mapping f can be approximated, in the $\mathcal{C}^0$ (resp. $\mathcal{C}^\infty$) topology, by regular mappings $X \to \mathbb{G}_{m,p}(\mathbb{F})$.

(iii) The mapping f is homotopic to a regular mapping $X \to \mathbb{G}_{m,p}(\mathbb{F})$.

13.3 Approximation of Mappings with Values in S^1, S^2 or S^4

Proof. The implications (ii) $\Rightarrow$ (iii) $\Rightarrow$ (i) are obvious. Let us prove (i) $\Rightarrow$ (ii). We deal with the $\mathcal{C}^0$ case, the $\mathcal{C}^\infty$ case being similar. The bundle $f^*(\gamma_{m,p}(\mathbb{F}))$ is a topological subbundle of the trivial bundle $X \times \mathbb{F}^m = \epsilon^m$. Denote by $i : f^*(\gamma_{m,p}(\mathbb{F})) \hookrightarrow \epsilon^m$ the inclusion. By assumption, there is an algebraic $\mathbb{F}$-vector bundle ξ and a topological isomorphism $\varphi : \xi \to f^*(\gamma_{m,p}(\mathbb{F}))$. Consider the composition $\psi = i \circ \varphi : \xi \to \epsilon^m$. Given $x \in X$, the morphism ψ maps the fibre ξ_x onto $\{x\} \times f(x) \subset \{x\} \times \mathbb{F}^m$. The morphism ψ therefore defines a continuous section σ of the $\mathbb{F}$-vector bundle $\text{Hom}(\xi, \epsilon^m)$. Since this bundle is algebraic (Proposition 12.1.8 (iii) for $\mathbb{F} = \mathbb{R}$), we can approximate σ by an algebraic section s (Theorem 12.3.1 for $\mathbb{F} = \mathbb{R}$). The section s, in turn, defines an algebraic morphism $h : \xi \to \epsilon^m$, which is injective if s is close enough to σ. Now define the mapping $g : X \to \mathbb{G}_{m,p}(\mathbb{F})$ by the property $h(\xi_x) = \{x\} \times g(x) \subset \{x\} \times \mathbb{F}^m$. The mapping g is regular (Proposition 3.4.7 for $\mathbb{F} = \mathbb{R}$), and it is clear that g approximates f if s approximates σ. $\square$

Corollary 13.3.2. *Let X be as in Theorem 13.3.1. If every topological $\mathbb{F}$-vector bundle of rank p over X is topologically isomorphic to an algebraic $\mathbb{F}$-vector bundle, then $\mathcal{R}(X, \mathbb{G}_{m,p}(\mathbb{F}))$ is dense in $\mathcal{C}^0(X, \mathbb{G}_{m,p}(\mathbb{F}))$ (resp. $\mathcal{C}^\infty(X, \mathbb{G}_{m,p}(\mathbb{F}))$).*

Example 13.3.3. (i) The set $\mathcal{R}(\mathbb{P}_k(\mathbb{R}), \mathbb{G}_{m,p}(\mathbb{R}))$ is dense in the space $\mathcal{C}^\infty(\mathbb{P}_k(\mathbb{R}), \mathbb{G}_{m,p}(\mathbb{R}))$. Indeed, every topological $\mathbb{R}$-vector bundle over $\mathbb{P}_k(\mathbb{R})$ is topologically isomorphic to an algebraic $\mathbb{R}$-vector bundle (Example 12.3.7 c)).

(ii) For the same reason, the set $\mathcal{R}(X, S^1)$ is dense in $\mathcal{C}^\infty(X, S^1)$ for every affine nonsingular compact real algebraic curve X (Theorem 12.5.1).

Theorem 13.3.4. *Let X be a compact affine real algebraic (resp. nonsingular) variety, and let $k = 1, 2$ or 4. Given a continuous (resp. $\mathcal{C}^\infty$) mapping $f : X \to S^k$, the following conditions are equivalent:*

(i) The mapping f can be approximated, in the $\mathcal{C}^0$ (resp. $\mathcal{C}^\infty$) topology, by regular mappings $X \to S^k$.

(ii) The mapping f is homotopic to a regular mapping $X \to S^k$.

Proof. Follows from Theorem 13.3.1 and the remarks made at the beginning of this section. $\square$

We shall now examine more closely regular mappings with values in S^1. Let X be a compact nonsingular affine real algebraic variety. Theorem 13.3.1 allows one to describe the closure $\overline{\mathcal{R}(X, S^1)}$ of $\mathcal{R}(X, S^1)$ in $\mathcal{C}^\infty(X, S^1)$ in terms of the group $H^1_{\text{alg}}(X, \mathbb{Z}/2)$. If γ is a nontrivial $\mathbb{R}$-line bundle over S^1, the first Stiefel-Whitney class $x = w_1(\gamma)$ of γ is the generator of $H^1(S^1, \mathbb{Z}/2)$. Consider $f \in \mathcal{C}^\infty(X, S^1)$. Theorems 13.3.1 and 12.4.6 imply that f belongs to $\overline{\mathcal{R}(X, S^1)}$ if and only if $f^*(x) = w_1(f^*(\gamma))$ is in $H^1_{\text{alg}}(X, \mathbb{Z}/2)$. In order to obtain a more precise result about the size of $\overline{\mathcal{R}(X, S^1)}$, it is convenient to introduce a group structure on $\mathcal{C}^\infty(X, S^1)$. This can be done by considering $S^1 = \{z \in \mathbb{C} \mid |z| = 1\}$ as the multiplicative group of complex numbers of norm 1 and endowing $\mathcal{C}^\infty(X, S^1)$ with the induced group structure. Clearly, $\mathcal{R}(X, S^1)$ and $\overline{\mathcal{R}(X, S^1)}$ are subgroups of $\mathcal{C}^\infty(X, S^1)$, and we can define the quotient group

$$\Gamma(X) = \mathcal{C}^\infty(X, S^1)/\overline{\mathcal{R}(X, S^1)}\,.$$

The group $\Gamma(X)$ measures the size of the set $\overline{\mathcal{R}(X, S^1)}$. In particular, $\mathcal{R}(X, S^1)$ is dense in $\mathcal{C}^\infty(X, S^1)$ if and only if $\Gamma(X) = 0$. We shall completely describe the structure of $\Gamma(X)$ in terms of $H^1_{\mathrm{alg}}(X, \mathbb{Z}/2)$ and deduce several interesting consequences.

Consider a compact $\mathcal{C}^\infty$ manifold M and the canonical monomorphism

$$r : H^1(M, \mathbb{Z}) \otimes \mathbb{Z}/2 \to H^1(M, \mathbb{Z}/2)\,.$$

Denote by $A(M)$ the image of r:

$$A(M) = r(H^1(M, \mathbb{Z}) \otimes \mathbb{Z}/2)\,.$$

The subgroup $A(M)$ of $H^1(M, \mathbb{Z}/2)$ plays an important role in the following theorem, which fully elucidates the structure of the group $\Gamma(X)$.

Theorem 13.3.5. *Let X be a compact connected nonsingular affine real algebraic variety. Then*

$$\Gamma(X) \simeq A(X)/(A(X) \cap H^1_{\mathrm{alg}}(X, \mathbb{Z}/2))\,.$$

Proof. Consider the homomorphism

$$\psi : \mathcal{C}^\infty(X, S^1) \ni f \longmapsto f^*(x) \in H^1(X, \mathbb{Z}/2)\,.$$

We claim that $\psi(\mathcal{C}^\infty(X, S^1)) = A(X)$. Indeed, the following diagram of group homomorphisms

$$\begin{array}{ccccc}
\mathcal{C}^\infty(X, S^1) & \xrightarrow{\varphi} & H^1(X, \mathbb{Z}) & \longrightarrow & H^1(X, \mathbb{Z}) \otimes \mathbb{Z}/2 \\
& \searrow^{\psi} & & \swarrow^{r} & \\
& & H^1(X, \mathbb{Z}/2) & &
\end{array}$$

is commutative, where φ is defined analogously to ψ (by replacing the generator x of $H^1(S^1, \mathbb{Z}/2)$ with a generator of $H^1(S^1, \mathbb{Z})$). Since φ is an epimorphism (cf. [169], p. 49), the claim follows.

As observed earlier, a mapping $f \in \mathcal{C}^\infty(X, S^1)$ is in $\overline{\mathcal{R}(X, S^1)}$ if and only if $f^*(x)$ belongs to $H^1_{\mathrm{alg}}(X, \mathbb{Z}/2)$. Therefore

$$\psi^{-1}\bigl(A(X) \cap H^1_{\mathrm{alg}}(X, \mathbb{Z}/2)\bigr) = \overline{\mathcal{R}(X, S^1)}\,,$$

and the homomorphism ψ induces an isomorphism

$$\Gamma(X) \longrightarrow A(X)/(A(X) \cap H^1_{\mathrm{alg}}(X, \mathbb{Z}/2))\,.$$

□

13.3 Approximation of Mappings with Values in S^1, S^2 or S^4

Corollary 13.3.6. *Let M be a compact connected C^∞ manifold of dimension greater than 1, and let*

$$\alpha(M) = \begin{cases} \mathrm{rank}(H^1(M,\mathbb{Z})) - 1 & \text{if M is nonorientable and} \\ & w_1(M) \in A(M), \\ \mathrm{rank}(H^1(M,\mathbb{Z})) & \text{otherwise.} \end{cases}$$

Then:

(i) *For each algebraic model X of M, one has $\Gamma(X) \simeq (\mathbb{Z}/2)^s$ for some integer s satisfying $0 \leq s \leq \alpha(M)$.*

(ii) *For each integer s satisfying $0 \leq s \leq \alpha(M)$, there exists an algebraic model X of M with $\Gamma(X) \simeq (\mathbb{Z}/2)^s$.*

Proof. (i) Let $\rho(M)$ be the rank of $H^1(M,\mathbb{Z})$ and let X be an algebraic model of M. Since $\Gamma(X) \simeq A(X)/(A(X) \cap H^1_{\mathrm{alg}}(X,\mathbb{Z}/2))$ and since the dimension of the $\mathbb{Z}/2$-vector space $A(X)$ is equal to $\rho(M)$, it is clear that $\Gamma(X) \simeq (\mathbb{Z}/2)^s$ for some s such that $0 \leq s \leq \rho(M)$.

Assume now that M is nonorientable and $w_1(M) \in A(M)$. Since $w_1(X)$ always belongs to $H^1_{\mathrm{alg}}(X,\mathbb{Z}/2)$, it follows that $A(X) \cap H^1_{\mathrm{alg}}(X,\mathbb{Z}/2)$ is of $\mathbb{Z}/2$-dimension greater than 0, that is,

$$\dim_{\mathbb{Z}/2}(A(X)/(A(X) \cap H^1_{\mathrm{alg}}(X,\mathbb{Z}/2))) \leq \rho(M) - 1,$$

and $\Gamma(X) \simeq (\mathbb{Z}/2)^s$, for some $0 \leq s \leq \rho(M) - 1$.

(ii) Given a subgroup G of $H^1(M,\mathbb{Z}/2)$ containing $w_1(M)$ we can find an algebraic model X of M such that

$$A(M)/(A(M) \cap G) \simeq A(X)/(A(X) \cap H^1_{\mathrm{alg}}(X,\mathbb{Z}/2))$$

(cf. Theorems 11.3.7 and 11.3.8). By Theorem 13.3.5, this implies (ii). □

The result described in Corollary 13.3.6 can be interpreted as the fact that $\overline{\mathcal{R}(X,S^1)}$ represents precisely $1/2^s$ of the total set $C^\infty(X,S^1)$, for an appropriate s depending only on $H^1_{\mathrm{alg}}(X,\mathbb{Z}/2)$.

Corollary 13.3.7. *Given a compact connected orientable C^∞ manifold M, the following conditions are equivalent:*

(i) *For each algebraic model X of M, the set of regular mappings $\mathcal{R}(X,S^1)$ is dense in $C^\infty(X,S^1)$.*

(ii) *The first Betti number of M is zero, or $\dim(M) = 1$.*

Proof. Follows from Corollary 13.3.6 if $\dim(M) \geq 2$, and from Example 13.3.3 (ii) if $\dim(M) = 1$. □

We now examine the group $\Gamma(X)$ for a compact connected nonsingular affine real algebraic surface X of genus g. If X is orientable, then $H^1(X,\mathbb{Z}/2) = A(X)$. For X nonorientable, one has

$$A(X) = \{v \in H^1(X,\mathbb{Z}/2) \mid v \cup v = 0\},$$

which is a $\mathbb{Z}/2$-vector hyperplane in $H^1(X,\mathbb{Z}/2)$. The Stiefel-Whitney class $w_1(X)$ of X does not belong to $A(X)$ if and only if X is nonorientable of odd genus. Denoting by $\beta(X)$ (resp. $\delta(X)$) the $\mathbb{Z}/2$-dimension of the vector space $H^1_{\text{alg}}(X,\mathbb{Z}/2)$ (resp. $A(X) \cap H^1_{\text{alg}}(X,\mathbb{Z}/2)$), we deduce from Corollary 13.3.6 the following values for $\Gamma(X)$:

$$\Gamma(X) \simeq \begin{cases} (\mathbb{Z}/2)^{2g-\beta(X)} & \text{if } X \text{ is orientable,} \\ (\mathbb{Z}/2)^{g-\beta(X)} & \text{if } X \text{ is nonorientable and } g \text{ is odd,} \\ (\mathbb{Z}/2)^{g-\delta(X)-1} & \text{if } X \text{ is nonorientable and } g \text{ is even.} \end{cases}$$

Corollary 13.3.8. *Given a compact connected C^∞ surface M, the following conditions are equivalent:*

(i) For each algebraic model X of M, the set $\mathcal{R}(X,S^1)$ of regular mappings is dense in $C^\infty(X,S^1)$.

(ii) M is homeomorphic to S^2 (the 2-sphere), or $\mathbb{P}_2(\mathbb{R})$ (the real projective plane), or the Klein bottle.

Proof. The invariant $\alpha(M)$ defined in Corollary 13.3.6 is zero if and only if M is one of the three surfaces listed in (ii). Hence, the corollary follows from Corollary 13.3.6. □

Studying regular mappings into S^2, we shall often use the fact that the group $V^1_{\mathbb{C}}(X)$ of isomorphism classes of topological $\mathbb{C}$-line bundles over X is canonically isomorphic to $H^2(X,\mathbb{Z})$.

Theorem 13.3.9. *Let X be a compact affine real algebraic (resp. nonsingular) variety, with $H^2(X,\mathbb{Z}) = 0$. Then $\mathcal{R}(X,S^2)$ is dense in $C^0(X,S^2)$ (resp. in $C^\infty(X,S^2)$).*

Proof. The assumption implies that every topological $\mathbb{C}$-line bundle over X is trivial. Thus the theorem follows from Corollary 13.3.2. □

The next result concerns the case of regular mappings $X \to S^k$, where $k = 1, 2$ or 4, and X is a real algebraic variety homeomorphic to S^n.

Theorem 13.3.10. *(i) For every positive integer n and $k = 1, 2$ or 4, the set $\mathcal{R}(S^n, S^k)$ is dense in $C^\infty(S^n, S^k)$.*

(ii) Let X be an affine real algebraic (resp. nonsingular) variety homeomorphic to S^n. Then $\mathcal{R}(X,S^k)$ is dense in $C^0(X,S^k)$ (resp. $C^\infty(X,S^k)$) in the following cases:

$$k = 1 \, ; \quad k = 2 \text{ and } n \neq 2 \, ; \quad k = 4 \text{ and } n \equiv 1,2,3 \text{ or } 7 \pmod 8 \, .$$

Proof. Case $k = 1$. The theorem follows from Corollary 13.3.7.

Case $k = 2$. If $n \neq 2$, the theorem is a consequence of Theorem 13.3.9. Now consider the assertion (i) for $n = 2$. Since S^2 is biregularly isomorphic to $\mathbb{P}_1(\mathbb{C})$ endowed with its real algebraic structure, and $V^1_{\mathbb{C}}(\mathbb{P}_1(\mathbb{C})) \simeq \mathbb{Z}$ is generated by the class of the universal $\mathbb{C}$-line bundle, which is algebraic (cf. Section 12.6), the assertion (i) for $n = 2$ follows from Corollary 13.3.2.

It should be stressed that the statement (ii) is not valid, in general, for $n = k = 2$ (cf. Theorem 13.3.15).

Case $k = 4$. (i) Every topological $\mathbb{H}$-vector bundle over S^n is topologically isomorphic to an algebraic $\mathbb{H}$-vector bundle [315]. It thus suffices to apply Corollary 13.3.2.

(ii) If $n \equiv 1, 2, 3$ or $7 \pmod 8$, every topological $\mathbb{H}$-vector bundle over S^n is stably trivial (cf. [174], p. 109). Hence, every topological $\mathbb{H}$-vector bundle over X is stably trivial, and, consequently, topologically isomorphic to an algebraic $\mathbb{H}$-vector bundle (Proposition 12.3.5). We then apply Corollary 13.3.2 once again. $\square$

It is a challenging open problem to find out whether there exists a couple of positive integers (n, k), $n \geq k$, $k \neq 1, 2, 4$, such that $\mathcal{R}(S^n, S^k)$ is dense in $\mathcal{C}^\infty(S^n, S^k)$.

Before examining further the mappings into S^2, it is convenient to consider a more general situation. Assume that X is a nonsingular compact connected oriented affine real algebraic variety of dimension n, and let

$$\mathrm{Deg}_\mathcal{R}(X) = \{k \in \mathbb{Z} \mid k = \deg(f), f \in \mathcal{R}(X, S^n)\}$$

be the set of topological degrees of regular mappings from X into S^n. Of course, $\mathrm{Deg}_\mathcal{R}(X)$ does not depend on the choice of the orientation on X and S^n. Recall that $V^1_{\mathbb{C}-\mathrm{alg}}(X)$ denotes the subgroup of $V^1_\mathbb{C}(X)$ of isomorphism classes represented by algebraic $\mathbb{C}$-line bundles.

Theorem 13.3.11. *Let X be an affine nonsingular real algebraic variety of dimension n. Assume that X is compact, connected and oriented, and n is either odd or $n = 2$ or 4. Then $\mathrm{Deg}_\mathcal{R}(X)$ is a subgroup of $\mathbb{Z}$.*

Proof. Case n odd. We shall prove in the next section that $\mathrm{Deg}_\mathcal{R}(X) = \mathbb{Z}$ or $2\mathbb{Z}$ for X of odd dimension (cf. Theorem 13.4.19).

Case $n = 2$. Let $[X, S^2]$ be the set of homotopy classes $[f]$ of $\mathcal{C}^\infty$ mappings f from the surface X into S^2. Let γ be the Hopf $\mathbb{C}$-line bundle over S^2 (in particular, γ is a generator of $V^1_\mathbb{C}(S^2)$). The Hopf classification theorem (cf. [308], p. 431) implies that the mappings

$$\varphi_1 : [X, S^2] \ni [f] \longmapsto f^*(\gamma) \in V^1_\mathbb{C}(X) \simeq H^2(X, \mathbb{Z})$$

and

$$\varphi_2 : [X, S^2] \ni [f] \longmapsto \deg(f) \in \mathbb{Z}$$

are bijective and that $\varphi_2 \circ \varphi_1^{-1} : V^1_\mathbb{C}(X) \to \mathbb{Z}$ is an isomorphism. By Theorem 13.3.1, the image by φ_1 of the set

$$\varphi_2^{-1}(\mathrm{Deg}_\mathcal{R}(X)) = \{[f] \in [X, S^2] \mid f \in \mathcal{R}(X, S^2)\}$$

is precisely the subgroup $V^1_{\mathbb{C}-\mathrm{alg}}(X)$ of $V^1_\mathbb{C}(X)$. It follows that

$$\mathrm{Deg}_\mathcal{R}(X) = \varphi_2 \circ \varphi_1^{-1}(V^1_{\mathbb{C}-\mathrm{alg}}(X))$$

is a subgroup of $\mathbb{Z}$. In particular, if $\mathrm{Deg}_{\mathcal{R}}(X) = b\mathbb{Z}$ for some integer b, then $V^1_{\mathbb{C}-\mathrm{alg}}(X) = bV^1_{\mathbb{C}}(X)$.

Case $n = 4$. The proof in this case is conceptually similar to the previous one. Instead of $\mathbb{C}$-line bundles we have to use the $\mathbb{H}$-line bundles $f^*(\gamma_\mathbb{H})$, where $\gamma_\mathbb{H}$ is the quaternionic Hopf line bundle over S^4. However, the details are more subtle and we refer the reader to the original paper [47]. □

It would be very interesting to decide whether $\mathrm{Deg}_{\mathcal{R}}(X)$ is always a subgroup of $\mathbb{Z}$, regardless of the dimension of X.

In each case where $\mathrm{Deg}_{\mathcal{R}}(X)$ is a subgroup of $\mathbb{Z}$, define $b(X)$ to be the unique nonnegative integer satisfying

$$\mathrm{Deg}_{\mathcal{R}}(X) = b(X)\mathbb{Z}.$$

Clearly, the invariant $b(X)$, if defined, contains all information about the homotopy classes represented by elements of $\mathcal{R}(X, S^{\dim(X)})$. In particular, $b(X) = 0$ if and only if each regular mapping from X into $S^{\dim(X)}$ is homotopically trivial. In dimension 2 or 4 (where, by Theorem 13.3.11, the invariant $b(X)$ is always defined) the closure of $\mathcal{R}(X, S^{\dim(X)})$ in $\mathcal{C}^\infty(X, S^{\dim(X)})$ has a particularly simple description:

$$\overline{\mathcal{R}(X, S^{\dim(X)})} = \{f \in \mathcal{C}^\infty(X, S^{\dim(X)}) \mid \deg(f) \in b(X)\mathbb{Z}\}.$$

Hence, in these dimensions, $\mathcal{R}(X, S^{\dim(X)})$ is dense in $\mathcal{C}^\infty(X, S^{\dim(X)})$ if and only if $b(X) = 1$.

As mentioned above, if X is of odd dimension, we always have $\mathrm{Deg}_{\mathcal{R}}(X) = \mathbb{Z}$ or $2\mathbb{Z}$, that is, $b(X) = 1$ or 2. The situation is completely different for varieties of even dimension (cf. Theorem 13.3.15 and Remark 13.3.16).

As observed in the proof of Theorem 13.3.11, we have the following.

Corollary 13.3.12. *Let X be a compact connected oriented nonsingular affine real algebraic surface. Then $V^1_{\mathbb{C}-\mathrm{alg}}(X) = b(X)V^1_{\mathbb{C}}(X)$.*

It follows that, for X as in Corollary 13.3.12, the set $\mathcal{R}(X, S^2)$ is dense in $\mathcal{C}^\infty(X, S^2)$ (resp. each regular mapping $X \to S^2$ is homotopically trivial) if and only if $V^1_{\mathbb{C}-\mathrm{alg}}(X) = V^1_{\mathbb{C}}(X)$ (resp. $V^1_{\mathbb{C}-\mathrm{alg}}(X) = 0$).

Example 13.3.13. Let $Y = S^1$ or, more generally, let Y be a connected nonsingular dividing real algebraic curve (cf. Chap. 11, Section 5). Then every regular mapping $Y \times Y \to S^2$ is homotopically trivial. Indeed, it follows from Theorem 12.6.14 that $V^1_{\mathbb{C}-\mathrm{alg}}(Y \times Y) = 0$, so $b(Y \times Y)$ is null (cf. [57] for a more general result).

Example 13.3.14. Let V be a nonsingular irreducible complex projective curve (resp. surface), and let $V_\mathbb{R}$ be the underlying real algebraic variety. Then $b(V_\mathbb{R}) = 1$ (resp. $b(V_\mathbb{R}) = 1$ or 2) [47].

13.3 Approximation of Mappings with Values in S^1, S^2 or S^4

The following theorem clarifies the structure of $\mathrm{Deg}_{\mathcal{R}}(X)$ in the case of orientable surfaces.

Theorem 13.3.15. *Let M be a compact connected orientable C^∞ surface and b a nonnegative integer. Then there exists an algebraic model X of M such that $b(X) = b$. In particular,*

$$\overline{\mathcal{R}(X, S^2)} = \{f \in C^\infty(X, S^2) \mid \deg(f) \in b\mathbb{Z}\} \ .$$

Reference for the proof: [55].

Remark 13.3.16. Concerning Theorem 13.3.15, it should be noted that, in general, the analogous statement is false for 4-dimensional manifolds. If M is a compact connected oriented C^∞ manifold of dimension 4 and $\sigma(M)$ is its signature, then M admits an algebraic model X with $b(X) = 0$ if and only if $\sigma(M) = 0$. For M with $\sigma(M) \neq 0$, the invariant $b(X)$ divides $6\sigma(M)$ [47]. We conjecture that a statement like Theorem 13.3.15 is valid for M if $\sigma(M) = 0$.

The proof of Theorem 13.3.15 is quite long. One of the crucial steps in this proof is the computation of the invariant $b(X)$ in the case where X is the product of two projective real cubic curves. We shall briefly describe this result, after recalling first the classification of connected projective nonsingular cubics.

Given a positive real number α, let $\tau_\alpha = \frac{1}{2}(1 + \alpha\sqrt{-1})$ and define

$$D_\alpha = \{[x:y:z] \in \mathbb{P}_2(\mathbb{R}) \mid y^2 z = 4x^3 - g_2(\tau_\alpha)xz^2 - g_3(\tau_\alpha)z^3\} \ ,$$

where, as usual, the $g_j(\tau_\alpha)$ are the numbers (real in this case) defined by

$$g_2(\tau_\alpha) = 60 \sum_{w \in \Lambda'} w^{-4}, \quad g_3(\tau_\alpha) = 140 \sum_{w \in \Lambda'} w^{-6} \ ,$$

and $\Lambda = \mathbb{Z} + \tau_\alpha \mathbb{Z}$ is a lattice in $\mathbb{C}$, $\Lambda' = \Lambda \setminus \{0\}$, cf. [175].

Each D_α is a nonsingular connected real cubic curve in $\mathbb{P}_2(\mathbb{R})$. Moreover, D_α and D_β are not biregularly isomorphic for $\alpha \neq \beta$ and every nonsingular connected real cubic curve in $\mathbb{P}_2(\mathbb{R})$ is biregularly isomorphic to some D_α. A similar description is known for cubics which are not connected, but will not be needed here.

It is possible to give explicit formulas for $b(D_\alpha \times D_\beta)$ as a function of (α, β), thereby clarifying completely the structure of the closure of $\mathcal{R}(C \times D, S^2)$ in $C^\infty(C \times D, S^2)$, for the product of arbitrary real cubic curves C, D in $\mathbb{P}_2(\mathbb{R})$.

Theorem 13.3.17. *Given two positive real numbers α and β, the following conditions are equivalent:*
  (i) *Every regular mapping from $D_\alpha \times D_\beta$ into S^2 is homotopically trivial.*
  (ii) *The product $\alpha\beta$ belongs to $\mathbb{R} \setminus \mathbb{Q}$.*

Reference for the proof: [55].

In particular, $b(D_\alpha \times D_\alpha) = 0$ if and only if $\alpha^2 \notin \mathbb{Q}$. We now examine the case where $\alpha^2 \in \mathbb{Q}$ (we omit the general case $\alpha\beta \in \mathbb{Q}$ in order to simplify the presentation, see [55]).

If p and q are positive integers, let (p,q) denote their greatest common divisor.

Theorem 13.3.18. *If* $\alpha = p\sqrt{d}/q$, *with* p, q, d *positive integers,* $(p,q) = 1$ *and* d *square free, then*

$$b(D_\alpha \times D_\alpha) = \begin{cases} q^2/(q,d) & \text{if } d \equiv 3 \pmod 4 \text{ and } pq \equiv 1 \pmod 2, \\ 2q^2/(q,d) & \text{if } d \equiv 1 \pmod 4 \text{ and } pq \equiv 1 \pmod 2, \\ 4q^2/(q,d) & \text{if } pqd \equiv 0 \pmod 2. \end{cases}$$

Reference for the proof: [55].

Combining Theorems 13.3.17 and 13.3.18, we obtain the following characterization of nonsingular connected real cubic curves C in $\mathbb{P}_2(\mathbb{R})$ for which $\mathcal{R}(C \times C, S^2)$ is dense in $\mathcal{C}^\infty(C \times C, S^2)$.

Corollary 13.3.19. *Given a connected nonsingular real cubic curve* C *in* $\mathbb{P}_2(\mathbb{R})$, *the following conditions are equivalent:*
  (i) $\mathcal{R}(C \times C, S^2)$ *is dense in* $\mathcal{C}^\infty(C \times C, S^2)$.
  (ii) C *is biregularly isomorphic to* D_α, *with* $\alpha = p\sqrt{d}$, *where* p, d *are positive integers,* p *odd and* $d \equiv 3 \pmod 4$.

The next corollary plays an essential role in the proof of Theorem 13.3.15.

Corollary 13.3.20. *Given a nonnegative integer* b, *there exists a connected nonsingular real cubic curve* C *in* $\mathbb{P}_2(\mathbb{R})$ *such that* $b(C \times C) = b$.

Proof. Indeed, for $b = 0$ it suffices to take $C = D_\alpha$, where α is an arbitrary positive real number with $\alpha^2 \notin \mathbb{Q}$ (cf. Theorem 13.3.17). For $b > 0$, we can take $C = D_\alpha$ with $\alpha = \sqrt{(4+3b)/b}$ (cf. Theorem 13.3.18). □

Quite complete results concerning the structure of the set $\mathcal{R}(Y_1 \times Y_2, S^2)$, where Y_1 and Y_2 are real algebraic curves of genus ≤ 2, are presented in [62].

To end this section, let us consider the case of regular mappings from nonorientable surfaces into S^2.

Theorem 13.3.21. *Let* X *be a nonorientable nonsingular connected compact affine real algebraic surface containing a nonsingular algebraic curve* Γ *such that*

$$(*) \qquad \#_2(\Gamma, \Gamma; X) = 1 \, .$$

Then $\mathcal{R}(X, S^2)$ *is dense in* $\mathcal{C}^\infty(X, S^2)$.

Furthermore, there exists a nonsingular algebraic curve $\Gamma \subset X$ *satisfying condition* $(*)$ *in each of the following cases:*
  (i) $H_1(X, \mathbb{Z}/2) = H_1^{\text{alg}}(X, \mathbb{Z}/2)$.
  (ii) *The genus of* X *is odd.*

Proof. Follows from Theorem 12.6.10 and Corollaries 12.6.11 and 13.3.2. □

In connection with the last theorem, observe that, given a compact connected nonorientable surface M of even genus, one can construct an algebraic model X of M such that every regular mapping from X into S^2 is homotopically trivial [47]. Combining this fact with Theorems 13.3.15 and 13.3.21, we obtain the following result.

Corollary 13.3.22. *Given a compact connected C^∞ surface M, the following conditions are equivalent:*

(i) For every algebraic model X of M, the set $\mathcal{R}(X, S^2)$ is dense in $C^\infty(X, S^2)$.

(ii) M is nonorientable of odd genus.

13.4 Homotopy Classes of Mappings into S^n Represented by Regular Mappings

Let (X, x_0) be a pointed space, where X is a topological space and $x_0 \in X$. Denote by $\pi^k(X) = \pi^k(X, x_0)$ the set of homotopy classes of continuous mappings $(X, x_0) \to (S^k, s_0)$. The natural mapping $\pi^k(X) \to [X, S^k]$, where $[X, S^k]$ is the set of homotopy classes of continuous mappings $X \to S^k$, is bijective. If $X = S^n$, then $\pi^k(S^n)$ is the underlying set of the nth homotopy group $\pi_n(S^k)$. Under appropriate conditions on X (for example, when X is a triangulable compact space of dimension $n < 2k - 1$), the set $\pi^k(X)$ can be endowed with the structure of an abelian group, called the kth *cohomotopy group of X*, useful for studying mappings $X \to S^k$. If $n < 2k - 1$, then $\pi_n(S^k) = \pi^k(S^n)$ as groups. The results we shall need concerning $\pi^k(X)$ can be found in [169], Chap. 7, Sections 5 and 12.

Throughout this section, X will always be a connected compact affine real algebraic variety of dimension n.

Notation 13.4.1. *We denote by $\pi^k_{\text{alg}}(X)$ (resp. $\pi^{\text{alg}}_n(S^k)$) the subset of $\pi^k(X)$ (resp. $\pi_n(S^k)$) of homotopy classes represented by regular mappings $X \to S^k$ (resp. $S^n \to S^k$).*

It is not known whether $\pi^k_{\text{alg}}(X)$ is always a subgroup of $\pi^k(X)$ for $n < 2k - 1$.

Theorem 13.4.2. *Let M be a compact connected orientable C^∞ hypersurface in $\mathbb{P}_{2n+1}(\mathbb{R})$. There exists a C^∞ diffeomorphism $h : \mathbb{P}_{2n+1}(\mathbb{R}) \to \mathbb{P}_{2n+1}(\mathbb{R})$, arbitrarily close to the identity mapping, such that:*

(i) The set $Y = h(M)$ is a nonsingular algebraic hypersurface of $\mathbb{P}_{2n+1}(\mathbb{R})$.

(ii) Every regular mapping from Y into S^{2n} is homotopically trivial, that is, $\pi^{2n}_{\text{alg}}(Y) = 0$.

Proof. By Theorem 12.6.12, there exists a C^∞ diffeomorphism h of $\mathbb{P}_{2n+1}(\mathbb{R})$, arbitrarily close to the identity mapping, such that $h(M) = Y$ is a nonsingular algebraic hypersurface and the group $\widetilde{K}_0(\mathcal{R}(Y,\mathbb{C}))$ is finite. It follows from Proposition 13.4.3 (ii), proved below, that, for such a Y, one has $\pi^{2n}_{\text{alg}}(Y) = 0$. $\square$

Proposition 13.4.3. *Let M be a compact connected oriented C^∞ manifold of dimension $2n$, and let $f : M \to S^{2n}$ be a C^∞ mapping. Assume that either*
  (i) *The homomorphism $\widetilde{K}(f) : \widetilde{K}(S^{2n}) \to \widetilde{K}(M)$, induced by f, is null,*
or
  (ii) *M is a nonsingular affine real algebraic variety such that the group $\widetilde{K}_0(\mathcal{R}(M,\mathbb{C}))$ finite, and f is regular.*
Then f is homotopically trivial.

Proof. We shall need a few facts about Chern classes (see [236] for the definition and basic properties). Recall that the nth-Chern class $c_n(\xi)$ of a topological $\mathbb{C}$-vector bundle ξ over M is an element of $H^{2n}(M,\mathbb{Z})$. If ξ is stably trivial, then $c_n(\xi) = 0$ for every $n > 0$. If $M = S^{2n}$, the converse holds true: a topological $\mathbb{C}$-vector bundle ξ over S^{2n} with $c_n(\xi) = 0$ is stably trivial [252]. We shall also need the formula

$$c_n(f^*(\xi)) = f^*(c_n(\xi)),$$

where ξ is a $\mathbb{C}$-vector bundle over S^{2n} and

$$f^* : H^{2n}(S^{2n}, \mathbb{Z}) \simeq \mathbb{Z} \longrightarrow H^{2n}(M, \mathbb{Z}) \simeq \mathbb{Z}$$

is the homomorphism induced by $f : M \to S^{2n}$ (f^* is essentially multiplication by $\deg(f)$).

We are going to apply these facts to prove the following: if ξ is a topological $\mathbb{C}$-vector bundle over S^{2n} and $f^*(\xi)$ is stably trivial, then either $\deg(f) = 0$ or ξ is stably trivial.

Indeed, assume that $f^*(\xi)$ is stably trivial, but ξ is not. Then

$$f^*(c_n(\xi)) = c_n(f^*(\xi)) = 0.$$

Since ξ is not stably trivial, $c_n(\xi) \neq 0$. Hence, $f^* = 0$ on $H^{2n}(S^{2n}, \mathbb{Z})$ and $\deg(f) = 0$. In particular, for every C^∞ mapping $f : M \to S^{2n}$, either $\widetilde{K}(f) = 0$ or $\widetilde{K}(f)$ is injective.

To prove (i), take a $\mathbb{C}$-vector bundle ξ over S^{2n} which is not stably trivial (such a ξ exists because $\widetilde{K}(S^{2n}) \simeq \mathbb{Z}$). If $\widetilde{K}(f) = 0$, then $f^*(\xi)$ is stably trivial and, hence, $\deg(f) = 0$.

(ii) Recall that each topological $\mathbb{C}$-vector bundle over S^{2n} is isomorphic to an algebraic one (cf. Example 12.6.1). If $f : M \to S^{2n}$ is regular, the image of $\widetilde{K}(f)$ is contained in a subgroup of $\widetilde{K}(M)$ isomorphic to $\widetilde{K}_0(\mathcal{R}(M,\mathbb{C}))$. This group is finite by assumption. Since $\widetilde{K}(f)$ is either injective or null, and $\widetilde{K}(S^{2n}) \simeq \mathbb{Z}$, it follows that $\widetilde{K}(f) = 0$ and therefore, by (i), $\deg(f) = 0$. $\square$

13.4 Homotopy Classes of Mappings into S^n 363

Remark 13.4.4. Theorem 13.4.2 is valid for a larger class of manifolds. For example, it suffices to assume that M is an orientable transversal intersection of k C^∞ hypersurfaces of $\mathbb{P}_{2n+1}(\mathbb{R})$, with k odd.

We now study the subset $\pi^k_{\text{alg}}(X)$ of the group $\pi^k(X)$ in the case where k is *odd* and $\dim(X) < 2k - 1$. The next main result is Theorem 13.4.13. We shall use the following topological lemma.

Lemma 13.4.5. *Let Y be a triangulable compact space of dimension $n < 2k - 1$. If $f : Y \to S^k$ and $\varphi : S^k \to S^k$ are continuous mappings, then $[\varphi \circ f] = (\deg(\varphi))[f]$ in the cohomotopy group $\pi^k(Y)$.*

Reference for the proof: [169], Chap. 7, Proposition 5.4, p.213.

Proposition 13.4.6. *Let $f, g : S^n \to S^k \subset \mathbb{R}^{k+1}$ be continuous mappings, with $n < 2k - 1$ and k odd. Then $h = -f + 2(f \cdot g)g$, where $f \cdot g = \sum_{i=1}^{k+1} f_i g_i$, is also a mapping from S^n into S^k, and the homotopy class $[h] \in \pi_n(S^k)$ is given by $[h] = -[f] + 2[g]$.*

Proof. We verify that h maps S^n into S^k by computing
$$\|h\|^2 = \|f\|^2 + 4(f \cdot g)^2 \|g\|^2 - 4(f \cdot g)^2 = 1 \ .$$

Since the homotopy class of h depends only on the homotopy classes of f and g, we may assume that f (resp. g) is constant on the upper (resp. lower) hemisphere of S^n, and that this constant is $(1, 0, \ldots, 0)$. We then have
$$h = (2g_1^2 - 1, 2g_1 g_2, \ldots, 2g_1 g_{k+1})$$
on the upper hemisphere. The term on the right-hand side is the composition $q \circ g$, where q is the quadratic form $S^k \to S^k$ defined by
$$q(x) = (x_1^2 - x_2^2 - \cdots - x_{k+1}^2, 2x_1 x_2, \ldots, 2x_1 x_{k+1}) \ .$$

The topological degree of q is 2, since k is odd. Using $\pi_n(S^k) = \pi^k(S^n)$ and Lemma 13.4.5, it follows that h represents $2[g]$ on the upper hemisphere of S^n. On the lower hemisphere, h can be written as
$$h = (f_1, -f_2, \ldots, -f_{k+1}) \ .$$

Hence, since k is odd, h represents $-[f]$ on the lower hemisphere (by using Lemma 13.4.5 once again). It follows that h represents $-[f] + 2[g]$, by definition of the group operation in the homotopy groups. □

Corollary 13.4.7. *If $\alpha, \beta \in \pi_n^{\text{alg}}(S^k)$, with $n < 2k - 1$ and k odd, then $-\alpha + 2\beta \in \pi_n^{\text{alg}}(S^k)$.*

Remark 13.4.8. If α and β in Corollary 13.4.7 are represented by polynomial mappings (resp. by forms), then $-\alpha + 2\beta$ is represented by a polynomial mapping (resp. a form). This is clear in the case of polynomial mappings. In the case of forms, observe that if f and g are forms of (algebraic) degree ℓ and m, respectively, then $h = -f + 2(f \cdot g)g = -\|g\|^2 f + 2(f \cdot g)g$ is a form of degree $\ell + 2m$.

Theorem 13.4.9. *If k is odd, the element of the group $\pi_k(S^k) \simeq \mathbb{Z}$ corresponding to the integer d can be represented by a $|d|$-form $S^k \to S^k$.*

Proof. The assertion is true if $d = \pm 1$ (take orthogonal mappings $S^k \to S^k$ in these cases), and also if $d = 2$ (take the form q of the proof of Proposition 13.4.6). The theorem then follows, for $k > 1$, from Corollary 13.4.7 and Remark 13.4.8. For $k = 1$, the theorem is obvious. □

Lemma 13.4.10. *Let X be a connected compact affine real algebraic variety of dimension n, and let k be an odd positive integer, with $n < 2k - 1$. Let $\varphi : X \to S^k$ be a continuous mapping, and $\varphi^* : \pi^k(S^k) \to \pi^k(X)$, the homomorphism induced by φ. If $\alpha, \beta \in \pi^k_{\text{alg}}(X) \cap \text{Im}(\varphi^*)$, then $-\alpha + 2\beta \in \pi^k_{\text{alg}}(X)$.*

Proof. Let $f_1, g_1 : S^k \to S^k$ be continuous mappings, and let $f_2, g_2 \in \mathcal{R}(X, S^k)$ be chosen such that $\alpha = [f_1 \circ \varphi] = [f_2]$ and $\beta = [g_1 \circ \varphi] = [g_2]$. If $h = -f_1 + 2(f_1 \cdot g_1)g_1$, then, by Proposition 13.4.6, $[h] = -[f_1] + 2[g_1]$ in $\pi_k(S^k) = \pi^k(S^k)$. It follows that $\varphi^*([h]) = -\varphi^*([f_1]) + 2\varphi^*([g_1]) = -\alpha + 2\beta$. Moreover, $\varphi^*([h]) = [h \circ \varphi] = [-f_2 + 2(f_2 \cdot g_2)g_2] \in \pi^k_{\text{alg}}(X)$. □

The next example plays an important role in the proof of Proposition 13.4.12.

Example 13.4.11. Let $\theta : \mathbb{P}_m(\mathbb{R}) \to S^m$ be the regular mapping defined by

$$\theta(x_1 : x_2 : \cdots : x_{m+1}) = \|x\|^{-2}\left(2x_1 x_{m+1}, \ldots, 2x_m x_{m+1}, \sum_{i=1}^{m} x_i^2 - x_{m+1}^2\right).$$

Then the following diagram commutes

$$\begin{array}{ccc} \mathbb{P}_m(\mathbb{R}) & \xrightarrow{\theta} & S^m \\ \varphi \uparrow & & \uparrow \\ \mathbb{R}^m & \xrightarrow{\Pi_N^{-1}} & S^m \setminus \{P_N\}, \end{array}$$

where $P_N = (0, \ldots, 0, 1)$ is the north pole of S^m, Π_N the stereographic projection from P_N (cf. Proposition 3.5.1) and φ the embedding defined by $\varphi(y_1, \ldots, y_m) = (y_1 : \cdots : y_m : 1)$. We may consider θ as an extension of Π_N^{-1}. This shows that $\deg(\theta) = 1$ for m odd, and $\deg_2(\theta) = 1$ (the topological degree modulo 2) for m even. It follows that $\pi^m_{\text{alg}}(\mathbb{P}_m(\mathbb{R})) = \pi^m(\mathbb{P}_m(\mathbb{R}))$ (for m odd we use Lemma 13.4.5 and Theorem 13.4.9).

Proposition 13.4.12. *Let X be a connected compact affine real algebraic variety of dimension n, and k, an odd positive integer satisfying $n < 2k - 1$. Then $2\alpha \in \pi^k_{\text{alg}}(X)$ for every $\alpha \in \pi^k(X)$.*

Proof. Let $\alpha = [f]$, where $f : X \to S^k$ is continuous. Consider the composition
$$g : X \xrightarrow{f} S^k \xrightarrow{j} \mathbb{R}^{k+1} \setminus \{0\} \xrightarrow{\pi} \mathbb{P}_k(\mathbb{R}) \xrightarrow{\theta} S^k,$$
where π is the canonical surjection, θ the regular mapping of Example 13.4.11, and j the inclusion. The Stone-Weierstrass theorem implies that $j \circ f$ is homotopic to a regular mapping $f' : X \to \mathbb{R}^{k+1} \setminus \{0\}$. Since k is odd, the topological degree of $\pi \circ j$ is 2, and, hence, the topological degree of $\theta \circ \pi \circ j$ is also 2. It follows, using Lemma 13.4.5, that $2\alpha = [g]$, and $[g]$ is represented by the regular mapping $\theta \circ \pi \circ f'$. □

Given a commutative group G, denote by $2G$ the subgroup
$$2G = \{\beta \in G \mid \exists \alpha \in G \ \beta = 2\alpha\}.$$

Theorem 13.4.13. *Let X be a connected compact affine real algebraic variety of dimension n, and k, an odd positive integer satisfying $n < 2k - 1$. Then*
 (i) $2\pi^k(X) \subset \pi^k_{\text{alg}}(X)$.
 (ii) *If $\pi^k(X)$ is cyclic and either infinite or finite of even order, then $\pi^k_{\text{alg}}(X) = \pi^k(X)$ or $\pi^k_{\text{alg}}(X) = 2\pi^k(X)$.*
 (iii) *If $\pi^k(X)$ is finite cyclic of odd order, then $\pi^k_{\text{alg}}(X) = \pi^k(X)$.*

Proof. (i) is a restatement of Proposition 13.4.12.

(ii) First consider the case where $\pi^k(X)$ is infinite cyclic, and identify its elements with the integers. Let $\varphi : X \to S^k$ be such that $[\varphi] = 1$. The homomorphism $\varphi^* : \pi^k(S^k) \to \pi^k(X)$ induced by φ is an isomorphism. By Lemma 13.4.10, it follows that, if $\alpha, \beta \in \pi^k_{\text{alg}}(X)$, then $-\alpha + 2\beta \in \pi^k_{\text{alg}}(X)$. Assume that $\pi^k_{\text{alg}}(X) \neq 2\pi^k(X) = 2\mathbb{Z}$. Thus there exists $f \in \mathcal{R}(X, S^k)$ with $[f]$ corresponding to an odd integer, say $[f] = 2\ell - 1$ in $\pi^k(X)$. Choose (cf. Theorem 13.4.9) a regular mapping $g : S^k \to S^k$ of topological degree $2\ell + 1$. By Lemma 13.4.5, we have $[g \circ f] = 4\ell^2 - 1$ in $\pi^k(X)$. Since $2\ell^2 \in \pi^k_{\text{alg}}(X)$, one has $1 = -[g \circ f] + 2(2\ell^2) \in \pi^k_{\text{alg}}(X)$. Hence, by applying again Lemma 13.4.5 and Theorem 13.4.9, we obtain $\pi^k_{\text{alg}}(X) = \pi^k(X)$.

The proof in the case where $\pi^k(X)$ is finite cyclic of even order is similar and is left to the reader.

(iii) If $\pi^k(X)$ is finite cyclic of odd order, then $\pi^k(X) = 2\pi^k(X)$, and the result follows from (i). □

Example 13.4.14. It is not known whether the case $\pi^k_{\text{alg}}(X) = 2\pi^k(X)$ in Theorem 13.4.13 (ii) can occur. Consider $X = S^p \times S^q$, where p and q are positive integers with $p < \rho(q)$ (ρ is the Hurwitz-Radon function, cf.

Notation 13.1.5). We shall show that $\pi^{p+q}_{\mathrm{alg}}(S^p \times S^q) = \pi^{p+q}(S^p \times S^q)$. It suffices to construct a polynomial mapping $f : S^p \times S^q \to S^{p+q}$ of topological degree 1. This can be done as follows. Let $F : \mathbb{R}^{p+1} \times \mathbb{R}^q \to \mathbb{R}^q$ be a normed bilinear form, and define $f(x,y) = z = (z_0, \ldots, z_{p+q})$, where $(1 - z_0) = (1 - x_0)(1 - y_0)/2$, $z_i = x_i(1 - y_0)/2$ for $i = 1, \ldots, p$, and $z_{p+j} = F_j(1 - x_0, x_1, \ldots, x_p, y_1, \ldots, y_q)/2$ for $j = 1, \ldots, q$. To check that f takes its values in S^{p+q}, it is convenient to set $x_0' = 1 - x_0$ and observe that the equation of S^p becomes $x_0'^2 + x_1^2 + \cdots + x_p^2 = 2x_0'$. The mapping f sends $S^p \vee S^q = (S^p \times \{*\}) \cup (\{*\} \times S^q) \subset S^p \times S^q$ onto $\{*\} \in S^{p+q}$ (where $*$ denotes the point $(1, 0, \ldots, 0)$ of the appropriate sphere), and $f|_{((S^p \times S^q) \setminus (S^p \vee S^q))}$ is a diffeomorphism onto $S^{p+q} \setminus \{*\}$. It follows that the topological degree of f is 1.

On the other hand, it is not known whether $\pi^5_{\mathrm{alg}}(S^2 \times S^3) = \pi^5(S^2 \times S^3)$. □

By replacing $\pi^k(S^n)$ with $\pi_n(S^k)$ in Theorem 13.4.13, one obtains the following result.

Theorem 13.4.15. *Let n and k be positive integers, with k odd and $n < 2k - 1$. Then*
  (i) $2\pi_n(S^k) \subset \pi^{\mathrm{alg}}_n(S^k)$.
  (ii) *If $\pi_n(S^k)$ is cyclic of even order, $\pi^{\mathrm{alg}}_n(S^k) = \pi_n(S^k)$ or $\pi^{\mathrm{alg}}_n(S^k) = 2\pi_n(S^k)$.*
  (iii) *If $\pi_n(S^k)$ is cyclic of odd order, $\pi^{\mathrm{alg}}_n(S^k) = \pi_n(S^k)$.*

Example 13.4.16. $\pi^{\mathrm{alg}}_{2p+14}(S^{2p+1}) = \pi_{2p+14}(S^{2p+1}) = \mathbb{Z}/3$ for every $p \geq 7$.

The previous theorem can be generalized for $k = 3$ or 7 as follows.

Proposition 13.4.17. *If $k = 3$ or 7, properties (i), (ii) and (iii) of Theorem 13.4.15 are satisfied for every positive integer n.*

Proof. It is known that, if $f : S^n \to S^k$ and $\varphi : S^k \to S^k$ are continuous mappings, with $k = 3$ or 7, and n an arbitrary positive integer, then $[\varphi \circ f] = (\deg(\varphi))[f]$ in $\pi_n(S^k)$ (cf. [337], Chap. 10, Corollary 8.4). Using this fact in place of Lemma 13.4.5, we proceed as in the proof of Theorem 13.4.13. □

Example 13.4.18. $\pi^{\mathrm{alg}}_9(S^3) = \pi_9(S^3) = \mathbb{Z}/3$; $\pi^{\mathrm{alg}}_{10}(S^3) = \pi_{10}(S^3) = \mathbb{Z}/15$.

Theorem 13.4.19. *Let X be an oriented nonsingular connected compact affine real algebraic variety of odd dimension n. Then, for each even integer p, there exists a regular mapping $X \to S^n$ of topological degree p. Moreover, if there exists a regular mapping $X \to S^n$ of odd topological degree, every continuous mapping $X \to S^n$ is homotopic to a regular mapping.*

Proof. One has $\pi^n(X) \simeq \mathbb{Z}$. If $n > 1$, the result follows from Theorem 13.4.13. For $n = 1$ one has in fact $\pi^1_{\mathrm{alg}}(X) = \pi^1(X)$ (cf. Example 13.3.3 (ii)). □

Remark 13.4.20. (i) It would be interesting to decide whether there exists an odd dimensional variety X, as in Theorem 13.4.19, with $\mathrm{Deg}_{\mathcal{R}}(X) = 2\mathbb{Z}$.

(ii) In Theorem 13.4.19, we can replace the assumption "X nonsingular" with the weaker one "X is a topological manifold".

We conclude this section with a simple and effective method of constructing elements of $\pi_{n+1}^{\mathrm{alg}}(S^{k+1})$: the suspension of forms $S^n \to S^k$.

Proposition 13.4.21. *The suspension $\Sigma h : S^{n+1} \to S^{k+1}$ of a d-form $h : S^n \to S^k$ is homotopic to a regular mapping, i.e., $[\Sigma h] \in \pi_{n+1}^{\mathrm{alg}}(S^{k+1})$.*

Proof. Let $u = (x, y) = (x_1, \ldots, x_{n+1}, y) \in S^{n+1}$. Define $\Sigma h : S^{n+1} \to S^{k+1}$ by
$$\Sigma h(u) = ((1-y)^d + (1+y)^d)^{-1}(2H(x), -(1-y)^d + (1+y)^d),$$
where $H : \mathbb{R}^{n+1} \to \mathbb{R}^{k+1}$ is a homogeneous polynomial mapping of degree d such that $H|_{S^n} = h$. An easy computation shows that Σh indeed maps S^{n+1} into S^{k+1}, and it is clear that Σh is the suspension of h. □

Corollary 13.4.22. *For every positive integer n, every continuous mapping $S^n \to S^n$ is homotopic to a regular mapping. In particular, $\pi_n(S^n) = \pi_n^{\mathrm{alg}}(S^n)$.*

Proof. For n odd the corollary follows from Theorem 13.4.9. Moreover, by the same theorem, each continuous mapping $S^n \to S^n$ (n odd) is homotopic to a form. Thus, for n even, the corollary follows from Proposition 13.4.21. □

Example 13.4.23. For every positive integer n of the form $n = 2p$ with p odd, one has $\pi_{n+1}(S^n) = \pi_{n+1}^{\mathrm{alg}}(S^n)$, $\pi_{n+2}(S^{n+1}) = \pi_{n+2}^{\mathrm{alg}}(S^{n+1})$ and $\pi_{n+2}(S^n) = \pi_{n+2}^{\mathrm{alg}}(S^n)$. Indeed, we have seen in Remark 13.2.13 that, if $\varphi_n : S^{n+\rho(n)-1} \to S^n$ is the Hopf form associated to the normed bilinear form $\mathbb{R}^n \times \mathbb{R}^{\rho(n)} \to \mathbb{R}^n$ given by the Hurwitz-Radon construction, then $[\varphi_n] \in \pi_{n+\rho(n)-1}(S^n)$ is different from zero. Since $\pi_{k+1}(S^k) \simeq \mathbb{Z}/2$ for $k \geq 3$ and $\rho(n) = 2$, it follows that $[\varphi_n]$ generates $\pi_{n+1}(S^n)$ for $n > 2$ and that $[\Sigma \varphi_n]$ generates $\pi_{n+2}(S^{n+1})$ for $n \geq 2$ (using the suspension theorem, cf. [169], p. 312). Since, by Proposition 13.4.21, $\Sigma \varphi_n \in \mathcal{R}(S^{n+2}, S^{n+1})$ and $[\varphi_n \circ \Sigma \varphi_n]$ generates $\pi_{n+2}(S^n) \simeq \mathbb{Z}/2$ for $n \geq 2$ (cf. [337], Theorem 2.8, p. 550), we obtain the desired result in all cases, except for $\pi_3(S^2) \simeq \mathbb{Z}$. However, the latter case is dealt with in Theorem 13.3.10.

Example 13.4.24. The generator of $\pi_8(S^5) = \mathbb{Z}/24$ is represented by the regular mapping $\Sigma \varphi_4 : S^8 \to S^5$, which is the suspension of the Hopf fibration $\varphi_4 : S^7 \to S^4$ (cf. [169], Theorem 16.4, p. 330). It follows, using Theorem 13.4.13 (ii), that $\pi_8(S^5) = \pi_8^{\mathrm{alg}}(S^5)$. Recall that, on the other hand, $\mathcal{P}(S^8, S^5)$ contains only constant mappings (cf. Theorem 13.1.9).

Finally, let us emphasize that the question whether $\pi_n(S^k) = \pi_n^{\mathrm{alg}}(S^k)$, for arbitrary n, k, remains open.

13.5 Regular or Polynomial Mappings from a Product of Spheres into a Sphere

We have seen in Example 13.3.13 that every regular mapping $S^1 \times S^1 \to S^2$ is homotopically trivial. The aim of this section is to prove a theorem generalizing this result.

Theorem 13.5.1. *Given positive integers $q_1, \ldots, q_n$, with $n \geq 2$, and $k = q_1 + \cdots + q_n$, the following properties are equivalent:*
  (i) *Every regular mapping $f : S^{q_1} \times \cdots \times S^{q_n} \to S^k$ is homotopically trivial.*
  (ii) *The integer k is even and at least two of the integers $q_1, \ldots, q_n$ are odd.*

In order to see, once again, the difference of behaviour of polynomial and regular mappings, it is useful to compare the last result with the next one.

Theorem 13.5.2. *Given positive integers $q_1, \ldots, q_n$, with $n \geq 2$, and $k = q_1 + \cdots + q_n$, the following properties are equivalent:*
  (i) *Every polynomial mapping $S^{q_1} \times \cdots \times S^{q_n} \to S^k$ is homotopically trivial.*
  (ii) *At least two of the integers $q_1, \ldots, q_n$ are odd.*

We shall prove both theorems after some preparation.

Theorem 13.5.3. *Let X be an orientable nonsingular connected compact affine real algebraic variety of odd dimension $n < 2k$. Then every regular mapping $X \times S^{2k-n} \to S^{2k}$ is homotopically trivial.*

Proof. We first prove the result for X of dimension $2k - 1$. Let $f : X \times S^1 \to S^{2k}$ be a regular mapping, and $i : X \to X \times S^1$, the inclusion defined by $i(x) = (x, a)$, where a is a given point of S^1. Consider the following commutative diagram of ring homomorphisms

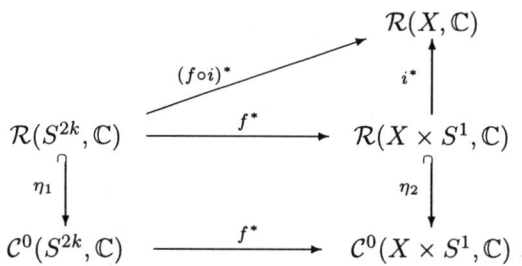

where η_1, η_2 are the inclusions, and $f^*, i^*, (f \circ i)^*$, the homomorphisms induced by $f, i, f \circ i$. The functor $\widetilde{K}_0$ induces the following commutative diagram

13.5 Mappings from a Product of Spheres into a Sphere

$$\begin{array}{ccc}
& & \widetilde{K}_0(\mathcal{R}(X,\mathbb{C})) \\
& \nearrow^{\widetilde{K}_0((f\circ i)^*)} & \uparrow^{\widetilde{K}_0(i^*)} \\
\widetilde{K}_0(\mathcal{R}(S^{2k},\mathbb{C})) & \xrightarrow{\widetilde{K}_0(f^*)} & \widetilde{K}_0(\mathcal{R}(X\times S^1,\mathbb{C})) \\
\downarrow^{\widetilde{K}_0(\eta_1)} & & \downarrow^{\widetilde{K}_0(\eta_2)} \\
\widetilde{K}_0(\mathcal{C}^0(S^{2k},\mathbb{C})) & \xrightarrow{\widetilde{K}_0(f^*)} & \widetilde{K}_0(\mathcal{C}^0(X\times S^1,\mathbb{C})) \\
\wr\| & & \wr\| \\
\widetilde{K}(S^{2k}) \simeq \mathbb{Z} & \xrightarrow{\widetilde{K}(f)} & \widetilde{K}(X\times S^1)
\end{array}$$

Observe that:
  (i) $\widetilde{K}_0(\eta_1)$ is an isomorphism (cf. Example 12.6.1),
  (ii) $\widetilde{K}_0(i^*)$ is an isomorphism (cf. Theorem 12.6.5),
  (iii) since $\dim(X) < 2k$, $f \circ i : X \to S^{2k}$ is homotopically trivial, and, hence, $\widetilde{K}_0((f \circ i)^*) = 0$.

It follows that $\widetilde{K}(f) = 0$, and therefore, by Proposition 13.4.3, f is homotopically trivial.

Now consider the general case, where $\dim(X) = n < 2k$, with n odd. Let $f : X \times S^{2k-n} \to S^{2k}$ be a regular mapping. Using Theorem 13.4.19, choose $\varphi : S^{2k-n-1} \times S^1 \to S^{2k-n}$ to be a regular mapping of topological degree 2, and consider the regular mapping

$$\psi = f \circ (\mathrm{Id}_X \times \varphi) : X \times S^{2k-n-1} \times S^1 \to S^{2k}\,.$$

By the first part of the proof, $\deg(\psi) = 0$. On the other hand, $\deg(\psi) = 2\deg(f)$. It follows that $\deg(f) = 0$, which completes the proof. □

Lemma 13.5.4. *Let p be an even positive integer. The polynomial mapping $\varphi : S^p \times S^q \to S^{p+q}$ defined by*

$$\varphi(x_0, x_1, \ldots, x_p, y_0, y_1, \ldots, y_q) = (x_0 y_0, x_1 y_0, \ldots, x_p y_0, y_1, \ldots, y_q)$$

is of topological degree 2.

Proof. Observe that $\varphi = \varphi \circ h$, where h is the symmetry

$$h(x_0, \ldots, x_p, y_0, \ldots, y_q) = (-x_0, \ldots, -x_p, -y_0, y_1, \ldots, y_q)\,,$$

of topological degree $(-1)^{p+2} = 1$. If $y_0 \neq 0$, then

$$\varphi^{-1}(\varphi(x,y)) = \{(x,y), h(x,y)\}$$

and the derivatives $d_{(x,y)}\varphi$ and $d_{h(x,y)}\varphi$ are orientation preserving isomorphisms (if the spheres are endowed with their standard orientation). Hence, φ is of topological degree 2. □

Proof of Theorem 13.5.1. (i) $\Rightarrow$ (ii) Assumption (i) and Theorem 13.4.19 imply that k is even. If all q_i are even, it follows from Lemma 13.5.4 that there is a regular mapping $S^{q_1} \times \cdots \times S^{q_n} \to S^k$ of topological degree 2^{n-1}, contradicting (i).

(ii) $\Rightarrow$ (i) Follows from Theorem 13.5.3. □

Proof of Theorem 13.5.2. (i) $\Rightarrow$ (ii) In order to construct a polynomial mapping $S^{q_1} \times \cdots \times S^{q_n} \to S^k$ of topological degree different from 0, where all but possibly one q_i are even, it suffices to apply Lemma 13.5.4 $k-1$ times.

(ii) $\Rightarrow$ (i) We shall prove the following stronger result:

Let X be a nonsingular algebraic subset of $\mathbb{R}^\ell$ of dimension m. Let p and q be odd positive integers and $k = m + p + q$. If X is compact, connected and orientable, every polynomial mapping $X \times S^p \times S^q \to S^k$ is homotopically trivial.

If k is even, the statement is a particular case of Theorem 13.5.3. We thus assume below that k is odd.

Let $f : X \times S^p \times S^q \to S^k$ be a polynomial mapping. We shall prove that $\deg(f) = 0$. Choose a polynomial mapping $F : \mathbb{R}^{\ell+p+1+q+1} \to \mathbb{R}^{k+1}$ such that $f(x,y,z) = F(x,y,z)$, for every $(x,y,z) \in X \times S^p \times S^q$. By replacing z with $h(z)$, where $h : S^q \to S^q$ is the 2-form

$$h(z) = (z_0^2 - z_1^2 - \cdots - z_q^2, 2z_0 z_1, \cdots, 2z_0 z_q),$$

we obtain a polynomial mapping

$$G = (G_1, \ldots, G_{k+1}) : \mathbb{R}^{\ell+p+1+q+1} \longrightarrow \mathbb{R}^{k+1}$$
$$(x,y,z) \longmapsto F(x,y,h(z)),$$

where each G_j is the sum of monomials which are all of even degree with respect to the variables $z_0, \ldots, z_q$. Multiplying each of these monomials by an appropriate power of $z_0^2 + \cdots + z_q^2$, we may assume that the G_j are homogeneous of the same degree r with respect to the variables $z_0, \ldots, z_q$ (some G_j may be identically 0).

Setting $g_f(x,y,z) = G(x,y,z)$, for $(x,y,z) \in X \times S^p \times S^q$, we obtain a well-defined polynomial mapping

$$g_f : X \times S^p \times S^q \to S^k,$$

satisfying $g_f(x,y,z) = f(x,y,h(z))$. Since $\deg(h) = 2$, we have $\deg(g_f) = 2 \deg(f)$. We shall now "suspend" g_f over S^q. For $u \in S^{q+1}$, set $u = (z,t) = (z_0, \ldots, z_q, t)$. Then define

$$\widetilde{g}_f : X \times S^p \times S^{q+1} \to S^{k+1}$$

by the formula

$$\widetilde{g}_f(x,y,u) = ((1+t)^r + (1-t)^r)^{-1}(2G(x,y,z), (1+t)^r - (1-t)^r),$$

for $(x, y, u) \in X \times S^p \times S^{q+1}$. It can be checked by an elementary computation that $\widetilde{g}_f$ is a well-defined regular mapping. Since, for each fixed $(x, y) \in X \times S^p$, the mapping $\widetilde{g}_f(x, y, \cdot) : S^{q+1} \to S^{k+1}$ is just the suspension of $g_f(x, y, \cdot) : S^q \to S^k$ (cf. Proposition 13.4.21), we have

$$\deg(\widetilde{g}_f) = \deg(g_f) = 2 \deg(f) \ .$$

Since $k + 1$ is even and p is odd, the mapping $\widetilde{g}_f$ is homotopically trivial by Theorem 13.5.3. It follows that $\deg(f) = 0$. □

Remark 13.5.5. It is not known whether there is always a regular mapping $S^{q_1} \times \cdots \times S^{q_n} \to S^{q_1 + \cdots + q_n}$ (with all q_i even) of topological degree 1. It is known for $q_1 = 2$, $q_2 = 4$, but is unknown for $q_1 = q_2 = 2$.

Bibliographic Notes. Theorems 13.1.9 and 13.1.11 are due to Wood [344]. The Hopf forms appear for the first time in [165]. Most results of Section 2 can be found (at least implicitly) in a paper by K.Y. Lam [203]. Numerous results concerning normed or nonsingular bilinear forms can be found in papers by Shapiro and Lam [202, 296]; these two papers also contain an excellent bibliography on the subject. Most results of Sections 3 to 5 can be found in papers by Bochnak and Kucharz [45, 46, 47, 48, 50, 52, 55, 57]. However, note that Theorems 13.3.1 and 13.3.4 (for $\mathbb{F} = \mathbb{R}$) are in Ivanov's paper [179], Proposition 13.4.6 (for $n = k$) and Theorem 13.4.9 are due to Wood [344], and Example 13.4.14 and Lemma 13.5.4 are due to Loday [212]. Theorem 13.5.2 (ii) $\Rightarrow$ (i) (in the case $n = 2$) and Theorem 13.5.3 (in the case $n = 2k - 1$ and f polynomial) are also due to Loday [212]. More results concerning the set $\mathcal{R}(X, S^k)$ can be found in the papers by Bochnak and Kucharz listed in the references.

14. Algebraic Models of $\mathcal{C}^\infty$ Manifolds

Abstract. The main result of this chapter is the Nash-Tognoli theorem stating that every compact $\mathcal{C}^\infty$ manifold is diffeomorphic to a nonsingular algebraic subset of $\mathbb{R}^p$, for some p. The proof is contained in Section 1. Section 2 contains some results concerning topological spaces homeomorphic to real algebraic sets.

14.1 Algebraic Models of $\mathcal{C}^\infty$ Manifolds

The aim of this section is to prove the Nash-Tognoli theorem which asserts that every compact $\mathcal{C}^\infty$ manifold is diffeomorphic to a nonsingular algebraic subset of $\mathbb{R}^p$, for some p. First we shall formulate the Thom isotopy theorem in the form needed for our purpose. Recall that two subsets A and B of a $\mathcal{C}^\infty$ manifold X are said to be diffeotopic if there exists a family $(F_t)_{t \in [0,1]}$ of diffeomorphisms of X, such that $F : [0,1] \times X \to X$ defined by $F(t,x) = F_t(x)$ is a $\mathcal{C}^\infty$ mapping, F_0 is the identity mapping of X and $F_1(A) = B$.

Theorem 14.1.1. *Let X and Y be $\mathcal{C}^\infty$ manifolds, with X compact and possibly with boundary ∂X. Let $Z \subset Y$ be a compact $\mathcal{C}^\infty$ submanifold. Let $f : X \to Y$ be a $\mathcal{C}^\infty$ mapping transverse to Z and such that $f^{-1}(Z) \cap \partial X = \emptyset$. Then there exists a neighbourhood Ω of f in $\mathcal{C}^\infty(X,Y)$ such that, for every $\mathcal{C}^\infty$ mapping $g : X \to Y$ in Ω, g is transverse to Z and the set $g^{-1}(Z) \subset X \setminus \partial X$ is a $\mathcal{C}^\infty$ submanifold diffeotopic to $f^{-1}(Z)$. Moreover, if V is a neighbourhood of $f^{-1}(Z)$ in X and Ω is small enough, the diffeomorphism of X sending $f^{-1}(Z)$ onto $g^{-1}(Z)$ can be chosen arbitrarily close to the identity, and equal to the identity on the set*

$$\{x \in X \mid f(x) = g(x)\} \cup (X \setminus V).$$

Reference for the proof: [1], p. 51.

We shall also need the following lemma concerning nonsingular points of algebraic sets (cf. Definition 3.3.9).

Lemma 14.1.2. *Let $W \subset \mathbb{R}^m$ and $Z \subset \mathbb{R}^n$ be algebraic sets, with Z nonsingular of dimension s. Let $f : W \to \mathbb{R}^n$ be a regular mapping. Let $x \in f^{-1}(Z)$*

be a point in W which is nonsingular in dimension r, and assume f transverse to Z at x. Then x is a point of the algebraic set $f^{-1}(Z)$, nonsingular in dimension $r + s - n$.

Proof. Since Z is nonsingular of dimension s, there exist $g_1, \ldots, g_{n-s}$ in $\mathcal{R}_{\mathbb{R}^n, f(x)}$ such that the rank of the jacobian matrix

$$\left[\frac{\partial g_i}{\partial y_j}\right] \quad (i = 1, \ldots, n-s \,;\, j = 1, \ldots, n)$$

is equal to $n - s$ at $f(x)$ and $\mathcal{R}_{Z, f(x)} = \mathcal{R}_{\mathbb{R}^n, f(x)}/(g_1, \ldots, g_{n-s})$ (cf. Proposition 3.3.10). Let $g = (g_1, \ldots, g_{n-s})$. The fact that f is transverse to Z at x means that the rank of the derivative $d(g \circ f)_x : T_x(W) \to \mathbb{R}^{n-s}$ is $n - s$. Then

$$\mathcal{R}_{f^{-1}(Z), x} = \mathcal{R}_{W, x}/(g_1 \circ f, \ldots, g_{n-s} \circ f)$$

is a regular local ring of dimension $r - (n - s)$, which proves the lemma. $\square$

The next theorem, due to Seifert [291], seems to be the first general result concerning the existence of algebraic models of $\mathcal{C}^\infty$ manifolds.

Theorem 14.1.3. *Let $V \subset \mathbb{R}^n$ be a nonsingular algebraic set, and $M \subset V$, a compact $\mathcal{C}^\infty$ submanifold whose normal bundle in V is trivial. Then M is diffeotopic to the union of some nonsingular connected components of an algebraic subset of V, by a diffeotopy arbitrarily close to the identity on V.*

Proof. We choose a compact tubular neighbourhood T of M in V, which is a $\mathcal{C}^\infty$ manifold with boundary. We also choose a $\mathcal{C}^\infty$ mapping $f : T \to \mathbb{R}^{v-m}$ (where $v = \dim(V)$ and $m = \dim(M)$) such that $0 \in \mathbb{R}^{v-m}$ is a regular value of f and $f^{-1}(0) = M$. We finally choose a neighbourhood Ω of f in $\mathcal{C}^\infty(T, \mathbb{R}^{v-m})$ as in the statement of Theorem 14.1.1 (with $Z = \{0\}$). By the Stone-Weierstrass theorem, there exists a polynomial mapping $g : V \to \mathbb{R}^{v-m}$ such that $g|_T \in \Omega$. Then, by Theorem 14.1.1, $g^{-1}(0) \cap T$ is diffeotopic to $M = f^{-1}(0)$ and, clearly, $g^{-1}(0) \cap T$ is the union of some nonsingular connected components of the algebraic set $g^{-1}(0) \subset V$. $\square$

The algebraic set $g^{-1}(0)$ in the proof of Theorem 14.1.3 has, in general, other connected components outside T. Usually, these extra components cannot be removed to make $g^{-1}(0) \cap T$ algebraic. For example, this is the case if the homology class $[M] \in H_m(V, \mathbb{Z}/2)$ does not belong to $H_m^{\text{alg}}(V, \mathbb{Z}/2)$ (cf. Section 11.3). However, we can get rid of the extra components in the following important case.

Theorem 14.1.4. *Let $M \subset \mathbb{R}^n$ be an orientable compact $\mathcal{C}^\infty$ submanifold of codimension ≤ 2. Then M is diffeotopic to a nonsingular algebraic set of $\mathbb{R}^n$, by a diffeotopy arbitrarily close to the identity.*

14.1 Algebraic Models of $\mathcal{C}^\infty$ Manifolds

Proof. The assumption implies that the normal bundle of M in $\mathbb{R}^n$ is trivial (cf. [225]). Consider M as embedded in S^n, and let $\varphi : S^n \to S^k$ ($k =$ codim(M)) be a $\mathcal{C}^\infty$ mapping such that $M = \varphi^{-1}(a)$ for some regular value $a \in S^k$. Then approximate φ by a regular mapping $\psi : S^n \to S^k$ (cf. Theorem 13.3.10). By Theorem 14.1.1, M and $\psi^{-1}(a)$ are diffeotopic if ψ is close enough to φ. □

A consequence of the previous theorem is that every $\mathcal{C}^\infty$ knot in $\mathbb{R}^3$ is diffeotopic to an algebraic knot.

We shall need the following technical result.

Proposition 14.1.5. *Let $A \subset K \subset \mathbb{R}^m$, $W \subset \mathbb{R}^n$, where A and W are nonsingular algebraic sets and K is a compact $\mathcal{C}^\infty$ submanifold, possibly with boundary. Let $f : K \to W$ be a $\mathcal{C}^\infty$ mapping such that $f|_A$ is regular. Then, for every neighbourhood Ω of f in $\mathcal{C}^\infty(K, W)$, there exist an algebraic set $Z \subset \mathbb{R}^m \times \mathbb{R}^n$ of dimension m, a regular mapping $h : Z \to W$ and a $\mathcal{C}^\infty$ mapping $\varphi : K \to \mathbb{R}^n$ such that:*

(i) *The image of the mapping $\sigma : K \to \mathbb{R}^m \times \mathbb{R}^n$, defined by $\sigma(x) = (x, \varphi(x))$, is contained in $\mathrm{Reg}(Z)$.*
(ii) *The mapping $h \circ \sigma : K \to W$ belongs to Ω.*
(iii) *For every $x \in A$, $\varphi(x) = 0$ and $h \circ \sigma(x) = f(x)$.*
(iv) *There exists $\epsilon > 0$ such that*

$$\sigma(K) = Z \cap (K \times \{y \in \mathbb{R}^n \mid \|y\| < \epsilon\}) .$$

Proof. Let W' be the intersection of W with an open ball containing $f(K)$. Choose an open tubular neighbourhood T of W' in $\mathbb{R}^n$ of radius ϵ, with the orthogonal retraction $\rho : T \to W'$. By Lemma 12.5.5, we can choose a polynomial mapping $g : \mathbb{R}^m \to \mathbb{R}^n$ such that $g|_K$ is arbitrarily close, in the $\mathcal{C}^\infty$ topology, to f (regarded as a mapping into $\mathbb{R}^n$) and $g|_A = f|_A$. In particular, we may assume that $g(K) \subset T$. Then $U = g^{-1}(T)$ is a semi-algebraic open neighbourhood of K in $\mathbb{R}^m$. For $x \in U$, set $\overline{\varphi}(x) = \rho(g(x)) - g(x)$ and $\overline{\sigma}(x) = (x, \overline{\varphi}(x))$. Denote $\varphi = \overline{\varphi}|_K$ and $\sigma = \overline{\sigma}|_K$.

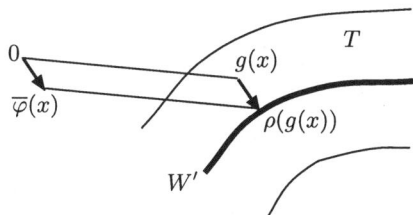

Figure 14.1.

Now, let Z be the Zariski closure of $\overline{\sigma}(U)$. Since $\overline{\sigma}$ is a semi-algebraic bijection, we have $\dim(\overline{\sigma}(U)) = \dim(U) = m$. Hence Z is of dimension m (cf. Proposition 2.8.2). Finally, set $h(x, y) = g(x) + y$ for $(x, y) \in Z \subset \mathbb{R}^m \times \mathbb{R}^n$.

It remains to verify properties (i)-(iv) of the statement. The verification of (iii) is straightforward, and (ii) is satisfied if g is chosen close enough to f (note that $h \circ \sigma = \rho \circ g$). Observe that Z is contained in the algebraic set $Z' \subset \mathbb{R}^m \times \mathbb{R}^n$ which is the inverse image of the total space

$$E = \{(w, y) \in W \times \mathbb{R}^n \mid y \text{ is orthogonal to } T_w(W)\}$$

of the normal bundle of W in $\mathbb{R}^n$ by the regular mapping

$$\begin{aligned} \gamma : \mathbb{R}^m \times \mathbb{R}^n &\longrightarrow \mathbb{R}^n \times \mathbb{R}^n \\ (x, y) &\longmapsto (g(x) + y, y) \, . \end{aligned}$$

Then observe that $\overline{\sigma}(U) = Z' \cap (U \times \{y \in \mathbb{R}^n \mid \|y\| < \epsilon\})$, which proves (iv). To prove (i), note first that E is nonsingular of dimension n, and γ is transverse to E at every point of $\overline{\sigma}(U)$ because $T_{(w,y)}(E) = T_w(W) \times (T_w(W))^\perp$ and the image of the derivative $d\gamma_{(x,y)}$ contains the diagonal of $\mathbb{R}^n \times \mathbb{R}^n$. By Lemma 14.1.2, $\overline{\sigma}(U)$ is contained in the subset of Z' of points nonsingular in dimension m. Now $Z \subset Z'$, $\dim(Z) = m$ and $\overline{\sigma}(U)$ is open both in Z and Z'. From these facts (i) follows. $\square$

Remark 14.1.6. The previous result is reminiscent of the Artin-Mazur description of Nash mappings (Theorem 8.4.4). Indeed, by applying this description to the Nash mapping $\rho \circ g|_U$ (with the same notation as in the proof of 14.1.5) we obtain directly a nonsingular algebraic set Z of dimension m, a Nash diffeomorphism σ from U onto an open subset of Z, and a regular mapping $h : Z \to W$ such that $h \circ \sigma|_K$ belongs to Ω and $h \circ \sigma|_A = f|_A$. Using Proposition 14.1.5, we obtain, moreover, that the image of A by σ is algebraic (and even that $\sigma(A) = A \times \{0\}$). This last result will be useful in the proof of Theorem 14.1.7, and does not follow from the Artin-Mazur description.

Theorem 14.1.7 (Generalized Nash's Theorem). *Let $M \subset \mathbb{R}^m$ be a connected compact C^∞ submanifold. Let $A \subset M$ be a nonsingular algebraic subset, and assume that some open neighbourhood U of A in M is an open subset of a nonsingular algebraic subset of $\mathbb{R}^m$. Then there exist a nonsingular connected component X of an algebraic subset of $\mathbb{R}^m \times \mathbb{R}^n$ (for some n) and a C^∞ diffeomorphism $\tau : M \to X$, such that $\tau(a) = (a, 0)$ for every $a \in A$. Moreover, X and τ can be chosen in such a way that τ is arbitrarily close to the inclusion mapping $M \hookrightarrow \mathbb{R}^m \times \mathbb{R}^n$ sending $x \in M$ to $(x, 0)$.*

Proof. Let k be the codimension of M in $\mathbb{R}^m$. Recall that

$$\gamma_{m,k} = (E_{m,k}, p_{m,k}, \mathbb{G}_{m,k})$$

is the universal vector bundle of rank k over the grassmannian $\mathbb{G}_{m,k}$, with $E_{m,k} = \{(g, v) \in \mathbb{G}_{m,k} \times \mathbb{R}^m \mid v \in g\}$. It follows from the description of $E_{m,k}$ given in Proposition 12.1.4 that we may consider $E_{m,k}$ as an algebraic subset of some $\mathbb{R}^n$.

Let K be a compact tubular neighbourhood of M in $\mathbb{R}^m$ (K is a $\mathcal{C}^\infty$ manifold with boundary), with the orthogonal retraction $\rho : K \to M$. Let $\nu : M \to \mathbb{G}_{m,k}$ be the Gauss mapping sending $x \in M$ to the normal space $\nu(x)$ to M at x. Since ρ is the orthogonal retraction, we have $\rho(y) - y \in \nu(\rho(y))$ for every $y \in K$. The formula $f(y) = (\nu(\rho(y)), \rho(y) - y)$ thus defines a $\mathcal{C}^\infty$ mapping $f : K \to E_{m,k}$. It is obvious that f is transverse to the zero section $\mathbb{G}_{m,k} \times \{0\} \subset E_{m,k}$ and $f^{-1}(\mathbb{G}_{m,k} \times \{0\}) = M$. Moreover, using the assumption concerning the neighbourhood U of A and applying Corollary 3.4.8, we see that $f|_A$ is a regular mapping.

Now, choose a neighbourhood Ω of f in $\mathcal{C}^\infty(K, E_{m,k})$ as in the statement of Theorem 14.1.1. Next, choose an algebraic set $Z \subset \mathbb{R}^m \times \mathbb{R}^n$, a regular mapping $h : Z \to E_{m,k}$ and a $\mathcal{C}^\infty$ mapping $\varphi : K \to \mathbb{R}^n$ satisfying properties (i)-(iv) of Proposition 14.1.5. In particular, the mapping $h \circ \sigma : K \to E_{m,k}$ belongs to Ω and, by Theorem 14.1.1, the set $(h \circ \sigma)^{-1}(\mathbb{G}_{m,k} \times \{0\})$ is a $\mathcal{C}^\infty$ submanifold of $K \setminus \partial K$ diffeotopic to $f^{-1}(\mathbb{G}_{m,k} \times \{0\}) = M$, the diffeotopy keeping A fixed (because $h \circ \sigma|_A = f|_A$). Denote $X = \sigma((h \circ \sigma)^{-1}(\mathbb{G}_{m,k} \times \{0\}))$. It is clear that M is diffeomorphic to X by a diffeomorphism τ (obtained by composing the diffeomorphism $M \to (h \circ \sigma)^{-1}(\mathbb{G}_{m,k} \times \{0\})$ with σ), such that $\tau(a) = (a, 0)$ for every $a \in A$. The set X is contained in the algebraic set $X' = h^{-1}(\mathbb{G}_{m,k} \times \{0\}) \subset Z$. By property (iv), we have $X = X' \cap ((K \setminus \partial K) \times \{y \in \mathbb{R}^n \mid \|y\| < \epsilon\})$, which shows that X is open in X'. Since, on the other hand, X is compact, it is a connected component of X'. Since $h|_{\sigma(K)}$ is transverse to $\mathbb{G}_{m,k} \times \{0\}$, Lemma 14.1.2 shows that X is contained in the subset of X' of points nonsingular in dimension $m - k$. The set X is therefore a nonsingular connected component of the Zariski closure of X. Finally, by construction of X and the diffeomorphism $\tau : M \to X$, we can choose τ arbitrarily close to the inclusion mapping $M \hookrightarrow \mathbb{R}^m \times \mathbb{R}^n$. $\square$

In the case $A = \emptyset$, the previous theorem is the Nash theorem.

Theorem 14.1.8 (Nash's theorem). *Let $M \subset \mathbb{R}^m$ be a compact connected $\mathcal{C}^\infty$ submanifold. Then M is diffeomorphic to a nonsingular connected component of an algebraic subset of $\mathbb{R}^m \times \mathbb{R}^n$ (for some n).*

The proof that the algebraic set in the Nash theorem can be chosen connected and, hence, diffeomorphic to M, requires the use of a deep result of differential topology. Recall that two compact $\mathcal{C}^\infty$ manifolds M_1 and M_2 of dimension m are said to be *cobordant* if their disjoint union is the boundary of a compact $\mathcal{C}^\infty$ manifold of dimension $m + 1$ (called a *cobordism* between M_1 and M_2). It was shown in [232] (cf. also [93]) that every compact $\mathcal{C}^\infty$ manifold is cobordant to a disjoint union of manifolds of the form

$$\mathbb{P}_{k_1}(\mathbb{R}) \times \cdots \times \mathbb{P}_{k_n}(\mathbb{R}) \times H_{s_1 q_1} \times \cdots \times H_{s_r q_r},$$

where, for $s \leq q$,

$$H_{sq} = \left\{ ((x_0 : \ldots : x_s), (y_0 : \ldots : y_q)) \in \mathbb{P}_s(\mathbb{R}) \times \mathbb{P}_q(\mathbb{R}) \mid \sum_{i=0}^{s} x_i y_i = 0 \right\}.$$

Since $\mathbb{P}_k(\mathbb{R})$ and H_{sq} are biregularly isomorphic to nonsingular real algebraic sets, one obtains therefore the following.

Theorem 14.1.9. *Every compact $\mathcal{C}^\infty$ manifold is cobordant to a nonsingular compact real algebraic set.*

Now we shall prove the main result of this section. In what follows, we identify $\mathbb{R}^k$ with the subspace $\mathbb{R}^k \times \{0\}$ of $\mathbb{R}^n$, when $n > k$.

Theorem 14.1.10 (Tognoli's theorem). *Let $M \subset \mathbb{R}^m$ be a compact $\mathcal{C}^\infty$ submanifold. Then M is diffeomorphic to a nonsingular algebraic subset of $\mathbb{R}^p$, for some $p \geq m$. Moreover, the diffeomorphism can be chosen arbitrarily close to the inclusion mapping $M \hookrightarrow \mathbb{R}^p$.*

Proof. Without loss of generality, we may assume M connected. Let A be a nonsingular compact algebraic set cobordant to M (cf. Theorem 14.1.9). Using the "collar neighbourhood theorem" (cf. [160], p. 113), we may assume the cobordism X contained in $\mathbb{R}^k \times [0, \infty[$, in such a way that $\partial X = M \cup A = X \cap (\mathbb{R}^k \times \{0\})$ and $X \cap (\mathbb{R}^k \times [0, \delta[) = (M \cup A) \times [0, \delta[$ for some $\delta > 0$. Let then Y be the "double" of X, i.e., the $\mathcal{C}^\infty$ boundaryless manifold defined by

$$Y = \{(v, r) \in \mathbb{R}^k \times \mathbb{R} \mid (v, r) \in X \text{ or } (v, -r) \in X\}.$$

Observe that the neighbourhood $A \times]-\delta, \delta[$ of A in Y is an open subset of the nonsingular algebraic set $A \times \mathbb{R}$. From Theorem 14.1.7 we obtain a nonsingular connected component Y' of some algebraic set $Z \subset \mathbb{R}^p$, with $p \geq k+1$, and a diffeomorphism $\tau : Y \to Y'$ such that $\tau(A) = A$. Moreover, Y' and τ can be chosen in such a way that τ is arbitrarily close to the inclusion mapping $Y \hookrightarrow \mathbb{R}^p$. Set $M' = \tau(M)$.

Let $g : Y' \to \mathbb{R}$ be a $\mathcal{C}^\infty$ function such that $g^{-1}(0) = M' \cup A$ and 0 is a regular value of g (for example, we can take g to be the composition of $\tau^{-1} : Y' \to Y$ with the "last coordinate" mapping: $Y \to R$). We now choose a regular function $h : Z \to \mathbb{R}$ such that $h|_{Y'}$ approximates g in $\mathcal{C}^\infty(Y', \mathbb{R})$, $h|_A = 0$ and $h^{-1}(0) \subset Y'$. This can be done in the following way: if $\overline{Z}$ is the Zariski closure of Z in $\mathbb{P}_p(\mathbb{R})$, approximate the function which coincides with g on Y' and is equal to 1 on $\overline{Z} \setminus Y'$ by a regular function $\overline{h} : \overline{Z} \to \mathbb{R}$ such that $\overline{h}|_A = 0$ (cf. Lemma 12.5.5), and take $h = \overline{h}|_Z$. The algebraic set $h^{-1}(0)$ is contained in Y'. By Theorem 14.1.1, if $h|_{Y'}$ is chosen close enough to g, then $h^{-1}(0)$ is a nonsingular algebraic set diffeotopic to $g^{-1}(0) = M' \cup A$ by a diffeotopy of Y' keeping A fixed. It follows that $h^{-1}(0) = M'' \cup A$, with M'' diffeotopic to M'. Since A and $M'' \cup A$ are nonsingular algebraic sets of the same dimension, it follows that M'' is an algebraic set (cf. Proposition 3.3.17). Moreover, the diffeomorphism $M \to M''$ can be chosen arbitrarily close to the inclusion mapping $M \hookrightarrow \mathbb{R}^p$ if we choose τ close enough to the inclusion

14.1 Algebraic Models of C^∞ Manifolds

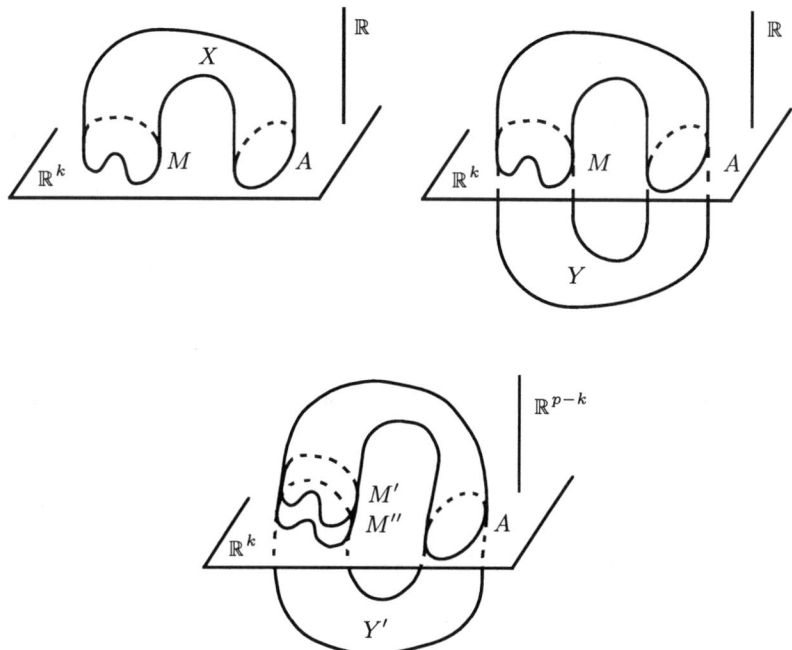

Figure 14.2.

mapping $Y \hookrightarrow \mathbb{R}^p$ and $h|_{Y'}$ close enough to g. This completes the proof. □

Remark 14.1.11. It is not known whether every compact submanifold $M \subset \mathbb{R}^p$ is diffeomorphic to a nonsingular algebraic subset of the *same* space $\mathbb{R}^p$. It can be proved that M has an algebraic model contained in $\mathbb{R}^{p+1}$ (cf. [9]).

Remark 14.1.12. There exists a projective version of Theorem 14.1.10, which says that every compact C^∞ manifold is diffeomorphic to a nonsingular projective real algebraic set. The proof relies on the same principles (cf. [186]).

Remark 14.1.13. The algebraic structure on a compact C^∞ manifold M provided by Theorem 14.1.10 is not unique (up to a biregular isomorphism). In fact, given a compact C^∞ manifold M, there exists an uncountable family $(X_i)_{i \in I}$ of algebraic models of M such that X_i and X_j are not birationally isomorphic if $i \neq j$ (cf. [51]).

Remark 14.1.14. A connected noncompact C^∞ manifold M is diffeomorphic to a nonsingular real algebraic set if and only if M is diffeomorphic to the interior of a compact C^∞ manifold with boundary (cf. [7]).

Remark 14.1.15. There is also a relative version of the Nash-Tognoli theorem [5, 32]. Let M be a compact C^∞ manifold and M_i, $i = 1, \ldots, k$, compact C^∞

submanifolds of M in general position. Then there exist a nonsingular real algebraic set V and a diffeomorphism $\varphi : M \to V$ such that $\varphi(M_i)$ is a nonsingular algebraic subset of V, for $i = 1, \ldots, k$.

Remark 14.1.16. There are versions of the Nash-Tognoli theorem in which one constructs algebraic models of $\mathcal{C}^\infty$ manifolds with additional structure. For example, given a compact $\mathcal{C}^\infty$ manifold M, there exists an algebraic model X of M such that every topological vector bundle over X is isomorphic to an algebraic one (cf. [33]). Other examples of such results were already mentioned in Chap. 11-13.

14.2 More about the Topology of Real Algebraic Sets

The answer to the question, which topological spaces are homeomorphic to real algebraic sets, is unknown. Below we indicate, without proof, some results describing classes of spaces homeomorphic to real algebraic sets. These results are based on the Nash Tognoli theorem 14.1.10. The main idea is as follows: we know that a compact $\mathcal{C}^\infty$ manifold is diffeomorphic to a real algebraic set; if we begin with a space having singularities, we may sometimes reduce the problem to the smooth case by a "topological desingularization". Next, we obtain the algebraic structure by using the relative version of the Nash-Tognoli theorem (Remark 14.1.15). Finally, we can find an algebraic set homeomorphic to the initial singular space by performing algebraic blowing downs (cf. Proposition 3.5.6). This method allows us to obtain a complete topological characterization of real algebraic sets with isolated singularities.

Theorem 14.2.1. *A topological space X is homeomorphic to a real algebraic set with isolated singularities if and only if X can be obtained from a compact $\mathcal{C}^\infty$ manifold Z with boundary $\partial Z = \cup_{i=1}^k M_i$ (disjoint union), where each M_i bounds a compact $\mathcal{C}^\infty$ manifold, by blowing down some of the M_i to points and removing the remaining M_i.*

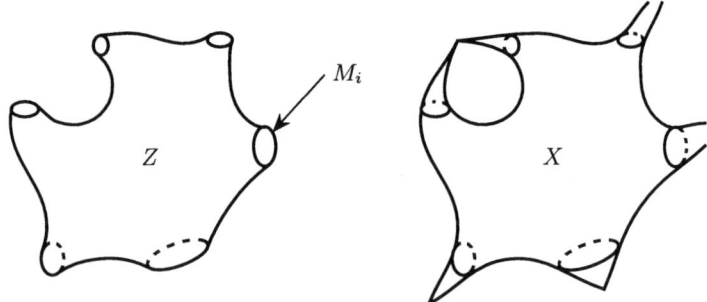

Figure 14.3.

Reference for the proof: [7].

The fact that the condition in Theorem 14.2.1 is necessary follows from the theorem on the resolution of singularities of [158]. The noncompact version of the Nash-Tognoli theorem, stated in Remark 14.1.14, is a consequence of Theorem 14.2.1.

The topological desingularization method applies to an important class of stratified spaces described in [6]. This class is large enough to contain all compact P.L. (piecewise linear) manifolds.

Theorem 14.2.2. *Every compact P.L. manifold is homeomorphic to a real algebraic set.*

Reference for the proof: [4, 6].

The following result is also obtained by the same technique.

Theorem 14.2.3. *Every compact real analytic set of dimension ≤ 3 is homeomorphic to a real algebraic set.*

Reference for the proof: [187].

It is not known whether Theorem 14.2.3 still holds true for higher dimensions. Some results concerning algebraicity of real analytic hypersurfaces can be found in [61].

Bibliographic Notes. The study of relations between $\mathcal{C}^\infty$ manifolds and real algebraic sets began in 1936 with Seifert [291], and resumed in 1952 with Nash [247], who proved Theorem 14.1.8. It is interesting to compare the technique and language of Nash's original paper with the contemporary presentation of his theorem. Nash's paper, like most of his work, was ahead of its time. The next was Wallace [335], who tried to prove that a connected compact $\mathcal{C}^\infty$ submanifold $M \subset \mathbb{R}^n$ is diffeomorphic to a nonsingular connected component of some algebraic subset of the same space $\mathbb{R}^n$. His proof was unfortunately not correct (the problem was solved later in [9]). However, his paper [335] has been very important because of the ideas it contained. He showed that the boundary of a compact $\mathcal{C}^\infty$ manifold is diffeomorphic to a nonsingular real algebraic set; the idea to apply cobordism originated in [335]. Finally, Tognoli proved Theorem 14.1.10 in [325].

Various results concerning algebraic models of topological spaces and related questions (some briefly presented in section 2) are contained in papers by Akbulut, Benedetti, Bochnak, King, Kucharz, Shiota and Tognoli, cited in the bibliography.

15. Witt Rings in Real Algebraic Geometry

Abstract. In the first section we define the Witt ring $W(A)$ of a commutative ring A and compare it with the group $K_0(A)$. In particular, if V is an algebraic set over a real closed field R, we show that the Witt ring $W(\mathcal{S}^0(V))$ coincides with $K_0(\mathcal{S}^0(V))$ ($\simeq KO(V)$ if $R = \mathbb{R}$). The second section is devoted to the result of Mahé concerning the separation of the semi-algebraically connected components of V by the signatures of elements of $W(\mathcal{P}(V))$. In the third section we prove that the morphism $W(\mathcal{P}(V))[1/2] \to W(\mathcal{S}^0(V))[1/2]$, induced by the inclusion $\mathcal{P}(V) \hookrightarrow \mathcal{S}^0(V)$, is surjective. This is part of the result of Brumfiel, which asserts that this morphism is actually an isomorphism.

15.1 K_0 and the Witt Ring

Throughout this chapter, A is a commutative ring in which 2 is invertible, R is a real closed field, and $V \subset R^n$ is an algebraic set. In practice A will usually be an R-algebra of functions on V (e.g. $A = \mathcal{P}(V), \mathcal{R}(V) \ldots$).

First we recall the definition of the group $K_0(A)$. We denote by $\mathrm{Proj}(A)$ the set of isomorphism classes of projective modules of finite type over A. The direct sum of modules induces an operation $\oplus$ on $\mathrm{Proj}(A)$. One defines an equivalence relation $\sim$ on $\mathrm{Proj}(A) \times \mathrm{Proj}(A)$ as follows: $(M, N) \sim (M', N')$ if there exists an integer $r \geq 0$ such that $M \oplus N' \oplus A^r \simeq M' \oplus N \oplus A^r$. The quotient $(\mathrm{Proj}(A) \times \mathrm{Proj}(A))/\sim$, equipped with the operation induced by

$$((M, N), (M', N')) \longmapsto (M \oplus M', N \oplus N'),$$

is a commutative group denoted by $K_0(A)$. We denote by $[M]$ the class of $(M, 0)$ in $K_0(A)$. Observe that every element of $K_0(A)$ can be written in the form $[M] - [N]$. We have $[M] = [N]$ if and only if there exists an integer $r \geq 0$ such that $M \oplus A^r \simeq N \oplus A^r$, and in this case M and N are said to be *stably isomorphic*. The class $[M]$ is called the *stable isomorphism class* of M. The group $K_0(A)$ contains $\mathbb{Z}$ (more precisely $\mathbb{Z} \cdot [A]$) as a subgroup, and the quotient $K_0(A)/\mathbb{Z}$ is isomorphic to the group $\widetilde{K}_0(A)$ considered in Chap. 12. We refer the reader to [234] for a systematic study of $K_0(A)$.

If $R = \mathbb{R}$ and $A = \mathcal{C}^0(V)$, there is a bijection between $\mathrm{Proj}(\mathcal{C}^0(V))$ and the set of isomorphism classes of topological vector bundles over V (cf. [314]).

Replacing projective modules of finite type with topological vector bundles in the above construction, we obtain a commutative group $KO(V)$, which is isomorphic to $K_0(\mathcal{C}^0(V))$.

A ring homomorphism $A \to A'$ induces a group homomorphism $K_0(A) \to K_0(A')$, and K_0 is a functor.

Assume now that A is an R-algebra of functions from V into R. Let $a \in V$ and let $\mathfrak{m}_a \subset A$ be the maximal ideal of all functions of A vanishing at a. If M is a projective module of finite type over A, the *rank of M at a* is, by definition, the dimension of the R-vector space $M/\mathfrak{m}_a M$. This rank depends only on the stable isomorphism class of M, and in this way we define a group homomorphism
$$\mathrm{rank} : K_0(A) \longrightarrow \mathbb{Z}^V$$
with values in the additive group $\mathbb{Z}^V$ of functions from V into $\mathbb{Z}$. The image of $K_0(\mathcal{P}(V))$ or $K_0(\mathcal{R}(V))$ (resp. $K_0(\mathcal{C}^0(V))$ if $R = \mathbb{R}$) by the rank homomorphism is the subgroup of functions on V which are locally constant with respect to the Zariski topology (resp. with respect to the euclidean topology). In particular, if V is irreducible, the image of $K_0(\mathcal{P}(V))$ consists of the constant functions, even if V has several semi-algebraically connected components.

Recall that, if $R = \mathbb{R}$, the inclusion $\mathcal{S}^0(V) \hookrightarrow \mathcal{C}^0(V)$ induces a bijection $\mathrm{Proj}(\mathcal{S}^0(V)) \to \mathrm{Proj}(\mathcal{C}^0(V))$ (Corollary 12.7.9) which, in turn, induces an isomorphism $K_0(\mathcal{S}^0(V)) \to K_0(\mathcal{C}^0(V))$. We are thus led to consider $K_0(\mathcal{S}^0(V))$ as the natural generalization of $KO(V)$, in the case of an arbitrary real closed field R.

We now introduce the Witt ring $W(A)$. We consider projective modules of finite type over A, equipped with a symmetric bilinear form. We say that a symmetric bilinear form $b : M \times M \to A$ (where M is a projective A-module of finite type) is *nondegenerate* if the linear mapping $h_b : M \to M^\vee$ from M into its dual, induced by b, is an isomorphism.

Definition 15.1.1. *A bilinear space over A is a pair (M, b), where M is a projective module of finite type over A and $b : M \times M \to A$ is a nondegenerate symmetric bilinear form.*

An isometry *between two bilinear spaces (M, b) and (M', b') over A is a linear isomorphism $\varphi : M \to M'$ such that $b'(\varphi(m), \varphi(n)) = b(m, n)$ for every $(m, n) \in M \times M$.*

If (M, b) and (M', b') are two bilinear spaces, their orthogonal sum *$(M, b) \perp (M', b') = (M \oplus M', b \perp b')$ is defined by*
$$(b \perp b')(x \oplus x', y \oplus y') = b(x, y) + b'(x', y'),$$
and their tensor product *$(M, b) \otimes_A (M', b') = (M \otimes_A M', b \otimes_A b')$ is defined by*
$$(b \otimes_A b')(x \otimes x', y \otimes y') = b(x, y)\, b'(x', y').$$

15.1 K_0 and the Witt Ring

Remark 15.1.2. a) A nondegenerate symmetric bilinear form on A^q can be identified with an invertible symmetric $q \times q$ matrix with entries in A. If this matrix is diagonal, with invertible entries $d_1, \ldots, d_q$ on the diagonal, the corresponding bilinear space will be denoted by $\langle d_1, \ldots, d_q \rangle$. If $d_1 = \ldots = d_q = d$, we set $\langle d_1, \ldots, d_q \rangle = q\langle d \rangle$.

b) Let $f : A \to B$ be a ring homomorphism, and (M, b), a bilinear space over A. We denote by

$$b \otimes_A B : (M \otimes_A B) \times (M \otimes_A B) \longrightarrow B$$

the symmetric B-bilinear form such that $(b \otimes_A B)(m_1 \otimes 1, m_2 \otimes 1) = f(b(m_1, m_2))$. Then $(M \otimes_A B, b \otimes_A B)$ is a bilinear space over B, denoted by $f^*(M, b)$.

c) We have seen that there is an equivalence between the category of projective modules of finite type over $\mathcal{R}(V)$ and the category of algebraic vector bundles over V (cf. Proposition 12.1.12). A bilinear space over $\mathcal{R}(V)$ can thus be identified with an algebraic vector bundle ξ over V, equipped with a nondegenerate symmetric bilinear form, that is, a regular function $b : \xi \oplus \xi \to R$ whose fibre $b_a : \xi_a \times \xi_a \to R$ at every point a of V is a nondegenerate symmetric bilinear form. We can also make a similar identification between bilinear spaces over $\mathcal{S}^0(V)$ (resp. $\mathcal{C}^0(V)$ if $R = \mathbb{R}$) and semi-algebraic (resp. topological) vector bundles equipped with a nondegenerate symmetric bilinear form, which is a continuous semi-algebraic (resp. continuous) function.

Definition 15.1.3. *A* hyperbolic space *over A is a bilinear space of the form $H(M) = (M \oplus M^\vee, \beta)$, where M is a projective module of finite type over A, $M^\vee$ is its dual and β is defined by $\beta(x \oplus \varphi, y \oplus \psi) = \psi(x) + \varphi(y)$.*

Two bilinear spaces (M, b) and (M', b') are said to be Witt-equivalent *if there exist hyperbolic spaces $H(N)$ and $H(N')$ such that $(M, b) \perp H(N)$ is isometric to $(M', b') \perp H(N')$. We denote by $[(M, b)]$ the Witt-equivalence class of (M, b).*

Theorem 15.1.4. *The set of Witt-equivalence classes of bilinear spaces over A, equipped with addition and multiplication defined by*

$$[(M, b)] + [(M', b')] = [(M, b) \perp (M', b')]$$
$$[(M, b)] \, [(M', b')] = [(M, b) \otimes_A (M', b')] ,$$

is a commutative ring. This ring is called the Witt ring *of A and is denoted by $W(A)$.*

Every ring homomorphism $f : A \to B$ induces a homomorphism of Witt rings $W(f) : W(A) \to W(B)$, defined by $W(f)([M, b]) = [f^*(M, b)]$.

Reference for the proof: [235], p. 14. The class of a hyperbolic space is, of course, the zero element of $W(A)$. Note that the negative of $[(M, b)]$ is $[(M, -b)]$. Indeed, there is an isometry $\varphi : (M, b) \perp (M, -b) \to H(M)$ defined by

$$\varphi(x \oplus y) = (x+y)/2 \oplus h_b((x-y)/2) \ .$$

The unit element of $W(A)$ is $[\langle 1 \rangle]$.

Definition 15.1.5. *Let A be an R-algebra of functions from V into R. Let a be a point of V, and (M,b), a bilinear space over A. The bilinear form b induces a nondegenerate symmetric bilinear form on the R-vector space $M/\mathfrak{m}_a M$, whose signature depends only on the Witt-equivalence class of (M,b). This defines a ring homomorphism*

$$\mathrm{sign} : W(A) \to \mathbb{Z}^V \ ,$$

called the signature homomorphism.

The following result will be useful in the study of the signature homomorphism.

If (M,b) is a bilinear space over $\mathcal{S}^0(V)$, U an open semi-algebraic subset of V and $\rho : \mathcal{S}^0(V) \to \mathcal{S}^0(U)$, the restriction homomorphism, we denote by $(M,b)|_U$ the bilinear space $\rho^*(M,b)$ over $\mathcal{S}^0(U)$.

Theorem 15.1.6. *Let (M,b) be a bilinear space over $\mathcal{S}^0(V)$. There exist a finite open semi-algebraic cover $V = \bigcup_{i=1}^k U_i$, and, for each $i = 1, \ldots, k$, integers $r_i, s_i \in \mathbb{N}$, such that $(M,b)|_{U_i}$ is isometric to the bilinear space $r_i \langle 1 \rangle \perp s_i \langle -1 \rangle$ over $\mathcal{S}^0(U_i)$.*

Proof. In order to prove the theorem, it will be more convenient to consider a semi-algebraic vector bundle ξ over V, equipped with a nondegenerate symmetric bilinear form $b : \xi \oplus \xi \to R$ which is continuous and semi-algebraic, using the identification of Remark 15.1.2 c). Let $\alpha \in \tilde{V}$. Without loss of generality, we may assume that there is an open semi-algebraic subset U of V such that $\alpha \in \tilde{U}$ and $\xi|_U = R^q \times U$. Then b induces a nondegenerate symmetric bilinear form

$$b_\alpha : k(\alpha)^q \times k(\alpha)^q \longrightarrow k(\alpha) \ .$$

We choose a basis $(e_1, \ldots, e_q)$ of $k(\alpha)^q$ such that $b_\alpha(e_i, e_j) = 0$ for $i \neq j$, $b_\alpha(e_i, e_i) = 1$ for $1 \leq i \leq r_\alpha$ and $b_\alpha(e_i, e_i) = -1$ for $r_\alpha < i \leq q$. By Proposition 7.3.4, there exist a semi-algebraic open subset U_α of U, such that $\alpha \in \tilde{U}_\alpha$, and continuous semi-algebraic functions $v_i : U_\alpha \to R^q$, such that $v_i(\alpha) = e_i$ for $i = 1, \ldots, q$. Shrinking U_α, we may assume that the functions constructed from v_i by the Gram-Schmidt orthogonalization process

$$w_1 = |b(v_1, v_1)|^{-1/2} v_1 \ ,$$

$$w_2 = |b(u_2, u_2)|^{-1/2} u_2 \ , \quad \text{where} \quad u_2 = v_2 - \frac{b(v_2, w_1)}{b(w_1, w_1)} w_1 \ ,$$

$$\cdots \quad \cdots \quad \cdots$$

$$w_q = |b(u_q, u_q)|^{-1/2} u_q \ , \quad \text{where} \quad u_q = v_q - \sum_{1 \leq i < q} \frac{b(v_q, w_i)}{b(w_i, w_i)} w_i \ ,$$

are well-defined on U_α, and satisfy $b(w_i, w_i) = 1$ for $1 \leq i \leq r_\alpha$ and $b(w_i, w_i) = -1$ for $r_\alpha < i \leq q$. The functions w_i induce an isometry between $(\xi, b)|_{U_\alpha}$ and the bilinear space $r_\alpha\langle 1\rangle \perp (q - r_\alpha)\langle -1\rangle$ over $\mathcal{S}^0(U_\alpha)$. Since the $\widetilde{U}_\alpha$ cover $\widetilde{V}$ and $\widetilde{V}$ is compact, we obtain a finite open semi-algebraic cover of V having the property stated in the theorem. □

Remark 15.1.7. If $R = \mathbb{R}$ and (M, b) is a bilinear space over $\mathcal{C}^0(V)$, there are an open cover $V = \bigcup_{i \in I} U_i$ and, for each $i \in I$, nonnegative integers r_i, s_i, such that $(M, b)|_{U_i}$ is isometric to the bilinear space $r_i\langle 1\rangle \perp s_i\langle -1\rangle$ over $\mathcal{C}^0(U_i)$. The proof is similar, but does not make use of the real spectrum.

Corollary 15.1.8. (i) *If A is a subring of $\mathcal{S}^0(V)$, the signature of a bilinear space over A is a mapping from V into $\mathbb{Z}$ which is constant on the semi-algebraically connected components of V.*

(ii) *If $R = \mathbb{R}$, the signature of a bilinear space over $\mathcal{C}^0(V)$ is constant on the connected components of V.*

Proof. (i) If $\rho : A \hookrightarrow \mathcal{S}^0(V)$ is the inclusion and (M, b) is a bilinear space over A, the signature of (M, b) is the same as the signature of $\rho^*(M, b)$. Hence, we may assume $A = \mathcal{S}^0(V)$. We use the notation of Theorem 15.1.6. If $a \in U_i$, the signature of (M, b) at a is $r_i - s_i$. Therefore, it is constant on an open semi-algebraic neighbourhood of a.

(ii) Follows in the same way using Remark 15.1.7. □

Theorem 15.1.6 also implies a "homotopy theorem" for bilinear spaces.

Corollary 15.1.9. (i) *Let (M, b) be a bilinear space over $\mathcal{S}^0(V \times [0,1])$. Denote by $\rho_t : \mathcal{S}^0(V \times [0,1]) \to \mathcal{S}^0(V)$, for $t = 0, 1$, the restriction homomorphisms induced by the inclusion mappings $V \simeq V \times \{t\} \hookrightarrow V \times [0,1]$. Then the two bilinear spaces $\rho_0^*(M, b)$ and $\rho_1^*(M, b)$ over $\mathcal{S}^0(V)$ are isometric.*

(ii) *In the case $R = \mathbb{R}$, the same statement holds with $\mathcal{S}^0$ replaced by $\mathcal{C}^0$.*

Proof. (i) We may assume that V is semi-algebraically connected (otherwise, we consider its semi-algebraically connected components). Then the r_i and s_i of Theorem 15.1.6 are all equal to r and s, respectively. We may thus view (M, b) as a semi-algebraic vector bundle over $V \times [0, 1]$ whose transition functions are required to take their values in the subgroup of $GL(r + s, R)$ consisting of the linear automorphisms preserving the symmetric bilinear form $r\langle 1\rangle \perp s\langle -1\rangle$. The result then follows from Lemma 11.7.4 and the standard argument similar to that used in [174], Theorem 9.8, p. 51.

(ii) We proceed in the same way, using Remark 15.1.7. Lemma 11.7.4 is not needed in this case. □

We shall now compare $K_0(\mathcal{S}^0(V))$ (or $K_0(\mathcal{C}^0(V))$ if $R = \mathbb{R}$) with $W(\mathcal{S}^0(V))$ (resp. $W(\mathcal{C}^0(V))$).

388   15. Witt Rings in Real Algebraic Geometry

Definition 15.1.10. *Let A be an R-algebra of functions from V into R, and (M,b), a bilinear space over A. Then b is said to be* positive definite *(resp.* negative definite*) if its signature is equal to the rank of M (resp. to $-(\operatorname{rank} M)$).*

Lemma 15.1.11. *If (M,b) is a bilinear space over $\mathcal{S}^0(V)$, there exists a decomposition $M \simeq M_+ \oplus M_-$ such that b is positive definite (resp. negative definite) on M_+ (resp. M_-). If $R = \mathbb{R}$, the same statement holds with $\mathcal{S}^0$ replaced by $\mathcal{C}^0$.*

Proof. For the proof, it will be more convenient to consider a semi-algebraic vector bundle ξ over V, equipped with a nondegenerate symmetric bilinear form $b : \xi \oplus \xi \to R$ which is continuous and semi-algebraic (cf. Remark 15.1.2 c)). We may assume that the rank and the signature of (ξ, b) are constant on V (otherwise, we replace V with its semi-algebraically connected components). By Theorem 15.1.6, there are a finite open semi-algebraic cover $V = \bigcup_{i=1}^{k} U_i$, and, for each $i = 1, \ldots, k$, an isometry

$$\varphi_i : (\epsilon_{U_i}^{r+s}, r\langle 1 \rangle \perp s\langle -1 \rangle) \longrightarrow (\xi, b)|_{U_i} \ .$$

The orthogonal sum $r\langle 1 \rangle \perp s\langle -1 \rangle$ gives a decomposition of $\xi|_{U_i}$ for each i, and we shall glue these decompositions together to obtain a decomposition $\xi = \xi_+ \oplus \xi_-$ into semi-algebraic subbundles, such that b is positive definite (resp. negative definite) on ξ_+ (resp. ξ_-). By induction on k, it is sufficient to consider the following situation: U and W are semi-algebraic open subsets of V such that $V = U \cup W$, there is an isometry $\varphi : (\epsilon_U^{r+s}, r\langle 1 \rangle \perp s\langle -1 \rangle) \to (\xi, b)|_U$ and there is a decomposition $\xi|_W = \pi \oplus \nu$ into semi-algebraic subbundles of ranks r and s, respectively, such that b is positive definite (resp. negative definite) on π (resp. ν).

For $a \in U \cap W$, we denote by $\sigma(a) \in \mathbb{G}_{r+s,r}(R)$ the subspace defined by

$$\{a\} \times \sigma(a) = \varphi^{-1}(\pi_a) \ .$$

The mapping $\sigma : U \cap W \to \mathbb{G}_{r+s,r}(R)$ is continuous, semi-algebraic, and the form $r\langle 1 \rangle + s\langle -1 \rangle$ is positive definite on $\sigma(a)$, for every $a \in U \cap W$. Recall (cf. subsection 3.4.2) that there is a biregular isomorphism ψ from $\mathbb{M}_{s,r}(R)$ onto the open subset of those V in $\mathbb{G}_{r+s,r}(R)$ such that $V \cap F = (0)$, where $F = \{0\} \times R^s \subset R^{r+s}$. The image by ψ of a matrix $\theta \in \mathbb{M}_{s,r}(R)$ is the subspace $\{(v, \theta \cdot v) \mid v \in R^r\}$. Let Ω be the *convex* open subset of $\mathbb{M}_{s,r}(R)$ consisting of those matrices θ for which $\|\theta \cdot v\| < \|v\|$, for every $v \in R^r \setminus \{0\}$, where $\| \ \|$ denotes the standard "euclidean norm" on R^r or R^s. Observe that the form $r\langle 1 \rangle \perp s\langle -1 \rangle$ is positive definite on $\psi(\theta)$ if and only if $\theta \in \Omega$. Let (λ, μ) be a semi-algebraic partition of unity on V, subordinate to the cover (U, W) (cf. Lemma 12.7.3). Denote by E the subspace $\psi(0) = R^r \times \{0\}$ of R^{r+s} and let ξ_+ be the semi-algebraic vector subbundle of rank r of ξ defined by:

$$(\xi_+)_a = \begin{cases} \varphi(\{a\} \times \psi(\mu(a)\psi^{-1}(\sigma(a)))) & \text{if } a \in U \cap W, \\ \varphi(\{a\} \times E) & \text{if } a \in U \setminus (U \cap W), \\ \pi_a & \text{if } a \in W \setminus (U \cap W). \end{cases}$$

By the convexity of Ω and the remark above, b is positive definite on ξ_+. In a similar way, we construct a semi-algebraic vector subbundle ξ_- of ξ having rank s, such that b is negative definite on ξ_-. Obviously, we have $\xi = \xi_+ \oplus \xi_-$. $\square$

Theorem 15.1.12. (i) *Let M be a projective module of finite type over $S^0(V)$. There exists a positive definite symmetric bilinear form b on M. Moreover, the isometry class of the bilinear space (M, b) does not depend on the choice of the positive definite symmetric bilinear form b. Hence, there is a canonical group homomorphism*

$$\Delta: K_0(S^0(V)) \to W(S^0(V)),$$

defined by $\Delta([M]) = [(M, b)]$.

(ii) *The homomorphism Δ is an isomorphism, and the following diagram is commutative:*

$$\begin{array}{ccc} K_0(S^0(V)) & \xrightarrow{\Delta} & W(S^0(V)) \\ & \searrow \text{rank} \quad \swarrow \text{sign} & \\ & \mathbb{Z}^V & \end{array}.$$

(iii) *If $R = \mathbb{R}$, the preceding statements hold with S^0 replaced by C^0.*

Proof. (i) We represent M as a direct factor of a free module of finite type: $M \oplus P \simeq (S^0(V))^q$. Let b be the restriction to M of the form $q\langle 1 \rangle$ on $(S^0(V))^q$. Then b is positive definite. Let b' be another positive definite form on M. The form $(1-t)b + tb'$ is positive definite and nondegenerate for every $t \in [0, 1]$. By Corollary 15.1.9, (M, b) is isometric to (M, b').

(ii) The equality sign $\circ \Delta =$ rank is clear from the construction of Δ. Now we construct the isomorphism ∇ which is the inverse of Δ. Using the notation of Lemma 15.1.11, we set $\nabla([(M, b)]) = [M_+] - [M_-]$. We have to check that ∇ is well-defined. Let $M \simeq M'_+ \oplus M'_-$ be another decomposition such that b is positive definite (resp. negative definite) on M'_+ (resp. M'_-). Since $M_+ \oplus M'_- \simeq M$, we have $M_+ \simeq M/M'_- \simeq M'_+$; similarly, $M_- \simeq M'_-$. To show that $\nabla([(M, b)])$ depends only on the Witt-equivalence class of (M, b), observe that $H(N)$ is isometric to $(N, \beta) \perp (N, -\beta)$, for every positive definite form β on N. Finally, it is clear that $\nabla \circ \Delta$ and $\Delta \circ \nabla$ are the identity mappings.

(iii) The proof is analogous to that for S^0. $\square$

Corollary 15.1.13. *If $R = \mathbb{R}$, all homomorphisms of the commutative diagram*

$$
\begin{array}{ccc}
K_0(\mathcal{S}^0(V)) & \longrightarrow & K_0(\mathcal{C}^0(V)) \simeq KO(V) \\
\Delta \downarrow & & \Delta \downarrow \\
W(\mathcal{S}^0(V)) & \longrightarrow & W(\mathcal{C}^0(V))
\end{array}
$$

are isomorphisms.

The decomposition property of Lemma 15.1.11 is valid for bilinear spaces over $\mathcal{R}(V)$ only when certain restrictions are placed on V. The following result gives a necessary and sufficient condition for the existence of such decompositions, in the case where $V \subset \mathbb{R}^n$ is compact and connected.

Theorem 15.1.14. *Let $V \subset \mathbb{R}^n$ be a compact connected algebraic set. The following properties are equivalent:*

(i) *The canonical homomorphism $K_0(\mathcal{R}(V)) \to K_0(\mathcal{C}^0(V))$ is an isomorphism.*

(ii) *Every topological vector bundle over V is topologically isomorphic to an algebraic vector bundle.*

(iii) *Every bilinear space (M,b) over $\mathcal{R}(V)$ admits a decomposition $M = M_+ \oplus M_-$ such that b is positive definite (resp. negative definite) on M_+ (resp. M_-).*

Proof. (i) $\Leftrightarrow$ (ii) is a straightforward consequence of Theorem 12.3.6.

(ii) $\Rightarrow$ (iii) Let (ξ, b) be a bilinear space over $\mathcal{R}(V)$. By Lemma 15.1.11, there is a decomposition $\xi = \eta_+ \oplus \eta_-$ into a Whitney sum of *topological* vector subbundles such that b is positive definite (resp. negative definite) on η_+ (resp. η_-). Let k and ℓ be the ranks of ξ and η_+, respectively. By the definition of algebraic vector bundles, there is an injective algebraic morphism $i : \xi \to \epsilon_V^n$, for some $n \in \mathbb{N}$. The morphism i induces a regular mapping $f : V \to \mathbb{G}_{n,k}(\mathbb{R})$ and a continuous mapping $g : V \to \mathbb{G}_{n,\ell}(\mathbb{R})$, defined by $i(\xi_a) = \{a\} \times f(a)$ and $i((\eta_+)_a) = \{a\} \times g(a)$, for every $a \in V$. Property (ii) and Theorem 13.3.1 imply that g can be approximated by a regular mapping $h_1 : V \to \mathbb{G}_{n,\ell}(\mathbb{R})$. For every $a \in V$, denote by $h(a) \subset \mathbb{R}^n$ the image of $h_1(a)$ by the orthogonal projection onto $f(a)$. If h_1 is close enough to g, we have $h(a) \in \mathbb{G}_{n,\ell}(\mathbb{R})$. Hence, h is a regular mapping from V to $\mathbb{G}_{n,\ell}(\mathbb{R})$. Let ξ_+ be the algebraic vector subbundle of ξ defined by $\xi_+ = i^{-1}(h^*(\gamma_{n,\ell}))$. If h_1 is close enough to g, h is close to g and b is positive definite on ξ_+. In the same way, we construct an algebraic vector subbundle ξ_- of ξ, having rank $k - \ell$, such that b is negative definite on ξ_-. Then ξ decomposes into $\xi_+ \oplus \xi_-$.

(iii) $\Rightarrow$ (ii) First we introduce some notation. We denote by $\mathrm{Sym}(n)$ the set of symmetric $n \times n$ matrices with entries in $\mathbb{R}$ whose determinant is nonzero, regarded as a Zariski open subset of $\mathbb{R}^{n(n+1)/2}$. We denote by $\mathrm{Sym}(n,k) \subset \mathrm{Sym}(n)$ the open subset of matrices whose signature is equal to $2k - n$. We shall identify a nondegenerate symmetric bilinear form $\mathbb{R}^n \times \mathbb{R}^n \to \mathbb{R}$ with its matrix in $\mathrm{Sym}(n)$. Let U be the set of $(\beta, F) \in \mathrm{Sym}(n) \times \mathbb{G}_{n,k}(\mathbb{R})$

such that β is positive definite on F. Observe that U is an open subset of $\mathrm{Sym}(n) \times \mathbb{G}_{n,k}(\mathbb{R})$.

Let ξ be a topological vector bundle over V. There exists a continuous mapping $f : V \to \mathbb{G}_{n,k}(\mathbb{R})$ such that $\xi \simeq_{\mathrm{top}} f^*(\gamma_{n,k})$. Let $b : V \to \mathrm{Sym}(n)$ be the continuous mapping defined by

$$b(x)(u,u) = \|f(x) \cdot u\|^2 - \|u - f(x) \cdot u\|^2 \, , \text{ for every } x \in V \, , \, u \in \mathbb{R}^n \, ,$$

where $f(x)$ is regarded as the matrix of an orthogonal projection, as in Theorem 3.4.4. We have $b(V) \subset \mathrm{Sym}(n,k)$ and $(b,f)(V) \subset U$. By the Stone-Weierstrass theorem, we can approximate b by a polynomial mapping $c : V \to \mathrm{Sym}(n)$ such that $c(V) \subset \mathrm{Sym}(n,k)$ and $(c,f)(V) \subset U$. By property (iii), there is a decomposition $\epsilon_V^n = \eta_+ \oplus \eta_-$, where η_+ (resp. η_-) is an algebraic vector subbundle of ϵ_V^n and c is positive definite (resp. negative definite) on η_+ (resp. η_-). Clearly, the rank of η_+ is k. The inclusion $\eta_+ \subset \epsilon_V^n$ induces a regular mapping $g : V \to \mathbb{G}_{n,k}(\mathbb{R})$ such that $(\eta_+)_a = \{a\} \times g(a)$, for every $a \in V$. The fibre of the projection $U \to \mathrm{Sym}(n)$ at $\beta \in \mathrm{Sym}(n,k)$ is

$$U_\beta = \{F \in \mathbb{G}_{n,k}(\mathbb{R}) \mid \beta \text{ is positive definite on } F\} \, .$$

The proof of Lemma 15.1.11 shows that U_β is homeomorphic to a convex open subset of $\mathbb{M}_{n-k,k}(\mathbb{R})$. It follows that (c,f) and (c,g) are homotopic in U, and, therefore, f and g are homotopic. This proves that ξ is isomorphic to the algebraic vector bundle $g^*(\gamma_{n,k})$. □

Corollary 15.1.15. *Assume that $V \subset \mathbb{R}^n$ is a compact connected algebraic set and the equivalent properties of Theorem 15.1.14 hold. Then there is a surjective homomorphism $\nabla_\mathcal{R} : W(\mathcal{R}(V)) \to K_0(\mathcal{R}(V))$ such that the following diagram is commutative:*

$$\begin{array}{ccc} K_0(\mathcal{R}(V)) & \longrightarrow & K_0(\mathcal{C}^0(V)) \simeq KO(V) \\ \uparrow \scriptstyle{\nabla_\mathcal{R}} & & \uparrow \scriptstyle{\nabla} \\ W(\mathcal{R}(V)) & \longrightarrow & W(\mathcal{C}^0(V)) \, . \end{array}$$

Proof. Define $\nabla_\mathcal{R}$ in the same way as ∇ was defined in the proof of Theorem 15.1.12. It is then clear that the diagram is commutative. The surjectivity of $\nabla_\mathcal{R}$ follows from the fact that every projective module of finite type over $\mathcal{R}(V)$ can be equipped with a positive definite symmetric bilinear form. This fact is proved as in Theorem 15.1.12 (i). □

The answer to the following question is not known: Are any two positive definite symmetric bilinear forms on the same projective module of finite type over $\mathcal{R}(V)$ isometric? A positive answer would imply that all morphisms in the diagram of Corollary 15.1.15 are isomorphisms.

15.2 Separation of Semi-algebraically Connected Components of Algebraic Sets by Signatures of Bilinear Spaces

Assume $R = \mathbb{R}$ and $V \subset \mathbb{R}^n$ is a bounded algebraic set. Let C be a connected component of V. By the Stone-Weierstrass theorem, there is a polynomial function on V which is positive on C and negative on $V \setminus C$. Neither assumption can be omitted, as the following examples show. The semi-algebraically connected components of an algebraic set V cannot, in general, be separated by signs of polynomials which do not vanish on V.

Example 15.2.1. Let $V = \{(x,y,z) \in \mathbb{R}^3 \mid z(y^2 - x^3 + x) = 1\}$. Denote by $\pi : \mathbb{R}^3 \to \mathbb{R}^2$ the projection defined by $\pi(x,y,z) = (x,y)$. Then π induces a biregular isomorphism from V onto the complement in $\mathbb{R}^2$ of the cubic curve given by the equation $f(x,y) = y^2 - x^3 + x = 0$. Denote by S_1, S_2 and S_3 the three connected components of the complement of the cubic curve, as indicated on Figure 15.1.

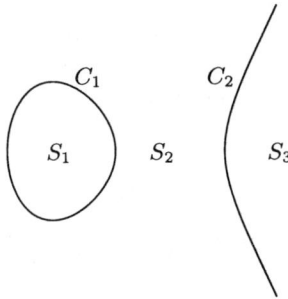

Figure 15.1.

Set $C_1 = \text{clos}(S_1) \cap \text{clos}(S_2)$ and $C_2 = \text{clos}(S_2) \cap \text{clos}(S_3)$. Suppose that there is a polynomial $P_1 \in \mathbb{R}[X,Y,Z]$ which is positive on $V \cap \pi^{-1}(S_1)$ and negative on $V \cap \pi^{-1}(S_2 \cup S_3)$. Substituting $1/f(X,Y)$ for Z in P_1 and multiplying by a suitable even power of $f(X,Y)$, we obtain a polynomial $P \in \mathbb{R}[X,Y]$ which is positive on S_1 and negative on $S_2 \cup S_3$. Hence, P vanishes on C_1 and we have $P = Qf^m$, where $m \geq 1$ and Q is not divisible by f. Therefore the zeros of Q are finitely many points on the curve $C_1 \cup C_2$. Then either m is even and the sign of P is the same on S_1, S_2 and S_3, or m is odd and P has opposite signs on S_2 and S_3. Both cases lead to a contradiction.

Example 15.2.2. Let $R = \mathbb{R}(t)^\wedge$ be the field of real Puiseux series. Set

$$f(X,Y) = (1 - X^2 + Y^2)X^2Y^2 - t \in R[X,Y]$$

and let V be the bounded algebraic subset $V \subset R^3$ given by the equation $z^2 = f(x,y)$. Denote by $\pi : R^3 \to R^2$ the projection defined by $\pi(x,y,z) =$

(x,y). The algebraic set V has four semi-algebraically connected components, which project onto the four semi-algebraically connected components of $S = \{(x,y) \in R^2 \mid f(x,y) \geq 0\}$. Let C be the semi-algebraically connected component of S contained in the set $K = \{(x,y) \in R^2 \mid x \geq 0,\ y \geq 0\}$. Suppose that there is a polynomial $P_1 \in R[X,Y,Z]$ which is positive on $V \cap \pi^{-1}(C)$ and negative on $V \setminus \pi^{-1}(C)$. Then $P_2 = P_1(X,Y,Z) + P_1(X,Y,-Z)$ has the same property, and contains only even powers of Z. Substituting $f(X,Y)$ for Z^2 in P_2, we obtain a polynomial $P \in R[X,Y]$ which is positive on C and negative on $S \setminus C$. Set $P = \sum_{(i,j) \in I} p_{i,j} X^i Y^j$. Multiplying by a suitable fractional power of t, we may assume that all Puiseux series $p_{i,j}$ have nonnegative order, and at least one $p_{i,j}$ has order 0. Hence, the polynomial $\overline{P} = \sum_{(i,j) \in I} p_{i,j}(0) X^i Y^j \in \mathbb{R}[X,Y]$ is nonzero. For every $(x,y) \in \mathbb{R}^2$, the Puiseux series $P(x,y)$ has nonnegative order, and $P(x,y)(0) = \overline{P}(x,y)$. Hence, $\overline{P}$ is nonnegative on $D \cap K$, and nonpositive on $D \cap (R^2 \setminus K)$, where D is the unit disk in R^2. Factor $\overline{P}$ as $\overline{P} = Y^e Q$, where Q is not divisible by Y. Choosing $x \in \mathbb{R}$ such that $Q(x,0) \neq 0$ and $0 < x < 1$ (resp. $-1 < x < 0$) and considering the sign of $\overline{P}$ on a small disk centred at $(x,0)$, we conclude that e must be even (resp. odd). This is impossible.

We shall show in this section that the semi-algebraically connected components of an algebraic set $V \subset R^n$ can be separated by signatures of bilinear spaces over $\mathcal{P}(V)$. More precisely, we shall prove the following result.

Theorem 15.2.3. *Let $V \subset R^n$ be an algebraic set and C a semi-algebraically connected component of V. There exist an integer $k \in \mathbb{N}$ and a bilinear space (M,b) over $\mathcal{P}(V)$, whose signature is equal to 2^k on C and 0 on $V \setminus C$.*

Remark 15.2.4. Recall (cf. Corollary 15.1.8) that the signature of a bilinear space is constant on the semi-algebraically connected components of V. In fact, the proof of Theorem 15.1.6 shows that the signature can be extended to a continuous mapping from $\widetilde{V} = \mathrm{Spec}_r(\mathcal{P}(V))$ into $\mathbb{Z}$. Theorem 15.2.3 then implies that every function of $C^0(\widetilde{V}, \mathbb{Z})$, multiplied by an appropriate power of 2, is the signature of some bilinear space over $\mathcal{P}(V)$. Equivalently, the cokernel of the homomorphism

$$\mathrm{sign} : W(\mathcal{P}(V)) \to C^0(\widetilde{V}, \mathbb{Z})$$

is a 2-*group*. This result is stated in [219] in a more general form: for every commutative ring A, one can define a "signature" homomorphism

$$\mathrm{sign} : W(A) \to C^0(\mathrm{Spec}_r(A), \mathbb{Z}) ,$$

and the cokernel of this homomorphism is a 2-group.

In this section and the next one, we shall use functions on V which can be constructed from polynomials by taking $1/f$ and $\sqrt{f}$ for positive functions f.

Notation 15.2.5. *We denote by $\mathcal{D}(V)$ the smallest subring of $\mathcal{S}^0(V)$ containing $\mathcal{P}(V)$ and such that*

$$f \in \mathcal{D}(V) \text{ and } f > 0 \text{ on } V \implies 1/f \in \mathcal{D}(V) \text{ and } \sqrt{f} \in \mathcal{D}(V).$$

Observe that every function $f \in \mathcal{D}(V)$ which has no zero on V has a multiplicative inverse in $\mathcal{D}(V)$, since $1/f = f/f^2$.

Proof of Theorem 15.2.3. It is not difficult to construct a bilinear space over $\mathcal{D}(V)$ whose signature separates C from $V \setminus C$. Indeed, by Mostowski's separation theorem 2.7.3, there is a function $f \in \mathcal{D}(V)$ which is positive on C and negative on $V \setminus C$. The function f is invertible in $\mathcal{D}(V)$, and the signature of the bilinear space $\langle 1, f \rangle$ is equal to 2 on C and 0 on $V \setminus C$. Hence, Theorem 15.2.3 is a consequence of the following theorem. □

Theorem 15.2.6. *Let (M_1, b_1) be a bilinear space over $\mathcal{D}(V)$. Denote by $i : \mathcal{P}(V) \hookrightarrow \mathcal{D}(V)$ the inclusion mapping. Then there exist an integer $k \in \mathbb{N}$ and a bilinear space (M, b) over $\mathcal{P}(V)$ such that the orthogonal sum $\underset{2^k}{\perp}(M_1, b_1)$ of 2^k copies of (M_1, b_1) is Witt-equivalent to $i^*(M, b)$. Equivalently, the cokernel of the homomorphism*

$$W(i) : W(\mathcal{P}(V)) \longrightarrow W(\mathcal{D}(V))$$

is a 2-group.

Theorem 15.2.6 will be proved in several steps. First, we reduce the proof to the case of nondegenerate symmetric bilinear forms on free modules of finite type, i.e., invertible symmetric matrices.

Lemma 15.2.7. *Let A be a commutative ring such that 2 is invertible in A. Denote by $W'(A)$ the subring of $W(A)$ consisting of the Witt-equivalence classes of bilinear spaces (M, b) over A, such that M is a free A-module of finite type. Then $2W(A) \subset W'(A)$.*

Proof. Let (M, b) be an arbitrary bilinear space over A. Recall that M can be represented as a direct factor of a free A-module: $M \oplus N \simeq A^q$. Then $(M, b) \perp (M, b)$ is Witt-equivalent to $(M, b) \perp (M, b) \perp H(N)$, and the module $M \oplus M \oplus N \oplus N^\vee$ is free and isomorphic to A^{2q} (recall that $N^\vee$ is isomorphic to N). This proves that $2[(M, b)] \in W'(A)$. □

The ring $\mathcal{D}(V)$ is constructed from $\mathcal{P}(V)$ by taking $1/f$ and $\sqrt{f}$ for positive functions f. We shall need a "formal" version of these operations. We introduce the following notation, which will be used in this section only. If A is a commutative ring, we denote by $\Sigma^{-1}A$ the ring of fractions of A with respect to the multiplicative subset

$$\Sigma = \{1 + a_1^2 + \cdots + a_\ell^2 \mid \ell \in \mathbb{N}, \ a_1, \ldots, a_\ell \in A\}.$$

If $a \in A$, we denote

15.2 Separation of Connected Components by Signatures

$$A\langle a^{1/2}\rangle = \bigcup_{m=1}^{\infty} A[Y_m]/(Y_m^{2^m} - a),$$

where $A[Y_m]/(Y_m^{2^m} - a)$ is considered as a subring of $A[Y_{m+1}]/(Y_{m+1}^{2^{m+1}} - a)$ by setting $Y_m = Y_{m+1}^2$.

Let $R \hookrightarrow K$ be a field extension, where K is real closed. Recall that every semi-algebraic function $h : V \to R$ has an extension $h_K : V_K \to K$ over K (cf. Definition 5.3.2). If $\psi : \mathcal{P}(V) \to K$ is an R-algebra homomorphism and $X = (X_1, \ldots, X_n)$ are the coordinate functions restricted to V, we denote by $\psi(X)$ the point $(\psi(X_1), \ldots, \psi(X_n)) \in K^n$. The point $\psi(X)$ belongs to the extension $V_K \subset K^n$, since it satisfies the equations defining V.

Lemma 15.2.8. *Consider a commutative diagram of ring homomorphisms*

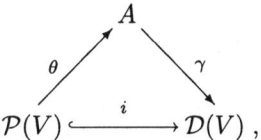

satisfying the following property:

(∗) *For every homomorphism $\varphi : A \to K$ into a real closed field K and for every $g \in A$, we have $\varphi(g) = (\gamma(g))_K(\varphi \circ \theta(X))$*

Let $f \in A$ be such that the function $\gamma(f)$ is positive on V. Then
(i) f is invertible in $\Sigma^{-1}A$,
(ii) γ can be extended in a unique way to a homomorphism

$$\gamma' : A' = (\Sigma^{-1}A)\langle f^{1/2}\rangle \to \mathcal{D}(V),$$

(iii) *the commutative diagram*

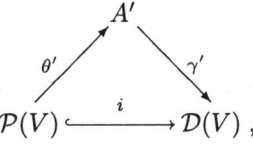

where θ' is the composition $\mathcal{P}(V) \xrightarrow{\theta} A \to A'$, satisfies property (∗).

Proof. (i) We claim that $\varphi(f) > 0$ for every homomorphism $\varphi : A \to K$ with K real closed. Indeed, $\varphi(f) = (\gamma(f))_K(\varphi \circ \theta(X))$ by property (∗) and $(\gamma(f))_K$ is positive on V_K by the Tarski-Seidenberg principle. Hence, by Proposition 4.4.1, f divides an element of the form $1 + a_1^2 + \cdots + a_\ell^2$ in A.

(ii) The homomorphism γ' is defined by $\gamma'(Y_m) = \sqrt[2^m]{\gamma(f)} \in \mathcal{D}(V)$.

(iii) If $\varphi : A' \to K$ is a homomorphism into a real closed field K, then $\varphi(Y_m) = \sqrt[2^m]{\varphi(f)} = (\gamma'(Y_m))_K(\varphi \circ \theta'(X))$. □

Lemma 15.2.9. *Let b be an invertible symmetric matrix with entries in $\mathcal{D}(V)$. There exist a commutative diagram of ring homomorphisms*

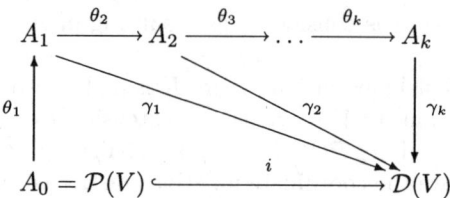

and an invertible symmetric matrix c with entries in A_k such that
  (i) *the image of c by γ_k is b,*
  (ii) *for $i = 0, \ldots, k-1$, the homomorphism $\theta_{i+1} : A_i \to A_{i+1}$ is of the form $A_i \to \Sigma^{-1} A_i$ or $A_i \to A_i \langle f^{1/2} \rangle$, where f is an invertible element of A_i.*

Proof. Starting from the diagram

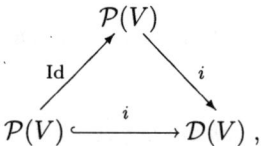

and applying the construction of Lemma 15.2.8 (ii) finitely many times, we obtain a commutative diagram of ring homomorphisms

(1)

$$\begin{array}{c} A \\ \theta \swarrow \quad \searrow \gamma \\ \mathcal{P}(V) \xhookrightarrow{i} \mathcal{D}(V) \,, \end{array}$$

such that the matrix b is the image by γ of a symmetric matrix c_1 with entries in A. Observe that Lemma 15.2.8 (iii) ensures that the construction can be iterated and the diagram (1) satisfies the property $(*)$ of this lemma. Applying Lemma 15.2.8 (i) to $f = (\det(c_1))^2$, we see that the image c of c_1 by $A \to \Sigma^{-1} A$ is an invertible matrix. Moreover, γ factors through $\Sigma^{-1} A$. □

We consider now the case of homomorphisms of the form $A \to \Sigma^{-1} A$.

Lemma 15.2.10. *Let A be a commutative ring such that 2 is invertible in A. The cokernel of the homomorphism $W'(A) \to W'(\Sigma^{-1} A)$ is a 2-group.*

Proof. Let b be an invertible symmetric $q \times q$ matrix with entries in $\Sigma^{-1} A$. Multiplying all entries of b by the square of a common denominator in Σ does not change the isometry class of b over $\Sigma^{-1} A$. Hence, we may assume that

(i) b is the image of a symmetric matrix b' with entries in A.

Moreover, since b is invertible,

(ii) there exists a symmetric matrix c' with entries in A such that $b' \cdot c' = s_\ell I_q$, where I_q is the $q \times q$ identity matrix, $s_\ell = 1 + g_1^2 + \cdots + g_\ell^2$ and $g_1, \ldots, g_\ell \in A$.

We claim that, for every matrix b satisfying properties (i) and (ii), the orthogonal sum $\underset{2^\ell}{\perp} b$ of 2^ℓ copies of b is isometric to the image of an invertible symmetric matrix with entries in A. The lemma follows from the claim.

We prove the claim by induction on ℓ. There is nothing to prove if $\ell = 0$. Assume $\ell > 0$ and the claim proved for $\ell - 1$. Set

$$b_1' = \begin{bmatrix} b' & g_\ell I_q \\ g_\ell I_q & c' \end{bmatrix} \quad \text{and} \quad c_1' = \begin{bmatrix} c' & -g_\ell I_q \\ -g_\ell I_q & b' \end{bmatrix}.$$

Since $b_1' \cdot c_1' = s_{\ell-1} I_{2q}$, it follows from the inductive assumption that the image of $\underset{2^{\ell-1}}{\perp} b_1'$ is isometric, over $\Sigma^{-1} A$, to the image of an invertible symmetric matrix with entries in A. If

$$\delta = \begin{bmatrix} I_q & -g_\ell I_q \\ 0 & b' \end{bmatrix},$$

then

$$^t\delta \cdot b_1' \cdot \delta = \begin{bmatrix} b' & 0 \\ 0 & s_{\ell-1} b' \end{bmatrix} = b' \otimes \begin{bmatrix} 1 & 0 \\ 0 & s_{\ell-1} \end{bmatrix}.$$

Hence, the image of b_1' is isometric, over $\Sigma^{-1} A$, to $b \otimes (\langle 1, s_{\ell-1} \rangle)$. Setting $\Lambda_0 = [1]$ and

$$\Lambda_p = \begin{bmatrix} \Lambda_{p-1} & -g_p I_{2^{p-1}} \\ g_p I_{2^{p-1}} & {}^t\Lambda_{p-1} \end{bmatrix}$$

for $1 \leq p < \ell$, we obtain ${}^t\Lambda_{\ell-1} \cdot \Lambda_{\ell-1} = s_{\ell-1} \cdot I_{2^{\ell-1}}$. This shows that $2^{\ell-1}\langle s_{\ell-1} \rangle$ is isometric to $2^{\ell-1}\langle 1 \rangle$. Hence $\underset{2^\ell}{\perp} b$ is isometric to $b \otimes (2^{\ell-1}\langle 1 \rangle \perp 2^{\ell-1}\langle s_{\ell-1} \rangle)$, which is isometric to the image of $\underset{2^{\ell-1}}{\perp} b_1'$. The claim follows. □

We now turn to the case of homomorphisms of the form $A \to A\langle f^{1/2} \rangle$. We assume that 2 and f are invertible in A. Let $A' = A[Y]/(Y^2 - f)$, $k : A \hookrightarrow A'$ the canonical injection, $\tau : A' \to A'$ the A-algebra involution defined by $\tau(Y) = -Y$, $t : A' \to A$ the A-linear homomorphism defined by $t(g) = g + \tau(g)$ for $g \in A'$. If $g = g_1 + g_2 Y$, where $g_1, g_2 \in A$, then $t(g) = 2g_1$. For every A'-module N, we denote by N_A the set N with its underlying structure of A-module.

Lemma 15.2.11. *Let (M, b) be a bilinear space over A'.*

(i) *$(M_A, t \circ b)$ is a bilinear space over A, denoted by $t_*(M, b)$.*

(ii) *The bilinear space $k^*(t_*(M, b))$ is isometric to $(M, b) \perp \tau^*(M, b)$.*

Proof. (i) Since $(A')_A \simeq A \oplus A$, it is clear that M_A is a projective A-module of finite type. It remains to prove that the symmetric A-bilinear form $t \circ b$ is nondegenerate, i.e. the A-linear homomorphism $h_{tob} : M_A \to (M_A)^\vee$, induced by $t \circ b$, is an isomorphism.

Let $\lambda : (M^\vee)_A \to (M_A)^\vee$ be the A-linear homomorphism defined by $\lambda(\varphi) = t \circ \varphi$, for $\varphi \in (M^\vee)_A$. If $\psi \in (M_A)^\vee$, define $\mu(\psi) : M \to A'$ by

$$\mu(\psi)(m) = \frac{1}{2}(\psi(m) + f^{-1}Y(\psi(Ym))) .$$

An easy computation shows that, for every $\psi \in (M_A)^\vee$ (resp. $\varphi \in (M^\vee)_A$), $\lambda(\mu(\psi)) = \psi$ (resp. $\mu(\lambda(\varphi)) = \varphi$). Hence, λ is an isomorphism. It follows that h_{tob}, which is the composition of $(h_b)_A : M_A \to (M^\vee)_A$ with λ, is an isomorphism.

(ii) We denote by τ^*M the set M with the structure of A'-module defined by

$$\begin{aligned} A' \times M &\longrightarrow M \\ (g, m) &\longmapsto \tau(g)m \end{aligned}$$

Observe that $\tau^*(M, b) = (\tau^*M, \tau \circ b)$. Let $\theta : M_A \otimes_A A' \to M \oplus \tau^*M$ be defined by

$$\theta(m_1 \otimes 1 + m_2 \otimes Y) = (m_1 + Ym_2,\ m_1 - Ym_2) .$$

Clearly, θ is bijective. Since

$$\theta(g(m_1 \otimes 1 + m_2 \otimes Y)) = (g(m_1 + Ym_2),\ \tau(g)(m_1 - Ym_2)) ,$$

θ is a A'-linear isomorphism. It remains to check that θ is an isometry from $k^*(t_*(M, b))$ onto $(M, b) \perp \tau^*(M, b)$. Since the elements $m \otimes 1$, for $m \in M_A$, generate $M_A \otimes_A A'$ as an A'-module, this follows from the equalities

$$((t \circ b) \otimes_A A')(m_1 \otimes 1, m_2 \otimes 1) = t(b(m_1, m_2)) = b(m_1, m_2) + \tau(b(m_1, m_2)) .$$

$\square$

Lemma 15.2.12. *Let A be a commutative ring such that 2 and f are invertible in A. Then the cokernel of the homomorphism $W'(A) \to W'(A\langle f^{1/2}\rangle)$ is a 2-group.*

Proof. Let b be an invertible symmetric matrix with entries in $A\langle f^{1/2}\rangle$. It is clear that b is the image of an invertible symmetric matrix c with entries in a subring $A_m = A[Y_m]/(Y_m^{2^m} - f)$. First assume that $m = 1$ and set $Y = Y_1$. By Lemma 15.2.11, $k^*(t_*(\langle 1, Y \rangle \otimes c))$ is isometric to $(\langle 1, Y \rangle \otimes c) \perp (\langle 1, -Y \rangle \otimes \tau^*(c))$. Since $Y = Y_1 = Y_2^2$ in $A\langle f^{1/2}\rangle$, the bilinear space over $A\langle f^{1/2}\rangle$ induced by $(\langle 1, Y \rangle \otimes c) \perp (\langle 1, -Y \rangle \otimes \tau^*(c))$ is Witt-equivalent to $\langle 1, 1 \rangle \otimes b$. Hence, $2[b]$ is the image of $[t_*(\langle 1, Y \rangle \otimes c)] \in W'(A)$. If $m > 1$, we observe that $A_m = A_{m-1}[Y_m]/(Y_m^2 - Y_{m-1})$ and we apply Lemma 15.2.11 m times as above to conclude that $2^m[b]$ belongs to the image of $W'(A)$. $\square$

Proof of Theorem 15.2.6. Let b be an invertible symmetric matrix with entries in $\mathcal{D}(V)$. By Lemma 15.2.9, there is a sequence of homomorphisms

$$\mathcal{P}(V) = A_0 \xrightarrow{\theta_1} A_1 \to \cdots \xrightarrow{\theta_k} A_k \xrightarrow{\gamma_k} \mathcal{D}(V),$$

such that $[b]$ belongs to the image of $W'(A_k)$ and each θ_i is either of the form $A_{i-1} \to \Sigma^{-1} A_{i-1}$ or of the form $A_{i-1} \to A_{i-1}\langle f^{1/2}\rangle$ with f invertible in A_{i-1}. Lemmas 15.2.10 and 15.2.12 show that, in both cases, the cokernel of the homomorphism $W'(A_{i-1}) \to W'(A_i)$ is a 2-group. Hence, there is an integer $s \in \mathbb{N}$ such that $2^s[b] \in W'(\mathcal{D}(V))$ belongs to the image of $W'(\mathcal{P}(V))$. By Lemma 15.2.7, it follows that the cokernel of the homomorphism $W(\mathcal{P}(V)) \to W(\mathcal{D}(V))$ is a 2-group. $\square$

15.3 Comparison between $W(\mathcal{P}(V))$ and $W(\mathcal{S}^0(V))$

In this section, $V \subset R^n$ is an algebraic set. The groups $K_0(\mathcal{P}(V))$ and $K_0(\mathcal{S}^0(V))$ ($\simeq KO(V)$ if $R = \mathbb{R}$) can be very different. For instance, if V is irreducible and has $s > 1$ semi-algebraically connected components, the images of these groups by the rank homomorphism are isomorphic to $\mathbb{Z}$ and $\mathbb{Z}^s$, respectively. On the other hand, we shall show that $W(\mathcal{P}(V))$ and $W(\mathcal{S}^0(V))$ are always closely related. More precisely, we shall prove the following result concerning the homomorphism $W(\mathcal{P}(V)) \to W(\mathcal{S}^0(V))$ induced by the inclusion $\mathcal{P}(V) \hookrightarrow \mathcal{S}^0(V)$.

Theorem 15.3.1. (i) *The cokernel of the homomorphism* $W(\mathcal{P}(V)) \to W(\mathcal{S}^0(V))$ *is a 2-group. Equivalently, the induced homomorphism*

$$W(\mathcal{P}(V))[1/2] \to W(\mathcal{S}^0(V))[1/2]$$

is surjective.

(ii) *If $R = \mathbb{R}$, the cokernel of the homomorphism*

$$W(\mathcal{P}(V)) \to W(\mathcal{C}^0(V)) \simeq KO(V)$$

is a 2-group. Equivalently, the induced homomorphism

$$W(\mathcal{P}(V))[1/2] \to KO(V)[1/2]$$

is surjective.

Remark 15.3.2. a) Theorem 15.3.1 is part of the result proved in [77]. The complete result asserts that $W(\mathcal{P}(V))[1/2] \to W(\mathcal{S}^0(V))[1/2]$ is an isomorphism. Moreover, this result is proved for an arbitrary ring A instead of $\mathcal{P}(V)$, with $\mathcal{S}^0(V)$ replaced by the ring of abstract continuous semi-algebraic functions on $\mathrm{Spec}_r(A)$.

b) Theorem 15.3.1 is an improvement of Theorem 15.2.3. Indeed, the following commutative diagram

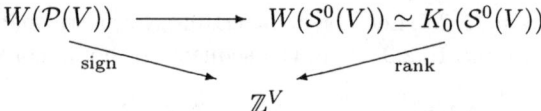

and the fact that the image of the rank homomorphism $K_0(\mathcal{S}^0(V)) \to \mathbb{Z}^V$ consists of all functions which are constant on the semi-algebraically connected components of V show that Theorem 15.2.3 follows from Theorem 15.3.1.

The proof of Theorem 15.3.1 is easy in the case where $R = \mathbb{R}$ and V is bounded. Let b be an invertible symmetric $q \times q$ matrix over $\mathcal{C}^0(V)$. We can consider b as a continuous mapping from V into the open subset $\mathrm{Sym}(q) \subset \mathbb{R}^{q(q+1)/2}$ of all symmetric $q \times q$ matrices β with entries in $\mathbb{R}$ such that $\det(\beta) \neq 0$. The Stone-Weierstrass theorem implies that b can be approximated by a polynomial mapping c which is homotopic to b in $\mathrm{Sym}(q)$. By Corollary 15.1.9, b and c are isometric over $\mathcal{C}^0(V)$. This proves that $W'(\mathcal{R}(V)) \to W'(\mathcal{C}^0(V))$ is surjective. Lemma 15.2.10 implies that the cokernel of the homomorphism $W'(\mathcal{P}(V)) \to W'(\mathcal{R}(V))$ is a 2-group. Using Lemma 15.2.7, Theorem 15.3.1 (ii) follows. Replacing $\mathcal{C}^0(V)$ with $\mathcal{S}^0(V)$, we obtain the statement (i) of this theorem (again under the assumptions $R = \mathbb{R}$ and V bounded). Observe that we have also proved that $2W(\mathcal{C}^0(V))$ is contained in the image of $W(\mathcal{R}(V))$.

The proof above does not apply to the case where V is not bounded or $R \neq \mathbb{R}$. The main idea of the proof in the general case remains the same, but we use $\mathcal{D}(V)$ as an intermediate step between $\mathcal{P}(V)$ and $\mathcal{S}^0(V)$, and the construction of the homotopy in $\mathrm{Sym}(q)$ is more complicated. For this construction, we need precise information concerning the geometry of $\mathrm{Sym}(q)$. This is the content of the following technical proposition.

We always regard the set $\mathrm{Sym}(q)$ of symmetric $q \times q$ matrices β with entries in R and such that $\det(\beta) \neq 0$ as an open subset of $R^{q(q+1)/2}$.

Proposition 15.3.3. *There exist*

  (i) *a mapping Γ from $\mathrm{Sym}(q)$ into the set of subsets of $\mathrm{Sym}(q)$ such that, for every $\beta \in \mathrm{Sym}(q)$, $\Gamma(\beta)$ is a convex open semi-algebraic neighbourhood of β,*

  (ii) *a finite number of points $\gamma_1, \ldots, \gamma_s$ of $\mathrm{Sym}(q)$ such that the sets*

$$D(\gamma_i) = \{\beta \in \mathrm{Sym}(q) \mid \gamma_i \in \Gamma(\beta)\},$$

for $i = 1, \ldots, s$, form an open semi-algebraic cover of $\mathrm{Sym}(q)$.

Example 15.3.4. For $q = 2$, we identify $\mathrm{Sym}(2)$ with

$$\{(a, b, c) \in R^3 \mid ac - b^2 \neq 0\}.$$

Thus, $\mathrm{Sym}(2)$ is the complement of a cone whose axis is the line given by $c = 0$ and $a = b$, and it has three connected components S_0, S_1 and S_2 consisting of the invertible symmetric matrices which are, respectively, positive definite,

15.3 Comparison between $W(\mathcal{P}(V))$ and $W(\mathcal{S}^0(V))$

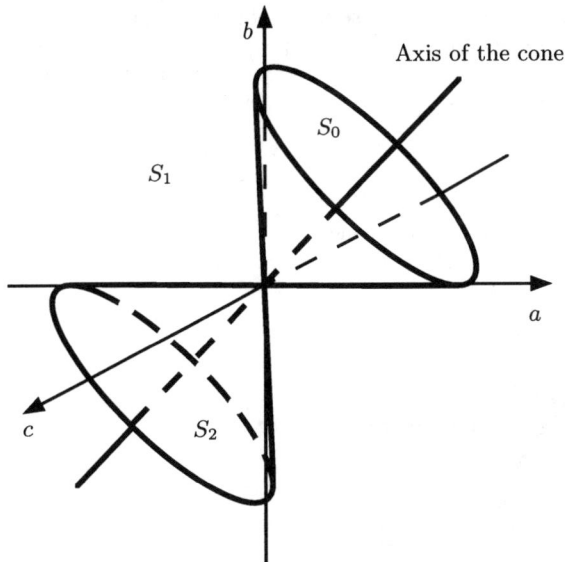

Figure 15.2.

of signature zero, and negative definite (cf. Figure 15.2). If $\beta \in S_0$ (resp. S_2), one sets $\Gamma(\beta) = S_0$ (resp. S_2). We now consider the case $\beta \in S_1$. First assume that $\beta = (a, b, 0)$, with $a > 0$ and $b < 0$. In this case one sets (cf. Figure 15.3)

$$\Gamma(\beta) = \{(a', b', c) \mid a' > 0, \ b' < 0, \ c \in R\}.$$

For an arbitrary $\beta \in S_1$, there is a unique rotation ρ, whose axis is the axis

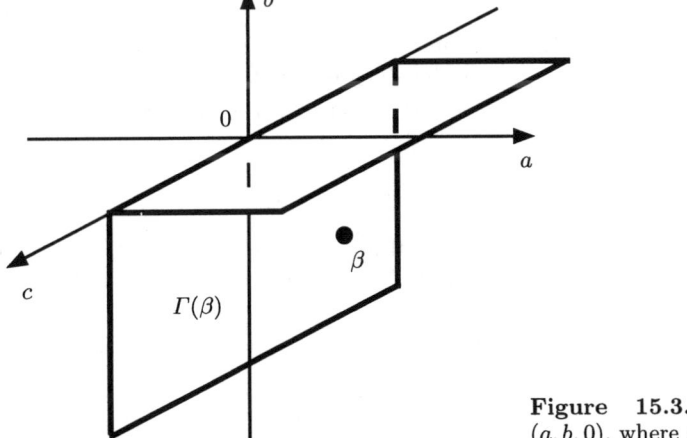

Figure 15.3. Case $\beta = (a, b, 0)$, where $a > 0$ and $b < 0$

of the cone, and such that $\rho(\beta)$ is of the form $(a, b, 0)$, with $a > 0$ and $b < 0$. One sets $\Gamma(\beta) = \rho^{-1}(\Gamma(\rho(\beta)))$.

Clearly, $\Gamma(\beta)$ is always a convex open semi-algebraic neighbourhood of β. It is also obvious that the six points $\gamma_1 = (1,1,0)$, $\gamma_2 = (1,-1,0)$, $\gamma_3 = (0,0,\sqrt{2})$, $\gamma_4 = (-1,1,0)$, $\gamma_5 = (0,0,-\sqrt{2})$, $\gamma_6 = (-1,-1,0)$ satisfy condition (ii) of the proposition (cf. Figure 15.4).

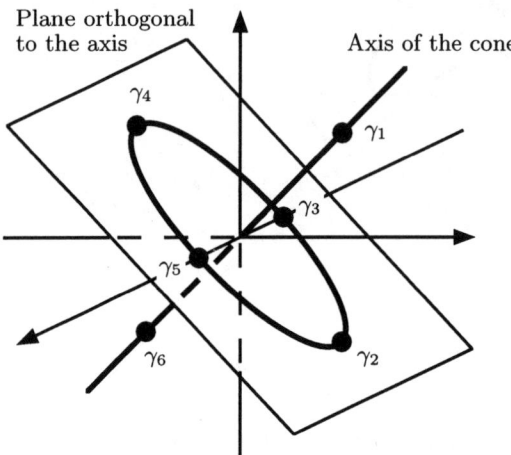

Figure 15.4.

Note that the rotations whose axes are the axis of the cone correspond, by the identification of Sym(2) with the complement of the cone, to the action on Sym(2) of elements σ of the orthogonal group $O(2,R)$ defined by $\beta \mapsto \sigma^{-1} \cdot \beta \cdot \sigma$. This example should help in understanding the proof of Proposition 15.3.3 for an arbitrary q.

Proof of Proposition 15.3.3. We denote by $\langle u_1, \ldots, u_q \rangle$ the diagonal $q \times q$ matrix with entries $u_1, \ldots, u_q$ on the diagonal. Let $\beta \in \text{Sym}(q)$. We can diagonalize β with respect to an orthonormal basis of R^q, i.e. there exists $\sigma \in O(q,R)$ such that:

$$\beta = \sigma^{-1} \cdot \langle a_1, \ldots, a_r, b_{r+1}, \ldots, b_q \rangle \cdot \sigma,$$

where $a_i > 0$ for $i = 1, \ldots, r$ and $b_i < 0$ for $i = r+1, \ldots, q$. Let Ω_r be the set of $q \times q$ matrices with entries in R of the form

$$\begin{bmatrix} A & C \\ {}^tC & B \end{bmatrix},$$

where A is a positive definite symmetric $r \times r$ matrix, B is a negative definite symmetric $(q-r) \times (q-r)$ matrix, and C is any $r \times (q-r)$ matrix. Set $\Gamma(\beta) = \Omega_r^\sigma = \sigma^{-1} \cdot \Omega_r \cdot \sigma$.

First we check that $\Gamma(\beta)$ does not depend on the choice of σ. Indeed, if $\sigma' \in O(q,R)$ is such that

15.3 Comparison between $W(\mathcal{P}(V))$ and $W(\mathcal{S}^0(V))$

$$\beta = \sigma'^{-1} \cdot \langle a'_1, \ldots, a'_{r'}, b'_{r'+1}, \ldots, b'_q \rangle \cdot \sigma',$$

where $a'_i > 0$ for $i = 1, \ldots, r'$ and $b'_i < 0$ for $i = r'+1, \ldots, q$, then $r' = r$ and $\rho = \sigma' \cdot \sigma^{-1}$ is of the form

$$\rho = \begin{bmatrix} \rho_1 & 0 \\ 0 & \rho_2 \end{bmatrix},$$

where $\rho_1 \in O(q, R)$ and $\rho_2 \in O(n-q, R)$. For such a ρ we have $\Omega_r = \rho^{-1} \cdot \Omega_r \cdot \rho$.

Next we check that $\Gamma(\beta) \subset \mathrm{Sym}(q)$. It is sufficient to prove that $\Omega_r \subset \mathrm{Sym}(q)$. If $\delta \in \Omega_r$, it is positive definite on a subspace of dimension r and negative definite on a subspace of dimension $q - r$. Therefore δ is nondegenerate (i.e. $\det(\delta) \neq 0$).

Since the positive definite symmetric $r \times r$ (resp. negative definite symmetric $(q-r) \times (q-r)$) matrices form a convex open semi-algebraic subset of $\mathrm{Sym}(r)$ (resp. $\mathrm{Sym}(q-r)$), Ω_r is a convex open semi-algebraic subset of $\mathrm{Sym}(q)$. Hence, $\Gamma(\beta)$ is a convex open semi-algebraic neighbourhood of β.

We now turn to part (ii) of the proposition. For r such that $0 \leq r \leq q$, set

$$\lambda_r = \langle \underbrace{1, \ldots, 1}_{r}, \underbrace{-1, \ldots, -1}_{q-r} \rangle.$$

If $\sigma \in O(q, R)$ and $\beta \in \mathrm{Sym}(q)$, we denote $\beta^\sigma = \sigma^{-1} \cdot \beta \cdot \sigma$. This defines an action of the orthogonal group $O(q, R)$ on $\mathrm{Sym}(q)$. For $\rho \in O(q, R)$, set

$$\Sigma(\rho) = \{\sigma \in O(q, R) \mid \lambda_r^\rho \in \Omega_r^\sigma\}.$$

Note that $\Sigma(\rho)$ is an open semi-algebraic subset of $O(q, R)$. We claim that there is a finite number of elements $\rho_1, \ldots, \rho_t$ in $O(q, R)$ (t depending on r) such that $O(q, R)$ is covered by the $\Sigma(\rho_j)$, for $j = 1, \ldots, t$. This is obvious if $R = \mathbb{R}$, since $O(q, \mathbb{R})$ is compact. We can even choose $\rho_1, \ldots, \rho_t$ in $O(q, \mathbb{R}_{\mathrm{alg}})$. For every real closed field R, these matrices $\rho_1, \ldots, \rho_t$ can be regarded as elements of $O(q, R)$, and the Tarski-Seidenberg principle implies that $O(q, R) = \bigcup_{i=1}^{t} \Sigma(\rho_i)$. For $0 \leq r \leq q$ and $1 \leq j \leq t$, we set

$$D(\lambda_r^{\rho_j}) = \{\beta \in \mathrm{Sym}(q) \mid \lambda_r^{\rho_j} \in \Gamma(\beta)\}.$$

If the signature of $\beta \in \mathrm{Sym}(q)$ is $2r-q$, then $\Gamma(\beta) = \Omega_r^\sigma$ for some $\sigma \in O(q, R)$, and therefore $\beta \in D(\lambda_r^{\rho_j})$ for some j, $1 \leq j \leq t$. If β' is close to β, then $\Gamma(\beta') = \Omega_r^{\sigma'}$ with σ' close to σ, and $\beta' \in D(\lambda_r^{\rho_j})$. This shows that $D(\lambda_r^{\rho_j})$ is open. We have proved that the $D(\lambda_r^{\rho_j})$ form a finite open semi-algebraic cover of $\mathrm{Sym}(q)$. $\square$

We recall that the ring $\mathcal{D}(V)$ was introduced in Notation 15.2.5.

Lemma 15.3.5. *Let $b: V \to \mathrm{Sym}(q)$ be a continuous semi-algebraic mapping. Then b is semi-algebraically homotopic to a mapping $c: V \to \mathrm{Sym}(q)$ such that all coordinates of c belong to $\mathcal{D}(V)$.*

Proof. We use the notation of Proposition 15.3.3. The sets $U_i = b^{-1}(D(\gamma_i))$, for $i = 1, \ldots, s$, form an open semi-algebraic cover of V. By Theorem 2.7.2, each U_i is of the form

$$U_i = \bigcup_{j=1}^{m} \bigcap_{k=1}^{p} \{x \in V \mid f_{i,j,k}(x) > 0\},$$

where $f_{i,j,k} \in \mathcal{P}(V)$. Set

$$\theta_i = \sum_{j=1}^{m} \prod_{k=1}^{p} (f_{i,j,k} + |f_{i,j,k}|), \quad \varphi_i = \theta_i \left/ \sum_{i=1}^{s} \theta_i \right. \quad \text{and} \quad d = \sum_{i=1}^{s} \varphi_i \gamma_i.$$

The mapping $d : V \to R^{q(q+1)/2}$ is continuous and semi-algebraic. For every $x \in V$, we have

$$\varphi_i(x) \neq 0 \Leftrightarrow x \in U_i \Leftrightarrow \gamma_i \in \Gamma(b(x)).$$

Since $\Gamma(b(x))$ is convex, $d(x) \in \Gamma(b(x))$ for every $x \in V$, and d is semi-algebraically homotopic to b in $\text{Sym}(q)$ by the homotopy $(t, x) \mapsto td(x) + (1-t)b(x)$.

For $x \in V \subset R^n$, set $\epsilon(x) = \eta(1 + \|x\|^2)^{-r}$, where $\eta \in R$ and $r \in \mathbb{N}$. Replacing $|f_{i,j,k}|$ with $\sqrt{f_{i,j,k}^2 + \epsilon^2}$ in the definition of d, we obtain a mapping $c : V \to R^{q(q+1)/2}$ such that all coordinates of c belong to $\mathcal{D}(V)$. By Proposition 2.6.2, we can choose η and r such that $\|c(x) - d(x)\|$ is smaller than the distance between $d(x)$ and $R^{q(q+1)/2} \setminus \text{Sym}(q)$, for every $x \in V$. Then c takes its values in $\text{Sym}(q)$, and is semi-algebraically homotopic to d (and, hence, to b) in $\text{Sym}(q)$. □

Proof of Theorem 15.3.1. (i) Recall that the cokernel of the homomorphism $W(\mathcal{P}(V)) \to W(\mathcal{D}(V))$ is a 2-group (Theorem 15.2.6). Lemma 15.3.5 and Corollary 15.1.9 imply that the homomorphism $W'(\mathcal{D}(V)) \to W'(\mathcal{S}^0(V))$ is surjective. By Lemma 15.2.7, it follows that the image of $W(\mathcal{D}(V))$ in $W(\mathcal{S}^0(V))$ contains $2W(\mathcal{S}^0(V))$. Hence, the cokernel of $W(\mathcal{P}(V)) \to W(\mathcal{S}^0(V))$ is a 2-group.

(ii) is a consequence of (i) and Corollary 15.1.13. □

Bibliographic Notes. This chapter is inspired mainly by the papers of Brumfiel [77] and Mahé [219]. We refer the reader to [235] or [188] for the study of Witt rings. The isomorphism between $W(C^0(V))$ and $KO(V)$ (cf. Theorem 15.1.12) is due to Lusztig and Gelfand Mishchenko [134]. Example 15.2.1 is due to Colliot-Thélène (cf. [219]). Lemma 15.2.10 can be found in [183]. The problem of separating connected components by signatures of bilinear spaces is posed in [188] by Knebusch, who solved the problem in the case of curves by using theorems of Witt [343]. Some partial answers (for smooth complete surfaces and abelian manifolds) are given in [92]. The general solution in the affine case appears in [219], and in the projective case in [168]. The results in [92] include a bound on the integer k such that $2^k \mathcal{C}^0(V,\mathbb{Z})$ is contained in the image of the "signature" homomorphism. A bound, depending only on the dimension of V, is given in [220] in the general case. This bound is obtained using the results of Bröcker and Scheiderer (cf. Theorem 6.5.1) and a "quantitative positivstellensatz", which gives a bound (depending only on $\dim(V)$) on the integer p such that every $f \in \mathcal{P}(V)$, which is positive on V, divides $1 +$ a sum of p squares in $\mathcal{P}(V)$. Improvements in the case of surfaces were obtained in [237]. The results in the third section are contained in [77].

Bibliography

1. Abraham, R., Robbin, J.: Transversal mappings and flows. New York: Benjamin (1967)
2. A'Campo, N.: Sur la première partie du seizième problème de Hilbert. Seminaire Bourbaki 1978/79, Lecture Notes in Math. **770**, 208-227. Berlin: Springer-Verlag (1980)
3. Acquistapace F., Andradas C., Broglia F.: Classification of obstructions for separation of semi-algebraic sets in dimension 3. Rev. Mat. Univ. Complutense Madrid **10** (Extra), 27-49 (1997)
4. Akbulut, S., King, H.: Real algebraic variety structures of P.L. manifolds. Bull. Amer. Math. Soc. **83**, 281-282 (1977)
5. Akbulut, S., King, H.: A relative Nash theorem. Trans. Amer. Math. Soc. **267**, 465-481 (1981)
6. Akbulut, S., King, H.: Real algebraic structures on topological spaces. Inst. Hautes Etudes Sci. Publ. Math. **53**, 79-162 (1981)
7. Akbulut, S., King, H.: The topology of real algebraic sets with isolated singularities. Ann. of Math. **113**, 425-446 (1981)
8. Akbulut, S., King, H.: The topology of real algebraic sets. Enseign. Math. (2), **29**, 221-261 (1983)
9. Akbulut, S., King, H.: On approximating submanifolds by algebraic sets and a solution of the Nash conjecture. Invent. Math. **107**, 87-98 (1992)
10. Akbulut, S., King, H.: Topology of real algebraic sets. Math. Sci. Research Institute Publ. **25**. Berlin: Springer-Verlag (1992)
11. Algorithms in Real Algebraic Geometry. Arnon, D., Buchberger, B. (Eds). J. Symbolic Comput. **5** (1988)
12. Alonso, M.E., Gamboa, J.M., Ruiz, J.: Ordres sur les surfaces réelles. C.R. Acad. Sci. Paris **298**, 17-19 (1984)
13. Andradas, C., Bröcker, L., Ruiz, J.: Constructible sets in Real Geometry. Ergeb. Math. Grenzgeb. (3) **33**. Berlin: Springer-Verlag (1996)
14. Aoki K., Nagachi, H.: On topological types of polynomial map germs of plane to plane. Memoirs of the School of Science and Engineering **44**, 133-156 Waseda University (1980)
15. Artin, E.: Über die Zerlegung definiter Funktionen in Quadrate. Hamb. Abh. **5**, 100-115 (1927). The collected papers of Emil Artin, 273-288. Reading: Addison-Wesley (1965)
16. Artin, E., Schreier, O.: Algebraische Konstruktion reeller Körper. Hamb. Abh. **5**, 85-99 (1926). The collected papers of Emil Artin, 258-272. Reading: Addison-Wesley (1965)
17. Artin, M.: On the solutions of analytic equations. Invent. Math. **5**, 277-291 (1968)
18. Artin, M.: Algebraic approximations of structures over complete local rings. Inst. Hautes Etudes Sci. Publ. Math. **36**, 23-58 (1969)

19. Artin, M., Mazur, B.: On periodic points. Ann. of Math. **81**, 82-99 (1965)
20. Atiyah, M.F., Macdonald, I.G.: Introduction to commutative algebra. Reading: Addison-Wesley (1969)
21. Baer, R.: Über nicht-archimedisch geordnete Körper (Beiträge zur Algebra 1). Sitz. Ber. der Heidelberger Akademie, 8. Abhandl. (1927)
22. Basu,S., Pollack,R., Roy, M.-F.: On the combinatorial and algebraic complexity of quantifier elimination. J. Assoc. Comput. Machin. **43**, 1002-1045, (1996)
23. Bass, H.: Algebraic K-theory. New York: Benjamin (1968)
24. Baum, P.F.: Quadratic maps and stable homotopy groups of spheres. Illinois J. Math. **11**, 586-595 (1967)
25. Becker, E.: Valuations and real places in the theory of formally real fields. Géométrie algébrique réelle et formes quadratiques. Lecture Notes in Math. **959**, 1-40. Berlin: Springer-Verlag (1982)
26. Becker, E.: On the real spectrum of a ring and its applications to semialgebraic geometry. Bull. Amer. Math. Soc. (NS) **15**, 19-60 (1986)
27. Becker, E., Köpping, E.: Reduzierte quadratische Formen und Semiordnungen reeller Körper, Abh. Math. Sem. Univ. Hamburg **46**, 143-177 (1977)
28. Bell, J.L., Slomson, A.R.: Models and ultraproducts: An introduction. Amsterdam: North Holland (1969)
29. Benedetti, R., Dedò, M.: The topology of two-dimensional real algebraic varieties. Ann. Mat. Pura Appl. (4) **127**, 141-171 (1981)
30. Benedetti, R., Dedò, M.: Counterexamples to representing homology classes by real algebraic subvarieties up to homeomorphism. Compositio Math. **53**, 143-151 (1984)
31. Benedetti, R., Shiota, M.: Finiteness of semialgebraic types of polynomial functions. Math. Z. **208**, 589-596 (1991)
32. Benedetti, R., Tognoli, A.: Théorèmes d'approximation en géométrie algébrique réelle. Sémin. sur la géométrie algébrique réelle, Publ. Math. Univ. Paris VII, **9**, 123-145 (1980). Approximation theorems in real algebraic geometry. Boll. Unione Mat. Ital., Suppl. 2, 209-228 (1980)
33. Benedetti, R., Tognoli, A.: On real algebraic vector bundles. Bull. Sci. Math. (2) **104**, 89-112 (1980)
34. Benedetti, R., Tognoli, A.: Remarks and counterexamples in the theory of real algebraic vector bundles and cycles. Géométrie algébrique réelle et formes quadratiques. Lecture Notes in Math. **959**, 198-211. Berlin: Springer-Verlag (1982)
35. Ben-Or, M.: Lower bounds for algebraic computation trees. Proc. 15th ACM Annual Symp. on Theory of Comput., 80-86 (1983)
36. Bhatwadekar, S.M., Ischebeck, F., Ojanguren, M., Schabhueser, G.: Strongly algebraic vector bundles over $\mathbb{R}^d$. Real analytic and algebraic geometry. Lecture Notes in Math. **1420**, 36-41. Berlin: Springer-Verlag (1990)
37. Białynicki-Birula, A., Rosenlicht, M.: Injective morphisms of real algebraic varieties. Proc. Amer. Math. Soc. **13**, 200-203 (1962)
38. Bierstone, E., Milman, P.: Composite differentiable functions. Ann. of Math. (2) **116**, 541-558 (1982)
39. Bierstone, E., Milman, P.: Relations among analytic functions. Ann. Inst. Fourier **37**, 187-239 (1987)
40. Bochnak, J.: Sur la factorialité des anneaux de fonctions de Nash. Comment. Math. Helv. **52**, 211-218 (1977)
41. Bochnak, J.: Algebraicity versus analyticity. Rocky Mountain J. Math.**14**, 863-880 (1984)
42. Bochnak, J., Buchner, M., Kucharz, W.: Vector bundles over real algebraic varieties. K-Theory **3**, 271-298 (1990), Erratum K-Theory **4**, 103 (1990)

43. Bochnak, J., Efroymson, G.: Real algebraic geometry and the Hilbert 17th problem. Math. Ann. **251**, 213-241 (1980)
44. Bochnak, J., Kucharz, W.: Local algebraicity of analytic sets. J. Reine. Angew. Math. **352**, 1-14 (1984)
45. Bochnak, J., Kucharz, W.: Algebraic approximation of mappings into spheres. Michigan Math. J. **34**, 119-125 (1987)
46. Bochnak, J., Kucharz, W.: Realization of homotopy classes by algebraic mappings. J. Reine. Angew. Math. **377**, 159-169 (1987)
47. Bochnak, J., Kucharz, W.: On real algebraic morphisms into even-dimensional spheres. Ann. of Math. **128**, 415-433 (1988)
48. Bochnak, J., Kucharz, W.: Algebraic models of smooth manifolds. Invent. Math. **97** 585-611 (1989)
49. Bochnak, J., Kucharz, W.: K-theory of real algebraic surfaces and threefolds. Math. Proc. Cambridge Philos. Soc **106**, 471-480 (1989)
50. Bochnak, J., Kucharz, W.: On vector bundles and real algebraic morphisms. Real analytic and algebraic geometry. Lecture Notes in Math. **1420**, 65-71. Berlin: Springer-Verlag (1990)
51. Bochnak, J., Kucharz, W.: Nonisomorphic algebraic models of a smooth manifold. Math. Ann. **290**, 1-2 (1991)
52. Bochnak, J., Kucharz, W.: Polynomial mappings from products of algebraic sets into spheres. J. Reine. Angew. Math. **417**, 135-139 (1991)
53. Bochnak, J., Kucharz, W.: Vector bundles on a product of real cubic curves. K-Theory **6**, 487-497 (1992)
54. Bochnak, J., Kucharz, W.: Algebraic cycles and approximation theorems in real algebraic geometry. Trans. Amer. Math. Soc. **337**, 463-472 (1993)
55. Bochnak, J., Kucharz, W.: Elliptic curves and real algebraic morphisms. J. Algebraic Geom. **2**, 635-666 (1993)
56. Bochnak, J., Kucharz, W.: Real algebraic hypersurfaces in complex projective varieties. Math. Ann. **301** 381-397 (1995)
57. Bochnak, J., Kucharz, W.: A characterization of dividing real algebraic curves. Topology **35**, 451-455 (1996)
58. Bochnak, J., Kucharz, W.: On homology classes represented by real algebraic varieties. Singularities Symposium (Krakow 1996). Banach Center Publications, Warsaw (1998)
59. Bochnak, J., Kucharz, W.: Topology of real algebraic varieties. Book in preparation
60. Bochnak, J., Kucharz, W., Shiota, M.: On equivalence of ideals of real global analytic functions and the 17th Hilbert problem. Invent. Math. **63**, 403-421 (1981)
61. Bochnak, J., Kucharz, W., Shiota, M.: On algebraicity of global real analytic sets and functions. Invent. Math. **70**, 115-156 (1982)
62. Bochnak, J., Kucharz, W., Silhol, R.: Morphisms, line bundles and moduli spaces in real algebraic geometry. Inst. Hautes Etudes Sci. Publ. Math. **86** (to appear)
63. Borel, A.: Injective endomorphisms of algebraic varieties. Arch. Math. **20**, 531-537 (1969)
64. Borel, A., Haefliger, A.: La classe d'homologie fondamentale d'un espace analytique. Bull. Soc. Math. France **89**, 461-513 (1961)
65. Borel, A., Moore, J.C.: Homology theory for locally compact spaces. Michigan Math. J. **7**, 137-159 (1960)
66. Bourbaki, N.: Algèbre commutative. Paris: Hermann (1961-1965)

67. Brakhage, H.: Topologische Eigenschaften algebraischer Gebilde über einem beliebigen reell-abgeschlossenen Konstantenkörper. Dissertation, Univ. Heidelberg (1954)
68. Bröcker, L.: Characterization of fans and hereditarily pythagorean fields. Math. Z. **151**, 149-163 (1976)
69. Bröcker, L.: Real spectra and distribution of signatures. Géométrie algébrique réelle et formes quadratiques. Lecture Notes in Math. **959**, 249-272. Berlin: Springer-Verlag (1982)
70. Bröcker, L.: Minimale Erzeugung von Positivbereichen. Geom. Dedicata **16**, 335-350 (1984)
71. Bröcker, L.: Spaces of orderings and semi-algebraic sets. Quadratic and Hermitian forms. CMS Conf. Proc.**4**, 231-248 (1984)
72. Bröcker, L.: On basic semi-algebraic sets. Exposition. Math. **9**, 289-334 (1991)
73. Brown, R.: Real places and ordered fields. Rocky Mountain J. Math.**1**, 633-636 (1971)
74. Bruhat, F., Cartan, H.: Sur les composantes irréductibles d'un sous-ensemble analytique réel. C.R. Acad. Sci. Paris **244**, 1123-1126 (1957)
75. Bruhat, F., Whitney, H.: Quelques propriétés fondamentales des ensembles analytiques réels. Comment. Math. Helv. **33**, 132-160 (1959)
76. Brumfiel, G.W.: Partially ordered rings and semi-algebraic geometry. Cambridge: Cambridge Univ. Press (1979)
77. Brumfiel, G.W.: Witt rings and K-theory. Rocky Mountain J. Math.**14**, 733-765 (1984)
78. Buresi, J., Mahé, L.: Reducing inequalities with bounds. Math. Z. (to appear)
79. Burghelea, D., Verona, A.: Local homological properties of analytic sets. Manuscripta Math. **7**, 55-66 (1972)
80. Carral, M., Coste, M.: Normal spectral spaces and their dimensions. J. Pure Appl. Algebra **30**, 227-235 (1983)
81. Cartan, H.: Variétés analytiques réelles et variétés analytiques complexes. Bull. Soc. Math. France **85**, 77-99 (1957)
82. Cassels, J.W.S.: On the representation of rational functions as sums of squares. Acta Arith. **9**, 79-82 (1964)
83. Cassels, J.W.S., Ellison, W.S., Pfister, A.: On sums of squares and on elliptic curves over functions fields. J. Number Theory **3**, 125-149 (1971)
84. Chillingworth, D.R.J., Hubbard, J.: A note on nonrigid Nash structure. Bull. Amer. Math. Soc. **77**, 429-431 (1971)
85. Choi, M.D., Dai, Z.D., Lam, T.Y., Reznick, B.: The Pythagoras number of some affine algebras and local algebras. J. Reine. Angew. Math. **336**, 45-82 (1982)
86. Choi, M.D., Lam, T.Y.: Extremal positive semi-definite forms. Math. Ann. **231**, 1-18 (1977)
87. Cohen, P.J.: Decision procedures for real and p-adic fields. Comm. Pure. Appl. Math. **22**, 131-151 (1969)
88. Collins, G.: Quantifier elimination for real closed fields by cylindrical algebraic decomposition. Autom. Theor. form. Lang.. Lect. Notes Comput. Sci. **33**, 134-183 (1975)
89. Colliot-Thélène, J.-L.: Variantes du Nullstellensatz réel et anneaux formellement réels. Géométrie algébrique réelle et formes quadratiques. Lecture Notes in Math. **959**, 98-108. Berlin: Springer-Verlag (1982)
90. Colliot-Thélène, J.-L.: The Noether-Lefschetz theorem and sums of 4 squares in the rational function field $\mathbb{R}(X,Y)$, Compositio Math. **86**, 235-243 (1993)
91. Colliot-Thélène, J.-L., Jannsen, U.: Sommes de carrés dans les corps de fonctions. C. R. Acad. Sci. Paris **312**, 759-762 (1991)

92. Colliot-Thélène, J.-L., Sansuc, J.-J.: Fibrés quadratiques et composantes connexes réelles. Math. Ann. **244**, 105-134 (1979)
93. Conner, R.J., Floyd, E.E.: Differential periodic maps. Ergeb. Math. Grenzgeb. (2) **33**. Berlin: Springer-Verlag (1964)
94. Coste, M.: Ensembles semi-algébriques et fonctions de Nash. Université Paris Nord (preprint) (1981)
95. Coste, M.: Ensembles semi-algébriques. Géométrie algébrique réelle et formes quadratiques. Lecture Notes in Math. **959**, 109-138. Berlin: Springer-Verlag (1982)
96. Coste, M.: Sous-ensembles algébriques réels de codimension 1. C.R. Acad. Sci. Paris **300**, 661-664 (1985)
97. Coste, M., Kurdyka, K.: On the link of a stratum in a real algebraic set. Topology **21**, 323-336 (1992)
98. Coste, M., Reguiat, M.: Trivialités en famille. Real Algebraic Geometry. Lecture Notes in Math. **1524**, 193-204. Berlin: Springer-Verlag (1992)
99. Coste, M., Roy, M.-F.: Topologies for real algebraic geometry. Various Publ. Ser. Aarhus Univ. **30**, 37-100 (1979)
100. Coste, M., Roy, M.-F.: La topologie du spectre réel. Contemp. Math. **8**, 27-59 (1982)
101. Coste, M., Ruiz, J., Shiota, M.: Approximation in compact Nash manifolds. Amer. J. of Math. **117**, 905-927 (1995)
102. Coste, M., Ruiz, J., Shiota, M.: Separation, factorization and finite sheaves on Nash manifolds. Compositio Math. **103** 31-62 (1996)
103. Coste, M., Shiota, M.: Nash triviality in families of Nash manifolds. Invent. Math. **108**, 349-368 (1992)
104. Cucker, F.: Sur les anneaux de sections globales du faisceau structural sur le spectre réel. Commun. Algebra **16**, 307-323 (1988)
105. Dai, Z. D., Lam T. Y.: Levels in algebra and topology, Comment. Math. Helv. **59**, 376-424 (1984)
106. Delfs, H.: The homotopy axiom in semialgebraic cohomology. J. Reine Angew. Math. **355**, 108-128 (1985)
107. Delfs, H.: Homology of locally semialgebraic spaces. Lecture Notes in Math. **1484**. Berlin: Springer-Verlag (1991)
108. Delfs, H., Knebusch, M.: Semialgebraic topology over a real closed field II: Basic theory of semialgebraic spaces. Math. Z. **178**, 175-213 (1981)
109. Delfs, H., Knebusch, M.: On the homology of algebraic varieties over real closed fields. J. Reine Angew. Math. **335**, 122-163 (1982)
110. Delfs, H., Knebusch, M.: Locally semialgebraic spaces. Lecture Notes in Math. **1173**. Berlin: Springer-Verlag (1985)
111. Delzell, C.N.: A continuous, constructive solution to Hilbert's 17th problem. Invent. Math. **76**, 365-384 (1984)
112. De Marco, G., Orsatti, A.: Commutative rings in which every prime ideal is contained in a unique maximal ideal. Proc. Amer. Math. Soc. **30**, 459-466 (1971)
113. Descartes, R.: Géométrie. (1636). A source book in Mathematics, 90-93. Massachussetts: Harvard Univ. Press (1969)
114. Dickmann, M.A.: Applications of model theory to real algebraic geometry. A survey. Methods in mathematical logic. Lecture Notes in Math. **1130**, 76-150. Berlin: Springer-Verlag (1985)
115. Dieudonné, J.: Foundations of modern analysis. New York: Academic Press (1960)
116. Dieudonné, J., Carrel, J.: Invariant theory, old and new. Adv. Math. **4**, 1-80 (1970)

117. van den Dries, L.: Artin-Schreier theory for commutative regular rings. Ann. Math. Logic **12**, 113-150 (1977)
118. van den Dries, L.: Some applications of a modeltheoretic fact to (semi-) algebraic geometry. Indagationes Math. **44**, 397-401 (1982)
119. van den Dries, L., Miller, C.: Geometric categories and o-minimal structures. Duke Math. J. **84**, 497-540 (1996)
120. Dubois, D.W.: Note on Artin's solution of Hilbert's 17th problem. Bull. Amer. Math. Soc. **73**, 540-541 (1967)
121. Dubois, D.W.: A nullstellensatz for ordered fields. Ark. Mat. **8**, 111-114 (1969)
122. Dubois, D.W.: Real commutative algebra I: places. Rev. Mat. Hisp.-Am. (4) **39**, 57-65 (1979)
123. Dubois, D.W., Efroymson, G.: Algebraic theory of real varieties I. Stud. and Essays Presented to Yu-Why Chen on his 60-th Birthday, 107-135 (1970)
124. Durfee, A.: Neighborhoods of algebraic sets. Trans. Amer. Math. Soc. **270**, 517-530 (1983)
125. Efroymson, G.: A Nullstellensatz for Nash rings. Pacific J. Math. **54**, 101-112 (1974)
126. Efroymson, G.: Substitution in Nash functions. Pacific J. Math. **63**, 137-145 (1976)
127. Efroymson, G.: The extension theorem for Nash functions. Géométrie algébrique réelle et formes quadratiques. Lecture Notes in Math. **959**, 343-357. Berlin: Springer-Verlag (1982)
128. Ehresmann, C.: Sur la topologie de certaines variétés algébriques réelles. J. Math. Pures Appl. (9) **16**, 69-100 (1937)
129. Eilenberg, S., Steenrod, N.: Foundations of algebraic topology. Princeton: Princeton Univ. Press (1952)
130. Evans, E.G.: Projective modules as fiber bundles. Proc. Amer. Math. Soc. **27**, 623-626 (1971)
131. Fossum, R.: Vector bundles over spheres are algebraic. Invent. Math. **8**, 222-225 (1969)
132. Fukuda, T.: Types topologiques des polynômes. Inst. Hautes Etudes Sci. Publ. Math. **46**, 87-106 (1976)
133. Gantmacher, F.R.: Théorie des matrices, vol. 2. Paris: Dunod (1966)
134. Gelfand, I.M., Mishchenko, A.S.: Quadratic forms over commutative group rings and the K-theory. Funct. Anal. Appl. **3**, 277-281 (1969)
135. Gibson, C.G., Wirthmüller, K., du Plessis, A.A., Looijenga, E.: Topological stability of smooth mappings. Lecture Notes in Math. **552**. Berlin: Springer-Verlag (1976)
136. Giesecke, B.: Simpliziale Zerlegung abzählbarer analytischer Räume. Math. Z. **83**, 177-213 (1964)
137. Godement, R.: Théorie des faisceaux. Paris: Hermann (1958)
138. Grigor'ev, D., Vorobjov, N.: Solving systems of polynomial inequalities in subexponential time. J. Symbolic Comput. **5** 37-64 (1988)
139. Gross, B., Harris, J.: Real algebraic curves. Ann. Sci. Ecole Norm. Sup. (4) **14**, 157-182 (1981)
140. Gudkov, D.A.: Construction of a new series of M-curves. Sov. Math., Dokl. **12**, 1559-1563 (1971)
141. Gudkov, D.A.: The topology of real projective algebraic varieties. Russian Math. Surveys **29**, No.4, 1-79 (1974)
142. Gunning, R., Rossi, H.: Analytic functions of several complex variables. Prentice-Hall Series in Modern Analysis. Englewood Cliffs, N.J.: Prentice-Hall, Inc. (1965)

143. Habicht, W.: Über die Zerlegung strikter definiter Formen in Quadrate. Comment. Math. Helv. **12**, 317-322 (1940)
144. Haefliger, A., Kosinski, A.: Un théorème de Thom sur les singularités des applications différentiables. Séminaire Henri Cartan **9** 1956/57, exposé 8
145. van Hamel, J.: Real algebraic cycles on complex projective varieties. Math. Z. **225**, 177-198 (1997)
146. Hardt, R.: Sullivan's local Euler characteristic theorem. Manuscripta Math. **12**, 87-92 (1974)
147. Hardt, R.: Triangulation of subanalytic sets and proper subanalytic maps. Invent. Math. **38**, 207-217 (1977)
148. Hardt, R.: Semi-algebraic local-triviality in semi-algebraic mappings. Amer. J. Math. **102**, 291-302 (1980)
149. Harnack, A.: Über die Vieltheiligkeit der ebenen algebraischen Kurven. Math. Ann. **10**, 189-198 (1876)
150. Hartshorne, R.: Algebraic geometry. Berlin: Springer-Verlag (1977)
151. Heintz, J., Roy, M.-F., Solerno, P.: Sur la complexité du Principe de Tarski-Seidenberg. Bull. Soc. Math. France **118**, 101-126 (1990)
152. Hermite, C.: Remarques sur le théorème de Sturm. C. R. Acad. Sci. Paris **36**, 52-54 (1853)
153. Hilbert, D.: Über die Darstellung definiter Formen als Summe von Formenquadraten. Math. Ann. **32**, 342-350 (1888). Ges. Abh. vol. 2, 154-161. New York: Chelsea Publishing Company (1965)
154. Hilbert, D.: Über die reellen Züge algebraischer Kurven. Math. Ann. **38**, 115-138 (1891). Ges. Abh. vol. 2, 415-436. New York: Chelsea Publishing Company (1965)
155. Hilbert, D.: Über ternäre definite Formen. Acta Math. **17**, 169-198 (1893). Ges. Abh. vol. 2, 345-366. New York: Chelsea Publishing Company (1965)
156. Hilbert, D.: Les principes fondamentaux de la géométrie. Paris: Gauthier-Villars (1900)
157. Hilbert, D.: Hermann Minkowski. Math. Ann. **68**, 445-471 (1910). Ges. Abh. vol. 3, 339-364. New York: Chelsea Publishing Company (1965)
158. Hironaka, H.: Resolution of singularities of an algebraic variety over a field of characteristic zero. Ann. of Math. **79**, 109-326 (1964)
159. Hironaka, H.: Triangulations of algebraic sets. Algebraic Geometry, Proc. Sympos. Pure Math. **29**, 165-185. Providence: Amer. Math. Soc. (1975)
160. Hirsch, M.: Differential topology. Berlin: Springer-Verlag (1976)
161. Hirzebruch, F.: Topological methods in algebraic geometry. Berlin: Springer-Verlag (1966)
162. Hochster, M.: Prime ideal structure in commutative rings. Trans. Amer. Math. Soc. **142**, 43-60 (1969)
163. Hodge, W.H.D., Pedoe, D.: Methods of algebraic geometry. Cambridge: Cambridge Univ. Press (1953)
164. Hollkott, A.: Finite Konstruktion geordneter algebraischer Erweiterungen von geordneten Grundkörpern, Dissertation, Univ. Hamburg (1941)
165. Hopf, H.: Über die Abbildungen von Sphären auf Sphären niedrigerer Dimension. Fund. Math. **25**, 427-440 (1935)
166. Hörmander, L.: On the division of distributions by polynomials. Ark. Mat. **3**, 555-568 (1958)
167. Hörmander, L.: The analysis of linear partial differential operators, vol. 2. Berlin: Springer-Verlag (1983)
168. Houdebine, J., Mahé, L.: Séparation des composantes connexes réelles dans le cas des variétés projectives. Géométrie algébrique réelle et formes quadratiques. Lecture Notes in Math. **959**, 358-370. Berlin: Springer-Verlag (1982)

169. Hu, S.T.: Homotopy theory. New York: Academic Press (1959)
170. Hubbard, J.: On the cohomology of Nash sheaves. Topology **11**, 265-270 (1972)
171. Huisman, J.: Real abelian varieties with complex multiplication. Thesis, Vrije Universiteit Amsterdam (1992)
172. Huisman, J.: The underlying real algebraic structure of complex elliptic curves. Math Ann. **294**, 19-35 (1992)
173. Huisman, J.: On real algebraic vector bundles. Math. Z. **219**, 335-342 (1995)
174. Husemoller, D.: Fibre bundles. Berlin: Springer-Verlag (1975)
175. Husemoller, D.: Elliptic curves. Berlin: Springer-Verlag (1987)
176. Itenberg, I.: Counter-examples to Ragsdale conjecture and T-curves. Contemp. Math. **182**, 55-72 (1995)
177. Itenberg, I.,Shustin, E.: Singular points and limit cycles of planar polynomial vector fields. Preprint. Rennes (1997)
178. Itenberg, I., Viro, O.: Patchworking algebraic curves disproves the Ragsdale conjecture. Math. Intelligencer, **18** (4), 19-28 (1996)
179. Ivanov, N.: Approximation of smooth manifolds by real algebraic sets. Russian Math. Surveys **37**, 1-59 (1982)
180. James, I.M.: Two problems studied by Heinz Hopf. Lectures on algebraic and differential topology. Lecture Notes in Math. **279**, 134-174. Berlin: Springer-Verlag (1972)
181. Johnstone, P.: Stone spaces. Cambridge: Cambridge Univ. Press (1983)
182. Kaplansky, I.: Hilbert's Problems. Lecture Notes, Chicago University (1977)
183. Karoubi, M.: Localisation de formes quadratiques 1. Ann. Sci. Ecole Norm. Sup. **7**, 359-403 (1974)
184. Khovanski, A.: On a class of systems of transcendental equations. Soviet. Math. Dokl. **22**, 762-765 (1980)
185. Khovansky, A.: Fewnomials. Transl. Math. Monogr. **88**. Providence, RI: American Mathematical Society (1991)
186. King, H.: Approximating submanifolds of real projective spaces by varieties. Topology **15**, 81-85 (1976)
187. King, H.: The topology of real algebraic sets. Singularities, Proc. Sympos. Pure Math. **40**, Part 1, 641-654. Providence: Amer. Math. Soc. (1983)
188. Knebusch, M.: Symmetric bilinear forms over algebraic varieties. Proc. Conf. Quadratic Forms, Kingston 1976, Queen's Pap. pure appl. Math. **46**, 103-283 (1977)
189. Knebusch, M.: An invitation to real spectra. Quadratic and Hermitian forms, Conf. Hamilton/Ont. 1983, CMS Conf. Proc. **4**, 51-105 (1984)
190. Knebusch, M.: Weakly semialgebraic spaces. Lecture Notes in Math. **1367**. Berlin: Springer-Verlag (1989)
191. Knebusch, M., Scheiderer, C.: Einführung in die reelle Algebra. Vieweg Studium **63**. Braunschweig: Friedr. Vieweg & Sohn (1989)
192. Kollàr, J.: Real algebraic surfaces, preprint, Univ. of Utah (1997)
193. Kreisel, G.: Sums of squares. Summaries Summer Inst. Symbolic Logic, Cornell Univ. 1957, 313-320 (1960)
194. Krivine, J.-L.: Anneaux préordonnés. J. Anal. Math. **12**, 307-326 (1964)
195. Krull, W.: Allgemeine Bewertungstheorie. J. Reine Angew. Math. **167**, 160-196 (1931)
196. Kucharz, W.: On homology of real algebraic sets. Invent. Math. **82**, 19-26 (1985)
197. Kucharz, W.: Vector bundles over real algebraic surfaces and threefolds. Compositio Math. **60**, 209-225 (1986)
198. Kucharz, W.: Algebraic morphisms into rational real algebraic surfaces, preprint, Univ. of New Mexico (1998)

199. Lafon, J.-P.: Séries formelles algébriques. C.R. Acad. Sci. Paris **260**, 3238-3241 (1965)
200. Lafon, J.-P.: Algèbre commutative: langages géométrique et algébrique. Paris: Hermann (1977)
201. Lam, K.Y.: Sectioning vector bundles over real projective spaces. Quart. J. Math. Oxford (2) **23**, 97-106 (1972).
202. Lam, K.Y.: Topological methods for studying the composition of quadratic forms. Quadratic and Hermitian forms, Conf. Hamilton/Ont. 1983, CMS Conf. Proc.4, 173-192 (1984)
203. Lam, K.Y.: Some new results on composition of quadratic forms. Invent. Math. **79**, 467-474 (1985)
204. Lam, T.Y.: The algebraic theory of quadratic forms. New York: Benjamin (1973)
205. Lam, T.Y.: Serre's conjecture. Lecture Notes in Math. **635**. Berlin: Springer-Verlag (1978)
206. Lam, T.Y.: An introduction to real algebra. Rocky Mountain J. Math.**14**, 767-814 (1984)
207. Lange, H., Birkenhake, C.: Complex abelian varieties. Grundlehren Math. Wiss. **302**. Berlin: Springer-Verlag (1992)
208. Landau, E.: Über die Darstellung definiter Funktionen durch Quadrate. Math. Ann. **62**, 272-285 (1906)
209. Lang, S.: The theory of real places. Ann. of Math. **57**, 378-391 (1953)
210. Lang, S.: Algebra. Reading: Addison-Wesley (1971)
211. Lazzeri, F., Tognoli, A.: Alcune proprietà degli spazi algebrici. Ann. Scuola Norm. Sup. Pisa **24**, 597-632 (1970)
212. Loday, J.-L.: Applications algébriques du tore dans la sphère et de $S^p \times S^q$ dans S^{p+q}. Algebraic K-theory II. Lecture Notes in Math. **342**, 79-91. Berlin: Springer-Verlag (1973)
213. Lønsted, K.: Vector bundles over finite CW-complexes are algebraic. Proc. Amer. Math. Soc. **38**, 27-31 (1973)
214. Łojasiewicz, S.: Sur le problème de la division. Studia Math. **18**, 87-136 (1959)
215. Łojasiewicz, S.: Ensembles semi-analytiques. Inst. Hautes Etudes Sci., (preprint) (1964)
216. Łojasiewicz, S.: Triangulation of semi-analytic sets. Ann. Scuola Norm. Sup. Pisa, Sci. Fis. Mat. (3) **18**, 449-474 (1964)
217. Lombardi, H.: Nullstellensatz réel effectif et variantes. C.R. Acad. Sci. Paris **310** 635-640 (1990)
218. Lombardi, H., Roy, M.-F.: Elementary constructive theory of ordered fields. Effective methods in algebraic geometry. Progress Math. **94**, 249-262 (1991)
219. Mahé, L.: Signatures et composantes connexes. Math. Ann. **260**, 191-210 (1982)
220. Mahé, L.: Théorème de Pfister pour les variétés et anneaux de Witt réduits. Invent. Math. **85**, 53-72 (1986)
221. Mahé, L.: Level and Pythagoras number of some geometric rings. Math. Z. **204**, 615-629 (1990), Erratum Math. Z. **209**, 481-483 (1992)
222. Mangolte, F.: Cycles algébriques sur les surfaces $K3$ réelles. Math. Z. **225**, 559-576 (1997).
223. Marinari, M.G., Raimondo, M.: Fibrati vettoriali su varietà algebriche definite su corpi non algebricamente chiusi. Boll. Un. Mat. Ital. A (5) **16**, 128-136 (1979).
224. Marshall, M.: Spaces of orderings and abstract real spectra. Lecture Notes in Math. **1636**. Berlin: Springer-Verlag (1996).

225. Massey, W.: On the normal bundle of a sphere embedded in Euclidean space. Proc. Amer. Math. Soc. **10**, 959-964 (1959)
226. Mather, J.: Notes on topological stability. Lecture Notes, Harvard University (1970)
227. Mather, J.: Stratifications and mappings. Dynamical Syst., Proc. Sympos. Univ. Bahia, Salvador 1971, 195-232 (1973)
228. Matsumura, H.: Commutative algebra. New York: Benjamin (1970)
229. McCrory, C., Parusinski, A.: Algebraically constructible functions. Ann. Sci. Ecole Norm. Sup (4) **30**, 527-552 (1997)
230. Milnor, J.: Morse theory. Princeton: Princeton Univ. Press (1963)
231. Milnor, J.: On the Betti numbers of real varieties. Proc. Amer. Math. Soc. **15**, 275-280 (1964)
232. Milnor, J.: On the Stiefel-Whitney numbers of complex manifolds and of spin manifolds. Topology **3**, 223-230 (1965)
233. Milnor, J.: Singular points of complex hypersurfaces. Princeton: Princeton Univ. Press (1968)
234. Milnor, J.: Introduction to algebraic K-theory. Princeton: Princeton Univ. Press (1971)
235. Milnor, J., Husemoller, D.: Symmetric bilinear forms. Berlin: Springer-Verlag (1973)
236. Milnor, J., Stasheff, J.: Characteristic classes. Princeton: Princeton Univ. Press (1974)
237. Monnier, J.-B.: About the image of the total signature map in the two dimensional case. Manuscripta Math. **93**, 143-161 (1997)
238. Mostowski, T.: Some properties of the ring of Nash functions. Ann. Scuola Norm. Sup. Pisa, Cl. Sci., (4) **3**, 245-266 (1976)
239. Mostowski, T.: Topological equivalence between analytic and algebraic sets. Bull. Acad. Polon. Sci. **32**, 393-400 (1984)
240. Motzkin, T.S.: The arithmetic-geometric inequality. Inequalities, O. Shisha ed., 205-224. New York: Academic Press (1967)
241. Motzkin, T.S.: The real solution set of a system of algebraic inequalities is the projection of a hypersurface in one more dimension. Inequalities II, O. Shisha ed., 251-254. New York: Academic Press (1970)
242. Mumford, D.: The red book of varieties and schemes. Lecture Notes in Math. **1358**. Berlin: Springer-Verlag (1988)
243. Munkres, J.R.: Elements of algebraic topology. Reading: Addison-Wesley (1984)
244. Nakai, I.: On topological types of polynomial mappings. Topology **23**, 45-66 (1984)
245. Narasimhan, R.: Introduction to the theory of analytic spaces. Lecture Notes in Math. **25**. Berlin: Springer-Verlag (1966)
246. Narasimhan, R.: Analysis on real and complex manifolds. Advanced Studies in Pure Mathematics **1**. Amsterdam: North-Holland Publishing Company (1968)
247. Nash, J.: Real algebraic manifolds. Ann. of Math. **56**, 405-421 (1952)
248. Natanzon, S.M.: Moduli spaces of real curves, Trans. Moscow Math. Soc. **1**, 233-272 (1980)
249. Oleinik, O.A.: Estimates of the Betti numbers of real algebraic hypersurfaces. Mat. Sb. (N.S.) **28** (70), 635-640 (1951)
250. Parusinski, A., Szafraniec, Z.: Algebraically constructible functions and signs of polynomials. Manuscripta Math. **93**, 443-456 (1997)
251. Pecker, D.: On Efroymson's extension theorem for Nash functions. J. Pure Appl. Algebra **37**, 193-203 (1985)

252. Peterson, F.: Some remarks on Chern classes, Ann. of Math. **69**, 414-420 (1959)
253. Petrovsky, I.: On the topology of real plane algebraic curves. Ann. of Math., **39**, 187-209 (1938)
254. Pfister, A.: Multiplikative quadratische Formen. Arch. Math. **16**, 363-370 (1965)
255. Pfister, A.: Zur Darstellung definiter Funktionen als Summe von Quadraten. Invent. Math. **4**, 229- 237 (1967)
256. Pfister, A.: Quadratic forms with applications to algebraic geometry and topology. London Math. Soc. Lecture Note Ser. **217**. Cambridge: Cambridge Univ. Press (1995)
257. Pourchet, Y.: Sur la représentation en sommes de carrés des polynômes à une indéterminée sur un corps de nombres algébriques. Acta Arith. **19**, 89-104 (1971)
258. Prestel, A.: Lecture on formally real fields. Lecture Notes in Math. **1093**. Berlin: Springer-Verlag (1984)
259. Procesi, C.: Positive symmetric functions. Adv. Math. **29**, 219-225 (1978)
260. Procesi, C., Schwarz, G.: Inequalities defining orbit spaces. Invent. Math. **81**, 539-554 (1985)
261. Quarez, R.: The idempotency of the real spectrum implies the extension theorem for Nash functions. Math. Z. (to appear)
262. Ragsdale, V.: On the arrangement of the real branches of plane algebraic curves. Amer. J. Math. **28**, 377-404 (1906)
263. Ramanakoraisina, R.: Complexité des fonctions de Nash. Commun. Algebra **17**, 1395-1406 (1989)
264. Raynaud, M.: Anneaux locaux henséliens. Lecture Notes in Math. **169**. Berlin: Springer-Verlag (1970)
265. Recio, T.: Una decomposición de un conjunto semi-algebráico. Actas del V congreso de la Agrupación de Matemáticos de Expresión Latina, Madrid (1978)
266. Renegar, J.: On the computational complexity and geometry of the first-order theory of the reals, parts I, II and III. J. Symbolic Comput. **13**, 255-352, (1992)
267. Risler, J.-J.: Une caractérisation des idéaux des variétés algébriques réelles. C.R. Acad. Sci. Paris **271**, 1171-1173 (1970)
268. Risler, J.-J.: Sur l'anneau des fonctions de Nash globales. Ann. Sci. Ecole Norm. Sup. (4) **8**, 365-378 (1975)
269. Risler, J.-J.: Le théorème des zéros en géométries algébrique et analytique réelle. Bull. Soc. Math. France **104**, 113-127 (1976)
270. Risler, J.-J.: Sur le 16ème problème de Hilbert: un résumé et quelques questions. Sémin. sur la géométrie algébrique réelle, Publ. Math. Univ. Paris VII, **9**, 11-25 (1980)
271. Risler, J.-J.: Sur l'homologie des surfaces algébriques réelles. Géométrie algébrique réelle et formes quadratiques. Lecture Notes in Math. **959**, 381-385. Berlin: Springer-Verlag (1982)
272. Risler, J.-J.: Complexité et géométrie réelle (d'après Khovanski). Séminaire Bourbaki Volume 1984/85. Astérisque **133/134**, 89-100 (1986)
273. Roberts, J.: Generic projections of algebraic varieties. Amer. J. Math. **93**, 191-214 (1971)
274. Robinson, A.: On ordered fields and definite functions. Math. Ann. **130**, 257-271 (1955)
275. Robinson, A.: Complete theories. Amsterdam: North Holland (1956)

276. Robinson, R.M.: Some definite polynomials which are not sums of squares of real polynomials. Notices Amer. Math. Soc. **16**, 554 (1969)
277. Robson, R.: Nash wings and real prime divisors. Math. Ann. **273**, 177-190 (1986)
278. Rockafellar, R.T.: Convex analysis. Princeton: Princeton Univ. Press (1970)
279. Rokhlin, V.: Congruences modulo 16 in Hilbert's sixteenth problem. Funct. Anal. Appl. **6**, 58-64 (1972)
280. Roy, M.-F.: Faisceau structural sur le spectre réel et fonctions de Nash. Géométrie algébrique réelle et formes quadratiques. Lecture Notes in Math. **959**, 406-432. Berlin: Springer-Verlag (1982)
281. Roy, M.-F.: Basic algorithms in real algebraic geometry : from Sturm theorem to the existential theory of reals. Lectures in Real Geometry, Expositions in Mathematics **23**, 1-67. Berlin, New York: de Gruyter (1996)
282. Ruiz, J.M.: On Hilbert's 17th problem and real nullstellensatz for global analytic functions. Math. Z. **190**, 447-459 (1985)
283. Sabbah, C.: Le type topologique éclaté d'une application analytique. Singularities, Proc. Sympos. Pure Math. **40**, Part 2, 433-440. Providence: Amer. Math. Soc. (1983)
284. Saliba, C.: Le théorème des zéros centraux. C.R. Acad. Sci. Paris **298**, 337-340 (1984)
285. Sander, T.: Existence and uniqueness of the real closure of an ordered field without Zorn's lemma. J. Pure Appl. Algebra **73**, 165-180 (1991)
286. Samuel, P.: Méthodes d'algèbre abstraite en géométrie algébrique. Berlin: Springer-Verlag (1967)
287. Scheiderer, C.: Stability index of real varieties. Invent. Math. **97**, 467-483 (1989)
288. Schmüdgen, K.: The K-moment problem for compact semi-algebraic sets. Math. Ann. **289**, 203-206 (1991)
289. Schwartz, N.: The basic theory of real closed spaces. Mem. Amer. Math. Soc. **397** (1989)
290. Seidenberg, A.: A new decision method for elementary algebra. Ann. of Math. **60**, 365-374 (1954)
291. Seifert, H.: Algebraische Approximation von Mannigfaltigkeiten. Math. Z. **41**, 1-17 (1936)
292. Seifert, H., Threlfall, W.: A textbook of topology. New York: Academic Press (1980)
293. Serre, J.-P.: Faisceaux algébriques cohérents. Ann. of Math. **61**, 197-278 (1955)
294. Serre, J.-P.: A course in arithmetic. Berlin: Springer-Verlag (1973)
295. Shafarevich, I.R.: Basic algebraic geometry. Berlin: Springer-Verlag (1974)
296. Shapiro, D.: Products of sums of squares. Exposition. Math. **2**, 235-261 (1984)
297. Shiota, M.: On the unique factorization property of the ring of Nash functions. Publ. Res. Inst. Math. Sci. **17**, 363-369 (1981)
298. Shiota, M.: Sur la factorialité de l'anneau des fonctions lisses rationnelles. C.R. Acad. Sci. Paris **292**, 67-70 (1981)
299. Shiota, M.: Real algebraic realization of characteristic classes. Publ. Res. Inst. Math. Sci. **18**, 995-1008 (1982)
300. Shiota, M.: Classification of Nash manifolds. Ann. Inst. Fourier **33**, 209-232 (1983)
301. Shiota, M.: Abstract Nash manifolds. Proc. Amer. Math. Soc. **96**, 155-162 (1986)
302. Shiota, M.: Nash manifolds. Lecture Notes in Math. **1269**. Berlin: Springer-Verlag (1987)

303. Shiota, M.: Piecewise linearization of subanalytic functions. II. Real analytic and algebraic geometry, Lecture Notes in Math. **1420**, 247-307. Berlin: Springer-Verlag (1990)
304. Shiota, M., Yokoi, M.: Triangulation of subanalytic sets and locally subanalytic manifolds. Trans. Amer. Math. Soc. **286**, 727-750 (1984)
305. Silhol, R.: A bound on the order of $H_{n-1}^{(a)}(X, \mathbb{Z}/2)$ on a real algebraic variety. Géométrie algébrique réelle et formes quadratiques. Lecture Notes in Math. **959**, 443-450. Berlin: Springer-Verlag (1982)
306. Silhol, R.: Real algebraic surfaces. Lecture Notes in Math. **1395**. Berlin: Springer-Verlag (1989)
307. Silhol, R.: Compactifications of moduli spaces in real algebraic geometry, Invent. Math. **107**, 151-202 (1992)
308. Spanier, E.H.: Algebraic topology. New York: McGraw-Hill (1966)
309. Steenrod, N.: The topology of fibre bundles. Princeton: Princeton Univ. Press (1951)
310. Stengle, G.: A Nullstellensatz and a Positivstellensatz in semialgebraic geometry. Math. Ann. **207**, 87-97 (1974)
311. Sturm, C.: Mémoire sur la résolution des équations numériques. Inst. France Sc. Math. Phys. **6** (1835)
312. Sullivan, D.: Combinatorial invariants of analytic spaces. Proc. Liverpool Singularities-Sympos. I, Lecture Notes in Math. **192**, 165-168. Berlin: Springer-Verlag (1971)
313. Sylvester, J.J.: On a theory of syzygetic relations of two rational integral functions, comprising an application to the theory of Sturm's function. Philos. Trans. Roy. Soc. London, **143** (1853)
314. Swan, R.: Vector bundles and projective modules. Trans. Amer. Math. Soc. **105**, 264-277 (1962)
315. Swan, R.: Topological examples of projective modules. Trans. Amer. Math. Soc. **230**, 201-234 (1977)
316. Swan, R.: K-theory of quadric hypersurfaces. Ann. of Math. **122**, 113-154 (1985)
317. Tarski, A.: Sur les ensembles définissables de nombres réels. Fund. Math. **17**, 210-239 (1931)
318. Tarski, A.: A decision method for elementary algebra and geometry. Prepared for publication by J.C.C. Mac Kinsey, Berkeley (1951)
319. Teichner, P.: 6-dimensional manifold without totally algebraic homology. Proc. Amer. Math. Soc. **123**, 2909-2914 (1995)
320. Thom, R.: Quelques propriétés globales des variétés différentiables. Comment. Math. Helv. **28**, 17-86 (1954)
321. Thom, R.: Un lemme sur les applications différentiables. Bol. Soc. Mat. Mexicana **1**, 59-71 (1956)
322. Thom, R.: La stabilité topologique des applications polynomiales. Enseign. Math. (2) **8**, 24-33 (1962)
323. Thom, R.: Sur l'homologie des variétés algébriques réelles. Differential and combinatorial topology, 255-265. Princeton: Princeton Univ. Press (1965)
324. Thom, R.: Structural stability and morphogenesis. An outline of a general theory of models. Reading, Mass.: W. A. Benjamin, Inc. (1975)
325. Tognoli, A.: Su una congettura di Nash. Ann. Scuola Norm. Sup. Pisa, Sci. Fis. Mat. (3) **27**, 167-185 (1973)
326. Tognoli, A.: Algebraic geometry and Nash functions. Institutiones Math. **3**. New York: Academic Press (1978)
327. Tougeron, J.-C.: Idéaux de fonctions différentiables. Berlin: Springer-Verlag (1972)

328. Tougeron, J.-C.: Solutions d'un système d'équations analytiques réelles et applications. Ann. Inst. Fourier **26**, 109-135 (1976)
329. Tougeron, J.-C.: Fonctions composées différentiables: cas algébrique. Ann. Inst. Fourier **30**, 51-74 (1980)
330. Varčenko, A.N.: Theorems on the topological equisingularity of families of algebraic varieties and families of polynomial mappings. Math. USSR Izv. **6**, 949-1008 (1972)
331. Viro, O.: Gluing of plane real algebraic curves and constructions of curves of degrees 6 and 7. Topology, Lecture Notes in Math. **1060**, 187-200. Berlin: Springer-Verlag (1984)
332. Viro, O.: Progress in the topology of real algebraic varieties over the last six years. Russian Math. Surveys, **41** (3), 55-82 (1986)
333. van der Waerden, B.L.: Topologische Begründung des Kalküls der abzählenden Geometrie. Math. Ann. **102**, 337-362 (1929)
334. Walker, R.: Algebraic curves. Princeton: Princeton Univ. Press (1950). Berlin: Springer-Verlag (1978)
335. Wallace, A.H.: Algebraic approximations of manifolds. Proc. London Math. Soc. **7**, 196-210 (1957)
336. Wallace, A.H.: Linear sections of algebraic varieties. Indiana Univ. Math. J. **20**, 1153-1162 (1971)
337. Whitehead, G.W.: Elements of homotopy theory. Berlin: Springer-Verlag (1978)
338. Whitney, H.: Elementary structure of real algebraic varieties. Ann. of Math. **66**, 545-556 (1957)
339. Whitney, H.: Local properties of analytic varieties. Differential and combinatorial topology, 205-244. Princeton: Princeton Univ. Press (1965)
340. Whitney, H.: Tangents to an analytic variety. Ann. of Math. **81**, 496-549 (1965)
341. Wilkie, A.: On the theory of the real exponential field. Illinois J. Math. **33**, 384-408 (1989)
342. Wilson, G.: Hilbert's sixteenth problem. Topology **17**, 53-73 (1978)
343. Witt, E.: Zerlegung reeller algebraischer Funktionen in Quadrate, Schiefkörper über reellen Funktionenkörpern. J. Reine. Angew. Math. **171**, 4-11 (1934)
344. Wood, R.: Polynomial maps from spheres to spheres. Invent. Math. **5**, 163-168 (1968)
345. Yiu, P.: Quadratic forms between spheres and the non-existence of sum of squares formulae. Math. Proc. Cambridge Philos. Soc. **100**, 493-504 (1986)
346. Yiu, P.: Quadratic forms between Euclidean spheres. Manuscripta Math. **83**, 171-181 (1994)
347. Yomdin, J.: The geometry of critical and near-critical values of differentiable mappings. Math. Ann. **264**, 495-515 (1983)
348. Yomdin, J.: Metric properties of semialgebraic sets and mappings and their applications in smooth analysis. Géométrie algébrique et applications III: Géométrie réelle. Systèmes différentiels et théorie de Hodge, Travaux en Cours **24**, 165-183 (1987)
349. Zariski, O., Samuel, P.: Commutative Algebra. Vol. 1. Graduate Texts in Math. **28**. Berlin: Springer-Verlag (1975)
350. Zariski, O., Samuel, P.: Commutative algebra. Vol. 2. Graduate Texts in Math. **29**. Berlin: Springer-Verlag (1976)

Index of Notation

Standard Notation

We denote by $\mathbb{N}$, $\mathbb{Z}$, $\mathbb{Q}$, $\mathbb{R}$, $\mathbb{C}$, $\mathbb{H}$ and $\mathbb{Z}/2$ the sets of nonnegative integers, integers, rational numbers, real numbers, complex numbers, quaternions and integers modulo 2, respectively.

If A is a ring, A/I is the quotient ring of A by the ideal I, A_I is the localization of A with respect to I if I is a prime ideal, $(a_1, \ldots, a_k)$ the ideal generated by the elements $a_1, \ldots, a_k$. We denote by $A[X]$ the ring of polynomials in the variable X with coefficients in A. If P is a polynomial in $A[X]$, $\deg(P)$ denotes its degree. The ring of polynomials in the variables $X_1, \ldots, X_n$ with coefficients in A is denoted by $A[X_1, \ldots, X_n]$ (or $A[X]$, if $X = (X_1, \ldots, X_n)$).

If R is a field, $R(X)$ denotes the field of rational functions in the variable X with coefficients in R, and $R[[X_1, \ldots, X_n]]$ (or $R[[X]]$) the ring of power series in the variables $X_1, \ldots, X_n$.

We denote by Id_A, or simply Id, the identity mapping of A.

The notation $\mathcal{C}^\infty$ stands for infinitely differentiable. If f is a differentiable mapping, $d_x f$ is its derivative at x.

The closure (resp. the interior) of a set A is denoted by $\operatorname{clos}(A)$ (resp $\operatorname{int}(A)$).

The open intervals in a totally ordered set are denoted:

$$]a,b[= \{x \mid a < x < b\}.$$

We also use the notation $[a, b[$ or $]a, b]$ for half-open intervals.

Special Notation

The following entries are listed in order of appearance.

a_+, a_-	ordering of $\mathbb{R}(X)$ for which $X > a$ (resp. $X < a$) and $(b - X)(b - a) > 0$, for every $b \neq a$ in $\mathbb{R}$	7
$\sum A^2$	set of sums of squares of the ring A	8
$P[(a_i)_{i \in I}]$	cone generated by the cone P and the elements a_i	8, 86
$A[\sqrt{a}]$	quotient of $A[X]$ by the ideal $(X^2 - a)$	9
$A[i]$	quotient of $A[X]$ by the ideal $(X^2 + 1)$	9
$\mathbb{R}_{\operatorname{alg}}$	field of real algebraic numbers	11

Index of Notation

$\mathbb{R}(X)^\wedge$, $\mathbb{C}(X)^\wedge$	field of real (resp. complex) Puiseux series	11
$v(f,g;a)$	number of sign changes in a Sturm sequence	12
$\lvert a \rvert$	absolute value of a in an ordered field	13
$\mathbb{R}(X)^\wedge_{\text{alg}}$	field of algebraic Puiseux series	16
$[K:F]$	degree of the extension K of F	17
$\text{sign}(a)$	sign of a, element of $\{-1,0,+1\}$	17
$\text{SIGN}_R(f_1,\ldots,f_s)$	table of signs of $f_1,\ldots,f_s$	18
$\mathcal{Z}(I)$, $\mathcal{Z}_V(I)$	zero set (in V) of the family I of functions	23, 62
$\mathcal{I}(S)$, $\mathcal{I}_A(S)$	ideal of polynomials (or ideal of functions in the ring A) vanishing on S	24, 62
$\lVert x \rVert$	$\sqrt{x_1^2 + \cdots + x_n^2}$, where $x=(x_1,\ldots,x_n)$	26
$B_n(x,r)$, $\overline{B}_n(x,r)$	open (resp. closed) ball with center x and radius r in R^n	26
$S^{n-1}(x,r)$	sphere with center x and radius r in R^n	26
S^{n-1}	standard sphere (with centre 0 and radius 1) in R^n	26
$\text{dist}(x,A)$	distance from x to A	29
$(\dot{S},\eta)$	Alexandrov compactification of S	42
$\mathcal{S}^0(A)$	ring of semi-algebraic continuous functions on A	44
$\mathcal{A}(R^n;U)$	subring of the ring of semi-algebraic functions on U used for the separation of closed semi-algebraic sets	48
$\mathcal{P}(A)$	ring of polynomial functions on A	50
$\text{clos}_{\text{Zar}}(A)$	Zariski closure of A	50
$\mathcal{K}(V)$	field of rational functions on V	50
$\dim(A_x)$	local dimension of A at x	53
$A^{(d)}$	set of points x of A such that $\dim(A_x)=d$	53
$T_x(A)$	tangent space to A at x	54
$\mathcal{S}^k(U,B)$	set of semi-algebraic mappings from U to B of class $\mathcal{C}^k$ (for $k=0,\ldots,\infty$)	54
$\mathcal{S}^k(U)$	ring of semi-algebraic functions from U to R of class $\mathcal{C}^k$ (for $k=0,\ldots,\infty$)	54
$\mathcal{N}(U)$	ring of Nash functions on U	55, 166
$\mathcal{N}_M$	sheaf of Nash functions on M	55, 171
$\mathbb{P}_n(K)$	projective space of dimension n over the field K	59
$\mathcal{P}(V,W)$	set of polynomial mappings from V to W	62
$\mathcal{R}(U)$	ring of regular functions on U	62
$\mathcal{R}(U,W)$	set of regular mappings from U to W	62
$\mathcal{R}_X$	sheaf of regular functions on X	63
$T_x^{\text{Zar}}(X)$	Zariski tangent space of X at x	65
$E^\vee$	dual of the vector space E	66
$\text{Sing}(V)$	set of singular points of V	69
$\text{Reg}(V)$	set of nonsingular points of V	69
$(x_0:\ldots:x_n)$	homogeneous coordinates in $\mathbb{P}_n(K)$	70
$\mathbb{P}\mathcal{Z}(P_1,\ldots,P_k)$	set of projective zeros of homogeneous polynomials $P_1,\ldots,P_k$	70

Index of Notation

$\mathbb{G}_{n,k}(K)$	grassmannian of vector subspaces of dimension k of K^n	71
$\mathbb{M}_{n-k,k}(R)$	set of $(n-k) \times k$ matrices with entries in R	71
$E(X,Y)$	blowing up of X with centre Y	78
$X \# Y$	connected sum of X and Y	82
$\sqrt[R]{I}$	real radical of the ideal I	85
$\sqrt[P]{I}$	P-radical of the ideal I	87
$\mathrm{supp}(P)$	support of the prime cone P	88
$k(\mathfrak{p})$	residue field at the prime ideal $\mathfrak{p}$	89
S_K, f_K	extension of the semi-algebraic set S (resp. semi-algebraic function f) to the real closed field K	98, 101
$\mathcal{L}(R)$	first-order language of ordered fields with parameters in R	99
$\mathrm{Fr}(A)$	field of fractions of an integral domain A	107
$GL(n,K)$	linear group of $n \times n$ invertible matrices with entries in K	111
$P_{n,m}$	set of nonnegative forms in n variables of degree m	111
$\Sigma_{n,m}$	subset of $P_{n,m}$ consisting of sums of squares	111
$p(A)$	Pythagoras number of the ring A	114
F^*	multiplicative group of invertible elements of F	114
$\langle a_1, \ldots, a_n \rangle$	diagonal quadratic form $\sum_{i=1}^n a_i X_i^2$	114, 385
$\varphi \simeq \psi$	equivalence between quadratic forms	114
$\varphi \perp \psi, \varphi \otimes \psi$	orthogonal sum and tensor product of quadratic forms	115
$\langle\langle a_1, \ldots, a_n \rangle\rangle$	Pfister form	117
$\mathcal{U}(g_1, \ldots, g_k)$	basic open semi-algebraic set	122
$\leq_\alpha$	ordering of the residue field $k(\mathrm{supp}(\alpha))$ induced by the prime cone α	134
$k(\alpha)$	real closure of the ordered field $(k(\mathrm{supp}(\alpha)), \leq_\alpha)$	134
$a(\alpha)$	image of a in $k(\alpha)$	134, 146
$\mathrm{Spec}_\mathrm{r}(A)$	real spectrum of A	134
$\widetilde{\mathcal{U}}(a_1, \ldots, a_n)$	basic open subset of $\mathrm{Spec}_\mathrm{r}(A)$	134
$\mathrm{Spec}_\mathrm{r}(f)$	mapping $\mathrm{Spec}_\mathrm{r}(B) \to \mathrm{Spec}_\mathrm{r}(A)$ induced by $f: A \to B$	136
$\widetilde{S}$	constructible subset corresponding, by the tilde operation, to the semi-algebraic set S	143
$\widetilde{f}$	mapping corresponding, by the tilde operation, to the semi-algebraic mapping f	145
$\mathcal{S}^0_{\widetilde{S}}$	sheaf of continuous semi-algebraic functions on $\widetilde{S}$	146
$X|_S$	restriction of the family X to S	149
X_t, X_α	fibre of the family X parametrized by R^p at $t \in R^p$, at $\alpha \in \widetilde{R^p}$	149
f_t, f_α	fibre of the family of mappings f parametrized by R^p at $t \in R^p$, at $\alpha \in \widetilde{R^p}$	149, 150
$\mathrm{Cent}(V)$	set of central points of V	158
$\widehat{A}$	completion of a local ring A	162
$R[[X]]_{\mathrm{alg}}$	ring of power series algebraic over the polynomials	165
k_A	residue field of a local ring A	185

Index of Notation

$^h A$	henselization of a local ring A	188
$V_{\mathbb{C}}$	complexification of V	190
$\mathcal{N}_{\widetilde{M}}$	sheaf of Nash functions on $\widetilde{M}$	192
$\mathcal{C}^\infty(M, N)$	set of $\mathcal{C}^\infty$ mappings from M into N,	200
$\mathcal{A}_k, \mathcal{B}_k, \mathcal{C}_k$	families of strata associated to a stratifying family of polynomials	208
$[a_0, \ldots, a_k]$	simplex with vertices $a_0, \ldots, a_k$	216
σ^0	open simplex	216
$\|K\|$	realization of the complex K	217
$\mathfrak{m}_B$	maximal ideal of the valuation ring B	246
λ_B, v_B, Γ_B	place, valuation and value group associated to the valuation ring B	246
$H_r(A, \Lambda)$	r-th homology group of A with coefficients in Λ	266
$b_r(A)$	r-th Betti number of A	266
$\chi(A)$	Euler-Poincaré characteristic of A	267
$\chi(A, A \setminus v)$	local Euler-Poincaré characteristic of A at v	267
$[V]$	fundamental class of V	271
$H_r^{\mathrm{alg}}(X, \mathbb{Z}/2)$	subgroup of $H_r(X, \mathbb{Z}/2)$ of homology classes represented by algebraic sets	272
$H_{\mathrm{alg}}^r(X, \mathbb{Z}/2)$	subgroup of cohomology classes Poincaré dual to algebraic homology classes	273
$w_r(M), w_r(\xi)$	r-th Stiefel-Whitney class of M (resp. of ξ)	274, 304
$H_r^{\mathrm{BM}}(V, \mathbb{Z}/2)$	r-th Borel-Moore homology group of V	278
$\check{H}^r(B, \Lambda)$	r-th Čech cohomology group of B with coefficients in Λ	291
$H_r(A, B; \Lambda)$	relative homology group with coefficients in Λ	293
$\simeq_{\mathrm{alg}}, \simeq_{\mathrm{top}}, \simeq_{\mathcal{C}^\infty}$	algebraic, topological and $\mathcal{C}^\infty$ isomorphism between vector bundles	298, 309
ϵ_X^n	trivial vector bundle of rank n on X	298
$f^*(\xi)$	induced vector bundle	298
$\xi\|_Y$	restriction of the vector bundle ξ to Y	298
$\xi \oplus \xi', \xi \otimes \xi'$	Whitney sum and tensor product of vector bundles	299
$\bigwedge^k \xi$	k-th exterior power of a vector bundle	299
$\xi^\vee$	dual vector bundle	299
$\mathrm{Hom}(\xi, \xi')$	vector bundle of morphisms from ξ to ξ'	299
$\mathcal{L}_{\mathrm{alg}}(\xi)$	sheaf of modules of algebraic sections of ξ	300
$\gamma_{n,k}$	universal vector bundle on the grassmannian	300
$\Gamma_{\mathrm{alg}}(\xi), \Gamma_{\mathrm{s.a.}}(\xi), \Gamma_{\mathrm{Nash}}(\xi)$	module of global algebraic, semi-algebraic and Nash sections of ξ	305, 332, 336
$\mathrm{Pic}(A)$	Picard group of A	306
$V_{\mathrm{alg}}^1(X), V^1(X), V_{\mathrm{Nash}}^1(M)$	group of isomorphism classes of algebraic, topological and Nash line bundles over X	306, 309, 337
$\mathrm{Cl}(A)$	divisor class group of the ring A	306
$\mathrm{Proj}(A)$	set of isomorphism classes of projective A-modules of finite type	309

Index of Notation   425

$\mathcal{C}^0(X)$	ring of continuous functions on X	309
$\widetilde{K}_0(A)$	reduced Grothendieck group of the ring A	309
$\widetilde{K}_0(f)$	homomorphism $\widetilde{K}_0(A) \to \widetilde{K}_0(B)$ induced by $f: A \to B$	310
$\widetilde{KO}(X)$	reduced Grothendieck group of stable equivalence classes of real vector bundles over X	310
$e(\xi)$	Euler class of ξ	323
$\widetilde{K}(X)$	reduced Grothendieck group of stable equivalence classes of complex vector bundles over X	326
$V_{\mathbb{C}}^1(X), V_{\mathbb{C}-\mathrm{alg}}^1(X)$	group of isomorphism classes of topological (resp. algebraic) $\mathbb{C}$-line bundles over X	326
$\#_2(Z, Z; Y)$	modulo 2 self-intersection number of Z in Y	329
$p*q, p\#q$	smallest k such that there exists a normed (resp. nonsingular) bilinear form $\mathbb{R}^p \times \mathbb{R}^q \to \mathbb{R}^k$	341, 346
ρ	Hurwitz-Radon function	341
$\pi_n(S^k)$	n-th homotopy group of S^k	350
$\overline{\mathcal{R}(X, S^p)}$	closure of $\mathcal{R}(X, S^p)$ in $\mathcal{C}^\infty(X, S^p)$	353
$\mathrm{Deg}_{\mathcal{R}}(X)$	set of topological degrees of regular mappings $X \to S^{\dim(X)}$	357
$\deg(f)$	topological degree of a continuous mapping $f: X \to S^{\dim(X)}$	357
$[f]$	homotopy class of f	357
$b(X)$	invariant defined by $\mathrm{Deg}_{\mathcal{R}}(X) = b(X)\mathbb{Z}$	358
$\pi^k(X)$	k^{th} cohomotopy group of X	361
$\pi_{\mathrm{alg}}^k(X)$	subset of $\pi^k(X)$ of homotopy classes represented by regular mappings $X \to S^k$	361
$\pi_n^{\mathrm{alg}}(S^k)$	subset of $\pi_n(S^k)$ of homotopy classes represented by regular mappings $S^n \to S^k$	361
$\widetilde{K}(f)$	homomorphism $\widetilde{K}(S^\ell) \to \widetilde{K}(X)$ induced by $f: X \to S^\ell$	362
$K_0(A)$	Grothendieck group of the ring A	383
$[M]$	stable isomorphism class of M	383
$KO(X)$	Grothendieck group of stable isomorphism classes of real vector bundles over X	384
$(M, b) \perp (M', b'), (M, b) \otimes_A (M', b')$	orthogonal sum and tensor product of bilinear spaces	384
$H(M)$	hyperbolic bilinear space	385
$[(M, b)]$	Witt-equivalence class of (M, b)	385
$W(A)$	Witt ring of A	385
sign	signature homomorphism	386
$(M, b)\|_U$	restriction of the bilinear space (M, b) to U	386
$\mathcal{D}(V)$	ring of functions on V constructed from polynomials by taking $1/f$ and $\sqrt{f}$ for positive functions f	394
$W'(A)$	subring of $W(A)$ consisting of classes $[(M, b)]$ such that M is free	394

Index

Alexandrov compactification
- algebraic - 77
- semi-algebraic - 42

Algebraic
- affine - variety 63, 72, 305
- homology 272, 318
- model 273, 317
- projective - set 70
- real - variety 64
- set 23

Algebraic power series 165

Approximation
- algebraic - 318

Archimedean 7
Artin's approximation theorem 171
Artin-Lang theorem 84
Artin-Mazur theorem 173, 376

Basic
- closed semi-algebraic set 46, 111, 259, 285
- generically - 107
- open semi-algebraic set 46, 122

Bertini's theorem
- semi-algebraic - 236

Betti numbers 266, 276, 284
Bezout's theorem 281, 287
Bilinear space 384
Birationnally equivalent 75
Biregular isomorphism 63
Blowing down 77
Blowing up 78, 319
Borel-Moore homology 278, 294
Bröcker-Scheiderer theorem 122, 405

Cartan's umbrella 60, 105, 237
Cellular decomposition
- semi-algebraic - 213
Central points 158, 250
Centre of a place 249
Chern class 327, 329, 362

Cobordism 377
Cohomology 291
- algebraic 273
Cohomotopy group 361
Complete semi-algebraic set 74

Cone
- of a field 8
- of a ring 86
- positive - of an ordered field 8
- prime - 88
- proper - 8, 86

Conic
- local - structure 225, 295

Connected
- semi-algebraically - 34
- semi-algebraically - components 34, 75, 285, 392
- semi-algebraically path - 42

Connected sum 82
Constructible
- subset 137
- topology 137

Convex
- envelope 249
- P- - ideal 87
- subring 247

Critical
- point 235
- value 235

Curve
- dividing - 288, 331
- M- - 288

Curve selection lemma 38
- Nash - 167, 240

Descartes's law of signs 14
Desingularization
- topological - 380
Diffeotopy 318, 373

Dimension
- local - 53

- of a constructible subset   155
- of a prime cone   157
- of a quadratic form   114
- of a ring   50
- of a semi-algebraic set   50
- of an ideal   65

Distance to a semi-algebraic set   29
Division theorem
- for Nash functions   169
- formal –   168

Divisor class group   306

Efroymson's extension theorem   200
Equivalence of quadratic forms   114
Etale mapping   162
Euler-Poincaré
- characteristic   267
- local – characteristic   267

Extension
- of a semi-algebraic mapping   101, 151
- of a semi-algebraic set   98, 150

Face of a simplex   216
Factoriality   169, 306, 308, 319, 327, 337
Family
- of Nash functions   202
- semi-algebraic – of mapping   149
- semi-algebraic – of sets   149

Fan   257
- trivial –   257

Filter   137
- prime –   138

Finiteness theorem   46, 131
Form   340
- nonsingular bilinear –   346
- normed bilinear –   340

Formula
- first-order –   28
- quantifier-free –   28, 99

Fundamental class   271, 278

Generization   140
Grassmannian   71, 272, 300, 352
Grothendieck group   310

Half-branches   232, 254
- at infinity   234
- of a Nash curve   234

Hardt's theorem   221
Harnack's theorem   286
Henselian local ring   185
Henselization   186

Hilbert's problem
- 16-th –   285
- 17-th –   103

Homogeneous coordinates   70
Homology
- algebraic –   272, 318
- Borel-Moore –   278, 294
- local –   295
- relative –   293
- totally algebraic –   272

Homotopy   291, 334, 361
Hopf form   340
Hurwitz-Radon function   341
Hyperbolic space   385

Ideal
- fractional –   306
- P-convex –   87
- P-radical –   87
- real –   84

Implicit function theorem
- semi-algebraic –   56

Integral closure   75, 163
Invertible modules   306
Irreducible
- algebraic set   50
- Nash set   180

Isolated singularities   380
Isometry between bilinear spaces   384

Knot   375

Line bundle
- algebraic –   306, 327

Local-étale   184
Łojasiewicz's inequality   44

M-curves   288
Mapping
- Nash –   57
- polynomial –   62
- regular –   62
- semi-algebraic –   28, 54

Model
- algebraic –   273, 317

Morphism
- algebraic – of vector bundles   298

Nash
- diffeomorphism   57
- equivalence   181
- function   48, 55
- mapping   57
- submanifold   57, 68, 227

– subset   178
– tubular neigbourhood   199
– wing   239
Nash's theorem   377
Nest   289
Nonsingular
– in dimension d 67
– point   66
– variety   67
Normalization   75
Nullstellensatz
– central –   159
– for Nash functions   179
– real –   84

Ordered field   7
Ordering
– compatible with a place   247
– of a field   7
Orthogonal sum
– of bilinear spaces   384
– of quadratic forms   115
Oval   286
– depth of an –   289
– even –   290
– odd –   290

Pfister form   117, 158
Picard group   306
Place   245
– centre of a –   249
– finite on a ring   249
– real –   247
Poincaré duality   273, 313
Positivstellensatz   92
– for Nash functions   178
– formal –   91
Preparation theorem
– for Nash functions   169
– formal –   168
Projective modules   302, 309, 334, 336, 383
Projective space   70
Pseudo-line   286
Puiseux series   11, 16, 148
Pythagoras number   114, 128, 129

Quadratic form
– anisotropic –   115
– isotropic –   115
– multiplicative –   118
– nondegenerate –   115, 124
– representing an element   115
Quantifier elimination   99

Quasi-monic   38, 207

Ramification index   252
Rank
– of a projective module   384
– of a vector bundle   298
Real
– algebraic numbers   11
– closed field   9
– closure   14
– field   9
– ideal   84
– Nullstellensatz   84
– place   247
– principal ideal   94
– radical   85
– spectrum   134
– valuation ring   247
Regular
– function   62
– germs of – functions   65
– local ring   66
– mapping   62, 64
– system of parameters   66, 246
Residual degree   252

Sard's theorem
– semi-algebraic –   235
Self-intersection number   329, 360
Semi-algebraic
– cellular decomposition   213
– closed and bounded – set   35
– complete – set   74
– family of mappings   149
– family of sets   149
– fibre of a – family   149
– function   28, 65
– generically equal – sets   107
– implicit function theorem   56
– mapping   28
– set   24
– subset of a real algebraic variety   65
– Tietze-Urysohn theorem   45
– topological type   230
– trivialization   221
Semi-algebraically
– connected   34
– connected components   34, 75, 154, 392
– homotopic   291
– path connected   42
– trivial   230
Separation theorem   49, 394

Sheaf
- locally free algebraic – 300
- of continuous semi-algebraic functions 146
- of Nash functions 192
- of regular functions 63, 64
Signature homomorphism 386
Simplex 216
- open – 217
Simplicial complex
- finite – 217
Slicing 33
Specialization 140
- chain 155
Stable equivalence classes 309, 310
Stable isomorphism classes 383
Stable under derivation 37
Stereographic projection 76, 364
Stiefel-Whitney class 274, 304, 312, 321, 353
Stone space 137
Stone-Weierstrass theorem 193, 308, 321
Stratification
- induced by a stratifying family 210
- Nash – 212
- satisfying Whitney's conditions 242
Stratifying family of polynomials 207
Stratum 210
- contiguous strata 269
- of a Nash stratification 212
Sturm sequence 12
Sturm's theorem 13
Submanifold
- Nash – 57, 227
Substitution theorem for Nash functions 176
Sums of squares 8, 103, 111
Support 88
Suspension of a form 367
Sylvester's theorem 12

Tangent space
- to a Nash submanifold 197
- Zariski – 65
Tarski-Seidenberg principle 17, 26, 28, 98
Tensor product
- of quadratic forms 115
- of bilinear spaces 384
- of vector bundles 299
Thom's lemma 37
Tietze-Urysohn theorem

- semi-algebraic – 45
Tilde operation 143
Tognoli's theorem 378
Topological type
- semi-algebraic – 230
Topology
- constructible – 137
- euclidean – 26, 65
- spectral – 134
- Zariski – 26, 64
Transfer principle 99
Triangulation 217
- of semi-algebraic functions 227
Trivial
- algebraically – vector bundle 298
Tsen-Lang theorem 120
Tubular neigbourhood
- Nash – 199

Ultrafilter 137
Uniformizing parameter 246

Valuation
- discrete – ring 246
- of a field 245
- real – ring 247
- ring 245
Value group 246
Variety
- affine algebraic – 63, 72, 305
- real algebraic – 64
Vector bundle
- algebraic – 302, 385
- algebraic section of a – 298
- induced – 298
- Nash – 335
- pre-algebraic – 297
- pre-Nash – 334
- rank of a – 298
- semi-algebraic – 332
- transition functions of a – 299
- universal – 300, 352

Whitney sum of vector bundles 299
Whitney's conditions 236, 237
Wing
- Nash – 239
Witt ring 385

Printing: Mercedesdruck, Berlin
Binding: Buchbinderei Lüderitz & Bauer, Berlin